Handbook of Functionalized Organometallics

Volume 1

Edited by Paul Knochel

Tamaru, Y. (Ed.)

Modern Organonickel Chemistry

2005, ISBN 3-527-30796-6

Bäckvall, J.-E. (Ed.)

Modern Oxidation Methods

2004, ISBN 3-527-30642-0

Murahashi, S.-I. (Ed.)

Ruthenium in Organic Synthesis

2004, ISBN 3-527-30692-7

Cornils, B., Herrmann, W. A. (Eds.)

Aqueous-Phase Organometallic Catalysis

Concepts and Applications

2004, ISBN 3-527-30712-5

de Meijere, A., Diederich, F. (Eds.)

Metal-Catalyzed Cross-Coupling Reactions

938 pages in 2 volumes with 29 figures and 61 tables

2004, ISBN 3-527-30518-1

Marek, I. (Ed.)

Titanium and Zirconium in Organic Synthesis

2002, ISBN 3-527-30428-2

Beller, M., Bolm, C. (Eds.)

Transition Metals for Organic Synthesis

Building Blocks and Fine Chemicals

2004, ISBN 3-527-30613-7

Cornils, B., Herrmann, W. A. (Eds.)

Applied Homogencous Catalysis with Organometallic Compounds

A Comprehensive Handbook in Three Volumes

2002, ISBN 3-527-30434-7

Handbook of Functionalized Organometallics

Applications in Synthesis

Volume 1

Edited by Paul Knochel

WILEY-VCH Verlag GmbH & Co. KGaA

Chem

0142455528

Editor

Prof. Paul Knochel
Department of Chemistry
Ludwig-Maximilians-Universität
Butenandtstraße 5–13
Haus F
81377 München
Germany

Library of Congress Card No.: applied for

British Library Cataloguing-in-Publication Data
A catalogue record for this book is available
from the British Library.

**Bibliographic information published by
Die Deutsche Bibliothek**
Die Deutsche Bibliothek lists this publication
in the Deutsche Nationalbibliografie; detailed
bibliographic data is available in the Internet at
<http://dnb.ddb.de>.

Printed in the Federal Republic of Germany.
Printed on acid-free paper.

Cover Design Grafik-Design Schulz, Fußgönheim
Typesetting Kühn & Weyh, Satz und Medien,
Freiburg
Printing betz-druck GmbH, Darmstadt
Bookbinding Litges & Dopf Buchbinderei GmbH,
Heppenheim

ISBN-13: 978-3-527-31131-6
ISBN-10: 3-527-31131-9

1/19/06

Contents

Organometallics. Paul Knochel
Copyright © 2005 WILEY-VCH Verlag GmbH & Co. KGaA, Weinheim
ISBN: 3-527-31131-9

Preface

Since the pioneering work of Frankland and Wurtz, organometallic intermediates have occupied a central position in organic synthesis. The chemical behavior of organometallic reagents depends greatly on the nature of the metal and on the carbon hybridization. Each metal has intrinsic chemical properties, which confer a specific reactivity for forming new carbon-carbon bonds to the organic moiety attached to it. The nature of the metal substituents (ligands) enables a modulation and adjustment to this reactivity of the organometallic to the organic substrate. Choosing the correct metal and ligand sphere to achieve any given transformation represents a major task for the synthetic chemist. During the course of the last thirty years, chemists have realized that this fine-tuning of the reactivity of organometallics has a number of synthetic advantages (selectivity, yields, reaction conditions, etc.). However, they have also noticed that a broad range of functionalities can be present in the organometallic intermediate itself and therefore these reagents allow for the preparation of polyfunctional molecules without the need for multiple protection and deprotection steps.

This book summarizes the synthetic knowledge available as of 2005 for preparing functionalized organometallics and the optimum conditions for their reacting with electrophilic species. It also covers main group and transition organometallics while outlining in detail the functional group compatibility for each class of organometallics in the various book chapters.

Organometallic chemistry is a field of chemistry that is constantly experiencing discoveries and is one of the motors of chemistry. Thus it can be expected that numerous new synthetic methods based on the use of functionalized organometallics will be added to the chemistry presented in this book within the next few years. An effort has been made to present the material in an attractive layout with many equations and numerous practical details, allowing for rapidly entry in the field. Therefore, this book is well suited for master and PhD students, for advanced undergraduate students, as well as industrial process and research chemists.

Munich, August 2005   *Paul Knochel*

Organometallics. Paul Knochel
Copyright © 2005 WILEY-VCH Verlag GmbH & Co. KGaA, Weinheim
ISBN: 3-527-31131-9

List of Authors

Oliver Baron
Department of Chemistry
Ludwig-Maximilians-Universität
Butenandtstrasse 5–13, Haus F
81377 München
Germany

Gerard Cahiez
Departement de Chimie, ESCOM
Université de Paris
13, Boulevard de l'Hautil
95092 Cergy-Pontoise
France

Helena Chechik-Lankin
Department of Chemistry
Technion-Israel Institute of Technology
Technion City
Haifa 32000
Israel

Karl-Heinz Dötz
Kékulé-Institut für Organische Chemie
und Biochemie
Gerhard-Domagk-Strasse 1
53121 Bonn
Germany

Francisco Foubelo
Departamento de Química Orgánica
Facultad de Ciencias and Instituto de
Síntesis Orgánica (ISO)
Universidad de Alicante
Apdo. 99
03080 Alicante
Spain

Eric Fouquet
Laboratoire de Chimie Organique et
Organometallique
Université de Bordeaux I
351, Cours de la Liberation
33405 Talence Cedex
France

Nina Gommermann
Department of Chemistry
Ludwig-Maximilians-Universität
Butenandtstrasse 5–13, Haus F
81377 München
Germany

Liu-Zhu Gong
Department of Chemistry
Ludwig-Maximilians-Universität
Butenandtstrasse 5–13, Haus F
81377 München
Germany

Organometallics. Paul Knochel
Copyright © 2005 WILEY-VCH Verlag GmbH & Co. KGaA, Weinheim
ISBN: 3-527-31131-9

Corinne Gosmini
Laboratoire d'Electrochimie,
Catalyse et Synthèse Organique
2 à 8, rue Henri-Dunant
B. P. 28
94320 Thiais
France

Agnès Herve
Laboratoire de Chimie Organique et
Organometallique
Université de Bordeaux I
351, Cours de la Liberation
33405 Talence Cedex
France

Tamejiro Hiyama
Department of Material Chemistry
Graduate School of Engineering
Kyoto University
Kyoto University Katsura
Nishikyo-ku
Kyoto 615-8510
Japan

Li-Fu Huang
Institute of Chemistry
Academia Sinica
Nangang
Taipei
Taiwan 125

Hiriyakkanavar Ila
Department of Chemistry
Ludwig-Maximilians-Universität
Butenandtstrasse 5–13, Haus F
81377 München
Germany

Florian F. Kneisel
Department of Chemistry
Ludwig-Maximilians-Universität
Butenandtstrasse 5–13, Haus F
81377 München
Germany

Paul Knochel
Department of Chemistry
Ludwig-Maximilians-Universität
Butenandtstrasse 5–13, Haus F
81377 München
Germany

Alexander Koch
Kékulé-Institut für Organische Chemie
und Biochemie
Gerhard-Domagk-Strasse 1
53121 Bonn
Germany

Felix Kopp
Department of Chemistry
Ludwig-Maximilians-Universität
Butenandtstrasse 5–13, Haus F
81377 München
Germany

Tobias J. Korn
Department of Chemistry
Ludwig-Maximilians-Universität
Butenandtstrasse 5–13, Haus F
81377 München
Germany

Arkady Krasovskiy
Department of Chemistry
Ludwig-Maximilians-Universität
Butenandtstrasse 5–13, Haus F
81377 München
Germany

Helena Leuser
Department of Chemistry
Ludwig-Maximilians-Universität
Butenandtstrasse 5–13, Haus F
81377 München
Germany

Tien-Yau Luh
Institute of Chemistry
Academia Sinica
Nangang
Taipei
Taiwan 125

Florence Mahuteau-Betzer
Departement de Chimie, ESCOM
Université de Paris
13, Boulevard de l'Hautil
95092 Cergy-Pontoise
France

Ilan Marek
Department of Chemistry
Technion-Israel Institute of Technology
Technion City
Haifa 32000
Israel

Seijiro Matsubara
Graduate School of Engineering
Kyoto University
Kyoutodaigaku-katsura
Nishikyo
Kyoto 615-8510
Japan

Jacques Périchon
Laboratoire d'Electrochimie,
Catalyse et Synthèse Organique
2 à 8, rue Henri-Dunant
B. P. 28
94320 Thiais
France

Sylvie Perrone
Department of Chemistry
Ludwig-Maximilians-Universität
Butenandtstrasse 5–13, Haus F
81377 München
Germany

Ioannis Sapountzis
Department of Chemistry
Ludwig-Maximilians-Universität
Butenandtstrasse 5–13, Haus F
81377 München
Germany

Masaki Shimizu
Department of Material Chemistry
Graduate School of Engineering
Kyoto University
Kyoto University Katsura
Nishikyo-ku
Kyoto 615-8510
Japan

G. Richard Stephenson
Wolfson Materials and Catalysis Centre
School of Chemical Sciences and
Pharmacy
University of East Anglia
Norwich NR4 7TJ
United Kingdom

Martin Werner
Kékulé-Institut für Organische Chemie
und Biochemie
Gerhard-Domagk-Strasse 1
53121 Bonn
Germany

Xiaoyin Yang
Department of Chemistry
Ludwig-Maximilians-Universität
Butenandtstrasse 5–13, Haus F
81377 München
Germany

Miguel Yus
Departamento de Química Orgánica
Facultad de Ciencias and Instituto de
Síntesis Orgánica (ISO)
Universidad de Alicante
Apdo. 99
03080 Alicante
Spain

1

Introduction

Paul Knochel and Felix Kopp

Achieving selectivity in chemical reactivity is one of the most important goals in synthetic chemistry. This holds especially true for organometallic compounds in which the carbon attached to the metallic center behaves as a nucleophile. The nature of the metal or of the metallic moiety (MetL$_n$) is exceedingly important for tuning this reactivity. An ionic character of the carbon–metal bond leads to a polarity of this bond and to a high reactivity towards electrophilic species. This high reactivity precludes the presence of many functional groups in organometallics like organolithiums. However, by lowering the reaction temperature and using a solvent of moderate polarity, it is also possible to prepare functionalized organolithiums. Especially interesting was the pioneering work of Parham, who showed that a bromine–lithium exchange can be readily performed at –100 °C in a THF-hexane mixture leading to the functionalized organolithium compound **1**.

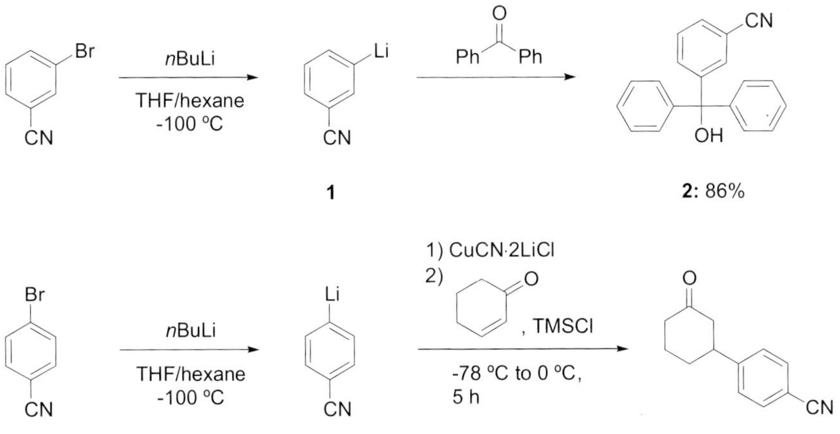

Scheme 1.1 Preparation of nitrile-functionalized aryllithiums by a low-temperature Br–Li exchange.

Organometallics. Paul Knochel
Copyright © 2005 WILEY-VCH Verlag GmbH & Co. KGaA, Weinheim
ISBN: 3-527-31131-9

Its reaction with benzophenone provides the desired alcohol **2** in 86% yield [1]. The need for keeping the highly reactive lithium intermediate **1** at −100 °C is a preparative limitation, which can be avoided by performing a transmetallation of the functionalized aryllithium to a less electropositive metal, such as zinc or copper. The resulting organometallic species is stable at much higher temperatures (0–25 °C), but still reactive enough to undergo efficient conjugated 1,4-addition. Thus, the cyano-substituted aryllithium **3**, readily prepared by a bromine–lithium exchange at −100 °C, reacts within a few minutes at this low temperature with the THF-soluble salt CuCN·2LiCl [2] and provides a functionalized copper reagent that reacts smoothly with cyclohexanone in the presence of TMSCl and provides the Michael-adduct **4** in 93% yield (Scheme 1.1) [3].

Even many sensitive functional groups, such as a ketone and an epoxide, can be present in a polar organometallic species, like an organolithium. Thus, the bicyclic bromide **5** undergoes an efficient Br–Li exchange reaction at −100 °C in THF/TMEDA affording the unstable polyfunctional lithium reagent **6**, which is trapped with benzaldehyde affording the benzylic alcohol **7** in 79% yield (Scheme 1.2) [4]. Low temperature allows the preparation of highly reactive organometallics, but only very active electrophiles react with these reagents at low temperature, which is a serious limitation. Furthermore, large-scale syntheses involving very low temperatures are difficult to set up and to apply in industry. An alternative constitutes a Barbier procedure [5], in which the reactive polyfunctional lithi-

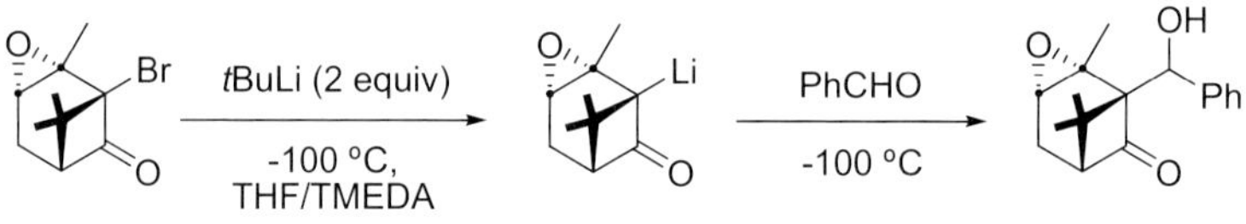

5   **6**   **7: 79%**

Scheme 1.2 Low-temperature preparation of an organolithium species bearing an epoxide and a ketone.

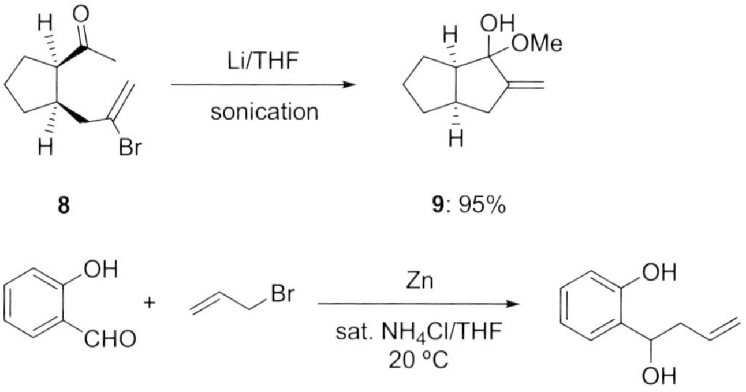

8   **9: 95%**

10   **11: 75%**

Scheme 1.3 Barbier reactions using polyfunctional substrates.

umorganic is directly generated in the presence of the electrophile. Thus, the sensitive bromoketone **8** can be converted *in situ* to the corresponding lithium species by the direct treatment with lithium metal when sonication is applied, leading to the bicyclic tertiary alcohol **9** in 95% yield [6].

Barbier reactions allow the generation of organometallic intermediates that are difficult to store or to generate in the absence of an electrophile. Thus, it was possible to perform the addition of an allylzinc moiety to an aldehyde, such as **10**, bearing a free hydroxy group in a 5:1 mixture of saturated aq. NH_4Cl/THF at room-temperature affording the product in 75% yield without the need of a protecting group (Scheme 1.3) [7]. New types of reactions, such as carbozincation reactions of alkynes using functionalized allylic zinc reagents can be readily accomplished. Thus, the reaction of a silylated propargyl alcohol with *t*-butyl(2-bromomethyl)acrylate (**12**) in the presence of zinc in THF at 45 °C under sonication provides the addition product **13** in 81% yield with excellent regioselectivity (Scheme 1.4) [8].

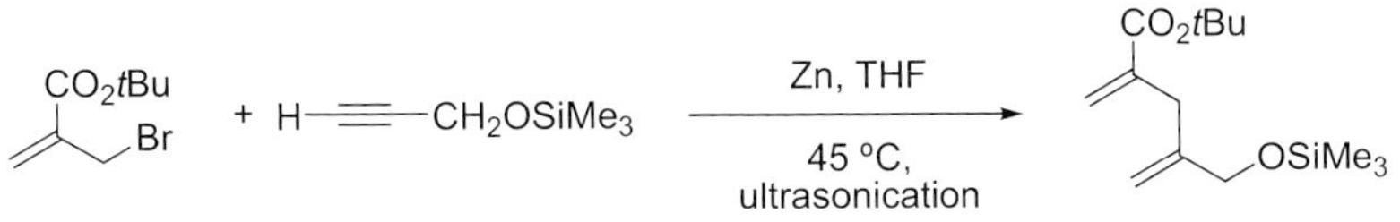

Scheme 1.4 Regioselective addition of a functionalized allylic bromide to alkynes.

The examples above show that the preparation of polyfunctional lithium reagents can be mastered in many cases, but often low temperature and very careful control of the reaction conditions have to be applied. The use of protecting groups is mandatory and reduces the overall yield of the reaction sequences including the protection and deprotection steps. The use of organometallics bearing more covalent carbon–metal bonds would avoid extreme reaction conditions. In 1988 it was found that organozinc iodides bearing numerous functional groups, such as an ester or a ketone can be readily prepared by the direct insertion of zinc dust at 45–50 °C. The resulting organozinc species has unfortunately a moderate reactivity and a transmetallation is required to adjust the intrinsic low reactivity of the zinc reagent. Therefore, the reaction of the functionalized alkyl iodide **14** with zinc dust produces within 4 h at 30 °C the corresponding alkylzinc iodide **15**. It is important in this reaction to keep the reaction temperature below 35 °C in order to avoid competitive deprotonation of the comparably acidic protons in the alpha position to the carbonyl group. Under these conditions, the desired zinc reagent can be prepared in high yield. The addition of $CuCN \cdot 2LiCl$ allows the preparation of the corresponding copper reagent **16** within a few minutes, which is readily acylated with PhCOCl providing the desired product **17** in 80% yield (Scheme 1.5) [2].

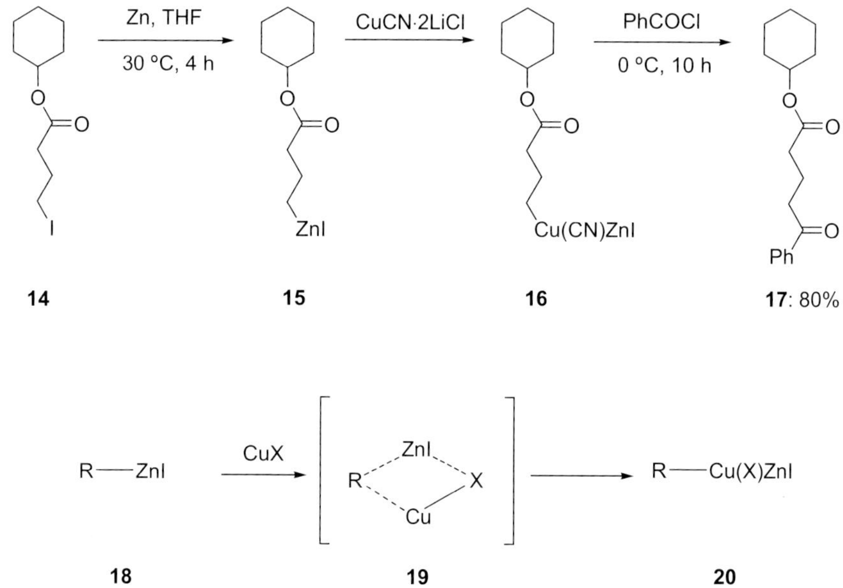

14   **15**   **16**   **17**: 80%

18   **19**   **20**

Scheme 1.5 Fine tuning of the reactivity of functionalized organometallic species by transmetallation.

The success of this transmetallation for adjusting the reactivity of the zinc organometallic relies on two main facts. Organozincs **18** undergo a series of transmetallations with transition metal salts due to the presence of empty p-orbitals of appropriate energy that facilitate 4-membered transition states such as **19**, leading to the copper–zinc species **20**. Secondly, the resulting copper reagent, although being thermodynamically more stable (more covalent carbon–copper bond) is also more reactive due to the presence of nucleophilic, nonbonding d-electrons that interact in an oxidative process with the electrophile and catalyze the formation of the new carbon–carbon bond. This concept is quite general and Negishi has shown the importance of sequential transmetallations for adjusting the reactivity in cross-coupling reactions [9]. Thus, the reaction of alkenylalane **21** with 3-iodo-toluene does not proceed directly. After one week at 25 °C less than 1% of the cross-coupling product is obtained. On the other hand, in the presence of zinc chloride rapid cross-coupling takes place within one hour leading to the styrene derivative **22** in 88% yield. This reaction takes place via an alkenylzinc species that is readily obtained by transmetallation from the alkenylalane **21** (Scheme 1.5). This example shows the importance of mixed metal catalysis. Recently, it was found that, while the cross-coupling of arylmagnesium derivatives with aryl iodides catalyzed by Fe(acac)$_3$ produces only homo-coupling products, the reaction of the corresponding magnesium cuprate **23** furnishes smoothly the desired cross-coupling product **24** in 75% yield (Scheme 1.6) [10].

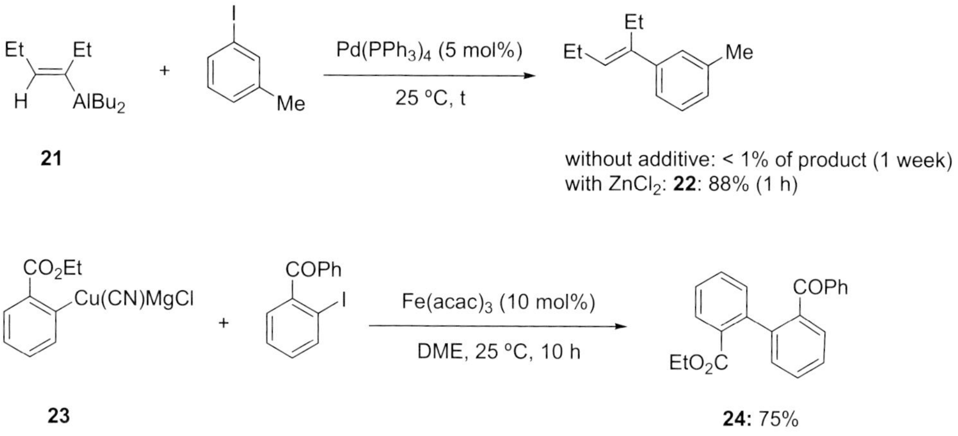

Scheme 1.6 Mixed metal catalysis for efficient cross-coupling reactions.

These general concepts can be applied to a number of organometallic compounds and over the years have allowed the development of an arsenal of synthetic methods for preparing polyfunctional organometallics and for selective reactions of these versatile intermediates. This book intends to cover the broad aspects of this chemistry that should find broad application in industry and universities.

References

1 (a) W. E. Parham, L. D. Jones, *J. Org. Chem.* **1976**, *41*, 1187; (b) W. E. Parham, L. D. Jones, *J. Org. Chem.* **1976**, *41*, 2704.

2 P. Knochel, N. C. P. Yeh, S. C. Berk, J. Talbert, *J. Org. Chem.* **1988**, *53*, 2390.

3 C. E. Tucker, T. N. Majid, P. Knochel, *J. Am. Chem. Soc.* **1992**, *114*, 3983.

4 (a) P. A. Wender, L. A. Wessjohann, B. Paschke, D. B. Rawlins, *Tetrahedron Lett.* **1995**; (b) P. A. Wender, T. E. Glass, *Synlett* **1995**, 516.

5 C. Blomberg, The Barbier Reaction and Related One-Step Processes, Springer-Verlag **1993**, Springer-Verlag Berlin, Heidelberg, New-York.

6 B. M. Trost, B. P. Coppola, *J. Am. Chem. Soc.* **1982**, *104*, 6879.

7 C. Einhorn, J. L. Luche, *J. Organomet. Chem.* **1987**, *322*, 177.

8 P. Knochel, J. F. Normant, *J. Organomet. Chem.* **1986**, *309*, 1.

9 (a) E. Negishi, T. Takahashi, S. Baba, D. E. van Horn, N. Okukado, *J. Am. Chem. Soc.* **1987**, *109*, 2393; (b) E. Negishi, S. Baba, *J. Chem. Soc., Chem. Commun.* **1976**, 596; (c) E. Negishi, N. Okukado, A. O. King, D. E. van Horn, B. I. Spiegel, *J. Am. Chem. Soc.* **1978**, *100*, 2254.

10 I. Sapountzis, W. Lin, C. C. Kofink, C. Despotopoulos, P. Knochel. *Angew. Chem. Int. Ed. Engl.* **2004**, in press.

2
Polyfunctional Lithium Organometallics for Organic Synthesis

Miguel Yus and Francisco Foubelo

2.1
Introduction

Multifunctional organic molecules can be achieved by reacting functionalized organolithium compounds [1] with electrophilic reagents, this fact makes these intermediates of relevant interest in synthetic organic chemistry [2]. Another remarkable fact concerning organolithium compounds, compared with other organometallic compounds is that they can be prepared following a great number of different methodologies [3]: hydrogen–lithium exchange (deprotonation), halogen–lithium exchange, transmetallation reactions, carbon–heteroatom bond cleavage and carbolithiation of multiple carbon–carbon bonds. However, the highly ionic character of the carbon–lithium bond [4] makes these compounds extremely reactive and also unstable, so the synthetic processes should be carried out in some cases under very mild reaction conditions.

Regarding the stability of functionalized organolithium compounds, it depends mainly on three factors, the most important one being the compatibility of the functional group with the carbon–lithium bond, so in some cases the functionality should be protected. The hybridization of the carbon atom bonded to the lithium atom is also important, so sp derivatives are more stable than the corresponding sp^2 and these are more stable than the sp^3 ones as it happens to the corresponding carbanions. Finally, the relative position of the functionality and the carbanionic center is also relevant, probably the β-functionalized derivatives being the most unstable species due to the existence of two consecutive carbon atoms with opposite polarities, the β-elimination process to give an olefin is extremely favored.

Stabilized organolithium compounds, such as lithium enolate intermediates [5], α-organolithium compounds bearing electron-withdrawing groups (RSO, RSO_2, CN, NO_2, etc.) and heteroatoms such as sulfur [6], selenium and phosphorus as well as functionalized aryllithium compounds will not be consider in depth in this chapter due to the limited length of this review.

The following study is ordered based on the relative position of the functionality and the carbanionic center, as well as on the hybridization of that center.

Organometallics. Paul Knochel
Copyright © 2005 WILEY-VCH Verlag GmbH & Co. KGaA, Weinheim
ISBN: 3-527-31131-9

2.2
α-Functionalized Organolithium Compounds

There are several examples of α-oxygen- and nitrogen-functionalized organo-lithium compounds with sp^2 and sp^3 hybridization in acyclic and cyclic systems of general structures I–IV. These compounds show a tendency to undergo α-elimination processes to generate carbene intermediates.

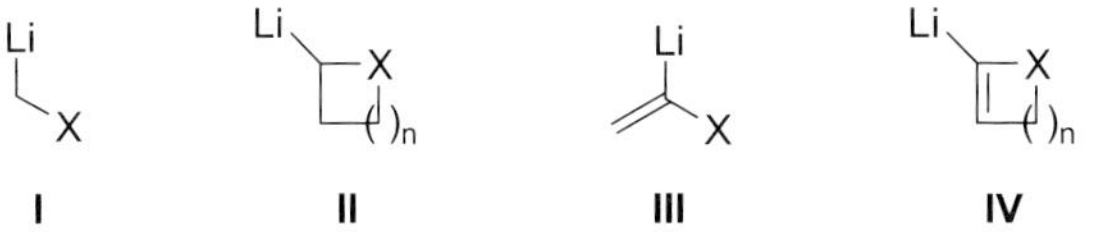

I   II   III   IV

2.2.1
sp^3-Hybridized α–Oxygenated Organolithium Compounds

sp^3-Hybridized α-oxygenated organolithium compounds of general structure I are accessible mainly through halogen–lithium exchange, tin–lithium exchange and by direct deprotonation of appropriate precursors, which means systems with hydrogens at a benzylic or allylic position. Lithiomethyl alkyl ether 2 has been prepared by chlorine–lithium exchange from the corresponding chloromethyl ether 1, lithium powder in the presence of a catalytic amount of 4,4′-di-*tert*-butylbiphenyl (DTBB) [7], or naphthalene, as well as biphenyl-supported polymers [8], being used as lithiating reagents (Scheme 2.1).

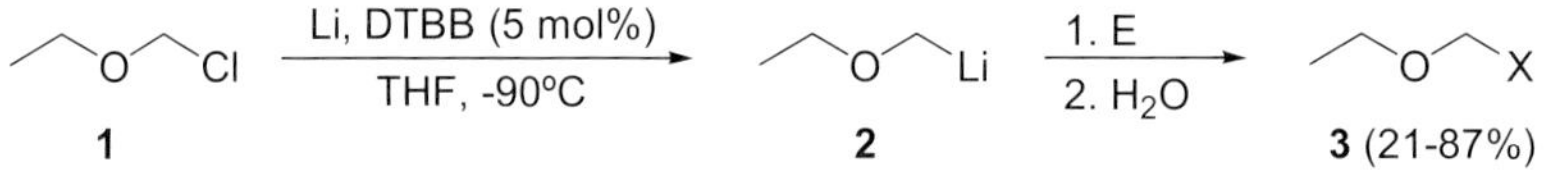

[E = RCHO, R_2CO, CO_2, PhCN, $PhCONMe_2$, CyNCO, PhN=CHPh]

Scheme 2.1

In the case of chiral lithiomethyl ethers 4 bearing the menthol unit, the lithiation was performed either in the presence of the electrophile (Barbier-reaction conditions) [9] at 0 °C or at –90 °C in a two-step process, which means lithiation followed by addition of the electrophile. Low stereoselectivity was observed in the case of using electrophiles with prostereogenic centers [10]. Similar results were obtained also in the case of preparing intermediates 4 and 5 through a tin–lithium exchange [11]. The O-chloroalkyl carbamates 6 were also lithiated by means of lithium in the presence of a catalytic amount of DTBB under Barbier-reaction conditions, because the step by step process at any temperature tried was not effective or gave poor yields. Final hydrolysis with lithium hydroxide or reduction with DIBALH gave 1,2-diols in the case of using carbonyl compounds as electrophiles [12].

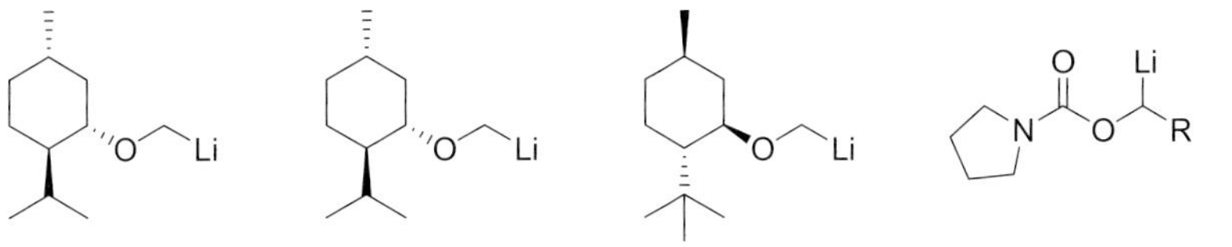

4   *ent*-**4**   **5**   **6**

It is possible to transform stereoselectively the anomeric position of a sugar unit into a organolithium compound by chlorine–lithium exchange using a stoichiometric amount of lithium naphthalene as the lithiating reagent [13]. The same kind of intermediates were prepared also by tin–lithium transmetallation [14] and sulfur–lithium exchange in phenyl sulfides and sulfones [15]. In the case of the dianion **8** derived from galactopyranose, the deprotonation of the amide moiety in compound **7** should be performed prior to heteroatom–lithium exchange. The reaction of this intermediate with electrophiles provided α-C-glycosides **9** (Scheme 2.2).

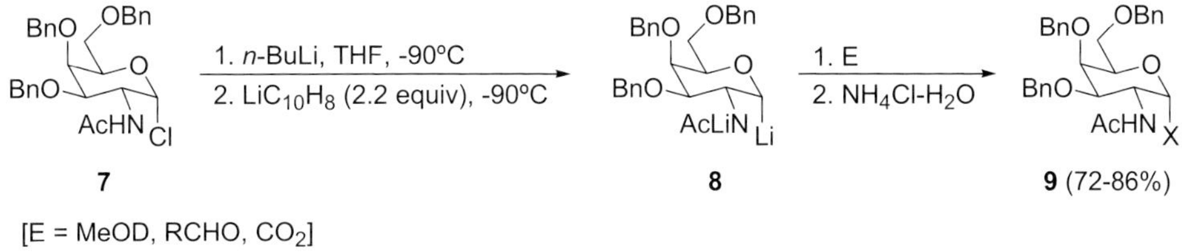

7   **8**   **9** (72-86%)

[E = MeOD, RCHO, CO$_2$]

Scheme 2.2

High diastereoselectivity was observed in the reaction of intermediate **11** (prepared by tin–lithium exchange from compound **10**, which derived from L-valine) with benzaldehyde to give the corresponding adducts in a 91:9 diastereomeric ratio. Final hydrolysis yields the 1,2-diol **13** and the chiral auxiliary **12** (Scheme 2.3) [16].

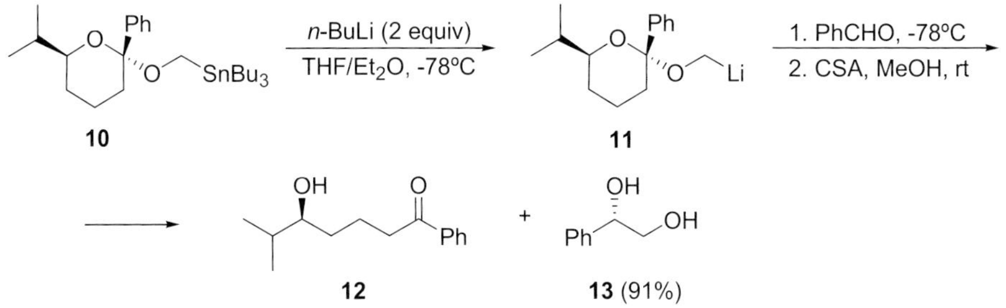

10   **11**

12   **13** (91%)

Scheme 2.3

Compounds **14**[11]-**18** have also been prepared from the corresponding stannanes. Surprisingly, cyclic α-alcoxy-β-aminoalkyllithium compounds **15** and **16** are stable at −78 °C, because, taking into account Baldwin's rules along with the microscopic reversibility principle, a β-elimination process (which would lead to a olefin) is the reverse of the n-*endo-trig* cyclization. Thus, the stability of intermediates **15** decreases with increasing ring size [17]. In the case of other α-stannylated hydroxycompounds, the corresponding transmetallation occurs only for their carbamate derivatives [18] but not in the case of *O*-MOM protected systems [19]. The reaction of intermediates **17** with aldehydes followed by reduction with AlH_3 gave 1,2-diols. The lithiation of enantiomerically enriched α-propargyloxy stannanes proceeds with complete inversion of the configuration and has been applied to study the Wittig rearrangement. In the case of compounds **18** the selectivity of [2,3] versus [1,2] rearrangement is dependent of the nature of the R^2 substituent on the acetylenic moiety [20].

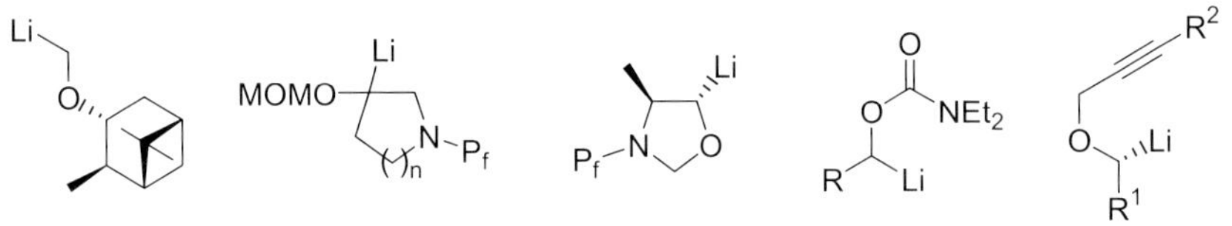

14   **15** (n = 0-2)   **16**   **17**   **18**

[P_f = 9-phenyl-9-fluorenyl]

The α-alkoxy alkyllithium derivative **20**, prepared from compound **19** by tin–lithium exchange with retention of the configuration, undergoes 5-*exo-trig* cyclization to form first the cyclic compound **21** and finally the *trans*-disubstituted tetrahydrofuran **22** through a β-elimination process (Scheme 2.4) [21].

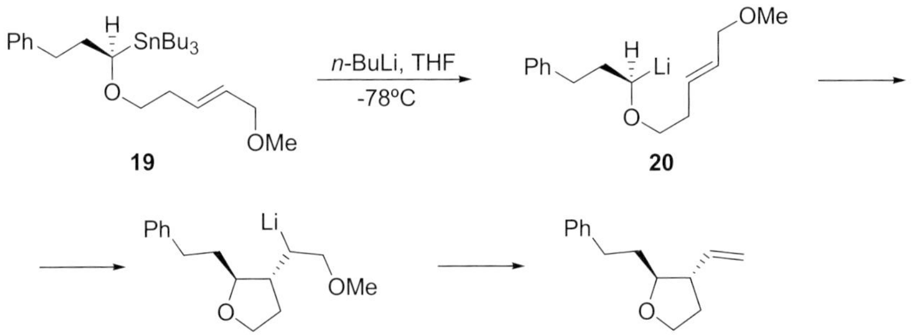

19   **20**

21   **22** (56%)

Scheme 2.4

The α- and β-configurated glycosyl anions **23–26** were prepared by tin–lithium exchange. *C*-Glycosilated aminoacids were obtained by reaction of intermediate **23** with lactams [14], meanwhile the reaction of compounds **24** and **25** with electrophiles led to α- and β-*C*-galactosides [22], respectively, intermediate **26** being used in the stereocontrolled synthesis of carba-*C*-disaccharides [23].

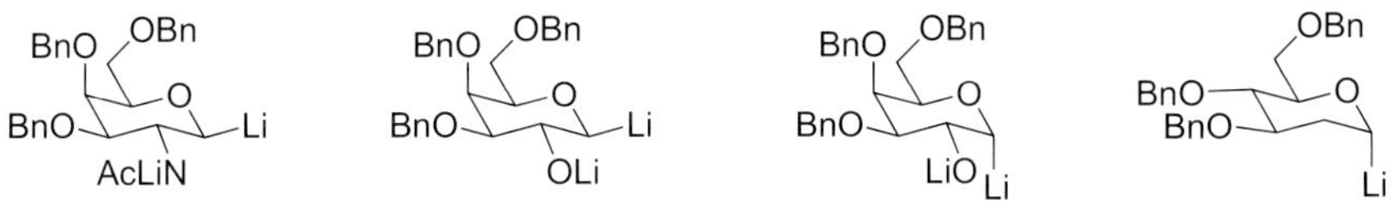

23   **24**   **25**   **26**

As commented above, this kind of functionalized organolithium compounds can also be prepared through deprotonation processes. The deprotonation can be also performed in a diastereoselective way, for instance, in the presence of (–)-sparteine [24]. Treatment of *O*-alkyl carbamate **27** with *s*-BuLi in the presence of (–)-sparteine at –78 °C gave the organolithium compound **28** with high ee, which upon carboxylation and acidic hydrolysis led to (*R*)-pantolactone **29** (Scheme 2.5) [25].

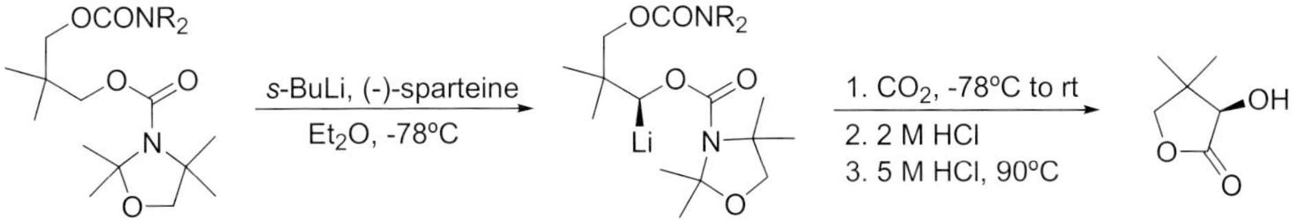

27   **28**   **29** (80%, >95% ee)

Scheme 2.5

The organolithium compound **30** was also prepared by an enantioselective de-protonation in the presence of (–)-sparteine. It is configurationally stable, but when reacting with electrophiles retention or inversion of its configuration is ob-served [26]. The deprotonation can also be performed in the presence of a chiral bisoxazolidine to give benzylic anions **31–33**, which reacted with electrophiles to yield the corresponding reaction products with good stereoselectivity. In this case, the stereoselectivity was explained by a dynamic thermodynamic mechanism instead of through an enantioselective deprotonation [27]. The *α*-lithiated *O*-pro-tected propargyl alcohol **34** was prepared by deprotonation with *n*-BuLi and used in the synthesis of a cyclic enediyne related to maduropeptin chromophor [28].

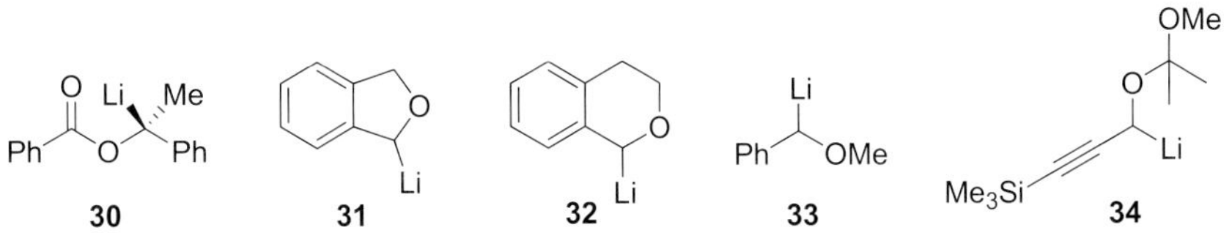

30   **31**   **32**   **33**   **34**

Lithiated epoxides [29] of general structure **II** (X = O, n = 0) were first postu-lated as intermediates by Cope, and many of them act as carbenoid species: they can undergo (a) *β*-elimination to give an enolate, (b) *α*-ring opening followed by

1,2-hydride shift to give an enolate or (c) α-ring opening followed by attack of an alkyllithium and subsequent elimination of lithium oxide affording an olefin. Only stabilized anions carrying electron-withdrawing or coordinating substituents, or heteroatomic, aromatic or unsaturated groups are stable enough to react with electrophiles. Lithiooxirane **35** was prepared by deprotonation with *s*-BuLi in the presence of TMEDA in THF at −98 °C and reacted with electrophiles [30] with retention of the configuration. The same reaction conditions were used in the preparation of oxazolinyloxiranyllithiums **36** [31], which reacted with nitrones to give, after hydrolysis and catalytic hydrogenation, α-epoxy-β-aminoacids [32]. Oxyranyllithium **37** was prepared by lithiation α to the silicon with *n*-BuLi in ether at −116 °C with retention of the configuration of the oxirane [33]. Ethynyl oxirane anionic species **38** was prepared by deprotonation with *n*-BuLi at the propargylic position and trapped with electrophiles to give trisubstituted oxiranes [34].

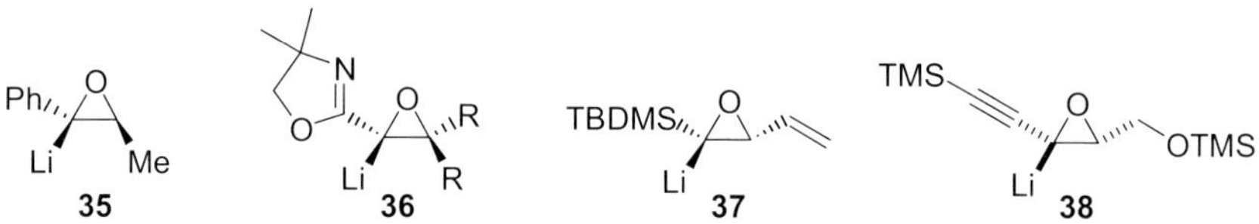

Nonstabilized oxiranyl anions were generated by lithiation of terminal epoxides in the presence of a diamine ligand [35], this methodology being applied to the asymmetric deprotonation of *meso*-epoxide **39** in the presence of (−)-sparteine. The resulting organolithium compound **40** reacted with different electrophiles to give compounds **41** in up to 86% ee (Scheme 2.6) [36].

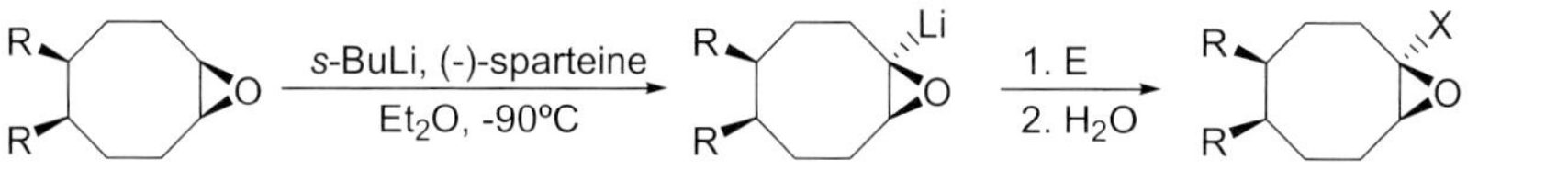

[E = CD$_3$OD, PhCHO, PhCONMe$_2$, EtCHO, Et$_2$CO, EtCONMe$_2$, EtOCOCl, *n*-Bu$_3$SnCl, Me$_3$SiCl, MeI]

Scheme 2.6

A α-oxygenated organolithium compound acts as a reaction intermediate in a spiroannulation reaction from 2-cyanotetrahydropyrans. The key feature of this method is the use of a nitrile to facilitate alkylation (to generate the corresponding precursor) and as a precursor of an alkyllithium reagent. Thus, reductive decyanation of compound **42** led to the intermediate **43** [37], which underwent intramolecular carbolithiation and, after carboxylation and reaction with CH$_2$N$_2$, gave spirocyclic ester **44** as a single diastereomer (Scheme 2.7) [38].

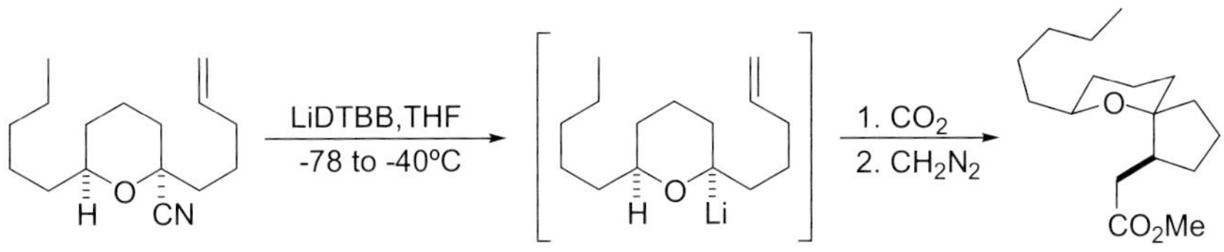

Scheme 2.7

2.2.2
sp²-Hybridized α-Oxygenated Organolithium Compounds

Simply vinyl ethers, such as compound **45**, and other vinyloxygenated systems undergo deprotonation with *t*-BuLi at low temperatures to give vinyl lithium derivatives of general structure **III** (X = OR) that react with different electrophiles [39]. For instance, the reaction of the intermediate **46** with chlorotrimethylgermane gave, after acidic hydrolysis, acylgermanane [40] (Scheme 2.8), compound **46** being an acyl anion equivalent [41]. The lithiation of 1-alkoxyallene **47** takes place more easily, using in this case *n*-BuLi as base at –78 °C in ether or THF as solvents (Scheme 2.8), intermediate **48** being stable at this temperature for several days [42]. Calculations and spectroscopic data suggest a 1,3-bridged structure for the anion **48** [43]. The reaction of this anion with α,β-unsaturated amides, followed by spontaneous Nazarov-type cyclization, gives methylene cyclopentenones [44]. The enantioselective variant of this synthesis was performed with the lithiated allene derived from D-glucose **49** [45].

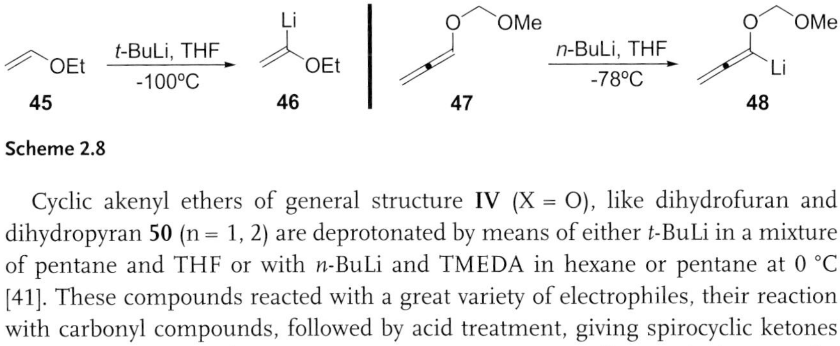

Scheme 2.8

Cyclic akenyl ethers of general structure **IV** (X = O), like dihydrofuran and dihydropyran **50** (n = 1, 2) are deprotonated by means of either *t*-BuLi in a mixture of pentane and THF or with *n*-BuLi and TMEDA in hexane or pentane at 0 °C [41]. These compounds reacted with a great variety of electrophiles, their reaction with carbonyl compounds, followed by acid treatment, giving spirocyclic ketones after suffering pinacol rearrangement [46]. Enantiomerically enriched spirocyclic compounds were obtained from the intermediate **51** (prepared from tin–lithium exchange) [47], or α-lithiated glucals **52** and **53** (prepared by direct deprotonation with *t*-BuLi) [48].

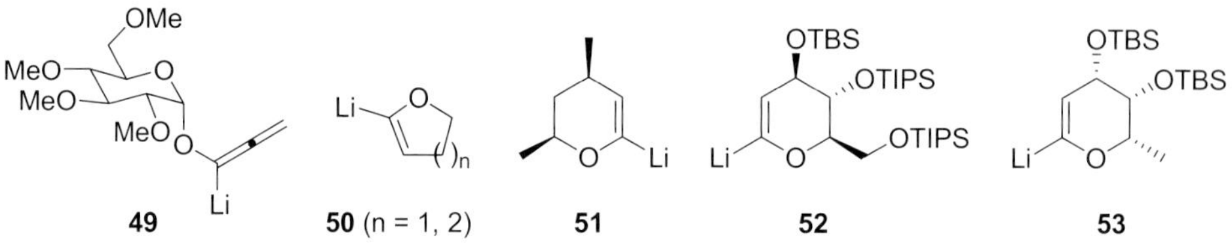

49   **50** (n = 1, 2)   **51**   **52**   **53**

Similarly, 2,3-dihydro-1,4-dioxin can be lithiated at a vinylic position with either *n*-BuLi at 0 °C or *t*-BuLi at −30 to −20 °C [43a]. Related benzocondensed systems **54** were lithiated with LDA at −78 °C to give the dianionic intermediates **55**, which were converted into lactones **56** upon successive reaction with acetaldehyde and propanoic acid [49] (Scheme 2.9).

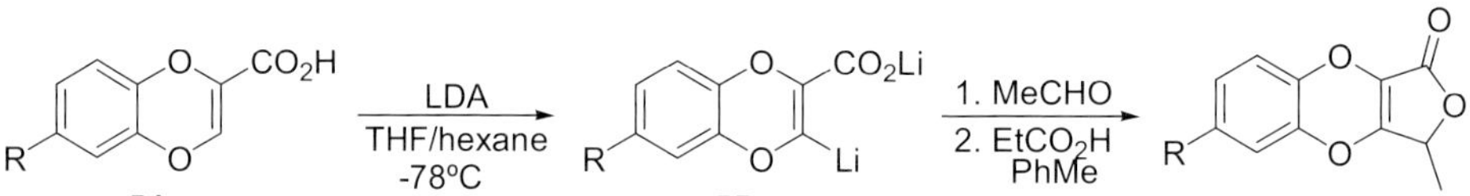

54   **55**   **56** (88-93%)

Scheme 2.9

2.2.3
sp³-Hybridized α-Nitrogenated Organolithium Compounds

There are two types of α-aminoalkyl organolithium compounds of general structures **I** (X = NRR′) and **II** (X = NR): those unstabilized (derived from tertiary alkylamines) and those dipole-stabilized.

The stabilized derivatives can be prepared by α-deprotonation of compounds where the nitrogen lone pair of electrons are involved in conjugation with unsaturated systems using strong bases [50]. This situation makes the nitrogen a more electron-withdrawing group, a stabilization of the organolithium compound occurring sometimes by intramolecular coordination of the metal with the functional group that contains the nitrogen atom. In the case of hindered amides α-lithiation to the amide nitrogen leads to stable organolithium compounds. Stereospecific deprotonation with *t*-BuLi of *N*-(α-methylbenzyl)-*N*-isopropyl-*p*-methoxybenzamide gave the corresponding configurationally stable tertiary benzyllithium derivative **57**, which underwent a stereospecific dearomatization-cyclization with >99% retention of the stereochemistry. This methodology was used in the synthesis of kainic acid derivatives [51]. Other stabilized α-nitrogenated organolithium compounds have been prepared by deprotonation of the appropriate precursors. The pyrazole derivative **58** was obtained by deprotonation with *n*-BuLi [2a], whereas in the case of the formamidine derivative **59** *t*-BuLi was used as base [52]. In the case of the lithiated cyclic *N*-nitrosoamine **60**, the deprotonation was performed with LDA to give exclusively the axial lithium derivative, which is stabilized by interaction with the LUMO of the adjacent π-system [53]. Quinuclidine N-oxide was lithiated with t-BuLi to give the intermediate **61** [54]. All these anionic systems **57–61** reacted with electrophiles to give functionalized molecules in good yields.

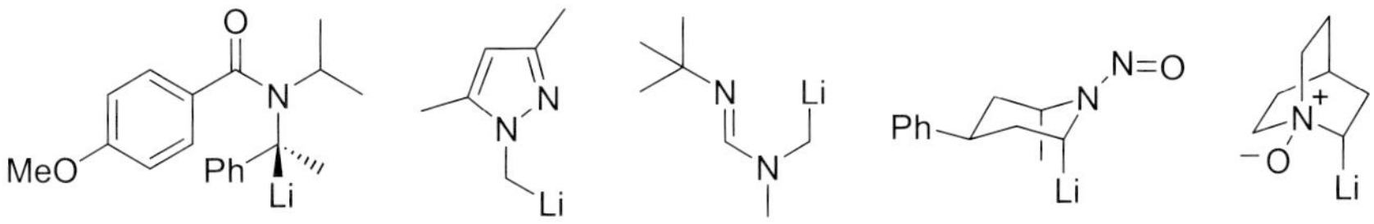

Dipole-stabilized α-aminoorganolithium compounds are less reactive than the unstabilized ones and for that reason more resistant to suffer racemization. Enantioselective deprotonation of N-Boc pyrrolidine **62** with s-BuLi and (–)-sparteine gave the chiral configurationally stable [55] organolithium compound **63**. Due to the fact that (–)-sparteine is only available in one enantiomeric form, O'Brien et al. reported the synthesis of a (+)-sparteine surrogate, which worked as a successful chiral ligand in the same processes as (–)-sparteine, so both enantiomers are available by choosing the right diamine ligand [56]. The enantio-determining step is the formation of the organolithium derivative, the reaction of this intermediate with electrophiles leading to chiral compounds with high stereoselectivity [57]. For instance, when 1-pyrrolidinecarbonyl chloride was used as electrophile, the corresponding amide **64** was obtained in 85% ee [58] (Scheme 2.10). Intermediate **63** can also be prepared by tin–lithium exchange from the corresponding enantiopure 2-(tributylstannyl)pyrrolidine because tin–lithium exchange takes place with retention of the configuration [59].

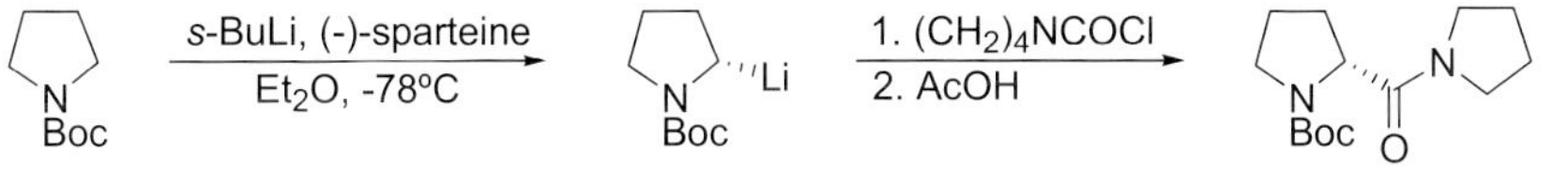

Scheme 2.10

The unchelated dipole-stabilized α-aminobenzyl organolithium compound **65** was prepared by deprotonation and used in the synthesis of the azaphenanthrene alkaloid eupolauramine [60]. The N-lithiomethylcarbamate **66** was prepared by DTBB-catalyzed lithiation of the corresponding chlorinated precursor in the presence of carbonyl compounds as electrophiles, the resulting adducts being transformed into the corresponding β-aminoalcohols [61]. Chiral N-lithiomethylpyrrolidine **67** was prepared through either a tin–lithium exchange with n-BuLi or by carbon–sulfur bond cleavage with lithium 4,4'-di-tert-butylbiphenyl (LiDTBB) from the corresponding prolinol derivatives. The reaction of the intermediate **67** with aldehydes gave β-aminoalcohols in good yields but poor diastereoselectivity [62]. The corresponding unchelated α-amino organolithium compound **68** was also prepared by tin–lithium transmetallation and reacted with electrophiles in moderate yields [63]. Reductive lithiation of a bicyclic oxazolidine with lithium in the presence of a catalytic amount of naphthalene (10 mol%) gave the benzylic organolithium compound **69**, which is stabilized by the alkoxide group. Epimerization at the benzylic position takes place rapidly and one of the diastereomers reacts faster than the other with alkyl halides to give mainly *syn*-products [64]. Nonstabilized aziridinyl anions can be prepared by direct deprotonation of aziridine borane complexes. Thus, deprotonation with *s*-BuLi of 1-(*tert*-butyldimethysililoxyethyl)-aziridine borane occurred *syn* to the boron substituent to give intermediate **70**, while lithiation of 1-(*tert*-butyldimethysililoxyethyl)-2-trimethylstannylaziridine borane occurred *anti* to the boron and tin, due probably to steric effects. When the deprotonation was performed in the presence of (–)-sparteine, stereoselectivity went up to 70% ee [65].

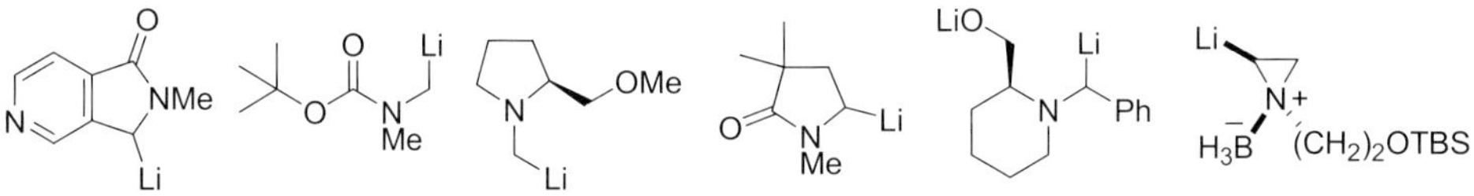

A trifluoromethyl stabilizing aziridinyl anion **72** was generated by deprotonation of the optically active *N*-tosyl-2-trifluoromethylaziridine **71** with *n*-BuLi. The organolithium intermediate **72** reacted with electrophiles to give the corresponding adducts **73** in good yields (Scheme 2.11), the whole reaction occurring with retention of the configuration at the stereogenic carbon center [66].

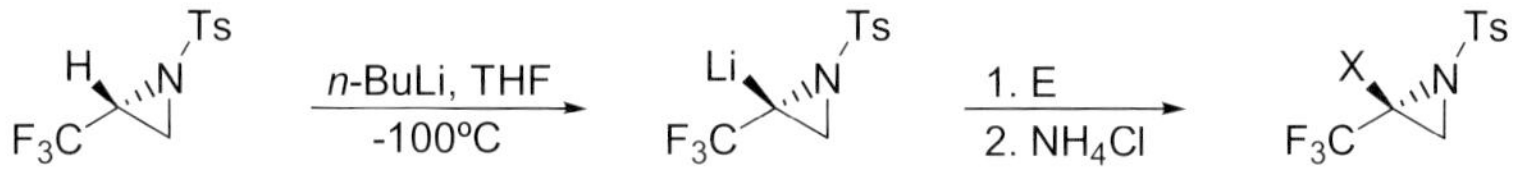

[E = PhCHO, Ph$_2$CO, PhCOMe, EtO$_2$CCHO, PhCOCl, MeOCOCl, EtCOCl, BnOCOCl]

Scheme 2.11

2.2.4
sp^2-Hybridized α-Nitrogenated Organolithium Compounds

Recently, different α-lithioenamines **75** have been prepared by chlorine–lithium exchange from the corresponding chloroenamines **74** and reacted with electrophiles to give functionalized enamines **76**. A mixture of lithium and a catalytic amount of DTBB was used as the lithiating reagent (Scheme 2.12) [67]. The process can be performed either step-by-step (lithiation-reaction with the electrophile) at –90 °C or under Barbier-reaction conditions at –40 °C. In the case of using carbonyl compounds as electrophiles, after acidic hydrolysis, α-hydroxyketones were obtained, intermediates **74** acting in this case as acyl anion equivalents [41].

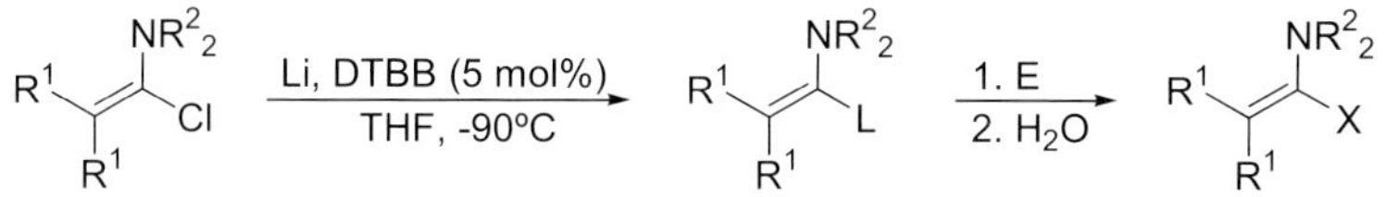

[E = D$_2$O, Me$_3$SiCl, PhCOCH=CHPh, CO$_2$, CyNCO]

Scheme 2.12

Some formamidines underwent lithiation of a vinylic position α to nitrogen, so for instance vinyllithium compound **77** was prepared by deprotonation with *t*-BuLi of the corresponding formamidine [68]. In the case of hydrazones, a double lithiation occurred leading, for instance, to intermediate **78**, where the second C-lithiation took place at the trigonal carbon atom α to the C=N bond [69]. Lithiated cyclic enamines **79** were prepared by tin–lithium transmetallation and coupled with different electrophiles [70]. Organolithium compounds with a nitrogen at the α-position and sp^2-hybridation can be prepared by direct metallation of nitrogen-containing aromatic heterocycles [71]. Thus, direct deprotonation of *N*-methylimidazol gave 2-lithio-*N*-methylimidazol **80** [72]. Direct lithiation of *N*-benzyloxypyrazole with *n*-BuLi gave compound **81**, which reacted with diethyl *N*-Boc-iminomalonate in the synthesis of *N*-hydroxypyrazole glycine derivatives [73]. 2-Lithiobenzothiazole **82** reacted with galactonolactone being used in saccharide chemistry [74]. Selective monobromo–lithium exchange from the corresponding dibromopyridine [75] gave intermediate **83**, which by reaction with a first electrophile and subsequent new lithiation and reaction with a second electrophile was used in the synthesis of a ceramide analog [76]. Using LiTMP as a base, organolithium compound **84** was prepared from 2-chloropyrazine and, after reaction with aldehydes, the resulting products were used in a route to the wheat-disease-impeding growth agent septorin [77]. The pyrazolopyrimidine lithium derivative **85** was prepared by tellurium–lithium transmetallation [78].

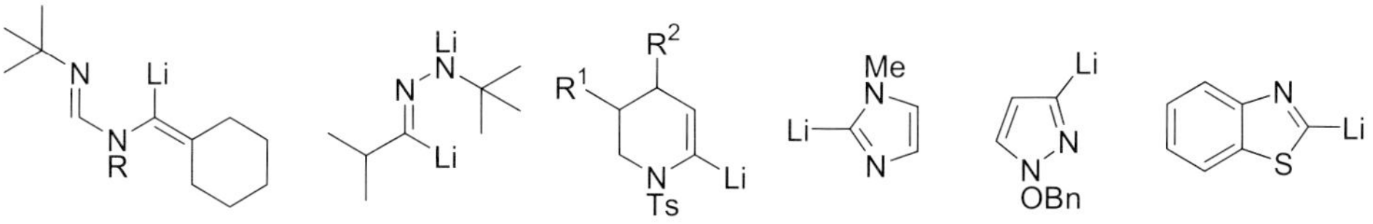

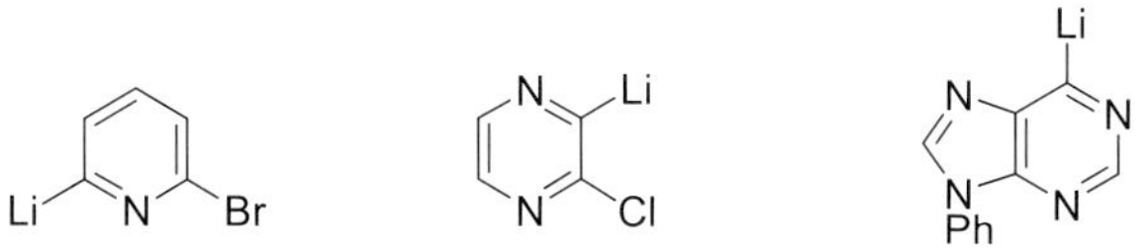

2.2.5
Other sp²-Hybridized α-Functionalized Organolithium Compounds

Two general strategies have been followed for preparing α-haloalkyllithiums: (a) α-deprotonation of vinyl halides with strong bases and (b) halogen–lithium exchange in 1,1-dihaloalkenes [43a,79]. In both cases the processes having to be carried out at low temperature (–100 °C). The deprotonation of (Z)-1-chloro-1,3-butadiene **86** with lithium 2,2,6,6-tetramethylpiperidine (LiTMP) at –90 °C gave (Z)-1-lithio-1-chloro-1,3-butadiene **87**. The reaction with benzaldehyde as electrophile took place with retention of the configuration of the double bond, giving compound **88** after hydrolysis (Scheme 2.13) [80].

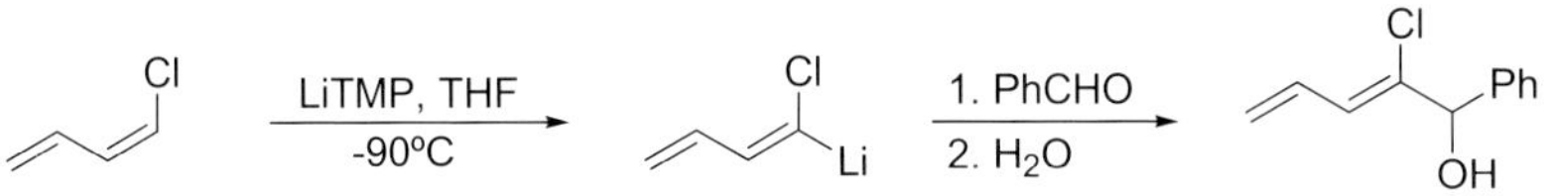

Scheme 2.13

Although vinyl bromides and iodides are useful precursors of vinyllihiums by reaction with alkyllithiums through a halogen–lithium exchange, in the case of chlorinated materials, deprotonation occurs preferentially. Thus, treatment of 1,1-diphenyl-2-chloroethylene with *n*-BuLi in THF at –100 °C gave the vinyllithium derivative **89** (deprotonation) [81], whereas under the same reaction conditions, chlorotrifluoroethylene gave intermediate **90**, (deprotonation is not possible) so a fluoro-stabilized vinyllithium compound by a chlorine–lithium exchange is formed [82].

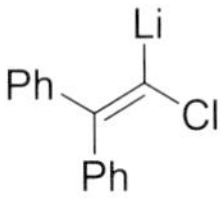

2.3
β-Functionalized Organolithium Compounds

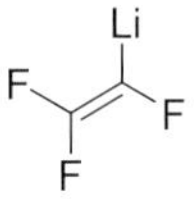

Organolithium compounds bearing a functional group at β-position are unstable species. They show a great tendency to undergo a β-elimination process to give olefins, the better leaving group the higher instability. The hybridization of the carbon atom attached to the metal could be sp³ (general structures **V** and **VI**) as well as sp² (general structures **VII–X**) in both cyclic and acyclic systems.

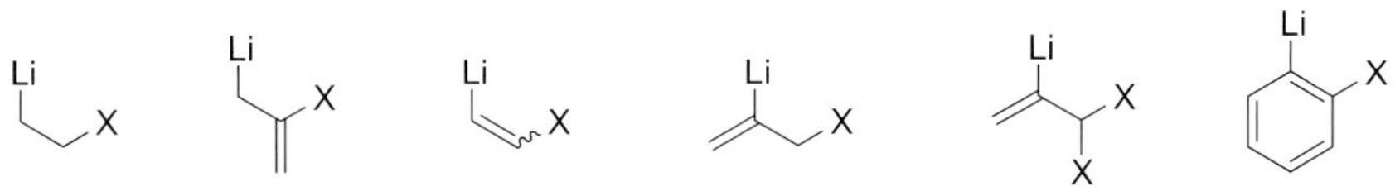

V VI VII VIII IX X

2.3.1
sp^3-Hybridized β-Functionalized Organolithium Compounds

One way to prevent the decomposition of β-functionalized organolithium compounds **V-X** (X = OR, NR_2) is by reducing the β-elimination process developing a negative charge on the heteroatom. These intermediates have been prepared following three general strategies: (a) mercury–lithium transmetallation, (b) chlorine–lithium exchange (in both cases oxygen and nitrogen deprotonation should be carried out first) and more recently by (c) reductive opening of epoxides and aziridines [83]. Deprotonation of organomercurials **91** with phenyllithium followed by mercury–lithium transmetallation with lithium metal gave organolithium intermediates **92**, which are stable at –78 °C and reacted effectively with electrophiles to give polyfunctionalized compounds **93** (Scheme 2.14) [84].

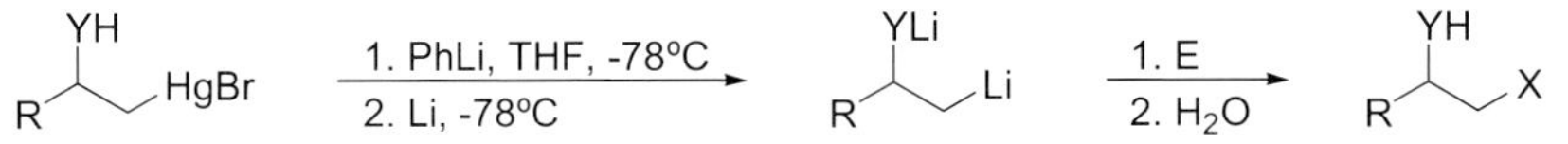

91 **92** **93** (32-90%)

[Y = O, NPh
E = H_2O, D_2O, CO_2, Me_3SiCl, R^1Br, Me_2S_2, R^1R^2CO, $R^1CH{=}NR^2$]

Scheme 2.14

Since enantiomerically pure epoxides are easily accessible or commercially available, chiral products can be obtained through a reductive cleavage followed by reaction with electrophiles. The DTBB-catalyzed lithiation of epoxide **94**, derived from D-glucose, led to the organolithium derivative **95**. In the case of unsymmetrically substituted epoxides, the regiochemistry of the ring opening always led to the most stable organolithium compounds, what is the case of the primary intermediate **95**. Its reaction with electrophiles followed by hydrolysis gave the corresponding 3*C*-substituted D-glucose derivatives **96** (Scheme 2.15) [85].

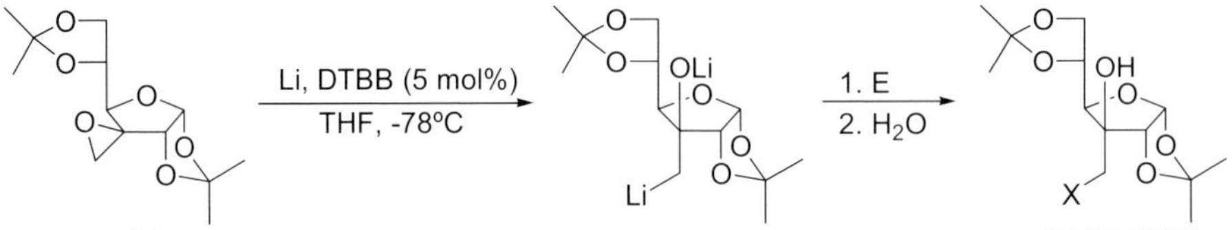

94 **95** **96** (20-75%)

[E = H_2O, D_2O, Me_3SiCl, CO_2, R_2CO]

Scheme 2.15

Enantiomerically pure intermediates **97** and **98** of type **V** (X = OLi) were prepared from the corresponding chiral chlorohydrines by deprotonation first with *n*-BuLi and subsequent chlorine–lithium exchange with lithium naphthalenide [86]. The reaction of these dianions with prostereogenic carbonyl compounds showed low diastereoselectivity. The same strategy was followed in order to prepare these intermediates but starting from β-(phenylsulfanyl)alcohols through a sulfur–lithium exchange [87]. Oxygen functionalized organolithium compounds **99** [88] and **100** [89] were prepared by reductive opening of the corresponding chiral epoxides with LiDTBB and used in the synthesis of calcitriol lactone and diterpene forskolin, respectively. Dianionic intermediate **101** was also prepared by DTBB-catalyzed lithiation of the corresponding epoxide derived from (–)-menthone and used in the synthesis of functionalized terpenes [90].

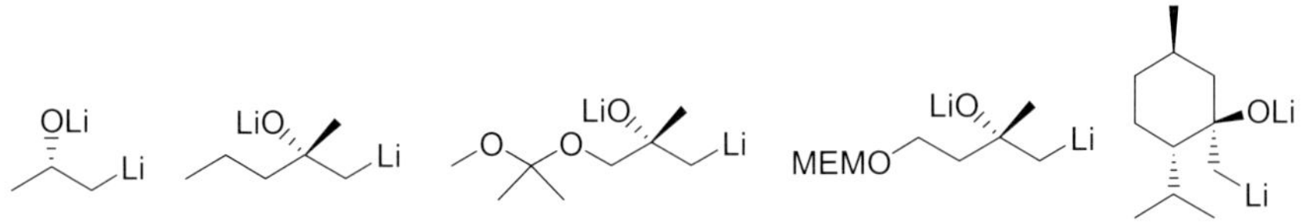

97 **98** **99** **100** **101**

Dilithium derivatives **102–105** were generated by reductive opening of epoxides derived from D-glucose, D-fructose, estrone and cholestanone, respectively, and trapped with different electrophiles. In this way, 6*C*-substituted 6-deoxy-D-glucose, 3*C*-substituted D-psycose [85], 17*C*-substituted-17β-estradiol and 3*C*-substituted-3α-cholestanol [91] derivatives were prepared.

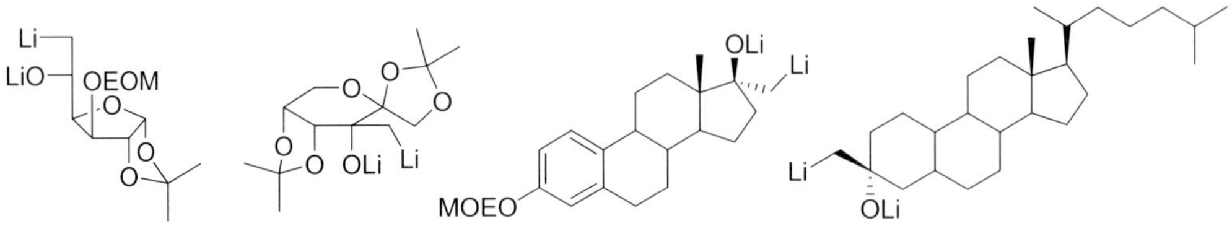

102 **103** **104** **105**

As has been described before, one of the most important problems to be overcome in the preparation of sp³-hybridized *β*-functionalized organolithium compounds is their decomposition by a *β*-elimination process to give olefins. Advantage of this reaction can be taken to prepare olefins starting from organomercurials, chlorohydrins and epoxides by performing the lithiation process at higher temperatures [92].

The reductive opening of aziridines **106** should be performed at –78 °C with an excess of lithium in the presence of a catalytic amount of an arene. A limitation in this methodology is that a phenyl or aryl group should be present as a substituent either at the nitrogen or at one of the carbon atoms of the aziridine ring. Regarding the regiochemistry of the process, the most stable primary organolihium intermediate **107** is always formed that by reaction with electrophiles and final hydrolysis led to the corresponding functionalized amines **108** (Scheme 2.16) [93].

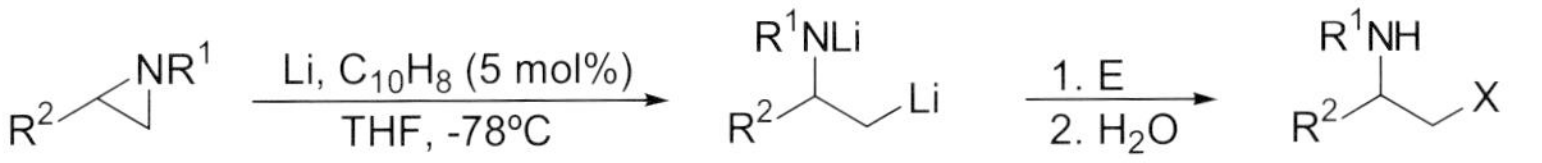

[E = H₂O, D₂O, RHal, Me₂S₂, RCHO, R³R⁴CO, (EtO)₂CO, CH₂=CHCO₂Et]

Scheme 2.16

The enantiomerically enriched *β*-nitrogenated organolithium compound **109** was prepared by reductive opening of the aziridines derived from (–)-ephedrine [93], whereas compounds **110** and **111** were generated from the corresponding chlorinated precursors through a DTBB-catalyzed lithiation by a chlorine–lithium exchange [94]. The same methodology but using lithium naphthanelide was employed to prepare the intermediate **112**. Compound **113** was formed by stereoselective deprotonation of the corresponding *N*-Boc-cyclopropylamines [96]. The reaction of all these intermediates **109**–**113** with electrophiles yielded chiral regioselectively functionalized nitrogenated compounds.

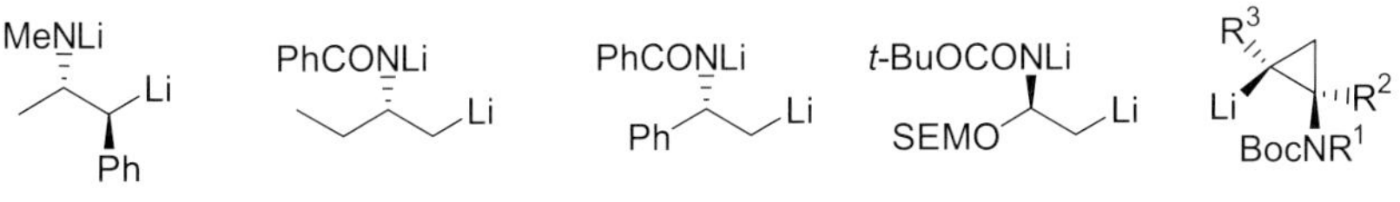

Although allylic organolithium compounds of the type **VI** decompose easily to give allenes, deprotontation of methyl isopropenyl ether with a mixture of *n*-BuLi/ *t*-BuOK at –78 °C gave the allylic intermediate **114**, which was trapped with electrophiles. At –30 °C compound **114** decomposes to give allene [97]. The organolithium derivative **115** was prepared by deprotonation with LDA at 0 °C and decomposed immediately to give the corresponding allene [98]. However, cyclic allylic organolithium compounds **116** [99] and **117** [100], which are more stable, were also prepared by direct deprotonation with LDA and *t*-BuLi, respectively, intermediate **116** being used in the synthesis of tetrahydropyranones.

114   **115**   **116**   **117**

2.3.2
sp²-Hybridized β-Functionalized Organolithium Compounds

Alkenyllithium compounds of type **VII** have been prepared mainly by halogen–lithium exchange. β-Ethoxyvinyllithium derivatives **118** and **119** were accessible from the corresponding vinyl bromides by stereospecific bromine–lithium exchange with *n*-BuLi at –78 °C. The (*Z*)-isomer **119** is much more stable than the (*E*)-isomer **118** because in this case an antiperiplanar elimination of lithium ethoxide is more favorable. In the case of (*Z*)- and (*E*)-diethoxyvinyllithiums **120** and **121**, they have been prepared by deprotonation with *t*-BuLi at –78 °C of the corresponding diethoxyethylenes. The (*Z*)-isomer **121** is also more stable than the (*E*)-isomer **120** due to the previously mentioned antiperiplanar β-elimination of lithium ethoxide. However, compounds **120** and **121** are much more stable than the monoethoxy substituted derivatives **118** and **119** due to the inductive effect of the oxygen atom at the α-position. All these intermediates reacted with electrophiles regio- and stereoselectively [101]. β-(Trimethylsilyl)vinyllithium **122**, generated from β-(trimethylsilyl)vinyl bromide with *t*-BuLi, reacted with β-alkenoyl acylsilanes to give eight-membered carbocycles [102]. Many vinyllithium compounds with other functional groups have been described in the literature, for instance, the cyclic β-bromoderivative **123** that was prepared from the corresponding dibromide through a selective monobromo–lithium exchange with *t*-BuLi at –78 °C [103]. In this case, decomposition through a synplanar process is obviously inhibited. Intermediates of type **X** are, in general, stable species and therefore useful functionalized organolithium reagents in order to prepare aromatic compounds. They have received much attention and are generated by direct deprotonation [71,104a] or by halogen–lithium exchange [75,104b]. For instance, the aryllithium **124** was prepared by iodo-lithium exchange with *n*-BuLi in toluene and used in the synthesis of antitumor antibiotics duocarmycins [105].

118   **119**   **120**   **121**   **122**   **123**   **124**

Dianions **126**, alkenyllithium compounds of type **VIII**, have been obtained by halogen–lithium exchange from the corresponding chlorinated or brominated precursors **125** using lithium naphthalenide; it was necessary to carry out a previous deprotonation with PhLi. The reaction of **126** with electrophiles gave regioselective functionalized allyl amines **127** (Scheme 2.17) [106].

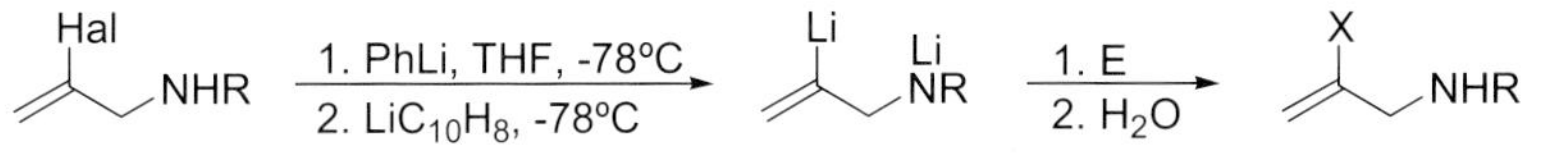

$$\begin{bmatrix} Hal = Cl, Br \\ E = H_2O, D_2O, Me_2S_2, R^1R^2CO, CH_2=CH_2CH_2Br \end{bmatrix}$$

Scheme 2.17

Vinyllithium derivatives **128–131** have been generated through a bromine–lithium exchange with *t*-BuLi. In the case of **128**, used in the synthesis of (–)-wodeshiol, a deprotonation of the alcohol functionality was performed prior lithiation [107]. Intermediates **129** were also used in the synthesis of triquinanes [102,108], meanwhile **130** acted as an intermediate in a synthetic route to (+)-pericosine B [109]. Dianionic species **131** showed an unexpected intramolecular carbometallation upon addition of TMEDA to give dilithiated dihydropyrroles, which finally reacted with different electrophiles [110].

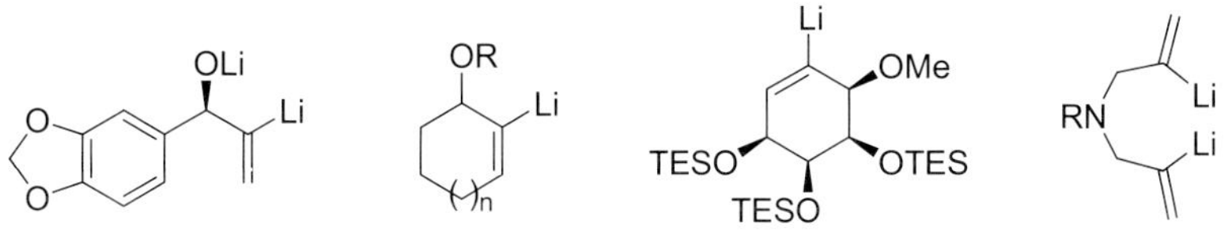

Vinyllithium acetals **133**, compounds of type **IX**, were prepared from the corresponding chlorinated precursor **132** by a DTBB-catalyzed lithiation at low temperature. They reacted with electrophiles to give compounds **134**. In the case of using chiral starting materials (**132**, R = Me), the reaction with prostereogenic carbonyl compounds took place with almost null stereoselectivity (Scheme 2.18) [111].

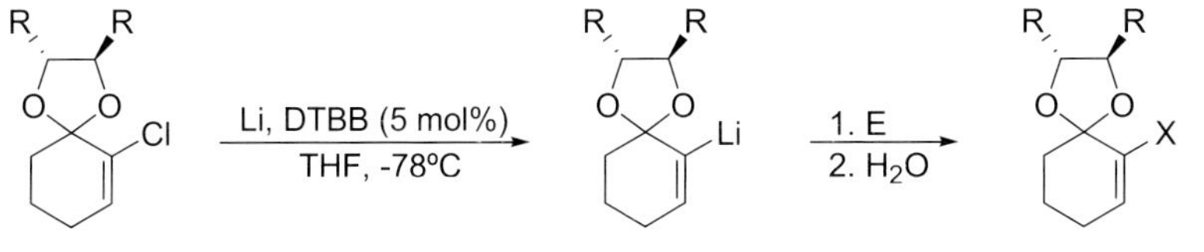

[E = H$_2$O, D$_2$O, Me$_3$SiCl, RCHO, R^{1}R^{2}CO]

Scheme 2.18

2.4
γ-Functionalized Organolithium Compounds

Organolithium compounds with a functional group at the γ position can be represented by a great number of general structures (**XI–XX**) depending on the hybridization of the carbon atoms bearing both the lithium and the functional group.

2.4.1
γ-Functionalized Alkyllithium Compounds

Functionalized organolithium compounds of type **XI** can be accessible through a large number of methodologies. They are accessible by halogen–, sulfur–, or selenium–lithium exchange, tin–lithium transmetallation, reductive opening of four-membered heterocycles and also by carbolithiation of cinnamyl systems.

Enantiomerically pure oxygen functionalized organolithium compounds **135–138** have been prepared by halogen–lithium exchange. The precursor of **135** was the corresponding chlorohydrin and lithium naphthalenide the lithiating reagent [86], meanwhile intermediates **136** [112], **137** [113] and **138** [114] were prepared by iodine–lithium exchange by means of *t*-BuLi. All of them reacted with electrophiles to yield polyfunctionalized compounds, compound **137** being used in the synthesis of scopadulcic acid A.

The carbolithiation of cinnamyl methyl ether **139** with an alkyllithium in the presence of TMEDA led to the corresponding benzylic organolithium intermediates **140**, which by reaction with electrophiles gave compounds **141** with high diastereoselectivity (Scheme 2.19) [115].

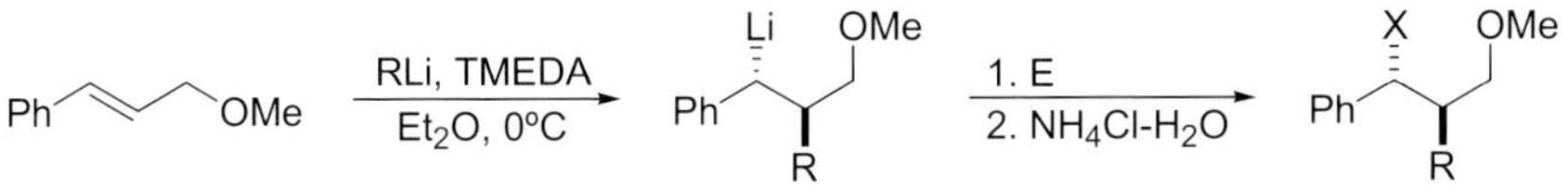

139 140 141 (59-68%)

[E = MeOH, MeOD, CO_2, $(MeS)_2$, MeI]

Scheme 2.19

When the carbolithiation of *tert*-butyl cinnamyl ether is performed in the presence of (–)-sparteine, chiral organolithium compounds **142** were obtained, so chiral 2-alkyl-3-phenylpropanols were the reaction products, after hydrolysis [116]. Dianionic species **143–145** were prepared by reductive opening of different heterocycles [83] by an arene-catalyzed lithiation and reacted with electrophiles to give polyfuncionalized molecules in good yields. Enantiomerically pure oxetanes [117,118] (for **143**), 4-phenyl-1,3-dioxolanes [119] (for **144**) and 2-phenyl-substituted azetidines or tietanes [120] (for **145**) were the starting heterocycles used. The reductive cleavage always gave the most stable intermediate: the primary alkyllithium compounds for the chiral oxetanes and the benzylic derivatives for the other. Although γ-halogenated organolithium compounds have a great tendency to undergo γ-elimination reactions, the norbornane derivative **146** (prepared from the brominated derivative by treatment with *t*-BuLi at –125 °C) could be trapped with carbon dioxide to give the corresponding carboxylic acid in 47% yield [121].

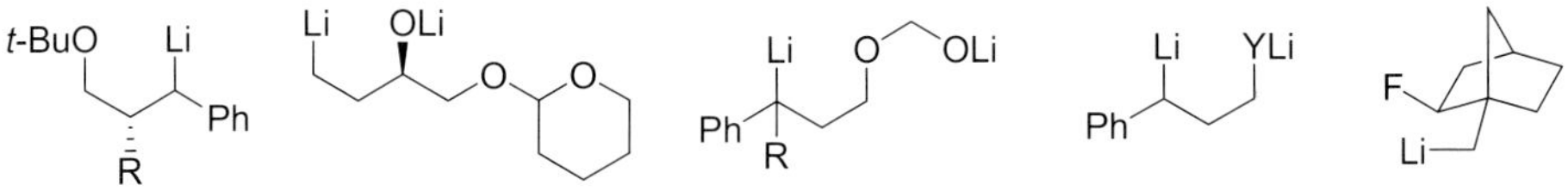

142 **143** **144** **145** (Y = S, NR) **146**

Masked lithium homoenolates of type **XII** are of interest in synthetic organic chemistry and can be considered as three-carbon homologating reagents with umpolung reactivity [122]. The lithiation of the β-chloro orthoester **147** with lithium in the presence of a catalytic amount of DTBB, under Barbier-reaction conditions, and using carbonyl compounds as electrophiles, followed by acidic hydrolysis, led to lactones **149** as reaction products, the masked lithium homoenolate **148** being proposed as a reaction intermediate (Scheme 2.20) [123].

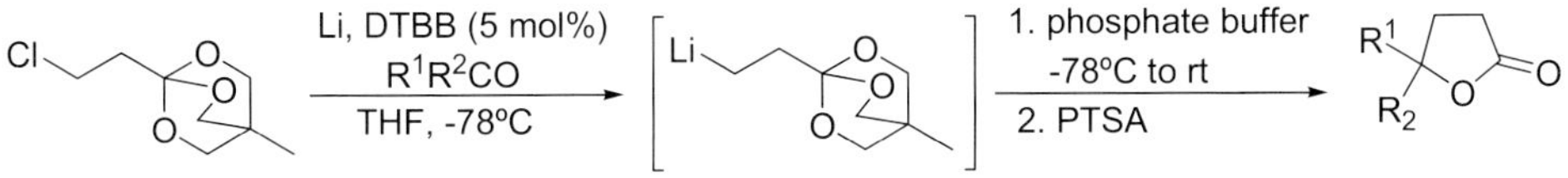

147 **148** **149** (37-45%)

Scheme 2.20

Cyclopentanone dioxolane lithium homoenolate **150** was prepared by reductive lithiation with LiDTBB of the corresponding β-penylsulfanyl derivative and alkylated with allylic bromides in the presence of copper salts [124]. The acyclic dioxane lithium homoenolate **151** was prepared by bromine–lithium exchange and used for the synthesis of highly functionalized porphyrins [125]. Carbolithiation of a cinnamyl acetal with *t*-BuLi in toluene gave the lithium homoenolate **152**, which reacted with electrophiles to give reaction products with high diastereoselectivity [126]. The chiral benzylic homoenolate **153** was prepared by deprotonation of the corresponding amide with 2 equiv of *s*-BuLi in the presence of TMEDA at –78 °C. First deprotonation deactivates the functionality and the second deprotonation is directed to the β-position of the carboxamide by the so-called complexinduced proximity effect (CIPE) [127]. In the reaction of intermediate **153** with electrophiles an almost 10:1 mixture of diastereomers was obtained [128]. Lithiated imines **154**, which were prepared by DTBB-catalyzed lithiation of the corresponding chlorinated derivatives, underwent intramolecular cyclization through an *endo-trig* process to yield 2-substituted pyrrolidines, after hydrolysis [129].

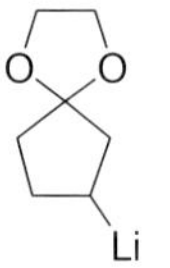

150 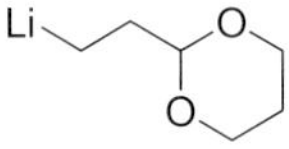   **151** 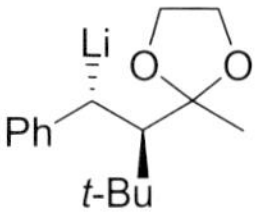   **152** 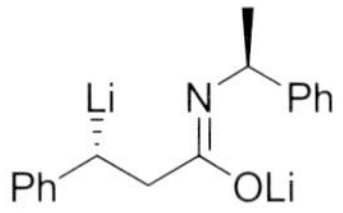   **153** 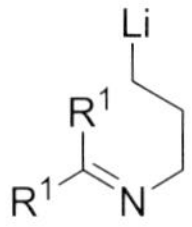   **154**

2.4.2
γ-Functionalized Allyllithium Compounds

Functionalized allyllithium compounds of type **XIII** are also homoenolate equivalents [122,130], but in their reaction with electrophiles sometimes it is not possible to control the regioselectivity. These compounds have been prepared mainly by either deprotonation or tin–lithium exchange. Deprotonation of (*E*)-cinnamyl-*N,N*-diisopropylcarbamate **155** with *n*-BuLi in the presence of (–)-sparteine in toluene gave a configurationally stable lithiated *O*-allyl carbamate (*epi*-**156**), which equilibrates at –50 °C to give the (*R*)-intermediate **156**. Whereas the reaction of these compounds with MeI and MeOTs gave the γ-attack, however acylation, silylation and stannylation took place at the α-position (Scheme 2.21) [131].

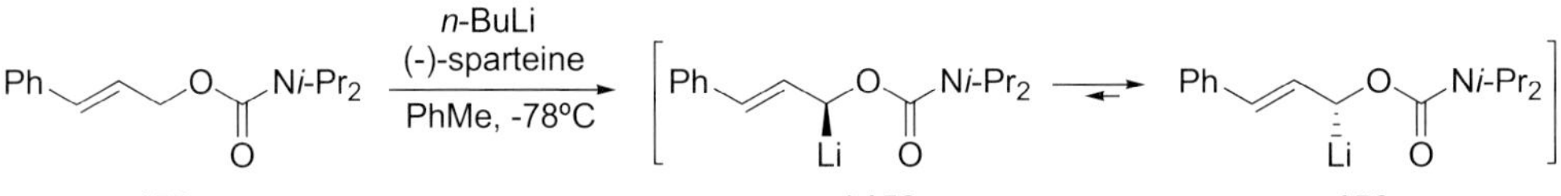

155   *epi*-**156**   **156**

Scheme 2.21

The allylic intermediate **157** was prepared by deprotonation with LDA and reacted with 1,2-dialkyloxiranes in the synthesis of parasorbic acid [132]. Direct deprotonation of silyl allyl ether with *s*-BuLi gave the allylic compound **158**, which reacted with electrophiles at the γ-position [133]. Tin–lithium transmetallation of a 3-stannylated enamine led to the intermediate **159**, which after reaction with electrophiles and acidic hydrolysis gave 3-funcionalized cyclohexenones [134]. The chiral *endo*-aminoallyllithium **160** was also prepared by tin–lithium transmetallation of the corresponding *O*-methylprolinol derivative, and alkylated to give after hydrolysis β-alkylated ketones [135]. Direct deprotonation with *t*-BuLi of *N*-methallylaniline gave the dianionic intermediate **161**, of type **XIV** [136].

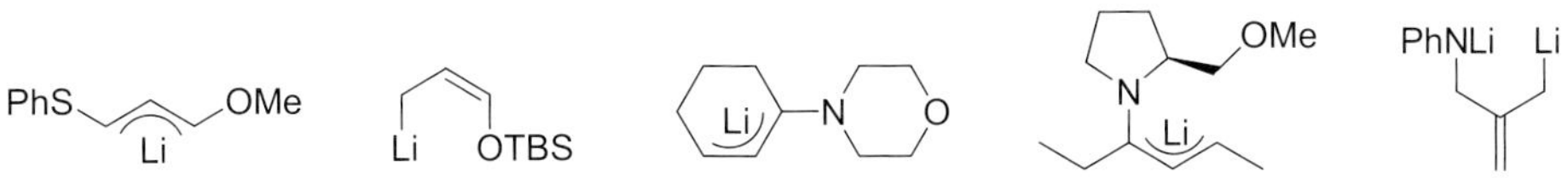

157	**158**	**159**	**160**	**161**

Starting from the symmetrical allylic diselenanyl compound **162**, and through a selenium–lithium exchange with *n*-BuLi, methylselenanyl methallyllithium compound **163** was prepared, which after reaction with electrophiles afforded products **164** (Scheme 2.22) [137].

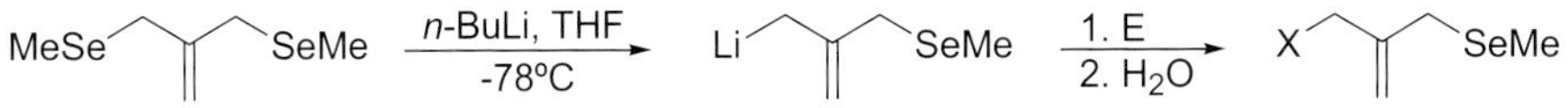

[E = RBr, R^{1}R^{2}CO]

Scheme 2.22

2.4.3
γ-Functionalized Benzyllithium Compounds

Functionalized benzyllithium compounds of type **XV** are prepared by proton abstraction at the benzylic position with appropriate bases. Trilithiated 2,6-dimethylphenol **165** was prepared by deprotonation with *n*-BuLi under hexane reflux and its tetrameric structure was determined by X-ray diffraction [138]. In the case of 2-methoxy benzyl ether, reductive cleavage of the benzylic carbon–oxygen bond by means of a naphthalene catalyzed lithiation at –10 °C in THF gave the benzyllithium derivative **166** [139], other benzyllithium compounds being prepared through this methodology. On the contrary, the benzyllithium derivatives **167** were generated by carbolithiation of *o*-vinyl substituted *N*-Boc protected aniline with *n*-BuLi in diethyl ether at –78 °C. Further reaction with DMF followed by final acidic hydrolysis yielded functionalized indoles in a one-pot process [140]. Dianions **168** [141] and **169** [142] were prepared by double deprotonation with LDA and *n*-BuLi, respectively, the first one being used in the synthesis of β-resorcylic acid derivatives [141].

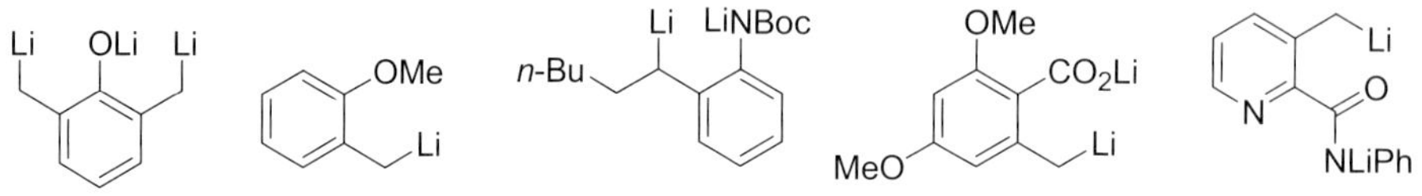

165   **166**   **167**   **168**   **169**

Benzylic organolithium compounds are in general configurationally unstable. However, the lateral lithiation of tertiary 2-alkyl-1-naphthamides, such as **170**, was stereoselective and yielded a single diastereomeric atropisomer **171** [143], which reacted with several electrophiles with retention, to give compounds **172** [144], except with trialkyltin halides for which an inversion of the configuration was observed (Scheme 2.23) [144a].

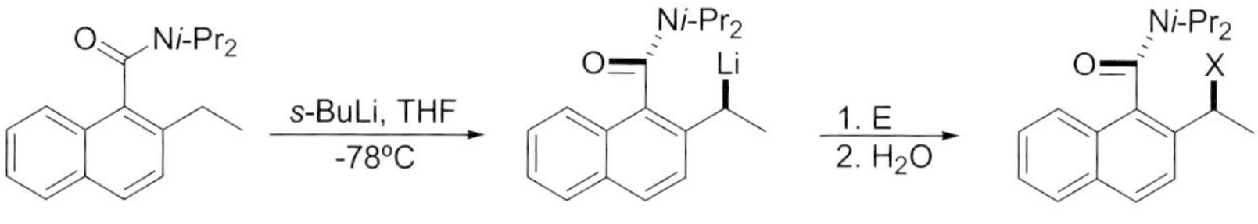

170   **171**   **172** (64–97%, >97:3 dr)

[E = MeOD, RX, RCHO, R^1R^2CO, $R^1CH=NR^2$, Me_3SiHal]

Scheme 2.23

2.4.4
γ-Functionalized Akenyllithium Compounds

Vinyllithium compounds of type **XVI** show *Z/E*-configuration and can be prepared by tin– or halide–lithium exchange, because deprotonation occurs mainly at the allylic position, except in very special cases. *O*-Silyl protected γ-lithiated allyl alcohol **173** has been prepared by iodine–lithium exchange with *t*-BuLi in hexane or THF at 0 °C, the process being totally stereospecific, so the resulting vilyllithium compound kept the configuration of the starting iodide [145]. The silylated compound **174** was prepared by tin–lithium transmetallation and reacted with cyclopentenones in the presence of zinc salts for the synthesis of prostaglandins [146]. Azasilacines and azagermacines were synthesized in a direct way from the trianion **175** [147], which was prepared by a sequential deprotonation at a vinylic position [136,148] followed by a tin–lithium transmetallation from (*Z*)-allyl-3-(tri-*n*-butyltin)allylamine. Reagent **176**, which derived from a protected diol, was prepared by bromine–lithium exchange and used in the synthesis of butenolide fugomycin [149]. The *N*-methoxyimine derived from (*Z*)-β-iodo acrolein was lithiated with *n*-BuLi in hexane at –78 °C to give vinyllithium intermediate **177**, which was trapped with isocyanates to give the corresponding amides [150].

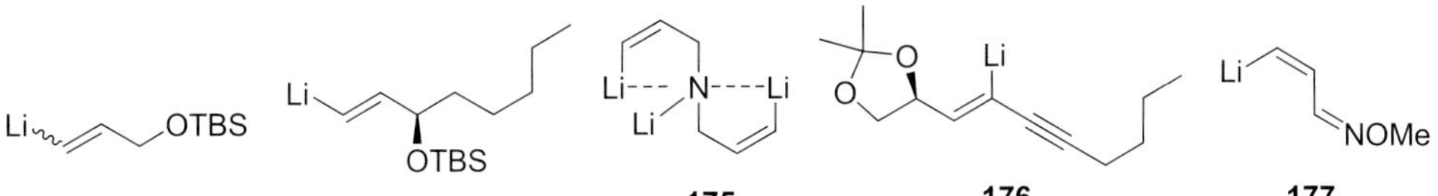

173 **174** **175** **176** **177**

β-Chloro *E-α,β*-unsaturated ketone acetals **178** underwent stereoselective DTBB-catalyzed lithiation to give vinyllithium derivatives **179**, which after reaction with electrophiles and final acidic hydrolysis led to regioselectively functionalized *α,β*-unsaturated carbonyl compounds **189**. In the reaction of enantiomerically enriched starting materials **180** (R^2 = Me) with prostereogenic carbonyl compounds, no diastereoselectivity was observed (Scheme 2.24) [111,151].

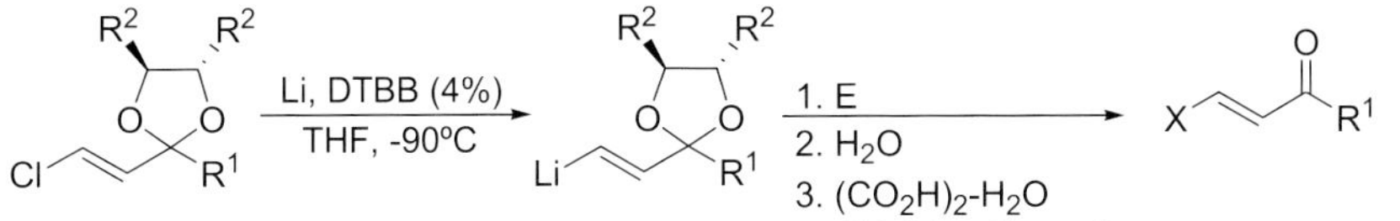

178 **179** **180** (43-90%)

[E = H_2O, D_2O, R^3R^4CO]

Scheme 2.24

Aryllithium compounds of type **XVIII** are generated mainly either by direct deprotonation [71,104] or by halogen–lithium exchange [75,105]. In the case *p*-methoxybenzaldehyde dimethyl acetal, *ortho* lithiation with *n*-BuLi (to give the intermediate **181**), followed by reaction with electrophiles and final acidic hydrolysis, led to polyfunctionalized benzaldehydes in high yields [152]. Vinyllithium derivatives **182** and **183** (of type **XIX)** have been prepared from the corresponding vinylic halides by treatment with *t*-BuLi. The silyl ether derivative **182** was used in the synthesis of the C1 alkyl side chains of Zaragozic acids A and C [153] and the intermediate **183** was involved in the total synthesis of marine metabolites (+)-calyculin A and (–)-calyculin B [154]. The chlorovinyllithium derivative **184** was accessible through a tin–lithium transmetallation and used in the synthesis of the marine sesquiterpenoid (±)-kelsoene [155]. Cyclic vinyllithium intermediate **185** was used in the construction of *seco*-taxanes, being prepared from the corresponding ketone trisylhydrazone applying the Shapiro reaction [156].

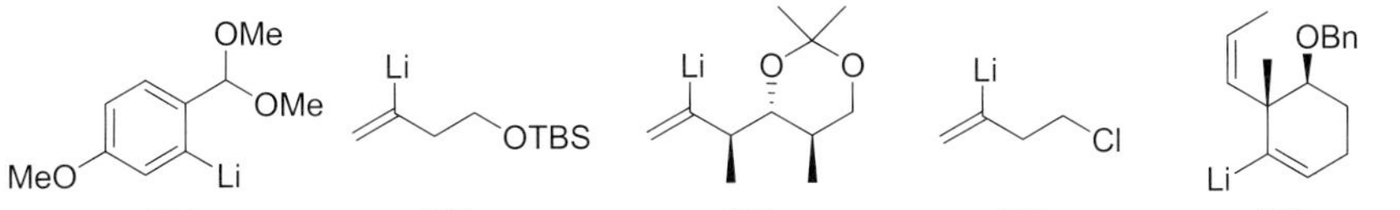

181 **182** **183** **184** **185**

Iron-catalyzed carbolithiation of the internal alkyne **186**, bearing an alkoxy group at the homopropargylic position, led to the vinyllithium compound **187**, which was trapped with electrophiles to give the corresponding functionalized olefins **188** with a defined stereochemistry and in high yields (Scheme 2.25) [157].

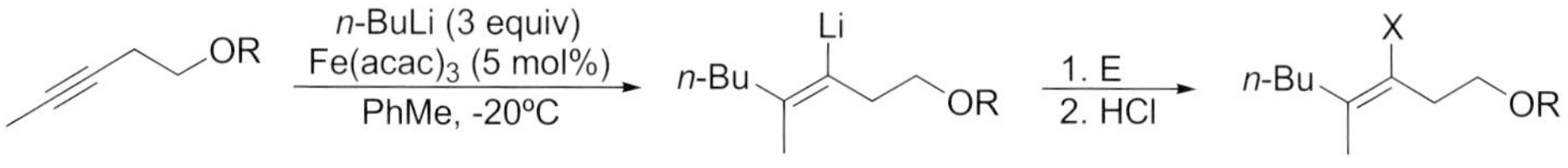

186 **187** **188** (69-96%)

[E = D_2O, Me_2SiHCl, R^1R^2CO]

Scheme 2.25

2.4.5 γ-Functionalized Alkynyllithium Compounds

Functionalized alkynyllithium compounds of type **XX** were prepared almost exclusively by deprotonation of the corresponding terminal alkynes and therefore, the functionality sometimes have to be protected in order to tolerate the presence of the base and to resist the conditions for the acetylenic proton to be removed. Functionalized lithium acetylide **189** was prepared by deprotonation of the corresponding *O*-protected propargyl alcohol and further alkylated with butadiene bis-epoxide in order to prepare dienediyens [158]. The alkynyl anion **190** was used in the synthesis of (–)-laulimalide, a microtube-stabilizing agent [159]. The intermediate **191**, prepared from Garner's aldehyde, reacted with aldehydes in high yields [160] and the orthoester reagent **192** acted as the acetylenic anion of propyolic acid and reacted with carbonyl compounds in good yields [161].

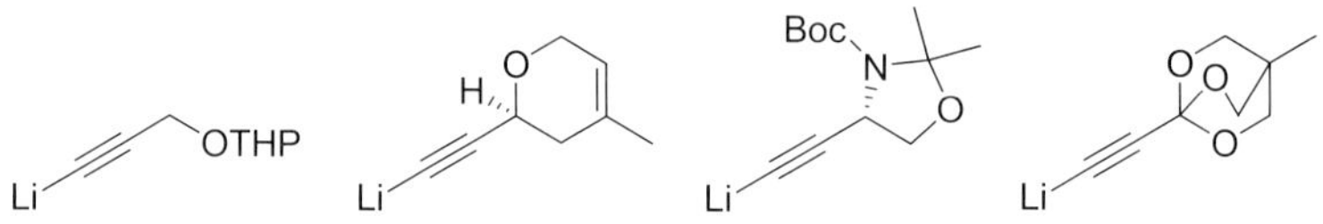

189 **190** **191** **192**

The lithium orthopropyolate **194** was prepared by a silicon–lithium exchange from the trimethylsilylacetylene derivative **193** and reacted with acetylenic aldehydes to give diacetylenic alcohols **195**, which were easily transformed into the corresponding ketones by treatment with manganese dioxide (Scheme 2.26) [162].

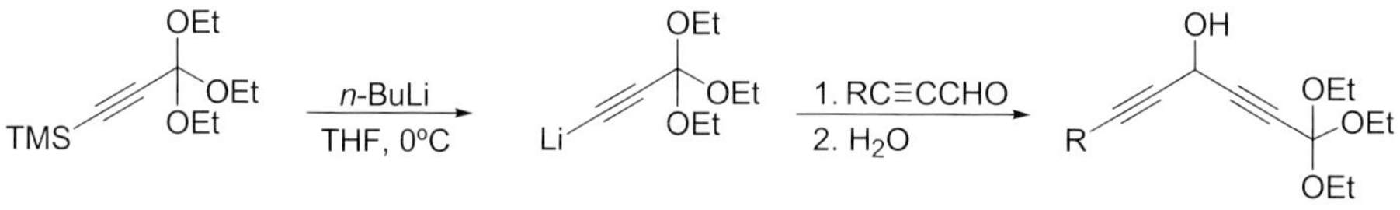

193 **194** **195** (74-80%)

Scheme 2.26

2.5 δ-Functionalized Organolithium Compounds

The number of possible structural devices increases as functionality gets further from the anionic center in functionalized organolithium compounds. So, δ-functionalized organolithium compounds can be classified according to the hybridization of the carbon atom bonded to the lithium in alkyllithium compounds (**XXI** and **XXII**), allylic and benzylic derivatives (**XXIII–XXV**) and alkenyl (**XXVI–XXVIII**) and alkynyl systems (**XXIX**).

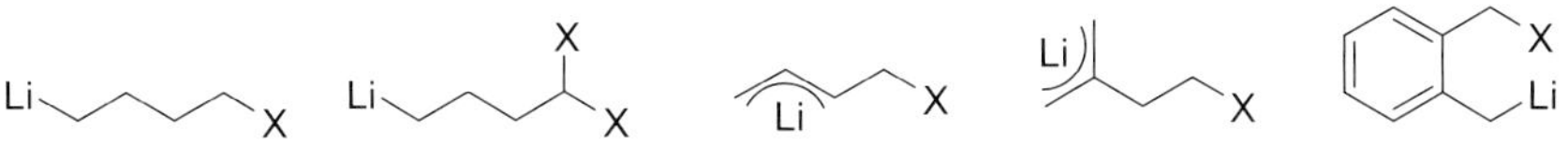

XXI **XXII** **XXIII** **XXIV** **XXV**

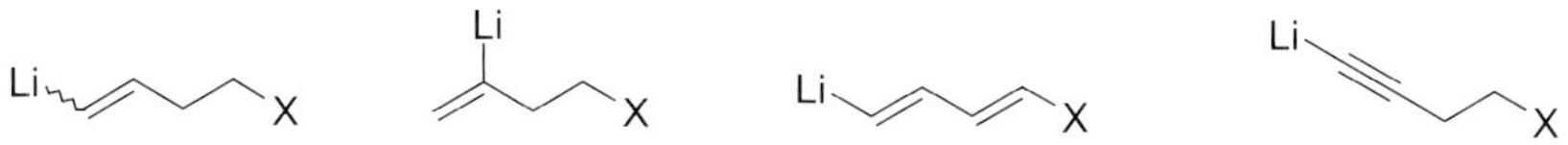

XXVI **XXVII** **XXVIII** **XXIX**

2.5.1 δ-Functionalized Alkyllithium Compounds

Different alcohols and protected alcohols (as hemiacetals, silyl, methoxymethyl or phenyl ethers) were lithiated at the δ-position to give the corresponding organolithium compounds. In the case of alcohols, a previous deprotonation of the hydroxyl functionality is required. The chiral intermediate **197** was prepared from the phenylsulfanyl derivative **196** first by deprotonation followed by carbon–sulfur bond cleavage with LiDTBB at low temperature. The reaction of the dianionic system **197** with γ- and δ-lactones in the presence of cerium(III) salts gave, after hydrolysis, spiroketal pheromones **198** (Scheme 2.27) [163].

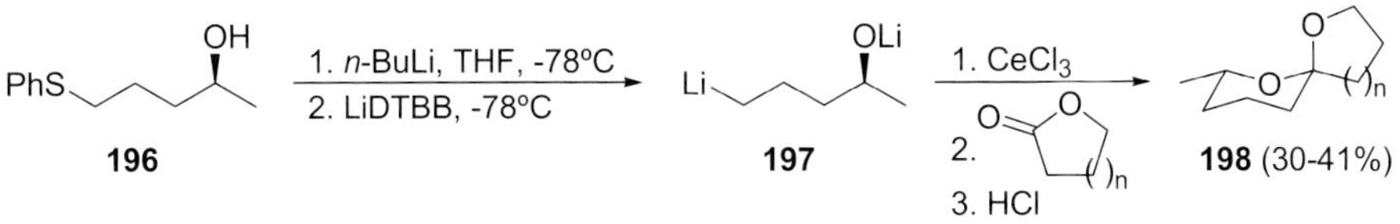

Scheme 2.27

Similarly to cinnamyl alcohols and ethers, which undergo carbolithiation by reaction with alkyllithiums [115,116], treatment of (*E*)-4-phenyl-3-buten-1-ol with two equivalents of *n*-BuLi in the presence of (–)-sparteine gave the organolithium compound **199**, which after hydrolysis afforded the corresponding alcohol with 72% ee [164]. The optically active intermediate **200** was prepared by bromine–lithium exchange with *t*-BuLi in ether and was used in the synthesis of the C28–C40 fragment of azaspiracids [165]. Masked bishomoenolates **201** and **202** were prepared by means of a DTBB-catalyzed lithiation of the corresponding γ-chloro orthoesters [166] and phenylsulfanyl derivative [167], respectively. Functionalized methyl esters were obtained when the lithiation leading to the intermediate **201** was performed in the presence of electrophiles followed by hydrolysis with methanol and a catalytic amount of *p*-toluenesulfonic acid [166]. The lithio-thio-acetal **202** reacted with *N*-phenethylimides, pyrroloisoquinolinones being obtained in good yields after acyliminium ion cyclization [167].

2.5.2
δ-Functionalized Allyl and Benzyllithium Compounds

Dianion **203** was prepared by double deprotonation of 3-methyl-3-buten-1-ol with *n*-BuLi in the presence of TMEDA and reacted with alkyl halides in the synthesis of different pheromones, the best results being obtained using diethyl ether as solvent [168]. However, the dianionic allylic isomer **204** was prepared through a tin–lithium transmetallation starting from the corresponding tri-*n*-butyltin derivative. The reaction of compound **204** with aldehydes took place at the internal position of the allylic system [169]. Reductive opening of heterocycles bearing the heteroatom at an allylic or benzylic position, as well as cyclic aryl ethers and thioethers, leads to functionalized organolithium compounds in a direct way [83]. The δ-amino functionalized organolithium compound **205** was prepared through an allylic carbon–nitrogen bond cleavage by a DTBB-catalyzed lithiation from *N*-phenyl-2,3-dihydropyrrole and reacted with electrophiles regioselectively [170]. The dilithium derivatives **206** and **207** have also been prepared from 2-phenyltetrahydrothiophene (Y = S) [171], *N*-isopropyl-2-phenylpyrrolidine (Y = NPh) [170], thiophthalan (Y = S) [172] and *N*-phenylisoindoline (Y = NPh) [170] under the same reductive reaction conditions. In all these cases, a benzylic carbon–heteroatom bond cleavage took place, their reaction with electrophiles leading to polyfunctionalized molecules in a regioselective manner.

In addition, the lithiation of phthalan **208** could be directed to the introduction of two different electrophiles at both benzylic positions in a sequential manner. Thus, after the reductive opening of compound **208**, the resulting dianionic intermediate **209** reacted with a first electrophile followed by hydrolysis to give the functionalized alcohols **211**. However, when, after the addition of the first electrophile, the resulting alcoholate **210** was allowed to react in the highly reductive reaction medium at room temperature, a new benzylic carbon–oxygen bond cleavage took place to give a new organolithium intermediate **212**. The addition of a second electrophile and final hydrolysis led to polyfunctionalized o-xylene derivatives **213** (Scheme 2.28) [173]. Functionalized benzyllithium compound **209** has found a wide applicability in organic synthesis. Depending on the electrophile it has been used in the synthesis of tetrahydroisoquinolines [174] and structurally modified sugars [85b], steroids [91b] and monoterpenes [175]. In the presence of different metallic salts, intermediate **209** participated in conjugate addition processes, acylation reactions, dimerization and alkylation with allylic systems, as well as in chemoselective additions to carbonyl compounds [176].

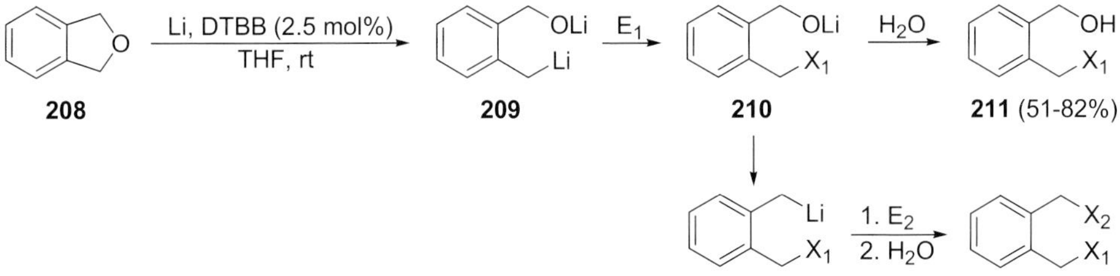

$[E_1, E_2 = H_2O, D_2O, CO_2, RCHO, R^1R^2CO, R^1CH=NR^2]$

Scheme 2.28

2.5.3
δ-Functionalized Alkenyllihium Compounds

Vinyllithium compounds with a functional group at the δ-position have been prepared stereospecifically from the corresponding vinylic precursors, mainly by halogen– or tin–lithium exchange. They are not accessible by direct deprotonation due to the lack of influence of the functional group, which is located too far away from the vinylic proton to be removed. The alkenyllithium derivative **214** (of type **XXVII**) was prepared from the corresponding ketone by means of a Shapiro reaction, acting as an intermediate in the synthesis of the antibiotic nodusmicin [177]. Meanwhile, the (E,E)-dienyllithium compound **215** (of type **XXVIII**) has been generated from 1-(tri-n-butylstannyl)-4-ethoxy-1,3-butadiene by a tin–lithium exchange and used in the synthesis of a polyene macrolide roflamycoin [178]. Reductive opening of benzofurane with lithium and a catalytic amount of DTBB at 0 °C gave the (Z)-organolithium derivative **216** that reacted with electrophiles in a regio- and stereoselective manner and was applied to the synthesis of substituted o-vinylphenol derivatives. In the case of using carbonyl compounds as electrophiles, the resulting diols gave substituted chromenes in high yields, after dehydration under acidic conditions [179].

2.5.4
δ-Functionalized Alkynyllithium Compounds

As previously commented for other acetylenic anions, functionalized alkynyl-lithium compounds of the type **XXIX** were prepared almost exclusively by depro-tonation of the corresponding terminal alkynes. Thus, the dihydropyran derivative **217** has been used for the convergent synthesis of *trans*-fused polytetrahydropyr-ans, which are present in marine toxins by coupling with triflates [180].

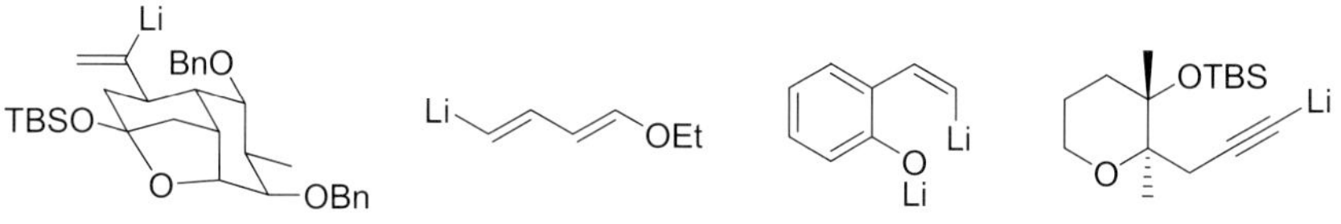

214	**215**	**216**	**217**

A different strategy towards δ-functionalized alkynyllithium compounds con-sists in the treatment of *gem*-dibromoalkenes (accessible from the corresponding aldehydes) with an excess of *n*-BuLi. Following this methodology, intermediate **219** was prepared from the dithiane derivative **218** and alkylated with methyl iodide to give the corresponding alkyne **220** (Scheme 2.29) [181].

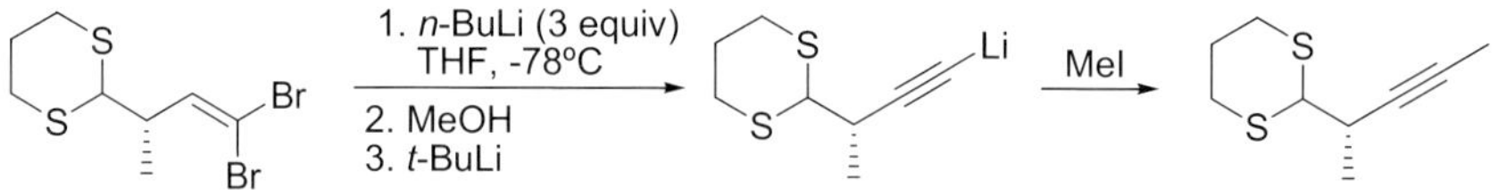

Scheme 2.29

2.6
Remote Functionalized Organolithium Compounds

The influence of the functional group on the reactivity and the stability of func-tionalized organolithium compounds decreases as it gets further from the carba-nionic center, so these compounds behave in many cases as normal organo-lithium compounds and can be prepared through classical methodologies. They can be classified according to the hybridization of the anionic carbon as in the pre-vious section.

2.6.1
Remote Functionalized Alkyllithium Compounds

In general, organolithium compounds bearing a leaving group at the ε-position undergo an intramolecular nucleophilic substitution to give the corresponding cyclic systems. Treatment of the iodinated methoxy compound **221** with two equivalents of *t*-BuLi in heptane at –78 °C gave the ε-functionalized organolithium compound **222**, which upon addition of TMEDA, suffered intramolecular S_N2' cyclization to give the vinylcyclopropane **223** in almost quantitative yield (Scheme 2.30) [182].

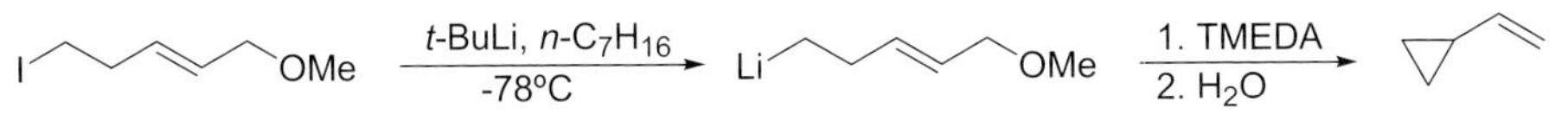

Scheme 2.30

On the other hand, the silyl-substituted vinyllithium **224**, prepared by bromine–lithium exchange, did not cyclize and reacted with 3-fluoro-3-buten-2-one to give the corresponding alcohol, which was used in the synthesis of (±)-dammarenediol [183]. The ε-oxido functionalized intermediate **225** was prepared from the corresponding chlorohydrine by a chlorine–lithium exchange with lithium naphthalenide after deprotonation, and reacted with iodoarenes to give the expected coupling products [184]. The organolithium compound **226**, containing a masked acylsilane functionality, was prepared from the corresponding iodinated precursor by treatment with *t*-BuLi, and reacted with substituted cyclopent-2-enones to give the expected 1,2-addition products. This strategy was used for the synthesis of isocarbacyclin, a stable analog of prostacyclin [185].

2.6.2
Remote Allyl and Benzyllithium Compounds

The allyllithium compound **227** was generated from the phenylsulfanyl precursors by means of lithium naphthalenide in THF at –78 °C in the presence of TMEDA and underwent a irreversible retro- [1,4]-Brook rearrangement to give an almost 4:1 mixture of *syn,trans* and *anti,trans* diastereomers [186].

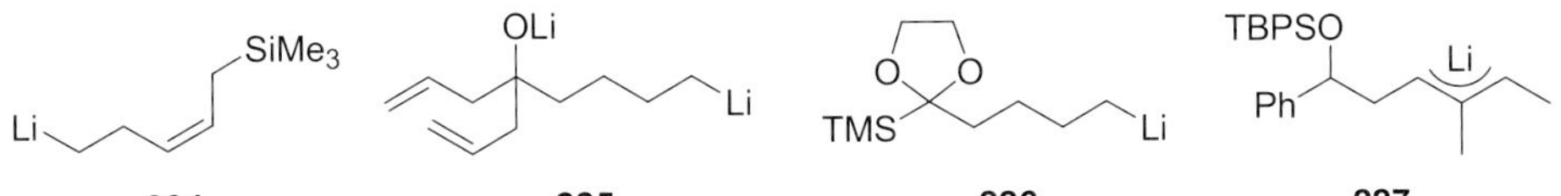

A diastereomeric mixture of isopentenyldimethylcyclopentanols **230** was obtained through a lithium-ene cyclization starting from the oxido allylic intermediate **229**, which was generated by LiDTBB carbon–sulfur bond cleavage of **228**, the lithium oxide unit facilitating the cyclization. The corresponding magnesium derivatives participated in the same cyclization process (Scheme 2.31) [187].

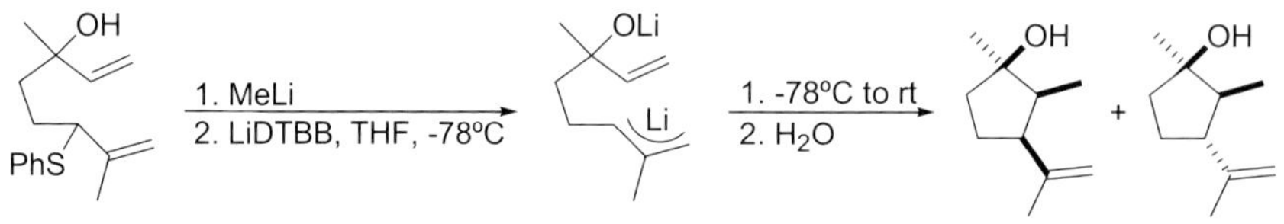

Scheme 2.31

Dianionic benzylic intermediates **231** [171], **232** (Y = O [188], Y = S [189], Y = NPh [170]), **233** (Y = O [190], Y = S [191], Y = NMe [190]), and **234** (Y = O [190], Y = S [190,192], Y = NMe [190]), were all prepared through a benzylic carbon–heteroatom bond cleavage from the corresponding heterocycles by an arene-catalyzed lithiation and reacted with electrophiles to give polyfunctionalized molecules in a regioselective manner. In a similar way to phthalan **208** (Scheme 2.28), after the reaction with a first electrophile intermediates **233** (Y = S) [191] and **234** (Y = O) [192], underwent a second benzylic carbon–heteroatom bond cleavage to give a new organolithium compound, which reacted with a second electrophile, yielding polyfunctionalized biphenyls and naphthalenes, respectively after hydrolysis.

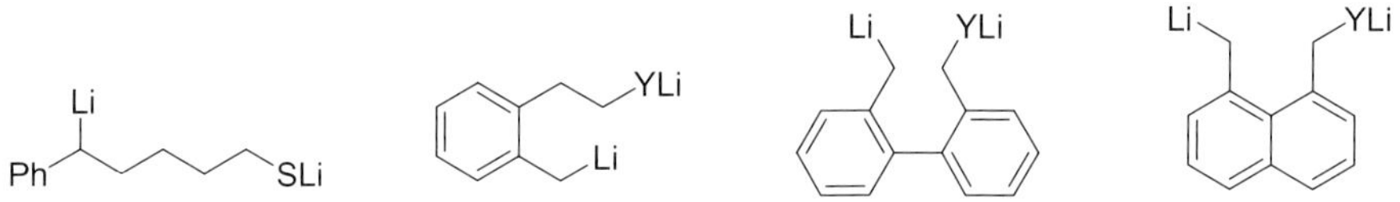

231   **232** (Y = O, S, NPh)   **233** (Y = O, NMe, S)   **234** (Y = O, NMe)

2.6.3
Remote Functionalized Alkenyl- and Alkynyllithium Compounds

As an example of a remote functionalized alkenyllithium compound, the alkoxy functionalized polyenyllithium **236** (prepared from **235** by bromine–lithium exchange with *t*-BuLi) reacted with carbonyl compounds to give after acidic hydrolysis the corresponding polyenic aldehydes **237** (Scheme 2.32) [193].

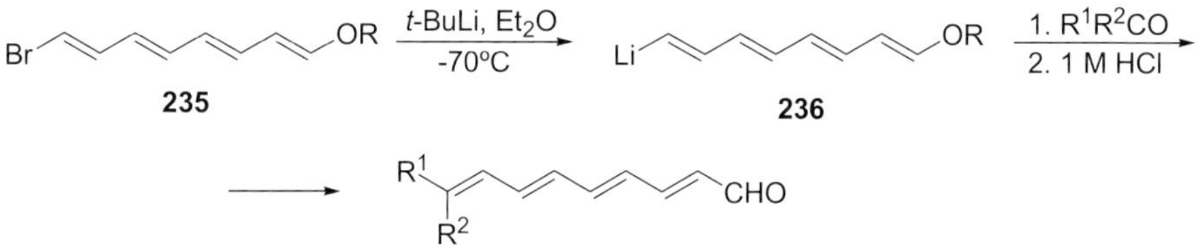

Scheme 2.32

Lithium (*Z*)-5-lithio-5-methyl-4-penten-1-olate **238** was also prepared by bromine–lithium exchange, after the deprotonation of the hydroxy unit, and acylated with amides in the synthesis of the enantiomer of natural epolactaene [194]. Dienyllithium **239** containing a masked carbonyl functionality (prepared by a bromine–lithium exchange) reacted with a lactam being used in the synthesis of trisporic acids [195]. Meanwhile, the lithium acetylide **240** with a remote oxide functionality was prepared by double deprotonation of the corresponding alkynol with *n*-BuLi and alkylated with 1-iodooctane, the resulting product being an intermediate in the synthetic route to the marine sponge natural products *R*-strongylodiols A and B [196]. Finally, the silyl enediyne lithium derivative **241** was prepared by deprotonation of the corresponding monoprotected enediyne with LiHMDS at –78 °C and used in the synthesis of antitumor agent (±)-calicheamicinone [197].

| **238** | **239** | **240** | **241** |

References

1 (a) C. Nájera, M. Yus, *Trends Org. Chem.* **1991**, *2*, 155–181. (b) C. Nájera, M. Yus, *Recent Res. Dev. Org. Chem.* **1997**, *1*, 67–96. (c) A. Boudier, L. O. Bromm, M. Lotz, P. Knochel, *Angew. Chem. Int. Ed.* **2000**, *39*, 4414–4435. (d) C. Nájera, M. Yus, *Current Org. Chem.* **2003**, *7*, 867–926.

2 (a) J. Clayden, *Organolithiums: Selectivity for Synthesis*, Pergamon, Oxford, **2002**. (b) D. M. Hodgson (Ed.), *Organolithiums in Enantioselective Synthesis*, Springer Verlag, Berlin, **2003**.

3 (a) M. Gray, M. Tinkl, V. Snieckus, in *Comprehensive Organometallic Chemistry II*, E. W. Abel, F. G. A. Stone, G. Wilkinson, A. Mckillop (Eds.), Pergamon, Oxford, **1995**, Vol. *11*, 2–81. (b) R. Bartsch, C. Drost, U. Klingebiel, in *Synthetic Methods of Organometallic and Inorganic Chemistry*, N. Auner, U. Klingebiel (Eds.), G. Thieme Verlag, Stuttgart, **1996**, Vol. *2*, 2–24. (c) F. Leroux, M. Schlosser, E. Zohar, I. Marek, in *The Chemistry of Organolithium Compounds (Part 1)*, Z. Rappoport, I. Marek (Eds.), Wiley, Chichester, **2004**, 435–494.

4 (a) A. Streitwieser, S. M. Bachrach, A. Dorigo, P. von R. Schleyer, in *Lithium Chemistry: A Theoretical and Experimental Overview*, A.-M. Sapse, P. von R. Schleyer (Eds.), Wiley, New York, **1995**, 2–15. (b) E. D. Jemmis, G. Gopakumar, in *The Chemistry of Organolithium Compounds (Part 1)*, Z. Rappoport, I. Marek (Eds.), Wiley, Chichester, **2004**, 2–6.

5 For a review, see: D. J. Berrisford, *Angew. Chem. Int. Ed. Engl.* **1995**, *34*, 178–180.

6 (a) D. Seebach, *Synthesis* **1969**, 17–36. (b) G. Boche, *Angew. Chem. Int. Ed. Engl.* **1989**, *28*, 277–297. (c) M. Yus, C. Nájera, F. Foubelo, *Tetrahedron* **2003**, *59*, 6147–6212. (d) A. B. Smith III, C. M. Adams, *Acc. Chem. Res.* **2004**, *37*, 365–377.

7 (a) A. Guijarro, M. Yus, *Tetrahedron Lett.* **1993**, *34*, 3487–3490. (b) A. Guijarro, B. Mancheño, J. Ortiz, M. Yus, *Tetrahedron* **1996**, *52*, 1643–1650.

8 (a) C. Gómez, S. Ruiz, M. Yus, *Tetrahedron Lett.* **1998**, *39*, 1397–1400. (b) C. Gómez, S. Ruiz, M. Yus, *Tetrahedron* **1999**, *55*, 7017–7026.

9 (a) C. Blomberg, *The Barbier Reaction and Related One-Step Processes*, Springer Verlag, Berlin, **1993**. (b) F. Alonso, M. Yus, *Recent Res. Dev. Org. Chem.* **1997**, *1*, 397–436.

10 J. Ortiz, A. Guijarro, M. Yus, *An. Quim. Int. Ed.* **1997**, *93*, 44–48.

11 V. L. Ponzo, T. S. Kaufman, *Can. J. Chem.* **1998**, *76*, 1338–1343.

12 (a) A. Guijarro, M. Yus, *Tetrahedron Lett.* **1996**, *37*, 5593–5596. (b) J. Ortiz, A. Guijarro, M. Yus, *Eur. J. Org. Chem.* **1999**, 3005–3012.

13 (a) O. Frey, M. Hoffmann, H. Kessler, *Angew. Chem. Int. Ed. Engl.* **1995**, *34*, 2026–2028. (b) F. Burkhart, H. Kessler, *Tetrahedron Lett.* **1998**, *39*, 255–256.

14 (a) M. Hoffmann, H. Kessler, *Tetrahedron Lett.* **1997**, *38*, 1903–1906. (b) F. Burkhart, M. Hoffmann, H. Kessler, *Angew. Chem. Int. Ed. Engl.* **1997**, *36*, 1191–1192. (c) M. A. Dechantsreiter, F. Burkhart, H. Kessler, *Tetrahedron Lett.* **1998**, *39*, 253–254. (d) B. Westermann, A. Walter, N. Diedrichs, *Angew. Chem. Int. Ed.* **1999**, *38*, 3384–3386.

15 (a) J.-M. Lancelin, L. Morin-Allory, P. Sinaÿ, *J. Chem. Soc., Chem. Commun.* **1984**, 355–356. (b) J.-M. Beau, P. Sinaÿ, *Tetrahedron Lett* **1985**, *26*, 6189–6192.

16 R. P. Smyj, J. M. Chong, *Org. Lett.* **2001**, *3*, 2903–2906.

17 M. I. Calaza, M. R. Paleo, F. J. Sardina, *J. Am. Chem. Soc.* **2001**, *123*, 2095–2096.

18 J. M. Chong, N. Nielsen, *Tetrahedron Lett.* **1998**, *39*, 9617–9620.

19 (a) W. C. Still, *J. Am. Chem. Soc.* **1978**, *100*, 1481–1487. (b) W. C. Still, C. Sreekumar, *J. Am. Chem. Soc.* **1980**, *102*, 1201–1202.

20 K. Tomooka, N. Komine, T. Nakai, *Synlett* **1997**, 1045–1046.

21 K. Tomooka, N. Komine, T. Nakai, *Tetrahedron Lett.* **1997**, *38*, 8939–8942.

22 F. Burkhart, M. Hoffmann, H. Kessler, *Tetrahedron Lett.* **1998**, *39*, 7699–7702.

23 R. Angelaud, Y. Landais, L. Parra-Rapado, *Tetrahedron Lett.* **1997**, *38*, 8845–8848.

24 For reviews, see: (a) V. K. Aggarwal, *Angew. Chem. Int. Ed. Engl.* **1994**, *33*, 175–177. (b) P. Beak, A. Basu, D. J. Gallagher, Y. S. Park, S. Thayumanavan, *Acc. Chem. Res.* **1996**, *29*, 552–560. (c) D. Hoppe, T. Hense, *Angew. Chem. Int. Ed. Engl.* **1997**, *36*, 2282–2316.

25 M. Paetow, H. Ahrens, D. Hoppe, *Tetrahedron Lett.* **1992**, *33*, 5323–5326.

26 (a) C. Derwing, D. Hoppe, *Synthesis* **1996**, 149–154. (b) F. Hammerschmidt, A. Hanninger, *Chem. Ber.* **1995**, *128*, 1069–1077.

27 (a) N. Komine, L.-F. Wang, K. Tomooka, T. Nakai, *Tetrahedron Lett.* **1999**, *40*, 6809–6812. (b) K. Tomooka, L.-F. Wang, N. Komine, T. Nakai, *Tetrahedron Lett.* **1999**, *40*, 6813–6816. (c) K. Tomooka, L.-F. Wang, F. Okazaki, T. Nakai, *Tetrahedron Lett.* **2000**, *41*, 6121–6125.

28 (a) J. Suffert, D. Toussaint, *Tetrahedron Lett.* **1997**, *38*, 5507–5510. (b) J. Suffert, B. Salem, P. Klotz, *J. Am. Chem. Soc.* **2001**, *123*, 1210712108.

29 For reviews, see: (a) T. Satoh, *Chem. Rev.* **1996**, *96*, 3303–3326. (b) D. M. Hodgson, E. Gras, *Synthesis* **2002**, 1625–1642.

30 V. Capriati, S. Florio, R. Luisi, I. Nuzzo, *J. Org. Chem.* **2004**, *69*, 3330–3335.

31 (a) A. Abbotto, V. Capriati, L. Degennaro, S. Florio, R. Luisi, M. Pierrot, A. Salomone, *J. Org. Chem.* **2001**, *66*, 3049–3058. (b) V. Capriati, S. Florio, R. Luisi, A. Salomone, *Org. Lett.* **2002**, *4*, 2445–2448.

32 R. Luisi, V. Capriati, L. Degennaro, S. Florio, *Org. Lett.* **2003**, *5*, 2723–2726.

33 C. Courillon, J.-C. Marié, M. Malacria, *Tetrahedron* **2003**, *59*, 9759–9766.

34 (a) G. P. Pale, J. Chuche, *Tetrahedron: Asymmetry* **1993**, *4*, 1991–1994. (b) S. Klein, J. H. Zhang, M. Holler, J.-M. Weibel, P. Pale, *Tetrahedron* **2003**, *59*, 9793–9802.

35 D. M. Hodgson, S. L. M. Norsikian, *Org. Lett.* **2001**, *3*, 461–463.

36 D. M. Hodgson, E. Gras, *Angew. Chem. Int. Ed.* **2002**, *41*, 2376–2378.

37 (a) S. D. Rychnovsky, J. P. Powers, T. J. LePage, *J. Am. Chem. Soc.* **1992**, *114*, 8375–8384. (b) D. J. Kopecky, S. D. Rychnovsky, *J. Org. Chem.* **2000**, *65*, 191–198. (c) S. D. Rychnovsky, T. Hata, A. I. Kim, A. J. Buckmelter, *Org. Lett.* **2001**, *3*, 807–810.

38 S. D. Rychnovsky, L. R. Takaoka, *Angew. Chem. Int. Ed.* **2003**, *42*, 818–820.

39 J. A. Soderquist, L. Castro-Rosario, in *Encyclopedia of Reagents for Organic Synthesis*, L. A. Paquette (Ed.), Wiley, London, **1995**, Vol. *5*, 3408–3410.

40 K. K. Banger, C. Birringer, R. U. Claessen, P. Lim, P. J. Toscano, J. T. Welch, *Organometallics* **2001**, *20*, 4745–4748.

41 For a review, see: C. Nájera, M. Yus, *Org. Prep. Proced. Int.* **1995**, *27*, 383–456.

42 W. B. Jang, H. Hu, M. M. Lieberman, J. A. Morgan, I. A. Stergiades, D. S. Clark, M. A. Tius, *J. Comb. Chem.* **2001**, *3*, 346–353.

43 For reviews, see: (a) M. Braun, *Angew. Chem. Int. Ed.* **1998**, *37*, 430–451. (b) R. W. Friesen, *J. Chem. Soc., Perkin Trans. 1* **2001**, 1969–2001.

45 P. E. Harrington, M. A. Tius, *Org. Lett.* **2000**, *2*, 2447–2450.

46 (a) L. A. Paquette, S. Brand, C. Behrens, *J. Org. Chem.* **1999**, *64*, 2010–2025. (b) L. A. Paquette, D. R. Owen, R. Todd Bibart, C. K. Seekamp, A. L. Kahane, J. C. Lanter, M. Alvarez Corral, *J. Org. Chem.* **2001**, *66*, 2828–2834. (c) L. A. Paquette, J. C. Lanter, D. R. Owen, F. Fabris, R. Todd Bibart, M. Alvarez Corral, *Heterocycles* **2001**, *54*, 49–53.

47 L. A. Paquette, J. C. Lanter, J. N. Johnston, *J. Org. Chem.* **1997**, *62*, 1702–1712.

48 L. A. Paquette, M. J. Kinney, U. Dullweber, *J. Org. Chem.* **1997**, *62*, 1713–1722.

49 C. Bozzo, M. D. Pujol, *Tetrahedron* **1999**, *55*, 11843–11852.

50 For reviews, see: (a) P. Beak, D. B. Reitz, *Chem. Rev.* **1978**, *78*, 275–316. (b) P. Beak, W. J. Zajdel, D. B. Reitz, *Chem. Rev.* **1984**, *84*, 471–523.

51 (a) J. Clayden, C. J. Menet, D. J. Mansfield, *Chem. Commun.* **2002**, 38–39. (b) J. Clayden, F. E. Knowles, C. J. Menet, *Tetrahedron Lett.* **2003**, *44*, 3397–3400.

52 A. I. Meyers, W. Ten Hoeve, *J. Am. Chem. Soc.* **1980**, *102*, 7125–7126.

53 (a) R. R. Fraser, T. B. Grindley, S. Passannanti, *Can. J. Chem.* **1975**, *53*, 2473–2480. (b) B. Renger, H. O. Kalinowski, D. Seebach, *Chem. Ber.* **1977**, *110*, 1866–1878.

54 D. H. R. Barton, R. Beugelmans, R. N. Young, *Nouv. J. Chim.* **1978**, *2*, 363–364.

55 K. B. Wiberg, W. F. Bailey, *J. Am. Chem. Soc.* **2001**, *123*, 8231–8238.

56 (a) M. J. Dearden, C. R. Firkin, J.-P. R. Hermet, P. O'Brien, *J. Am. Chem. Soc.* **2002**, *124*, 11870–11871. (b) J.-P. R. Hermet, D. W. Porter, M. J. Dearden, J. R. Harrison, T. Koplin, P. O'Brien, J. Parmene, V. Tyurin, A. C. Whitwood, J. Gilday, N. M. Smith, *Org. Biomol. Chem.* **2003**, *1*, 3977–3988.

57 (a) S. T. Kerrick, P. Beak, *J. Am. Chem. Soc.* **1991**, *113*, 9708–9710. (b) R. K. Dieter, C. M. Topping, K. R. Chandupatla, K. Lu, *J. Am. Chem. Soc.* **2001**, *123*, 5132–5133.

58 J. R. Harrison, P. O'Brien, *Synth. Commun.* **2001**, *31*, 1155–1160.

59 R. E. Gawley, G. Barolli, S. Madan, M. Saverin, S. O'Connor, *Tetrahedron Lett.* **2004**, *45*, 1759–1761.

60 C. Hoarau, A. Couture, H. Cornet, E. Deniau, P. Grandclaudon, *J. Org. Chem.* **2001**, *66*, 8064–8069.

61 (a) A. Guijarro, J. Ortiz, M. Yus, *Tetrahedron Lett.* **1996**, *37*, 5597–5600. (b) J. Ortiz, A. Guijarro, M. Yus, *Tetrahedron* **1999**, *55*, 4831–4842.

62 I. Coldham, S. Colman, M. M. S. Lang-Anderson, *J. Chem. Soc., Perkin Trans. 1* **1997**, 1481–1490.

63 D. M. Iula, R. E. Gawley, *J. Org. Chem.* **2000**, *65*, 6196–6201.

64 U. Azzena, L. Pilo, E. Piras, *Tetrahedron Lett.* **2001**, *42*, 129–131.

65 (a) E. Vedejs, J. T. Kendall, *J. Am. Chem. Soc.* **1997**, *119*, 6941–6942. (b) E. Vedejs, A. S. Bhanu Prasad, J. T. Kendall, J. S. Russel, *Tetrahedron* **2003**, *59*, 9849–9856.

66 Y. Yamauchi, T. Kawate, T. Katagiri, K. Uneyama, *Tetrahedron* **2003**, *59*, 9839–9847.

67 M. Yus, J. Ortiz, C. Nájera, *ARKIVOC* **2002**, *2*, 38–47.

68 A. I. Meyers, P. D. Edwards, T. R. Bailey, G. E. Jagdmann, *J. Org. Chem.* **1985**, *50*, 1019–1026.

69 R. M. Adlington, J. E. Baldwin, J. C. Bottaro, M. W. D. Perry, *J. Chem. Soc., Chem. Commun.* **1983**, 1040–1041.

70 M. D. B. Fenster, B. O. Patrick,
G. R. Dake, *Org. Lett.* **2001**, *3*,
2109–2112.

71 For a review, see: R. Chinchilla,
C. Nájera, M. Yus, *Chem. Rev.* **2004**, *104*,
2667–2722.

72 K. Worm, F. Chu, K. Matsumoto,
M. D. Best, V. Lynch, E. V. Anslyn,
Chem. Eur. J. **2003**, *9*, 741–747.

73 P. Calí, M. Begtrup, *Tetrahedron* **2002**,
58, 1595–1605.

74 A. Dondoni, A. Marra, M. Mizuno,
P. P. Giovannini, *J. Org. Chem.* **2002**, *67*,
4186–4199.

75 For reviews, see: (a) N. Sotomayor,
E. Lete, *Curr. Org. Chem.* **2003**, *7*, 1–26.
(b) C. Nájera, J. M. Sansano, M. Yus,
Tetrahedron **2003**, *59*, 9255–9303.

76 J. Chun, L. He, H.-S. Byun, R. Bittman,
J. Org. Chem. **2000**, *65*, 7634–7640.

77 C. Fruit, A. Turck, N. Plé, L. Mojovic,
G. Quéguiner, *Tetrahedron* **2001**, *57*,
9429–9435.

78 (a) O. Sugimoto, M. Sudo, K. Tanji,
Tetrahedron Lett. **1999**, *40*, 2139–2140.
(b) O. Sugimoto, M. Sudo, K. Tanji,
Tetrahedron **2001**, *57*, 2133–2138.

79 M. Shimizu, T. Kurahashi, T. Hiyama,
J. Synth. Org. Chem. Jpn. **2001**, *59*,
1062–1069.

80 A. Kasatkin, R. J. Whitby, *Tetrahedron
Lett.* **1997**, *38*, 4857–4860.

81 G. Köbrich, H. Trapp, I. Hornke, *Tetrahedron Lett.* **1964**, 1131–1136.

82 F. Tellier, R. Sauvêtre, J. F. Normant,
Y. Dromzee, Y. Jeannin, *J. Organomet.
Chem.* **1987**, *331*, 281–298.

83 (a) F. Foubelo, M. Yus, *Rev. Heteroatom
Chem.* **1997**, *17*, 73–107. (b) M. Yus,
F. Foubelo, *Targets Heterocycl. Syst.* **2002**,
6, 136–171. (c) M. Yus, *Pure Appl. Chem.*
2003, *75*, 1453–1475.

84 (a) J. Barluenga, F. J. Fañanás, M. Yus,
G. Asensio, *Tetrahedron Lett.* **1978**,
2015–2016. (b) J. Barluenga,
F. J. Fañanás, M. Yus, *J. Org. Chem.*
1979, *44*, 4798–4801. (c) J. Barluenga,
F. J. Fañanás, J. Villamaña, M. Yus,
J. Org. Chem. **1982**, *47*, 1560–1564.

85 (a) T. Soler, A. Bachki, L. R. Falvello,
F. Foubelo, M. Yus, *Tetrahedron: Asym-
metry* **1998**, *9*, 3939–3943. (b) T. Soler,
A. Bachki, L. R. Falvello, F. Foubelo,

M. Yus, *Tetrahedron: Asymmetry* **2000**,
11, 493–517.

86 C. Nájera, M. Yus, D. Seebach, *Helv.
Chim. Acta* **1984**, *67*, 289–300.

87 (a) F. Foubelo, A. Gutiérrez, M. Yus,
Tetrahedron Lett. **1997**, *38*, 4837–4840.
(b) F. Foubelo, A. Gutiérrez, M. Yus,
Synthesis **1999**, 503–514.

88 R. E. Conrow, *Tetrahedron Lett.* **1993**, *34*,
5553–5554.

89 D. Behnke, L. Hennig, M. Findeisen,
P. Welzel, D. Müller, M. Thormann,
H.-J. Hofmann, *Tetrahedron* **2000**, *56*,
1081–1095.

90 (a) A. Bachki, F. Foubelo, M. Yus, *Tetrahedron: Asymmetry* **1995**, *6*, 1907–1910.
(b) A. Bachki, F. Foubelo, M. Yus, *Tetrahedron: Asymmetry* **1996**, *7*, 2997–3008.

91 (a) L. R. Falvello, F. Foubelo, T. Soler,
M. Yus, *Tetrahedron: Asymmetry* **2000**,
11, 2063–2066. (b) M. Yus, T. Soler,
F. Foubelo, *Tetrahedron: Asymmetry*
2001, *12*, 801–810.

92 (a) J. Barluenga, M. Yus, P. Bernad,
J. Chem. Soc., Chem. Commun. **1978**,
847. (b) J. Barluenga, M. Yus,
J. M. Concellón, P. Bernad, *J. Chem.
Research (S)* **1980**, 41. (c) J. Barluenga,
M. Yus, J. M. Concellón, P. Bernad,
J. Org. Chem. **1981**, *46*, 2721–2726.
(d) J. Barluenga, M. Yus, J. M. Concellón,
P. Bernad, *J. Org. Chem.* **1983**, *48*,
609–611.

93 (a) J. Almena, F. Foubelo, M. Yus,
Tetrahedron Lett. **1993**, *34*, 1649–1652.
(b) J. Almena, F. Foubelo, M. Yus, *J. Org.
Chem.* **1994**, *59*, 3210–3215.

94 (a) F. Foubelo, M. Yus, *Tetrahedron Lett.*
1994, *35*, 4831–4834. (b) F. Foubelo,
M. Yus, *Tetrahedron: Asymmetry* **1996**, *7*,
2911–2922.

95 M. N. Kenworthy, J. P. Kilburn,
R. J. K. Taylor, *Org. Lett.* **2004**, *6*, 19–22.

96 Y. S. Park, P. Beak, *Tetrahedron* **1996**, *52*,
12333–12350.

97 F. Taherirastgar, L. Brandsma, *Chem.
Ber.* **1997**, *130*, 45–48.

98 P. Langer, M. Döring, D. Seyferth,
H. Görls, *Chem. Eur. J.* **2001**, *7*, 573–584.

99 A. Armstrong, F. W. Goldberg,
D. A. Sandham, *Tetrahedron Lett.* **2001**,
42, 4585–4587.

100 J. Tholander, J. Bergman, *Tetrahedron*
1999, *55*, 12595–12602.

101 M. Schlosser, H. Wei, *Tetrahedron* **1997**, *53*, 1735–1742.

102 K. Takeda, H. Haraguchi, Y. Okamoto, *Org. Lett.* **2003**, *5*, 3705–3707.

103 C. Yip, S. Handerson, G. K. Tranmer, W. Tam, *J. Org. Chem.* **2001**, *66*, 276–286.

104 For reviews, see: (a) W. E. Parham, C. K. Bardher, *Acc. Chem. Res.* **1982**, *15*, 300–305. (b) P. Beak, V. Snieckus, *Acc. Chem. Res.* **1982**, *15*, 306–312.

105 K. Yamada, T. Kurokawa, H. Tokuyama, T. Fukuyama, *J. Am. Chem. Soc.* **2003**, *125*, 6630–6631.

106 J. Barluenga, F. Foubelo, F. J. Fañanás, M. Yus, *J. Chem. Soc., Perkin Trans. 1* **1989**, 553–557.

107 X. Han, E. J. Corey, *Org. Lett.* **1999**, *1*, 1871–1872.

108 (a) L. A. Paquette, L. H. Kuo, A. T. Hamme, R. Kreuzholz, J. Doyon, *J. Org. Chem.* **1997**, *62*, 1730–1736. (b) J.-K. Ergüden, H. W. Moore, *Org. Lett.* **1999**, *1*, 375–378.

109 T. J. Donohoe, K. Blades, M. Helliwell, M. J. Waring, N. J. Newcombe, *Tetrahedron Lett.* **1998**, *39*, 8755–8758.

110 (a) J. Barluenga, R. Sanz, A. Granados, F. J. Fañanás, *J. Am. Chem. Soc.* **1998**, *120*, 4865–4866. (b) F. J. Fañanás, A. Granados, R. Sanz, J. M. Ignacio, J. Barluenga, *Chem. Eur. J.* **2001**, *7*, 2896–2907.

111 A. Bachki, F. Foubelo, M. Yus, *Tetrahedron* **1997**, *53*, 4921–4934.

112 J. Aigner, E. Gössinger, H. Kählig, G. Menz, K. Pflugseder, *Angew. Chem. Int. Ed.* **1998**, *37*, 2226–2228.

113 M. E. Fox, C. Li, J. P. Marino, L. E. Overman, *J. Am. Chem. Soc.* **1999**, *121*, 5467–5480.

114 S. P. Götzö, D. Seebach, J.-J. Sanglier, *Eur. J. Org. Chem.* **1999**, 2533–2544.

115 C. Mück-Lichtenfeld, H. Ahlbrecht, *Tetrahedron* **1999**, *55*, 2609–2642.

116 S. Norsikian, I. Marek, S. Klein, J. F. Poisson, J. F. Normant, *Chem. Eur. J.* **1999**, *5*, 2055–2068.

117 B. Mudryk, T. Cohen, *J. Org. Chem.* **1991**, *56*, 5760–5761.

118 A. Bachki, L. R. Falvello, F. Foubelo, M. Yus, *Tetrahedron: Asymmetry* **1997**, *8*, 2633–2643.

119 U. Azzena, L. Pilo, *Synthesis* **1999**, 664–668.

120 (a) J. Almena, F. Foubelo, M. Yus, *Tetrahedron* **1997**, *53*, 5563–5572. (b) J. Almena, F. Foubelo, M. Yus, *Tetrahedron* **1994**, *50*, 5775–5782.

121 L. Garamszegi, M. Schlosser, *Angew. Chem. Int. Ed.* **1998**, *37*, 3173–3175.

122 For a review, see: I. Kuwajima, E. Nakamura, *Top. Curr. Chem.* **1990**, *115*, 1–39.

123 I. M. Pastor, M. Yus, *Tetrahedron Lett.* **2001**, *42*, 1029–1032.

124 S. Zhu, T. Cohen, *Tetrahedron* **1997**, *53*, 17607–17624.

125 X. Feng, M. O. Senge, *J. Chem. Soc., Perkin Trans. 1* **2001**, 1030–1038.

126 N. Bremand, I. Marek, J. F. Normant, *Tetrahedron Lett.* **1999**, *40*, 3383–3386.

127 For reviews, see: (a) P. Beak, A. I. Meyers, *Acc. Chem. Res.* **1986**, *19*, 356–363. (b) M. C. Whisler, S. MacNeil, V. Snieckus, P. Beak, *Angew. Chem. Int. Ed.* **2004**, *43*, 2206–2225.

128 D. J. Pippel, M. D. Curtis, H. Du, P. Beak, *J. Org. Chem.* **1998**, *63*, 2–3.

129 M. Yus, T. Soler, F. Foubelo, *J. Org. Chem.* **2001**, *66*, 6207–6208.

130 For reviews. see: (a) H. Ahlbrecht, U. Beyer, *Synthesis* **1999**, 365–390. (b) A. R. Katritzky, M. Piffl, H. Lang, E. Anders, *Chem. Rev.* **1999**, *99*, 665–722.

131 K. Behrens, R. Fröhlich, O. Meyer, D. Hoppe, *Eur. J. Org. Chem.* **1998**, 2397–2403.

132 R. Tiedemann, F. Narjes, E. Schaumann, *Synlett* **1994**, 594–596.

133 M. Lombardo, S. Spada, C. Trombini, *Eur. J. Org. Chem.* **1998**, 2361–2364.

134 H. Ahlbrecht, P. Weber, *Synthesis* **1992**, 1018–1025.

135 H. Ahlbrecht, R. Schmidt, U. Beyer, *Eur. J. Org. Chem.* **1998**, 1371–1377.

136 J. Barluenga, F. J. Fañanás, F. Foubelo, M. Yus, *J. Chem. Soc., Chem. Commun.* **1988**, 1135–1136.

137 A. Krief, W. Dumont, *Tetrahedron Lett.* **1997**, *38*, 657–660.

138 S. Harder, M. Lutz, A. Streitwieser, *J. Am. Chem. Soc.* **1995**, *117*, 2361–2362.

139 U. Azzena, S. Carta, G. Melloni, A. Sechi, *Tetrahedron* **1997**, *53*, 16205–16212.

140 C. M. Coleman, D. F. O'Shea, *J. Am. Chem. Soc.* **2003**, *125*, 4054–4055.

141 Q. Yang, H. Toshima, T. Yoshihara, *Tetrahedron* **2001**, *57*, 5377–5384.

142 M. Poirier, F. Chen, C. Bernard, Y.-S. Wong, G. G. Wu, *Org. Lett.* **2001**, *3*, 3795–3798.

143 J. Clayden, J. H. Pink, *Tetrahedron Lett.* **1997**, *38*, 2561–2564.

144 (a) J. Clayden, J. H. Pink, *Tetrahedron Lett.* **1997**, *38*, 2565–2568. (b) J. Clayden, J. H. Pink, S. A. Yasin, *Tetrahedron Lett.* **1998**, *39*, 105–108. (c) J. Clayden, N. Westlund, F. X. Wilson, *Tetrahedron Lett.* **1999**, *40*, 3331–3334. (d) J. Clayden, M. N. Kenworthy, L. H. Youssef, M. Helliwell, *Tetrahedron Lett.* **2000**, *41*, 5171–5175.

145 R. Polt, D. Sames, J. Chruma, *J. Org. Chem.* **1999**, *64*, 6147–6158.

146 K. Furuta, K. Tomokiyo, T. Satoh, Y. Watanabe, M. Suzuki, *Chembiochem* **2000**, 283–286.

147 (a) J. Barluenga, R. González, F. J. Fañanás, *Tetrahedron Lett.* **1992**, *33*, 831–834. (b) J. Barluenga, R. González, F. J. Fañanás, M. Yus, F. Foubelo, *J. Chem. Soc., Perkin Trans. 1* **1994**, 1069–1077.

148 J. Barluenga, F. J. Fañanás, F. Foubelo, M. Yus, *Tetrahedron Lett.* **1988**, *29*, 4859–4862.

149 M. Braun, J. Rahematpura, C. Bühne, T. C. Paulitz, *Synlett* **2000**, 1070–1072.

150 K. Kuramochi, H. Watanabe, T. Kitahara, *Synlett* **2000**, 397–399.

151 A. Bachki, F. Foubelo, M. Yus, *Tetrahedron Lett.* **1994**, *35*, 7643–7646.

152 H. P. Plaumann, B. A. Keay, R. Rodrigo, *Tetrahedron Lett.* **1979**, 4921–4924.

153 J. Bach, M. Galobardes, J. García, P. Romea, C. Tey, F. Urpí, J. Vilarrasa, *Tetrahedron Lett.* **1998**, *39*, 6765–6768.

154 (a) A. B. Smith III, G. K. Friestad, J. J.-W. Duan, J. Barbosa, K. G. Hull, M. Iwashima, Y. Qiu, P. G. Spoors, E. Bertounesque, B. A. Salvatore, *J. Org. Chem.* **1998**, *63*, 7596–7597. (b) A. B. Smith III, G. K. Friestad, J. Barbosa, E. Bertounesque, K. G. Hull, M. Iwashima, Y. Qiu, B. A. Salvatore, P. G. Spoors, J. J.-W. Duan, *J. A. Chem. Soc.* **1999**, *121*, 10468–10477.

155 E. Piers, A. Orellana, *Synthesis* **2001**, 2138–2142.

156 D. Bourgeois, J.-Y. Lallemand, A. Pancrazi, J. Prunet, *Synlett* **1999**, 1555–1558.

157 M. Hojo, Y. Murakami, H. Aihara, R. Sakuragi, Y. Baba, A. Hosomi, *Angew. Chem. Int. Ed.* **2001**, *40*, 621–623.

158 W. Pitsch, B. König, *Synth. Commun.* **2001**, *31*, 3135–3139.

159 A. K. Glosh, Y. Wang, J. T. Kim, *J. Org. Chem.* **2001**, *66*, 8973–8982.

160 X. Serrat, G. Cabarrocas, S. Rafel, M. Ventura, A. Linden, J. M. Villalgordo, *Tetrahedron: Asymmetry* **1999**, *10*, 3417–3430.

161 P. Ducray, H. Lamotte, B. Rousseau, *Synthesis* **1997**, 404–406.

162 J. E. Baldwin, G. J. Pritchard, R. E. Rathmell, *J. Chem. Soc., Perkin Trans. 1* **2001**, 2906–2908.

163 H. Liu, T. Cohen, *J. Org. Chem.* **1995**, *60*, 2022–2025.

164 S. Klein, I. Marek, J. F. Poisson, J. F. Normant, *J. Am. Chem. Soc.* **1995**, *117*, 8853–8854.

165 J. Aiguade, J. Hao, C. J. Forsyth, *Tetrahedron Lett.* **2001**, *42*, 817–820.

166 (a) I. M. Pastor, M. Yus, *Tetrahedron Lett.* **2001**, *42*, 1029–1032. (b) M. Yus, R. Torregrosa, I. M. Pastor, *Molecules* **2004**, *9*, 330–348.

167 (a) I. Manteca, N. Sotomayor, M. J. Villa, E. Lete, *Tetrahedron Lett.* **1996**, *37*, 7841–7844. (b) I. Manteca, B. Etxarri, A. Ardeo, S. Arrasate, I. Osante, N. Sotomayor, E. Lete, *Tetrahedron* **1998**, *54*, 12361–12378.

168 (a) G. Cardillo, M. Contento, S. Sandri, *Tetrahedron Lett.* **1974**, 2215–2216. (b) K. H. Yong, J. A. Lotoski, J. M. Chong, *J. Org. Chem.* **2001**, *66*, 8248–8251.

169 D. Behnke, S. Hamm, L. Henning, P. Welzel, *Tetrahedron Lett.* **1997**, *38*, 7059–7062.

170 J. Almena, F. Foubelo, M. Yus, *Tetrahedron* **1996**, *52*, 8545–8564.

171 J. Almena, F. Foubelo, M. Yus, *Tetrahedron* **1997**, *53*, 5563–5572.

172 J. Almena, F. Foubelo, M. Yus, *J. Org. Chem.* **1996**, *61*, 1859–1863.

173 J. Almena, F. Foubelo, M. Yus, *Tetrahedron* **1995**, *51*, 3351–3364.

174 F. Foubelo, C. Gómez, A. Gutiérrez, M. Yus, *J. Heterocycl. Chem.* **2000**, *37*, 1061–1064.

175 M. Yus, B. Moreno, F. Foubelo, *Synthesis* **2004**, 1115–1118.

176 (a) I. M. Pastor, M. Yus, *Tetrahedron Lett.* **2000**, *41*, 1589–1592. (b) I. M. Pastor, M. Yus, *Tetrahedron* **2001**, *57*, 2365–2370. (c) I. M. Pastor, M. Yus, *Tetrahedron* **2001**, *57*, 2371–2378. (d) M. Yus, I. M. Pastor, J. Gomis, *Tetrahedron* **2001**, *57*, 5799–5805. (e) M. Yus, J. Gomis, *Tetrahedron Lett.* **2001**, *42*, 5721–5724. (f) M. Yus, J. Gomis, *Eur. J. Org. Chem.* **2002**, 1989–1995.

177 E. Auer, E. Gössinger, M. Graupe, *Tetrahedron Lett.* **1997**, *38*, 6577–6580.

178 M. J. Dabdoub, V. B. Dabdoub, P. G. Guerrero Jr., C. C. Silveira, *Tetrahedron* **1997**, *53*, 4199–4218.

179 (a) M. Yus, F. Foubelo, J. V. Ferrández, *Eur. J. Org. Chem.* **2001**, 2809–2813. (b) M. Yus, F. Foubelo, J. V. Ferrández, A. Bachki, *Tetrahedron* **2002**, *58*, 4907–4915.

180 K. Fujiwara, H. Morishita, K. Saka, A. Murai, *Tetrahedron Lett.* **2000**, *41*, 507–508.

181 A. B. Smith III, S. A. Lodise, *Org. Lett.* **1999**, *1*, 1249–1252.

182 W. F. Bailey, Y. Tao, *Tetrahedron Lett.* **1997**, *38*, 6157–6158.

183 W. S. Johnson, W. R. Bartlett, B. A. Czeskis, A. Gautier, C. H. Lee, R. Lemoine, E. J. Leopold, G. R. Luedtke, K. J. Bankcroft, *J. Org. Chem.* **1999**, *64*, 9587–9595.

184 J. F. Gil, D. J. Ramón, M. Yus, *Tetrahedron* **1993**, *49*, 4923–4938.

185 T. Mandai, S. Matsumoto, M. Kohama, M. Kawada, J. Tsuji, S. Saito, T. Moriwake, *J. Org. Chem.* **1990**, *55*, 5671–5673.

186 C. Gibson, T. Buck, M. Noltemeyer, R. Brückner, *Tetrahedron Lett.* **1997**, *38*, 2933–2936.

187 (a) D. Cheng, K. R. Knox, T. Cohen, *J. Am. Chem. Soc.* **2000**, *122*, 412–413. (b) D. Cheng, S. Zhu, T. Cohen, *J. Am. Chem. Soc.* **2001**, *123*, 30–34.

188 J. Almena, F. Foubelo, M. Yus, *Tetrahedron* **1995**, *51*, 3365–3374.

189 J. Almena, F. Foubelo, M. Yus, *J. Org. Chem.* **1996**, *61*, 1859–1862.

190 U. Azzena, S. Demartis, L. Pilo, E. Piras, *Tetrahedron* **2000**, *56*, 8375–8382.

191 M. Yus, F. Foubelo, *Tetrahedron Lett.* **2001**, *42*, 2469–2472.

192 F. Foubelo, B. Moreno, M. Yus, *Tetrahedron* **2004**, *60*, 4655–4662.

193 D. Soullez, G. Plé, L. Duhamel, *J. Chem. Soc., Perkin Trans. 1* **1997**, 1639–1645.

194 S. Marumoto, H. Kogen, S. Naruto, *Tetrahedron* **1999**, *55*, 7145–7156.

195 M. R. Reeder, A. I. Meyers, *Tetrahedron Lett.* **1999**, *40*, 3115–3118.

196 J. E. D. Kirkham, T. D. L. Courtney, V. Lee, J. E. Baldwin, *Tetrahedron Lett.* **2004**, *45*, 5645–5647.

197 I. Churcher, D. Hallett, P. Magnus, *J. Am. Chem. Soc.* **1998**, *120*, 10350–10358.

3
Functionalized Organoborane Derivatives in Organic Synthesis

Paul Knochel, Hiriyakkanavar Ila, Tobias J. Korn, and Oliver Baron

3.1
Introduction

Organoboranes have quite covalent carbon–boron bonds, which are compatible with a broad range of functional groups [1]. Thus, numerous highly functionalized boron derivatives can be prepared by various synthetic methods (hydroboration, transmetallation, cross-coupling). The synthetic utility of the resulting organoboranes is enhanced through transition-metal catalysis especially by the use of palladium complexes (Suzuki–Miyaura cross-coupling) [2]. In this chapter, we wish to describe the recent developments of polyfunctional organoboron compounds in organic synthesis. In particular, we will focus on methods for the preparation of polyfunctional aryl, heteroaryl, alkenyl, alkynyl, allylic and alkyl boron derivatives and try in each case to demonstrate the synthetic utility of the organoboron intermediates.

3.2
Preparation and Reaction of Functionalized Aryl and Heteroaryl Boranes

3.2.1
Preparation from Polar Organometallics

The preparation of functionalized aryl boronic reagents can be achieved by directed metallation followed by a transmetallation of aryllithiums with organoboron compounds. Thus, Caron and Hawkins have described a directed *ortho*-metallation of aryl neopentyl esters such as **1** for the synthesis of substituted *ortho*-boronyl neopentyl benzoates using lithium diisopropylamide (LDA) as the base and B(O*i*Pr)$_3$ as an *in situ* trap [3]. The crude boronic acids obtained by acidic hydrolysis were subsequently treated with ethanolamine and converted to stable diethanolamine complexes such as **2**. This methodology allows the preparation of a new class of boronic acids with *ortho*-carbonyl substituents and other functionalities

Organometallics. Paul Knochel
Copyright © 2005 WILEY-VCH Verlag GmbH & Co. KGaA, Weinheim
ISBN: 3-527-31131-9

especially bromine, which is not possible by a standard lithium–halogen exchange reaction (Scheme 3.1).

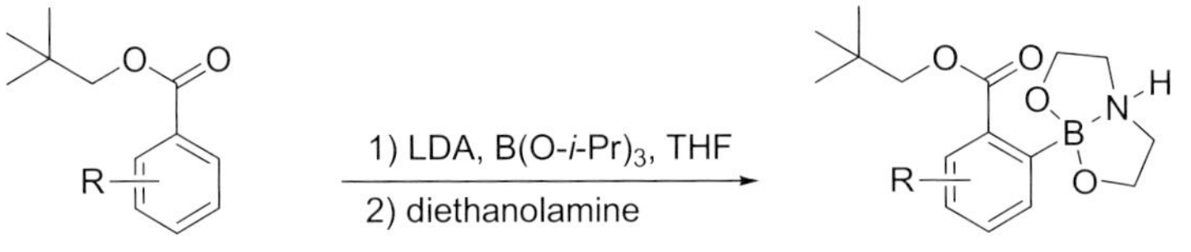

1: R = 4-Cl; 2-Br; 4-Br; 3-F;
2-CF$_3$; 4-CF$_3$; 4-OMe

2: 70 - 93 %

Scheme 3.1 Preparation of boronic esters via metallation.

Furthermore, Vedsø and Begtrup have demonstrated that *ortho*-lithiation, *in situ* borylation using lithium 2,2,6,6-tetramethylpiperidide (LTMP) in combination with triisopropylborate, is highly efficient and represents an experimentally straightforward preparation of *ortho*-substituted arylboronic esters such as **3** [4]. The mild reaction conditions allow the presence of functionalities such as ester or cyano groups or halogen substituents that are usually not compatible with the conditions used in directed metallation of arenes (Scheme 3.2).

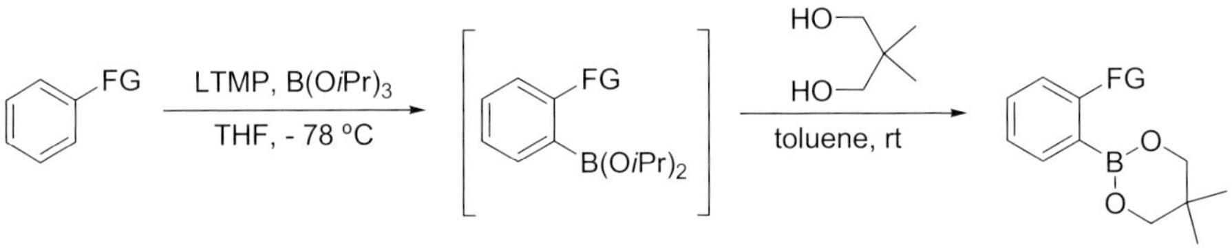

FG = CO$_2$Et; CN; F; Cl

3: 61 - 98 %

Scheme 3.2 *Ortho*-lithiation and borylation of functionalized aromatics.

Morin has reported the preparation of 4-mercaptophenyl boronic acid as a potential melanoma-seeking agent suitable for boron neutron capture therapy [5]. This synthesis involves a metallation-boration sequence starting from 4-bromo-*t*-butyldimethylsilyl thioether **4**. The parent 4-mercaptophenyl boronic acid, generated by boration could subsequently be S-alkylated yielding 4-carboxymethylthiophenyl boronic acid **5** after hydrolysis of the *t*-butyl ester without the need of protecting the boronic acid group leading to the corresponding acid **6** (Scheme 3.3).

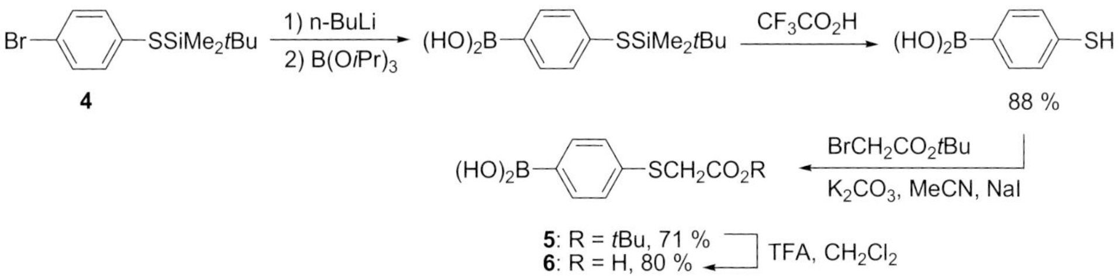

Scheme 3.3 Synthesis of a functionalized boronic acid via a Br/Li-exchange.

Although, the reaction of organolithiums with boronic esters (B(OR)$_3$) is an excellent method for preparing polyfunctional boronic esters, its scope and practicality is limited by the high reactivity of the intermediate organolithiums. They are compatible with a limited number of functional groups and their generation often requires low reaction temperature [6,7]. Organomagnesium derivatives have a much better functional group compatibility and are readily available via a halogen–magnesium exchange reaction [8]. Recently, it was shown that by using the mixed Li/Mg-species: *i*-PrMgCl·LiCl [9], the performance of bromine–magnesium exchange occurs under mild conditions and is compatible with a broad range of functional groups like a nitrile, amide or an ester. The resulting Grignard reagents display also an enhanced reactivity due to the complexation with lithium chloride that confer them a magnesiate character making the aryl moiety more nucleophilic. Thus, the reaction of 1,2-dibromobenzene **7** with *i*-PrMgCl·LiCl furnishes the corresponding Grignard reagent **8** that reacts with 2-methoxy-4,4,5,5-tetramethyl-1,3,2-dioxaborolane **9** affording the corresponding ester **10** in 85% yield. The functionalized aryl halides **11** and **12** give rise, under similar reaction conditions, to the arylmagnesium species **13** and **14** that react with the borate **9**, yielding the boronic esters **15** and **16** in 89–91% yield (Scheme 3.4) [10].

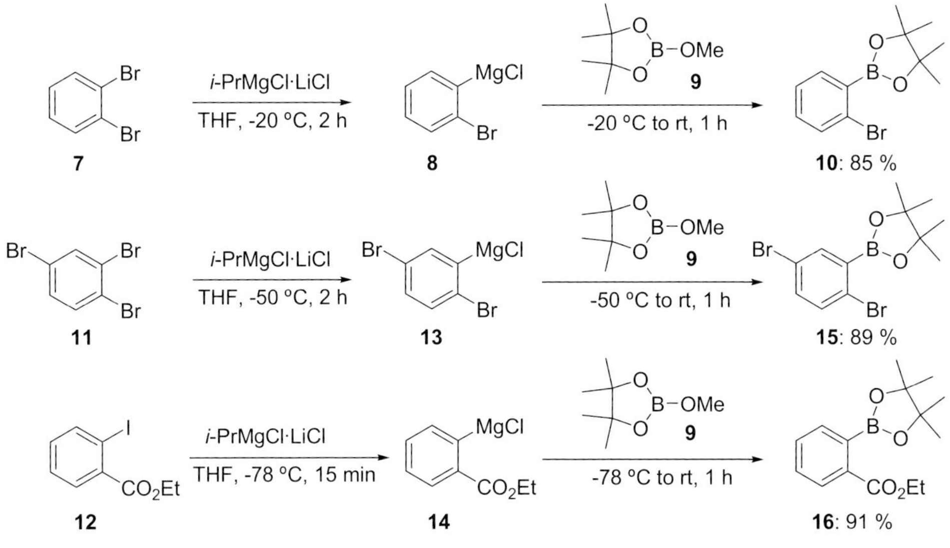

Scheme 3.4 Preparation of boronic esters from arylmagnesium reagents.

Functionalized boronic esters can be alternatively prepared from readily available iodoaryl boronic esters such as **17** and **18**. I/Mg-exchange with *i*-PrMgCl·LiCl provides a bimetallic B/Mg-species such as **19** and **20** that react with a range of electrophiles (aldehyde, allyl bromide, acid chloride, 3-iodo-2-cyclohexenone) providing the products **21** in good yields (Scheme 3.5) [11].

The Pd-catalyzed reaction of such polyfunctional boronic esters like **21b** with various aryl halides provides the desired cross-coupling products like **22** in high yields [10,11]. Interestingly, a one-pot reaction allowing the selective reaction with two electrophiles is possible. Thus, the treatment of the *meta*-iodophenyl boronic pinacol ester **18** with *i*-PrMgCl·LiCl followed by a transmetallation to the copper derivative and subsequent reaction with 2-methyl-3-iodocyclohexenone provides the intermediate boronic ester **23** that after a Suzuki–Miyaura cross-coupling, furnishes the heterocyclic product **24** in 52% yield (Scheme 3.6) [11].

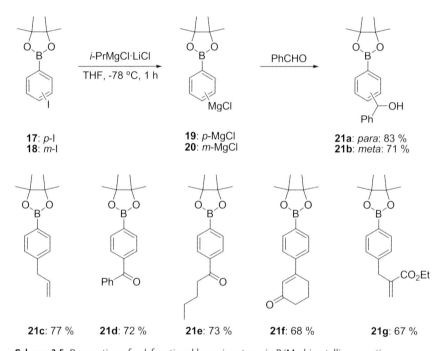

17: *p*-I
18: *m*-I

19: *p*-MgCl
20: *m*-MgCl

21a: *para*: 83 %
21b: *meta*: 71 %

21c: 77 %   **21d**: 72 %   **21e**: 73 %   **21f**: 68 %   **21g**: 67 %

Scheme 3.5 Preparation of polyfunctional boronic esters via B/Mg-bimetallic aromatics.

Finally, this approach can be extended to various heterocyclic systems allowing the preparation of polyfunctional boronic esters such as **25–27** (Scheme 3.) [11]. The boronic esters **26** and **27** have been converted to the polyfunctional heterocycles like- **28** and **29** in excellent yields using the Suzuki–Miyaura cross-coupling reaction (Scheme 3.7).

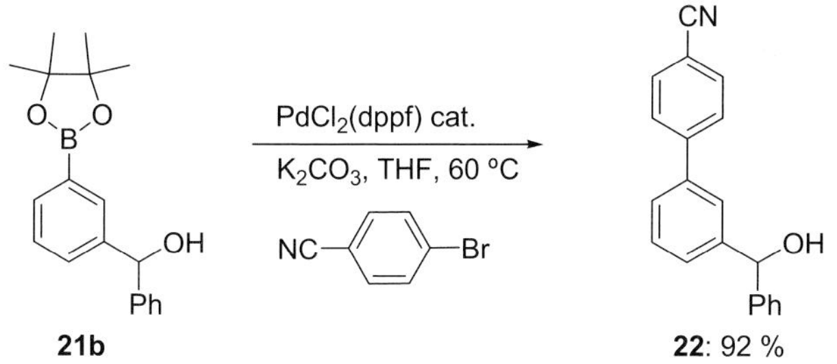

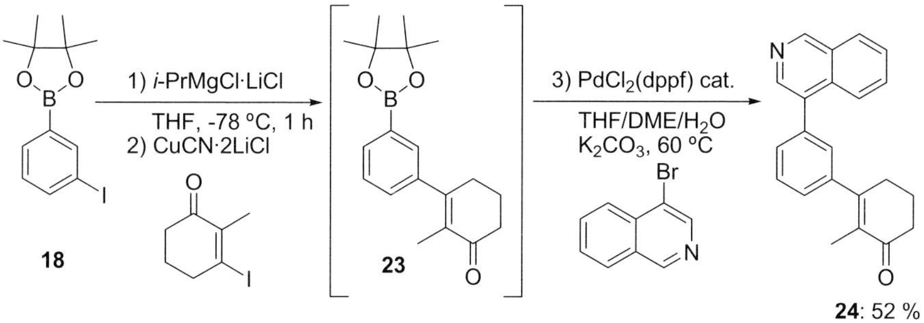

Scheme 3.6 Successive copper-catalyzed cross-coupling and Suzuki–Miyaura cross-coupling reaction.

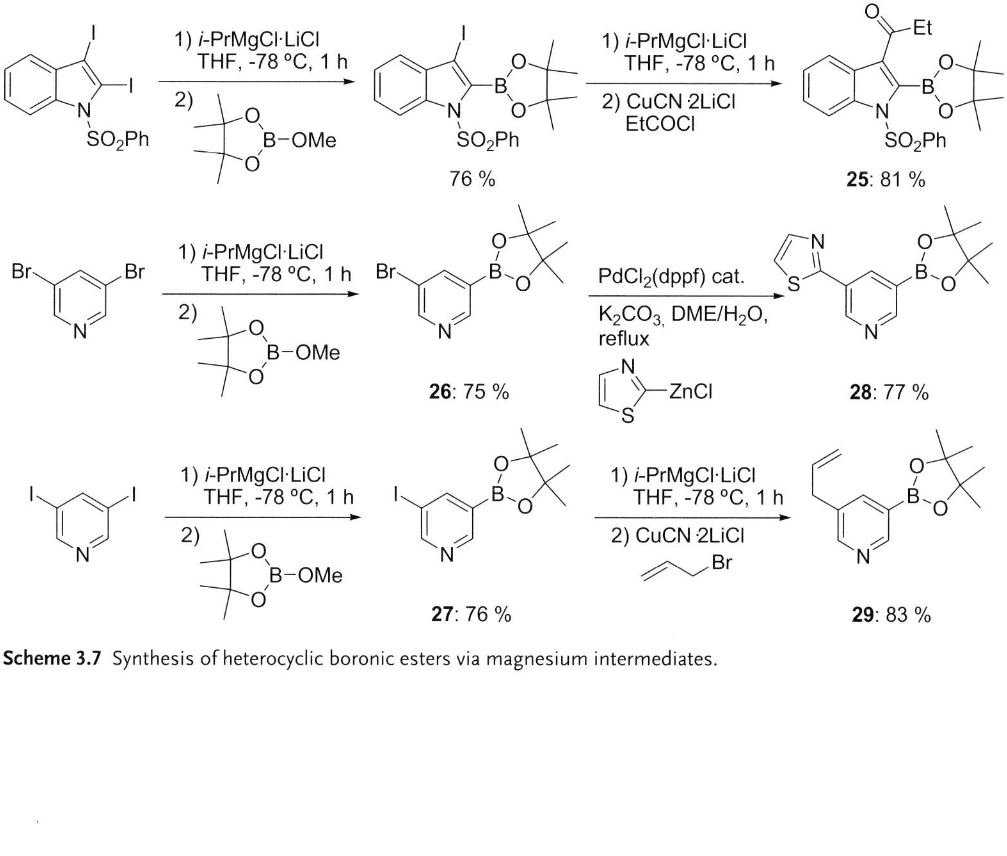

Scheme 3.7 Synthesis of heterocyclic boronic esters via magnesium intermediates.

3.2.2
Preparation from Aryl Halides and Sulfonates by Cross-coupling

A direct synthesis of substituted aryl boronic esters has been recently reported by Masuda [12,13]. Thus, coupling of pinacol borane **30** with aryl halides or triflates in the presence of a catalytic amount of PdCl$_2$(dppf) and a base like triethylamine allows the preparation of aryl boronates having a variety of functional groups such as carbonyl, cyano, nitro, and acylamino of type **31** in high yields. The product distribution **31** versus **32** (reduced product) is strongly dependent on the choice of the base employed. In the presence of triethylamine as a base, selective formation of boronates **31** was observed with negligible amount of the reduced aromatic compound **32** (Scheme 3.8).

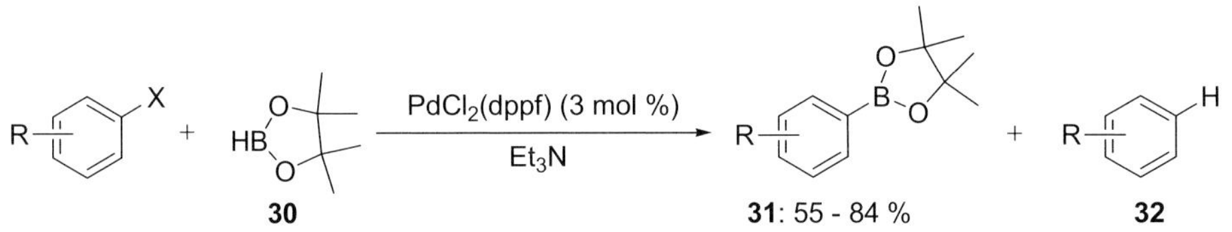

R = 4-Cl, 4-CN, 4-NMe$_2$, 4-NHAc, 4-CH$_2$CN, 4-COMe, 4-CO$_2$Me, 3-COMe; X = Br, I, OTf

Scheme 3.8 Preparation of aryl boronic esters by palladium-catalyzed borylation.

Baudoin has prepared a range of sterically crowded *ortho*-substituted functionalized aryl boronates of type **33** as advanced intermediates for the synthesis of rhazinilam analogs [14]. This was accomplished by a palladium-catalyzed borylation of *ortho*-substituted phenyl halides with pinacol borane (**30**) using a sterically crowded phosphine ligand **34** that significantly improved the yields of borylated products in comparison with dppf. With halides such as 2-bromonitrobenzene and 2-bromoacetophenone, however only little or no pinacol boronates were obtained (Scheme 3.9).

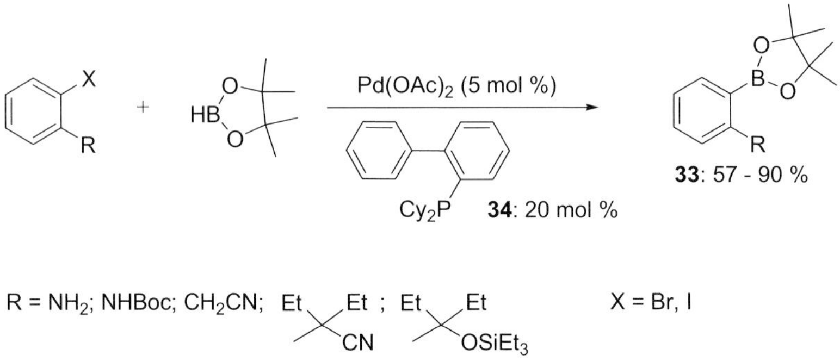

R = NH$_2$; NHBoc; CH$_2$CN; Et$\diagdown$Et ; Et$\diagdown$Et   X = Br, I
                               CN   OSiEt$_3$

Scheme 3.9 Synthesis of aryl boronic esters by palladium-catalyzed borylation with pinacolborane.

Begtrup and Vedsø have developed a new gramme scale synthesis for 2- and 3-substituted 3- or 3-thienylboronic acids and esters **35** from 2,3-dihalothiophenes using a combination of halogen–metal exchange followed by trapping with an electrophile and palladium-catalyzed borylation of the resulting 2- (or 3-) bromo-3- (or 2-) substituted thiophenes **36** [15]. Under optimized conditions the borylation of bromothiophenes were performed in THF at 40 °C using Pd(P(*t*Bu)₃)₂ as catalyst with Et₃N as base and pinacol borane (**30**) as borylation agent, providing in general, the desired 2,3-substituted thienylboronic acids and esters in good to excellent yields. The borylation protocol is compatible with a range of functional groups, however strongly electron withdrawing substituents decreased the stability of the thienylboronic acids and esters that were used as such without isolation in the subsequent coupling (Scheme 3.10).

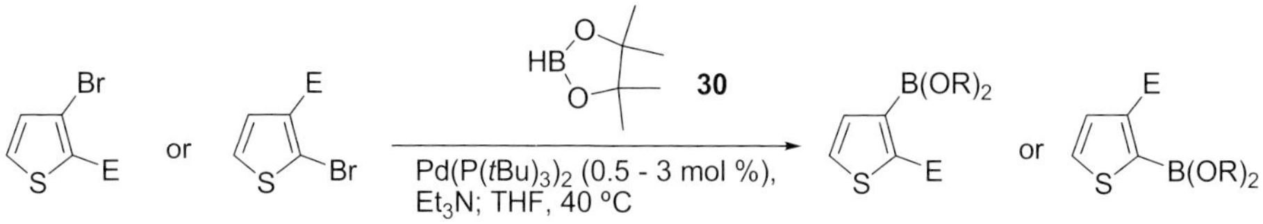

36: E = CO₂Et; CONH*t*Bu;
CHO; CN; COMe; NO₂; SiMe₃

35: 79 - 91 % (crude yields)
43 - 88 % (isolated yields)

Scheme 3.10 Synthesis of heteroaryl boronic esters by palladium-catalyzed borylation.

The cross-coupling reaction between *bis*(pinacol)diborane **37** and aryl halides or triflates in the presence of PdCl₂(dppf) and a base like potassium acetate was first reported by Miyaura and coworkers [16,17]. It is an efficient and direct method for the preparation of functionalized aryl boronates. In contrast to more traditional organometallic approaches involving transmetallation between organomagnesium/lithium reagents and boronate esters, Miyaura's procedure enables facile access to boronic acid derivatives in the presence of sensitive functionalities such as an ester, aldehyde, ketone, cyano, nitro or halogens. A heteroaryl triflate from 7-hydroxycoumarine (**38**) gave the corresponding boronate **39** in 84% yield (Scheme 3.11).

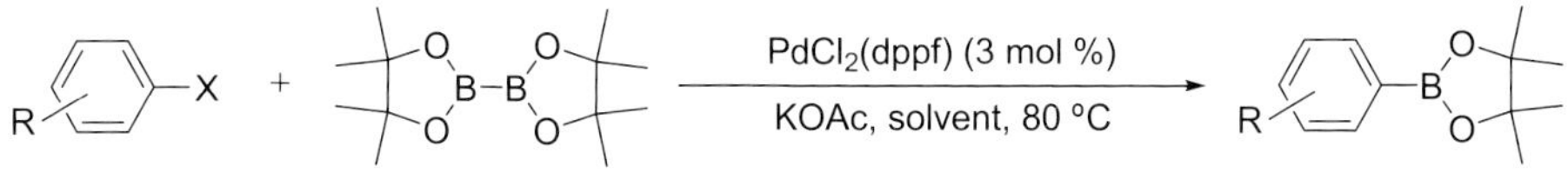

R = 4-COMe; 4-CO₂Me; 4-CHO; 4-CN; 4-SMe; 4-Br; 4-I; 4-NO₂; 2-NO₂; 2-OMe
X = Br, I, OTf

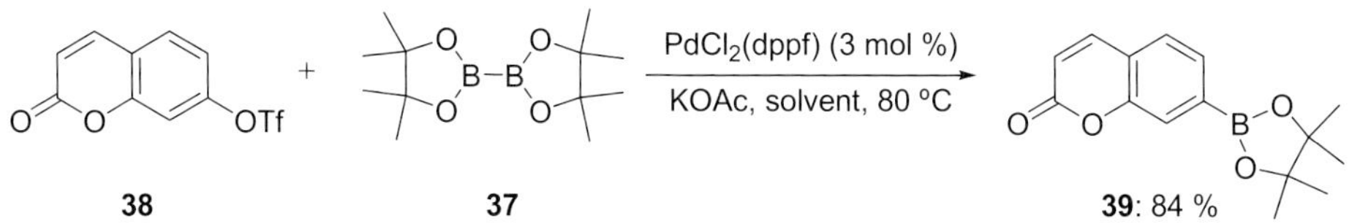

38   **37**   **39**: 84 %

Scheme 3.11 Synthesis of aryl boronates by palladium- catalyzed borylation of aryl halides with *bis*(pinacol)diborane (**37**).

In a subsequent paper, Miyaura demonstrated that several chloroarenes of type **40** are also efficient substrates for the cross-coupling reaction of *bis*(pinacol)diborane (**37**) in the presence of Pd(dba)$_2$, the sterically crowded ligand tricyclohexylphosphine (Cy$_3$P) and potassium acetate as a base leading to pinacol boronic esters of type **41** in 72–84% yield [18]. The reaction is compatible with a range of functional groups in the chloroarene. The catalyst system is also shown to be effective for analogous couplings with aryl bromides and triflates shortening significantly the reaction time for arylation (6–7 h) compared with previous procedures catalyzed by PdCl$_2$(dppf) in DMSO (Scheme 3.12).

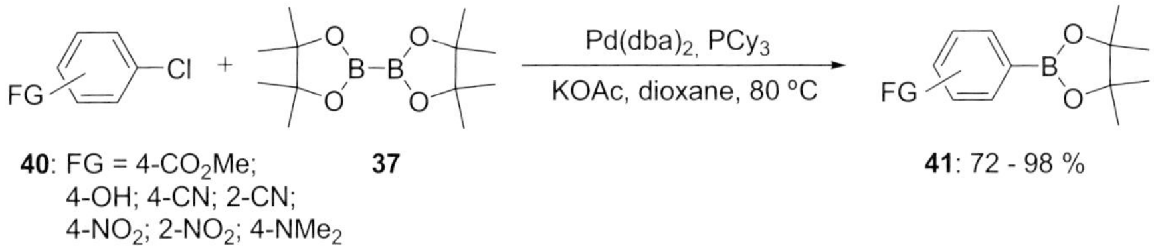

40: FG = 4-CO$_2$Me; **37** **41**: 72 - 98 %
4-OH; 4-CN; 2-CN;
4-NO$_2$; 2-NO$_2$; 4-NMe$_2$

Scheme 3.12 Coupling of chloroarenes with *bis*(pinacol)diborane (37).

Fürstner has described a synthesis of aryl boronates of type **42** bearing electron-withdrawing groups by cross-coupling of the respective *para*-substituted chloroarenes **43** with *bis*(pinacol)diborane (**37**) in the presence of a catalyst formed *in situ* from Pd(OAc)$_2$ and the imidazolium chloride **44** [19]. A particularly noteworthy aspect of this reaction is the very significant rate acceleration, if the borylation is carried out with microwave heating. This allows the reduction of the overall reaction time from several hours to 10–20 min without affecting the yields (Scheme 3.13).

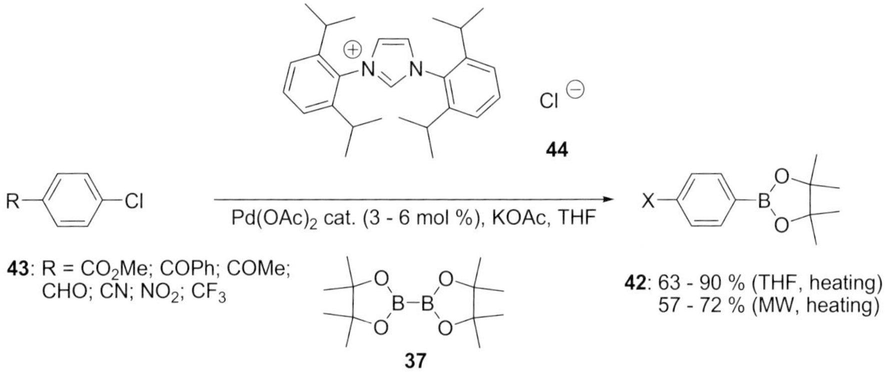

44

43: R = CO$_2$Me; COPh; COMe; Pd(OAc)$_2$ cat. (3 - 6 mol %), KOAc, THF **42**: 63 - 90 % (THF, heating)
CHO; CN; NO$_2$; CF$_3$ 57 - 72 % (MW, heating)

37

Scheme 3.13 Cross-coupling using a Pd-catalyst bearing carbene ligand.

Zhang has developed a modification of Miyaura's aryl boronate synthesis by using a ligandless palladium catalyst [20]. Pd(OAc)$_2$, a much cheaper palladium catalyst is found to be highly effective for such coupling reactions with aryl bromides bearing electron-withdrawing groups such as COMe, CN, CO$_2$Me and NO$_2$.

However, aryl bromides substituted with electron-donating groups such as NMe$_2$ and OMe poorly undergo the palladium-acetate-catalyzed cross-coupling. This modified procedure is advantageous compared to the original Miyaura synthesis in ease of work-up, catalyst removal and low catalyst cost (Scheme 3.14).

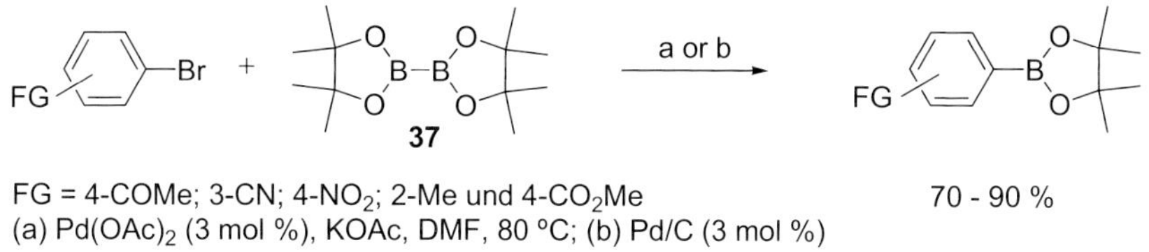

FG = 4-COMe; 3-CN; 4-NO$_2$; 2-Me und 4-CO$_2$Me   70 - 90 %
(a) Pd(OAc)$_2$ (3 mol %), KOAc, DMF, 80 °C; (b) Pd/C (3 mol %)

Scheme 3.14 Ligandless palladium-catalyzed cross-coupling.

Wang has synthesized boronic acids derived from nitrophenols such as **45** and **46** with possible use as a recognition and signalling unit for the construction of polyboronic acid sensors [21]. Synthesis of boronic acid **45** involves cross-coupling of 2-bromo-4-nitroanisole with **37** followed by hydrolysis of boronate ester and deprotection of the methoxy group, whereas the dinitrophenyl boronic acid **46** is prepared by nitration of 2-methoxyboronic acid and subsequent cleavage of the methoxy ether with BBr$_3$. The resulting functionalized boronic acids are both obtained in 90% yield (Scheme 3.15).

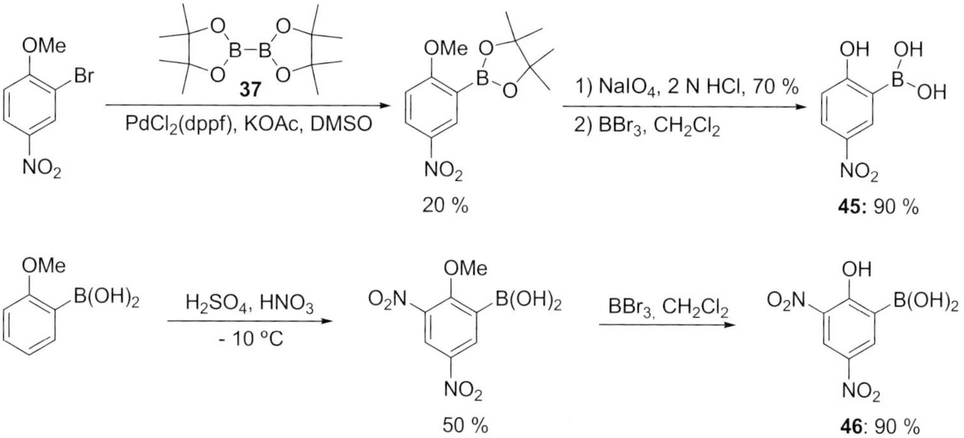

Scheme 3.15 Synthesis of functionalized aryl boranes by palladium-catalyzed cross-coupling or nitration.

Yamamoto has reported the synthesis of (4-boronylphenyl)alanine (BPA), used clinically for treatment of malignant melanoma and brain tumors in neutron cap-ture therapy by Pd-catalyzed coupling of triflate **47** with the diboron derivative **48** [22]. The boronic ester **49** could be easily cleaved by hydrogenolysis to give *L*-BPA **50** in 74% yield, whereas the corresponding pinacol ester yielded mixture of products on hydrolysis with $NaIO_4$ (Scheme 3.16). On the other hand, Morin has reported the synthesis of L-BPA **50** by Pd-catalyzed cross-coupling of 4-iodophenylalanine with *bis*(pinacol)diborane (**37**) [23].

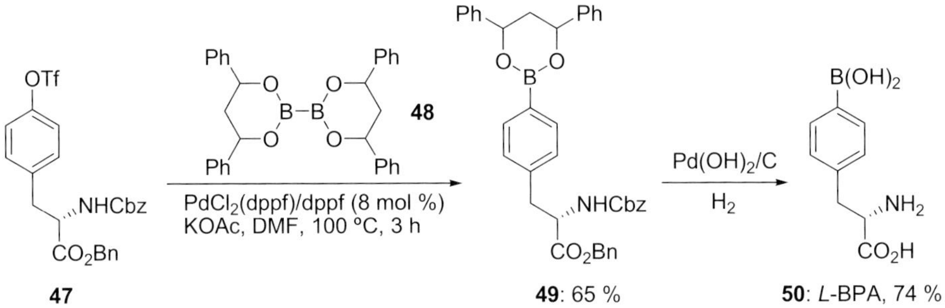

Scheme 3.16 Synthesis of (4-boronylphenyl)alanine.

3.2.3
Synthesis of Functionalized Aryl Boranes by Catalytic Aromatic C–H Borylation

Direct borylation of aliphatic and aromatic hydrocarbons provides a more efficient and convenient route to alkyl and aryl boronic compounds, because of their wide availability and low cost [24]. Hartwig has reported the C–H coupling of benzene with *bis*(pinacol)diborane (**37**) catalyzed by $Cp*Re(CO)_3$ [25] or $Cp*Rh(\eta^4\text{-}C_6Me_6)$ [26] under photoirradiation or by using temperatures above 150 °C. Similar reactions with pinacol borane (**30**) in the presence of a $(\eta^5\text{-}C_9H_7)Ir(COD)\text{-}dppe(\text{-}dmpe)$ [27] or a $(Cp*RhCl)_2$ [28] have also been developed by Smith. Clearly, the synthetic utility of the catalytic borylation hinges on the demonstration of functional-group tolerance and regioselective activation for substituted arenes. It was shown in initial studies [29,30] that both Ir and Rh precatalysts gave an approximately statistical distribution of *meta* and *para* isomers in borylation of mono-substituted arenes, demonstrating that borylation regioselectivity is sterically controlled for the most substituted arenes.

Thus, aromatic borylation by pinacol borane with Hartwig's catalyst $Cp*Rh(\eta^4\text{-}C_6Me_6)$ has been amply demonstrated by Smith and coworkers [29]. Borylation of several 1,3-substituted aromatic species ranging from electron rich $(1,3\text{-}(NMe_2)_2C_6H_4)$ to electron deficient groups $(1,3\text{-}(CF_3)_2C_6H_4)$ in cyclohexane at 150 °C yields 5-borylated arenes, whereas 1,2-disubstituted arenes such as veratrole were selectively borylated at the 4-position. Selective borylation in 3-position of N-triisopropylsilyl-pyrrole has also been demonstrated providing a valuable heteroaryl borane reagent for cross-coupling reaction (Scheme 3.17).

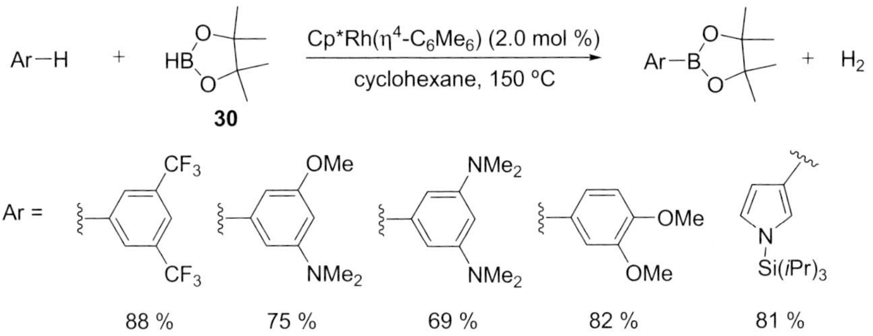

Scheme 3.17 Direct borylation of aromatics with pinacol borane (30).

Miyaura has demonstrated that the iridium complex derived from the reaction of [Ir(OMe)(COD)$_2$] with 4,4′-di-*ter*-butyl-2,2′-bipyridine is a highly active catalyst for aromatic C-H borylation by *bis*(pinacol)diborane (**37**) allowing for the first time, room-temperature borylation with stoichiometric amount of arenes [31]. The reactions are tolerated with a wide range of functionalities such as MeO, I, Br, Cl, CO$_2$Me, CN and CF$_3$ yielding aryl boronates in highly regioselective fashion. Thus, the reaction occurs regioselectively with disubstituted arenes and 1,2-, 1,3- and 1,4-dichlorobenzenes are yielding a single product. The borylation of 1,3- disubstituted benzenes bearing electron donating or electron withdrawing groups occurs only at the *meta* position (Scheme 3.18).

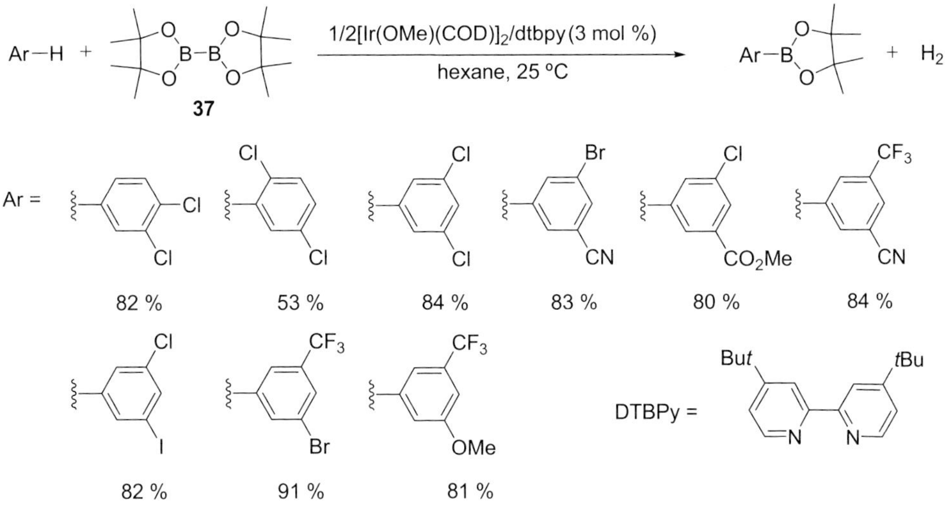

Scheme 3.18 Synthesis of functionalized aryl boronic esters by transition-metal-catalyzed C-H borylation of arenes.

The [IrCl(COD)]$_2$dtbpy system was also found to be effective for borylation of five-membered aromatic heterocycles such as thiophene, furan, pyrrole and their benzo fused derivatives yielding 2-borylated products exclusively in high yields, whereas six-membered heterocycles including pyridines and quinolines selectively gave 3-borylated products (Scheme 3.19) [32]. Regioselective synthesis of 2,5-*bis*-borylated heteroaromatics was also achieved by using equimolar amounts of substrates and *bis*(pinacol)diborane (**37**) (Scheme 3.19). The regiochemistry can be tuned by varying the steric hindrance of the substituents in triisopropylsilylpyrrole and the corresponding indole derivative that gave selectively 3-borylated products. The 2,6-dichloropyridine is shown to undergo γ-borylation (Scheme 3.19) [27].

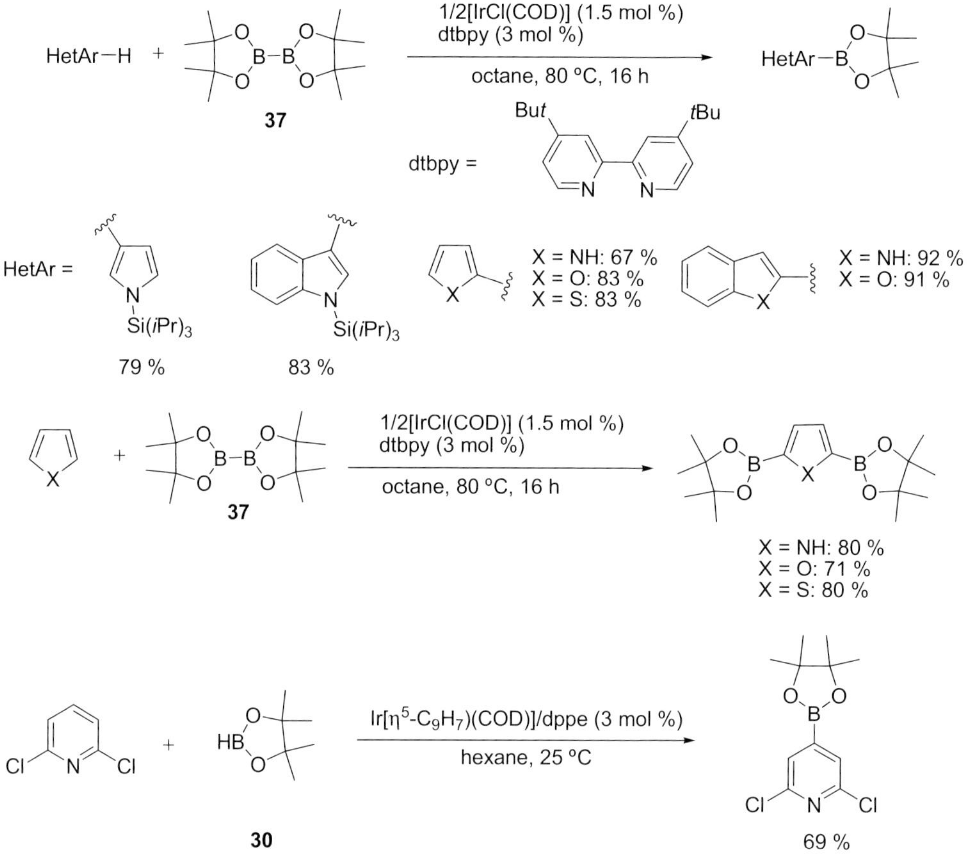

Scheme 3.19 Synthesis of heteroaryl boronates by transition-metal-catalyzed borylation of heteroarenes via C–H activation.

3.2.4
Synthesis of Functionalized Trifluoroborates and their Palladium-catalyzed Suzuki–Miyaura Cross-coupling Reactions

Molander has demonstrated in a detailed study on Suzuki coupling reactions using aryl and heteroaryl trifluoroborates that these fluoroborates are more robust, easier to handle and less prone to protodeboronation[33]. A wide array of electron-withdrawing and electron-donating groups such as fluoro-, acetyl-, nitro-, trifluoromethyl and methoxy groups are tolerated in aryl trifluoroborates used in Suzuki cross-coupling reactions both under ligandless[2] and ligand-added catalytic protocols[34]. In general, lower loading of catalyst, lower temperature and shorter reaction time as well as lack of inert atmosphere is needed in these reactions (Scheme 3.20).

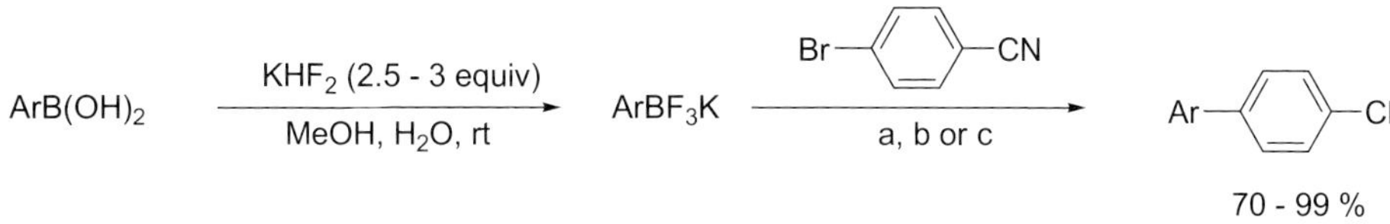

Ar = 4-MeOC$_6$H$_4$; 3-MeOC$_6$H$_4$; 4-FC$_6$H$_4$; 4-MeCOC$_6$H$_4$; 3-NO$_2$C$_6$H$_4$; 3,5-(CF$_3$)$_2$C$_6$H$_3$; 2,6-F$_2$C$_6$H$_3$

(a) Pd(OAc)$_2$ (0.5 mol %), K$_2$CO$_3$ (3 equiv), MeOH reflux; (b) Pd(OAc)$_2$, (0.5 mol %), Ph$_3$P (0.5 mol %), K$_2$CO$_3$ (3 equiv), MeOH reflux; (c) PdCl$_2$(dppf)·CH$_2$Cl$_2$ (0.5 mol %), Et$_3$N (3 equiv), EtOH reflux

Scheme 3.20 Synthesis and cross-coupling of functionalized potassium aryl trifluoroborates.

Batey has reported Pd-catalyzed cross-coupling reactions of tetraalkylammonium aryl trifluoroborate salts bearing functional groups such as acetyl, nitro and chloro with functionalized aryl bromides under mild conditions[35]. These tetraalkylammonium organofluoroborates are prepared from the respective boronic acids using counter ion exchange protocol. They are air and moisture stable and are soluble in various organic solvents (Scheme 3.21).

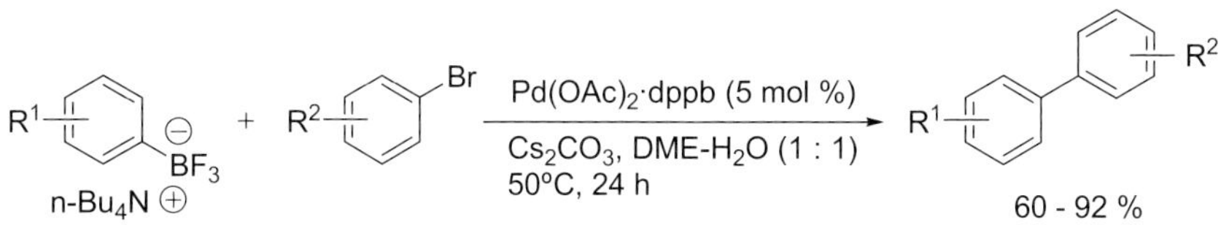

R^1 = 4-MeCO, 3-NO$_2$, 4-Cl, 3-Cl
R^2 = 4-CHO, 3-CHO, 4-MeCO

Scheme 3.21 Palladium-catalyzed cross-coupling of functionalized tetraalkylammonium aryl trifluoroborates.

Potassium pentafluorophenyl and trifluorovinyl trifluoroborates containing electron-poor substituents such as **51** undergo facile cross-coupling with *p*-substituted iodoarenes such as **52** in the presence of Pd(OAc)$_2$/2 PPh$_3$ and a stoichiometric amount of silver oxide yielding fluorinated cross-coupled products of type **53** in very good yields (Scheme 3.22) [36].

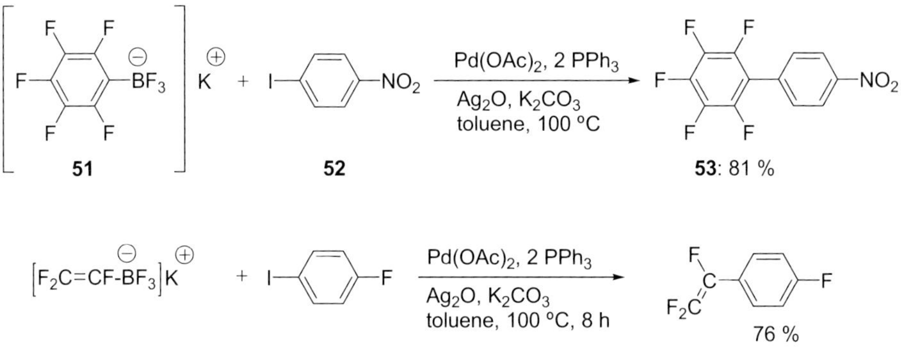

Scheme 3.22 Palladium-catalyzed cross-coupling reactions of perfluoroorgano trifluoroborates.

3.2.5
Palladium-catalyzed Suzuki–Miyaura Cross-coupling Reactions of Functionalized Aryl and Heteroaryl Boronic Esters

Buchwald has reported that the sterically crowded ligand o-*bis*(*t*-butyl)phosphino-biphenyl (**54**) is a very effective ligand for Pd-catalyzed Suzuki cross-coupling for a wide array of aryl chlorides with substituted boronic acids at room temperature (Scheme 3.23, Eq. 1) [37]. Simultaneously, Fu reported a versatile method for Suzuki cross-coupling of aryl chlorides (which are otherwise not reactive under normal Suzuki conditions) using the sterically crowded and electron-rich trialkyl-phosphane P(*t*-Bu)$_3$ (Scheme 3.23, Eq. 2) [38]. Deactivated and sterically hindered aryl chlorides were suitable substrates for this catalytic system. Fu subsequently demonstrated that KF is a more effective additive than Cs$_2$CO$_3$ that allowed Suzuki coupling of activated aryl chlorides including heteroaryl chlorides to proceed at room temperature [39]. The Pd/P(*t*-Bu)$_3$-based catalytic system exhibits a highly unusual reactivity profile that has unprecedented selectivity for the coupling of an aryl chloride in preference to an aryl triflate (Scheme 3.23, Eq. 3).

Despite a large variety of functional groups that can be tolerated in aryl chlorides, only a limited number of substituted aryl boronic acids with functionalities like MeCO, CF$_3$ and OMe have been used as coupling partners in these reactions (Scheme 3.23).

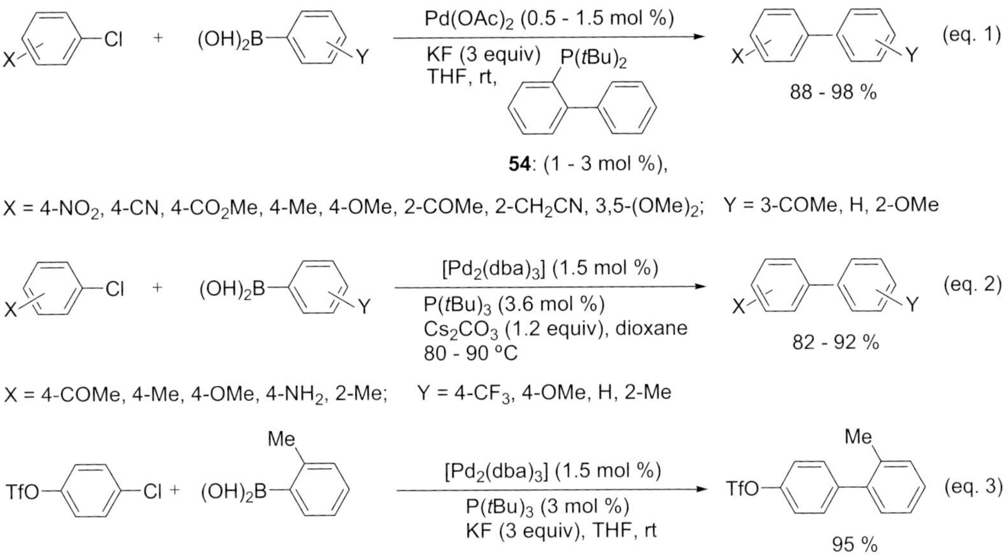

$$X = 4\text{-}NO_2, 4\text{-}CN, 4\text{-}CO_2Me, 4\text{-}Me, 4\text{-}OMe, 2\text{-}COMe, 2\text{-}CH_2CN, 3,5\text{-}(OMe)_2; \quad Y = 3\text{-}COMe, H, 2\text{-}OMe$$

$$X = 4\text{-}COMe, 4\text{-}Me, 4\text{-}OMe, 4\text{-}NH_2, 2\text{-}Me; \quad Y = 4\text{-}CF_3, 4\text{-}OMe, H, 2\text{-}Me$$

Scheme 3.23 Suzuki cross-coupling of aryl chlorides with functionalized boronic esters.

A heavily functionalized atropisomeric biphenyl derivative (designed for use as liquid crystal dopant) has been recently synthesized, although in low yield, with Suzuki coupling as the key step [40]. The coupling reaction is complicated by rapid hydrolytic deboronation of the sterically crowded electron-deficient boronate ester **55**. Rigorously anhydrous conditions are required to avoid the deboronation step (Scheme 3.24).

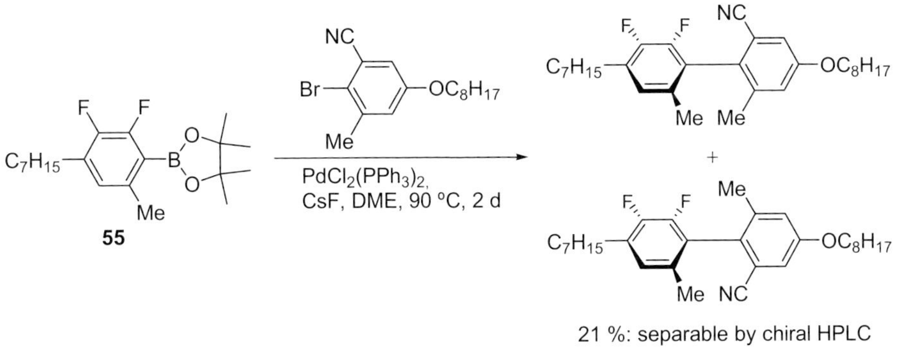

Scheme 3.24 Synthesis of sterically crowded biphenyls by Suzuki coupling.

A new synthesis of biologically important 1,4-benzodiazepines and 3-amino-1,4-benzodiazepines (**56**) using a Pd-catalyzed cross-coupling reaction of imidoyl chlorides with a variety of functionalized organoboronates or boronic acids as the key step has been described by Nadin et al. [41]. Examples of C–C bond formations using Pd(0)-catalyzed reactions of imidoyl halides or triflates are surprisingly rare (Scheme 3.25).

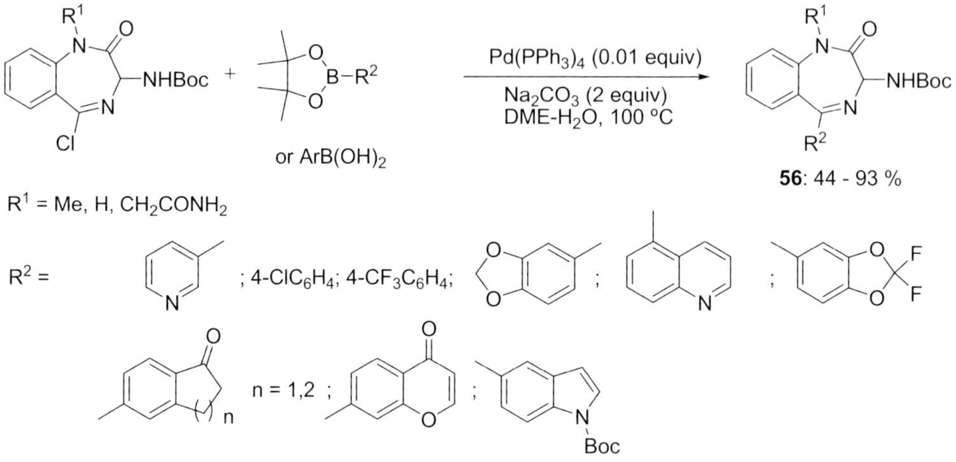

R^1 = Me, H, CH_2CONH_2

R^2 =

4-ClC$_6$H$_4$; 4-CF$_3$C$_6$H$_4$;

n = 1,2 ;

Scheme 3.25 Cross-coupling of functionalized aryl boronic esters with heterocyclic imidoyl chlorides.

An efficient large-scale chromatography free synthesis of cathepsin K inhibitor **57** has been recently reported involving Suzuki coupling of aryl bromide **58** with unprotected piperizinoaryl boronic acid **59** [42]. The residual palladium generated in the Suzuki coupling was efficiently removed from crude **57** via simple extractive work-up using lactic acid (Scheme 3.26).

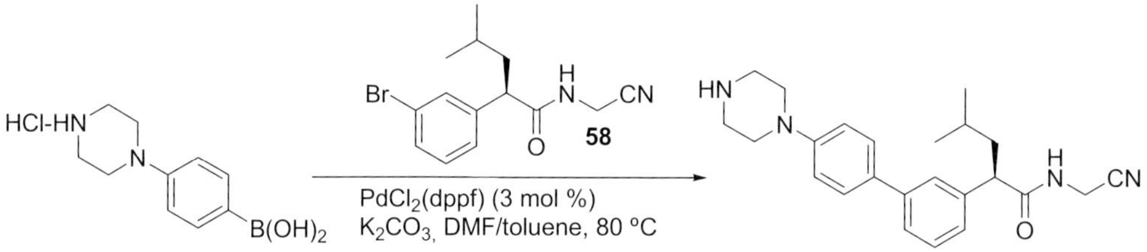

Scheme 3.26 Synthesis of potent cathepsin K inhibitor.

The Suzuki–Miyaura reaction has also found applications in nucleoside chemistry. Thus, Shaughnessy has reported an aqueous-phase modification of unprotected halonucleosides involving cross-coupling of either 8-bromodeoxyguanosine (or guanosine) and 8-bromodeoxyadenosine (or adenosine) with various substituted aryl boronic acids using a catalytic system derived from palladium acetate and water soluble *tris*(3-sulfonatophenyl)phosphine (TPPTS; **60**) in a 2:1 water-acetonitrile mixture giving 8-aryl adducts of the corresponding nucleosides in excellent yields [43]. This coupling protocol has also been extended to 5-iodo-2′-deoxyuridine (5-IDU) (Scheme 3.27).

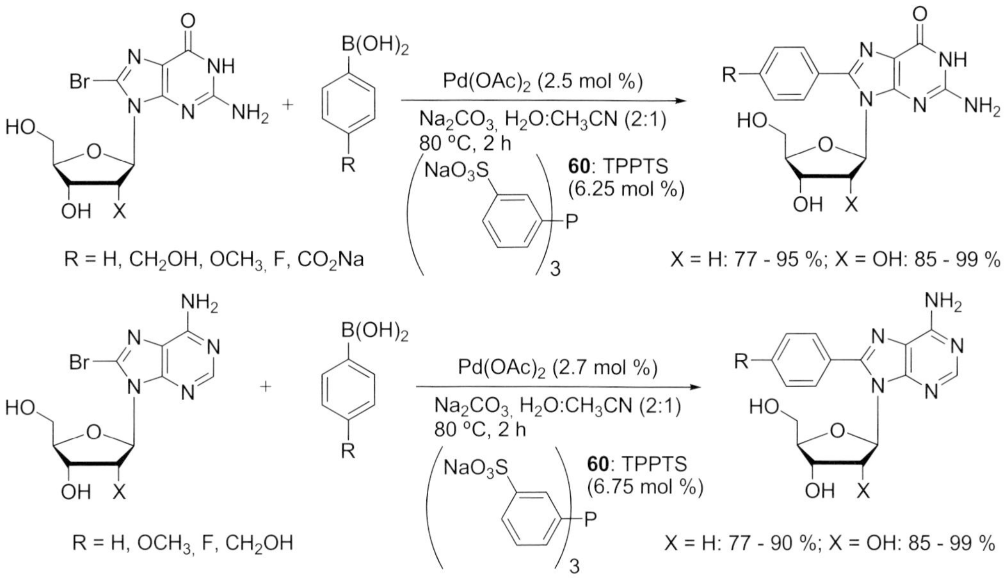

Scheme 3.27 Suzuki–Miyaura cross-coupling of unprotected halonucleosides.

Firooznia has reported the synthesis of 4-substituted phenylalanine derivatives via cross-coupling of protected (4-pinacolylboron)phenylalanine derivatives such as **61** with aryl and alkenyl iodides, bromides and triflates [44]. They have further shown that BOC derivatives of (4-pinacolylboron)phenylalanine ethyl ester **61** or the corresponding boronic acids undergo Suzuki–Miyaura reactions with a number of aryl chlorides in the presence of PdCl$_2$(PCy)$_3$ or NiCl$_2$(dppf), respectively providing diverse sets of 4-substituted phenylalanine derivatives of type **62** [45]. This strategy has also been used for the synthesis of enantiomerically enriched 4-substituted phenylalanine derivatives (Scheme 3.28) [46].

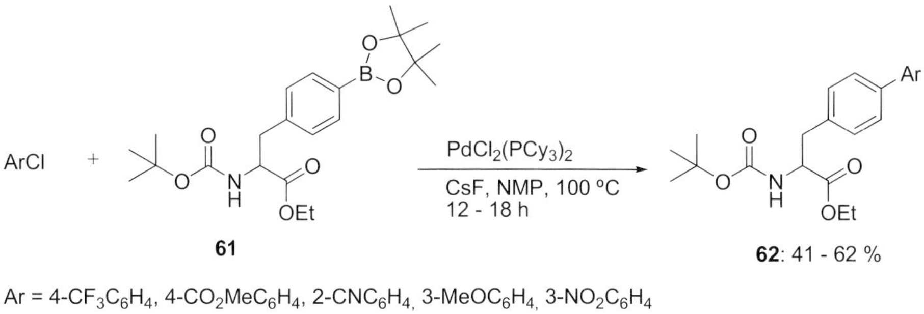

Ar = 4-CF$_3$C$_6$H$_4$, 4-CO$_2$MeC$_6$H$_4$, 2-CNC$_6$H$_4$, 3-MeOC$_6$H$_4$, 3-NO$_2$C$_6$H$_4$

Scheme 3.28 Synthesis of substituted phenylalanine derivatives by Suzuki–Miyaura cross-coupling.

Palladium-mediated cross-coupling of 2-pyrone-5-boronate (**63**), prepared by coupling of 5-bromo-2-pyrone with pinacol borane (**30**), with a range of androsterone-derived alkenyl triflates **64** affords bufadienolide type steroids **65** in high yields (Scheme 3.29) [47].

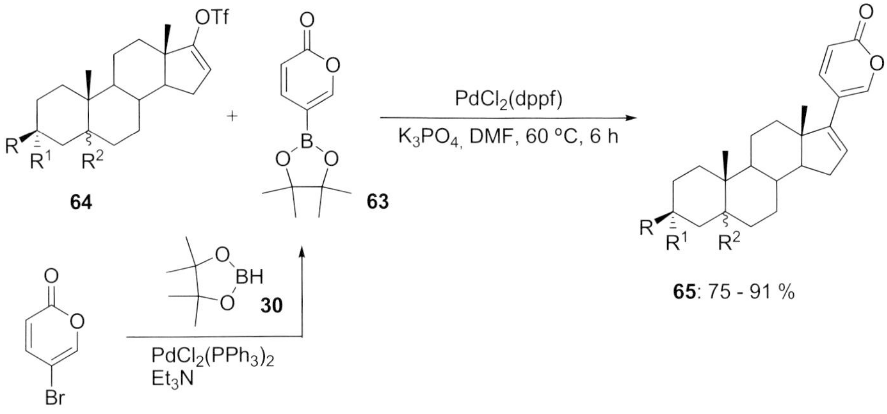

Scheme 3.29 Suzuki cross-coupling of heteroaryl boranes.

An efficient and flexible two-step synthesis of nemertelline (**66**), a quaterpyridine neurotoxin isolated from a hoplonemertine sea worm, involving regioselective Suzuki cross-coupling of chloropyridinyl boronic acids to give 2,2′-dichloro-3,4′-bipyridine (**67**) on the multigram scale followed by its coupling with excess of pyridin-3-yl boronic acid has been recently described by Rault (Scheme 3.30) [48].

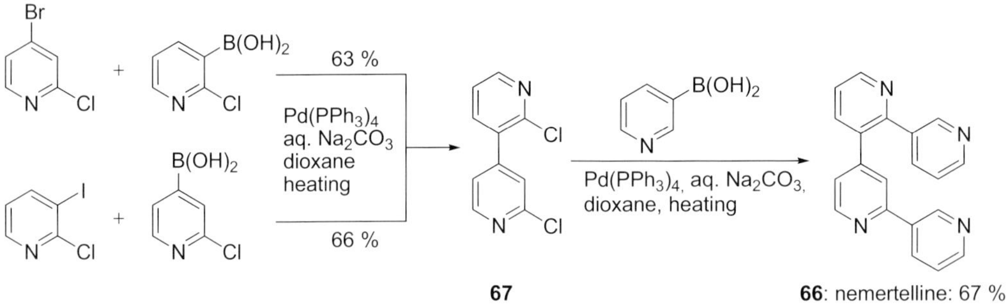

Scheme 3.30 Selective Suzuki coupling of functionalized heteroaryl boronic acids.

Williams has reported the synthesis of biaryl moiety of proteasome inhibitors TMC-95/A/B by Suzuki cross-coupling of boronic ester **68** with indolyl iodide **69** to give precursor **70** in 90% yield [49]. The boronic ester **68** was obtained from the tyrosine derivative **71** with *bis*(pinacol)diborane (**37**), Pd(dppf)Cl$_2$ and KOAc via the Miyaura protocol (Scheme 3.31).

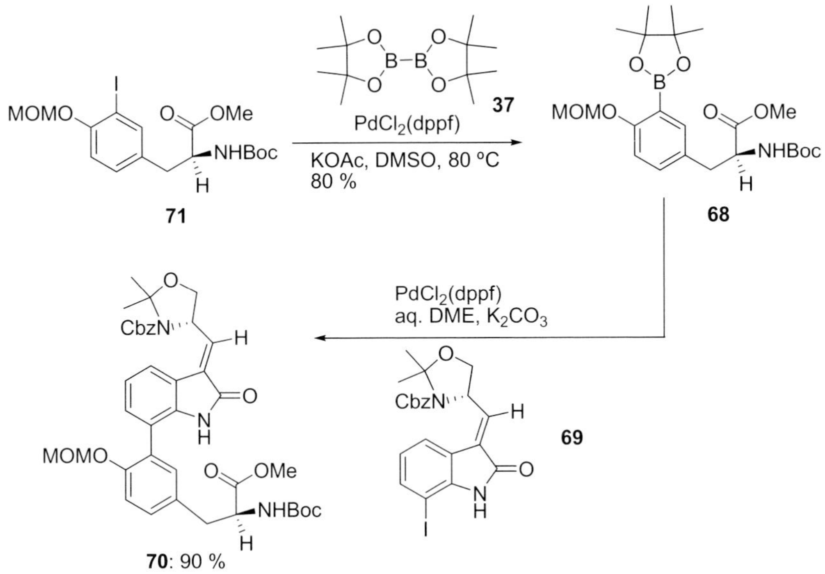

Scheme 3.31 Synthesis of polyfunctional heterocycles via Suzuki–Miyaura cross-couplings.

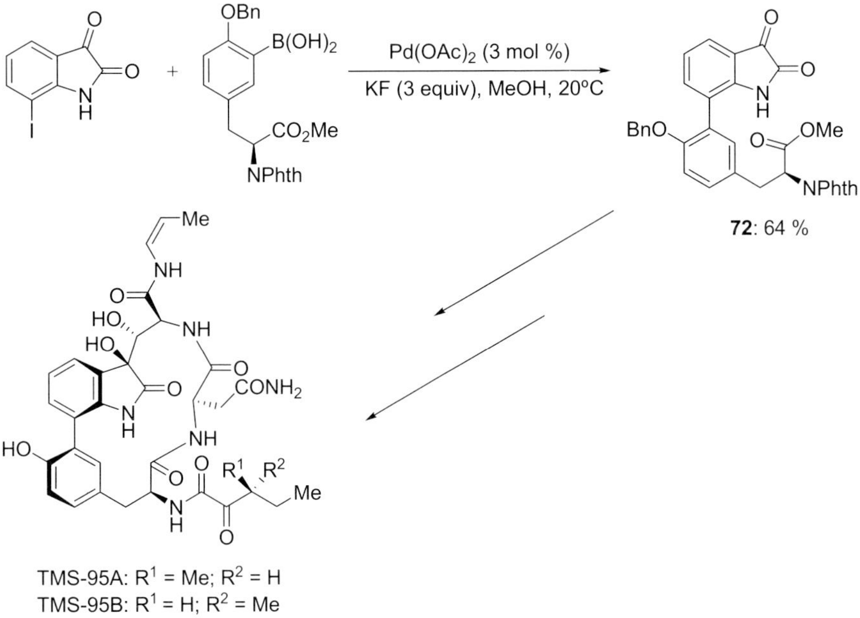

TMS-95A: R^1 = Me; R^2 = H
TMS-95B: R^1 = H; R^2 = Me

Scheme 3.32 Cross-coupling of isatin derivatives.

The biaryl moiety of proteasome inhibitor TMC-95 has been prepared via a ligandless Pd(OAc)$_2$-catalyzed Suzuki-coupling reaction of 7-iodoisatin with sterically hindered tyrosine-derived aryl boronic acid using potassium fluoride as a base in 64% yield [50]. It should be noted that the product **72** was obtained previously only in a yield of 19% under standard Suzuki-coupling conditions (Scheme 3.32) [51].

The macrocyclic core of diazonamide A, a cytotoxic marine natural product has been prepared by Vedejs via Suzuki coupling of boronic acid **73** and the triflate **74** yielding interconverting mixture of two atropisomers that on treatment with LDA at –23 °C affords the macrocyclic ketone **75** in 57% yield (Scheme 3.33) [52].

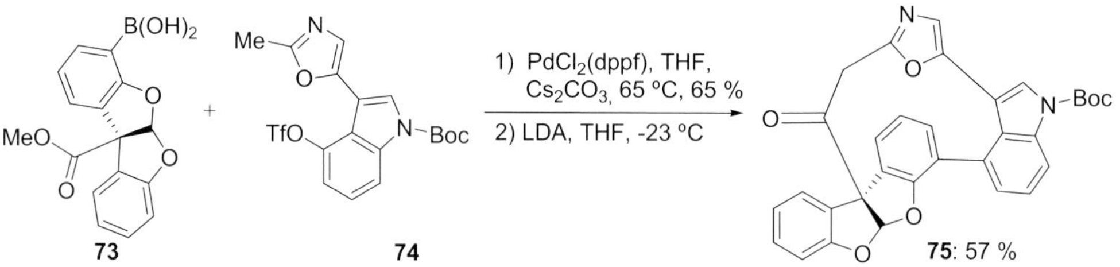

Scheme 3.33 Macrocyclic heterocycle synthesis.

The synthesis of anti-MRSA carbapenam has been reported as a multigram scale synthesis involving Suzuki–Miyaura cross-coupling between carbapenam triflate **76** and the highly functionalized aryl boronate salt **77** as the key step by Merck scientists yielding polyfunctional product **78** in 60% yield over four steps. It highlights the versatility and efficiency of the Suzuki–Miyaura cross-coupling reaction (Scheme 3.34) [53].

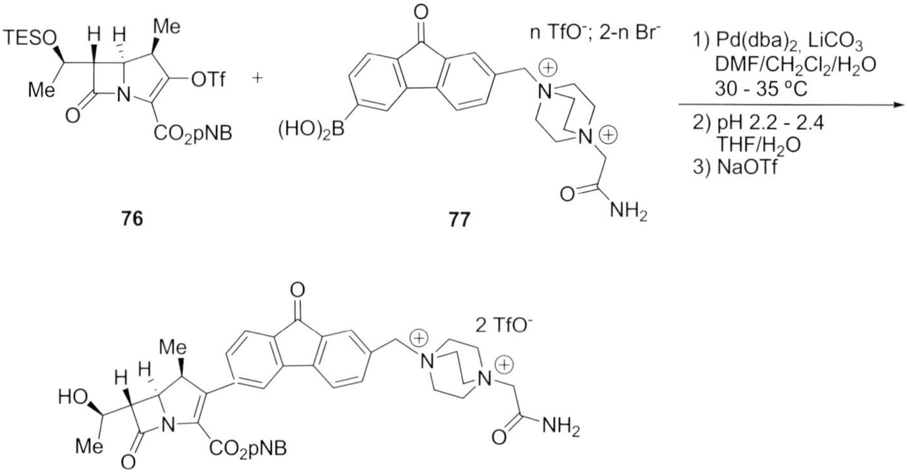

Scheme 3.34 Synthesis of a carbapenam derivative via Suzuki–Miyaura cross-coupling.

Kozikowski has reported the synthesis of a novel class of spirocyclic cocaine analogs **79a** and **b** by using Suzuki cross-coupling of the *ortho*-functionalized aryl boronic acids **80** and **81** and an bicyclic enol triflate **82** as the key step followed by further transformations of the resulting cross-coupled products **83** and **84**, respectively (Scheme 3.35) [54].

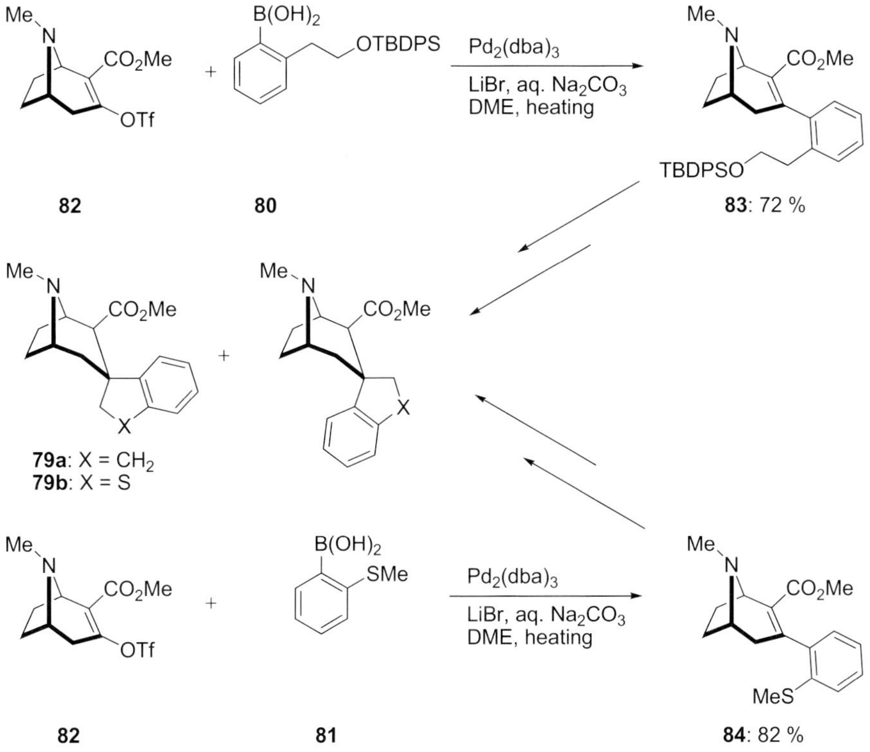

Scheme 3.35 Synthesis of biologically important natural products and their analogs.

The first total synthesis of biologically significant *bis*-indole alkaloid dragmacidin D (**85**) has been reported by Stoltz [55]. The key steps in this synthesis involve a series of thermally and electronically modulated palladium-catalyzed Suzuki cross-coupling reactions of highly functionalized indole-3-boronic acid and ester derivatives furnishing the core structure of this guanidine and amino imidazole marine natural product (Scheme 3.36).

A novel macrocyclization procedure involving two distinct cross-coupling manifolds in a domino fashion has been reported by Zhu for the synthesis of biphenomycin model **86** [56]. Thus, treatment of linear *bis*-iodide with *bis*(pinacol)diborane (**37**) in the presence of Pd(dppf)$_2$Cl$_2$ under defined conditions affords the biphenyl macrocyclic compound **86** in 45% yield through a Miyaura aryl boronic ester formation followed by its intramolecular Suzuki cross-coupling. The diiodide containing a free phenol function (R=H) gave the macrocycle (R=H) in only 22–25% yield under these conditions (Scheme 3.37).

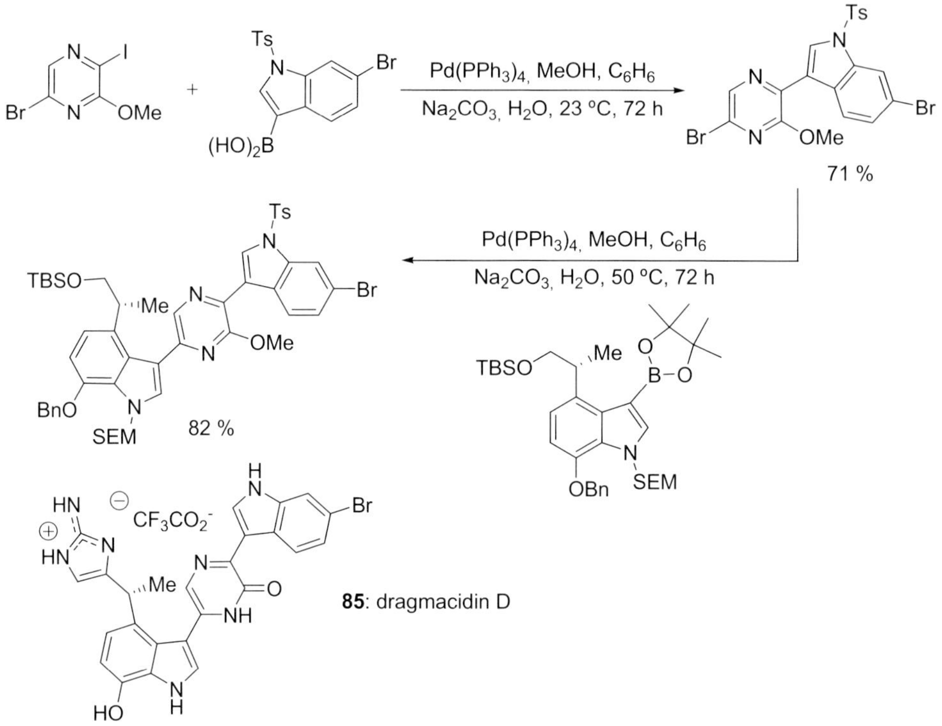

Scheme 3.36 Synthesis of *bis*-indole compounds via Suzuki–Miyaura cross-coupling.

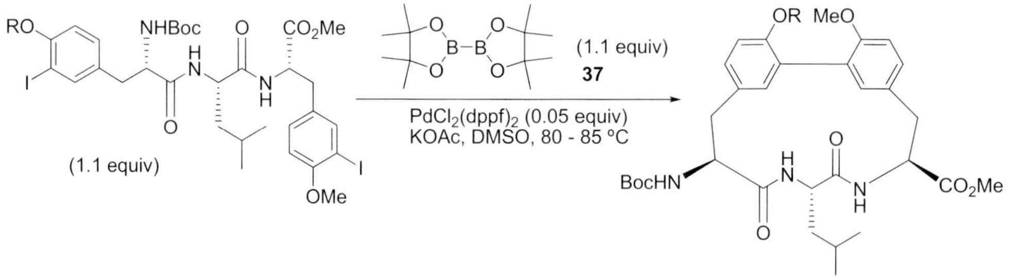

Scheme 3.37 Macrocyclization using the Suzuki–Miyaura reaction.

Vaultier has developed a solid-phase synthesis of macrocyclic systems by intramolecular Suzuki–Miyaura aryl-aryl macrocyclization of polymer ionically bound borates **87** obtained by trapping of the respective aryl boronic acids by an ammonium hydroxide form Dowex® ion exchanger resin (D-OH), leading to macrocycles of type **88** in 16–22% yield (Scheme 3.38) [57].

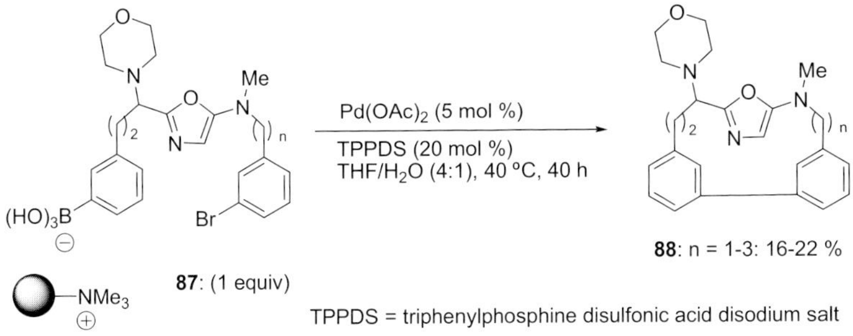

Scheme 3.38 Solid-phase intramolecular macrocyclization.

Dehaen has developed several approaches for the synthesis of highly soluble rod-like diketopyrrolopyrrole oligomers of specified lengths by employing Pd-catalyzed Suzuki cross-coupling reactions of *bis*-boronate **89** and the brominated 1,4-dioxo-3,6-diphenylpyrrolo[3,4-*c*]pyrroles (DPP) **90** and **91** [58]. An example of a convergent approach for the synthesis of trimer of DPPS (**92**) is shown. These compounds are useful for the construction of organic light emitting devices (Scheme 3.39).

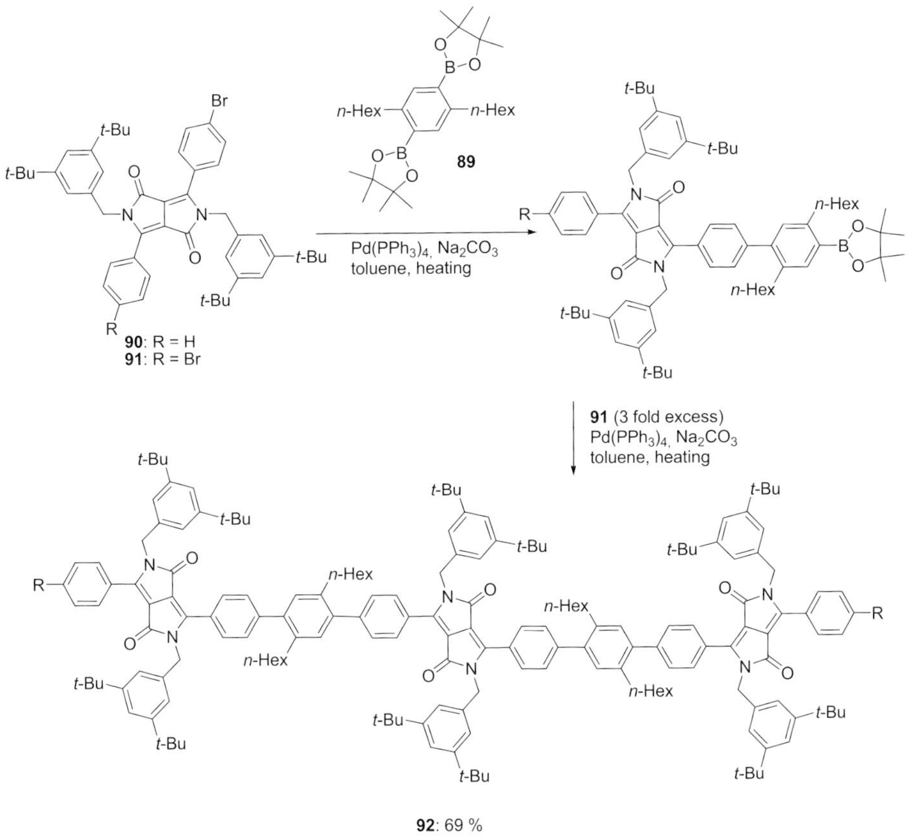

Scheme 3.39 Synthesis of highly functionalized polymers and oligomers.

3.2.6
Copper-mediated Carbon–Heteroatom-Bond-forming Reactions with Functionalized Aryl Boronic Acids

Cu(II)-mediated N-arylation of amines and azoles with aryl boronic acids has been reviewed recently [59]. Pfizer chemists have demonstrated that N-arylpyrroles can be prepared from aryl boronic acids and electron-deficient pyrroles via Cu(OAc)$_2$ mediated coupling reaction at room temperature in air [60]. These reaction conditions are compatible with a variety of functional groups on boronic acids, but are sensitive to steric hindrance. The method could be successfully applied for the synthesis of compound **93** bearing a 4-cyanoaryl group which is a pivotal intermediate in the synthesis of MMP inhibitor AG 3433 (Scheme 3.40).

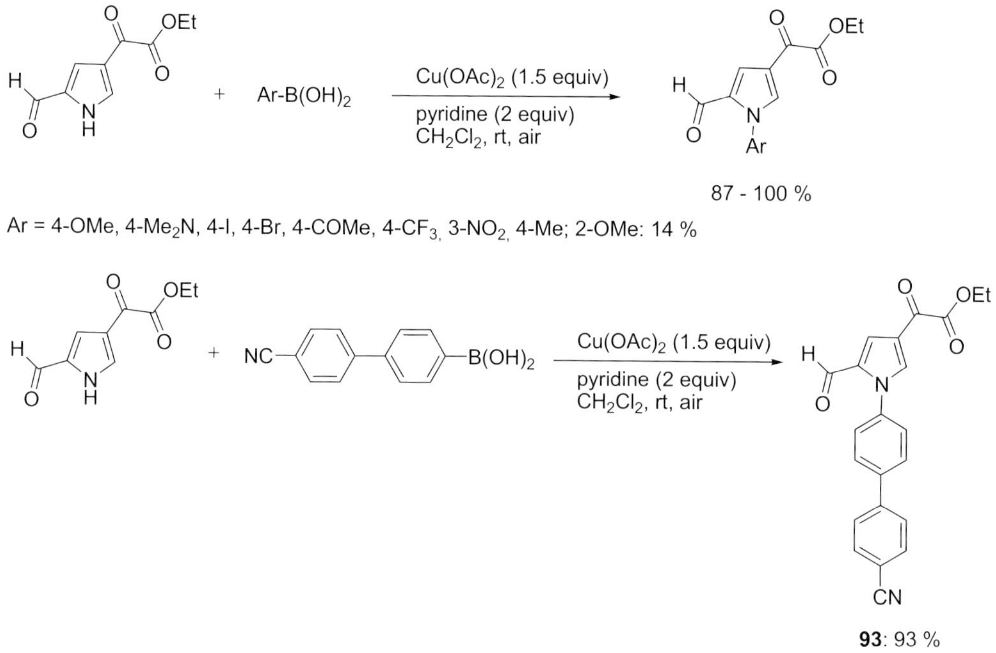

Ar = 4-OMe, 4-Me$_2$N, 4-I, 4-Br, 4-COMe, 4-CF$_3$, 3-NO$_2$, 4-Me; 2-OMe: 14 %

Scheme 3.40 Copper-mediated N-arylation of amines.

The copper-catalyzed amination developed by Buchwald is an effective method for catalytic coupling of aryl boronic acids with amines [61]. Yudin has recently shown that N-arylation of aziridines is possible under modification of Buchwald's method and is successful with a range of functionalized aryl boronic acids (Scheme 3.41) [62].

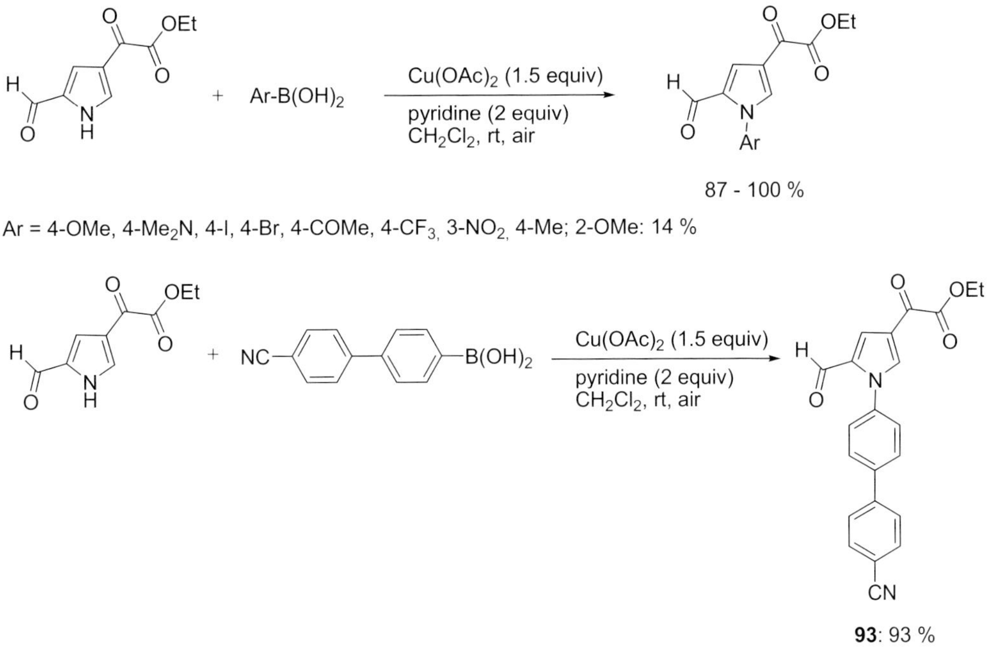

Ar = 4-OMe, 4-Me$_2$N, 4-I, 4-Br, 4-COMe, 4-CF$_3$, 3-NO$_2$, 4-Me; 2-OMe: 14 %

Scheme 3.41 Copper-mediated N-arylation of aziridine.

An unprecedented copper-mediated cross-coupling of N-hydroxyphthalimide and aryl boronic acids yielding aryloxyamines **94** in high yields has been reported by Kelly [63]. The reaction proceeds with both electron-rich and electron-deficient aryl boronic acids and works well in the presence of additional functional groups including halides, esters, ether, nitrile and aldehyde. The phthalimide group can be removed using hydrazine affording the corresponding free aryloxyamines of type **94** (Scheme 3.42).

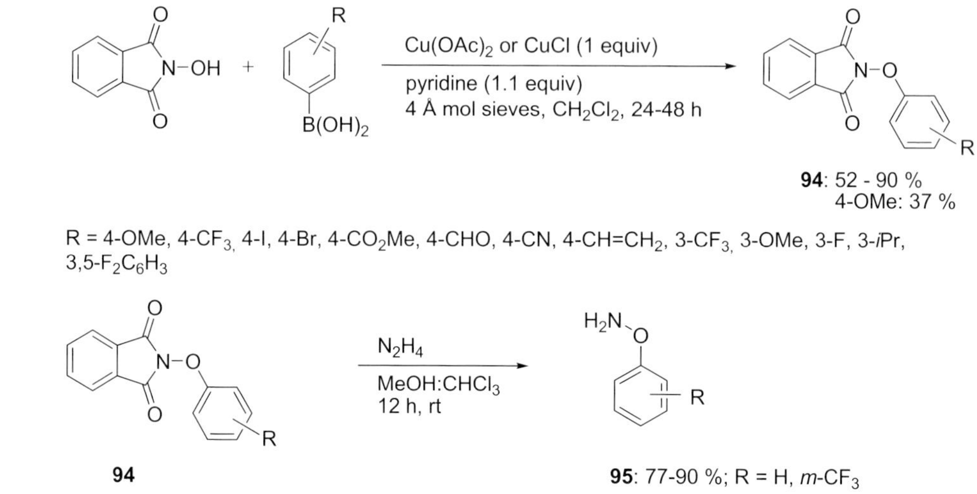

94: 52 - 90 %
4-OMe: 37 %

R = 4-OMe, 4-CF$_3$, 4-I, 4-Br, 4-CO$_2$Me, 4-CHO, 4-CN, 4-CH=CH$_2$, 3-CF$_3$, 3-OMe, 3-F, 3-*i*Pr,
3,5-F$_2$C$_6$H$_3$

Scheme 3.42 Copper-mediated O-arylation of N-hydroxyimides.

Evans has developed a direct synthesis of diaryl ethers while investigating the coupling of functionalized phenolic tyrosine derivatives **96** and **97** [64]. The desired diaryl ethers were obtained in good to excellent yields with pyridine as base and no observed racemization. The thyroxine intermediate **98** was obtained in 81% yield under these conditions using a mixture of pyridine/Et$_3$N (1:1) (5 equiv) as a base (Scheme 3.43).

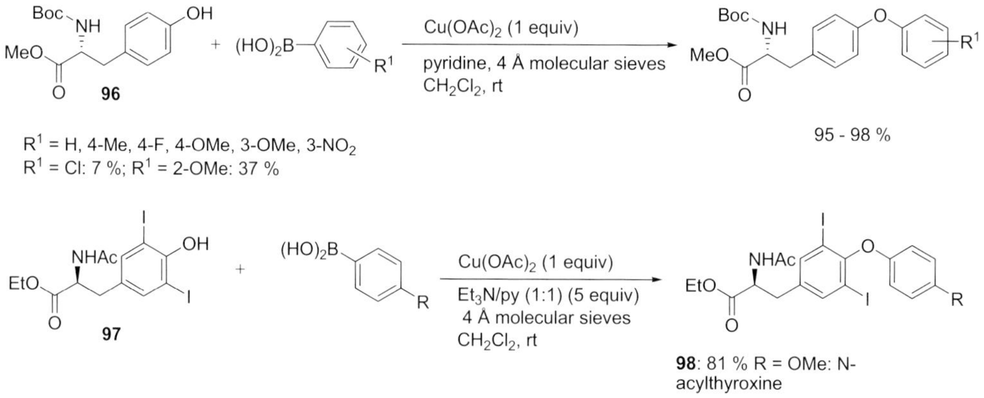

R^1 = H, 4-Me, 4-F, 4-OMe, 3-OMe, 3-NO$_2$
R^1 = Cl: 7 %; R^1 = 2-OMe: 37 %

95 - 98 %

98: 81 % R = OMe: N-
acylthyroxine

Scheme 3.43 Copper-promoted arylation of phenols.

Evans has also developed an intramolecular version of the Cu(II)-assisted boronic acid O-arylation reaction and has applied it to the synthesis of macrocyclic biphenyl ether hydroxamic acid inhibitors of collagenase 1 and gelatinases A and B [65]. The reaction proceeds under sufficiently mild conditions to accommodate chemical functionalities commonly used in peptidomimetics synthesis (Scheme 3.44).

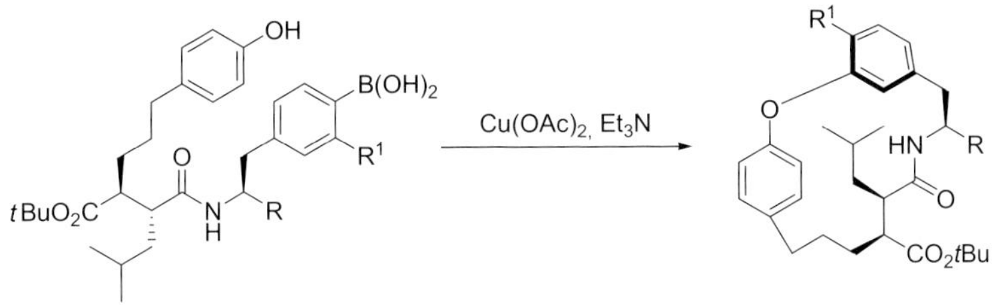

R = CO_2Me, R^1 = H: 54%
R = CONHMe, R^1 = H: 52 %
R = R^1 = H: 43 %
R = CO_2Me, R^1 = OMe: 52 %

Scheme 3.44 Synthesis of macrocyclic biaryl ethers.

Evans has elaborated a total synthesis of tiecoplanin aglycon (**99**) based on the coupling reaction of aryl boronic acid **100** and the phenol **101** [66]. In the key step, the diaryl ether **102** was obtained in 80% yield with no detection of epimerization at any of the three stereogenic centers. An elegant sequence of reactions subsequently led to the completion of the synthesis of **99** (Scheme 3.45).

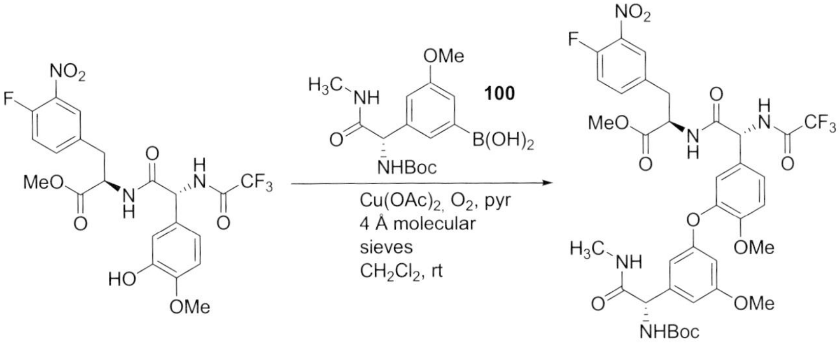

101 **102**: 80 %

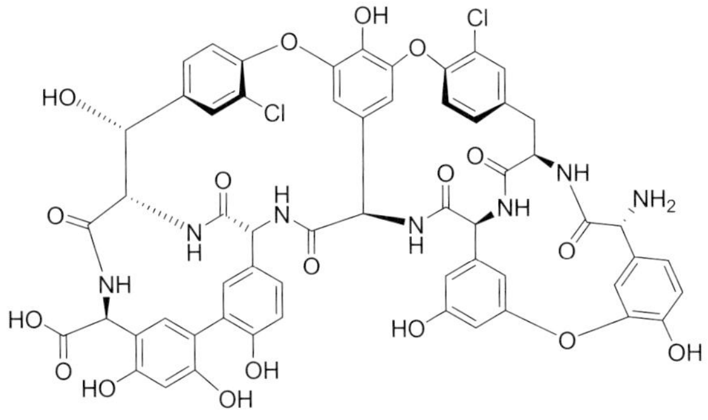

99: teicoplanin aglycon

Scheme 3.45 Copper-promoted O-arylation of phenols.

Guy has explored copper-mediated S-arylation of thiols and aryl boronic acids [67]. Earlier studies revealed that the reactions were slow for S-arylation under the conditions developed previously for N- and O-arylation reactions because of a significant disulfide formation. However, it was shown later that the reaction of a wide range of electronically diverse aryl boronic acids with a range of thiolate substrates proceeded well when heated at 155 °C in DMF affording cross-coupled products in good yields. Similarly cysteine phenyl sulfide **103** and an arylthio glycoside **104** were also prepared in 50–80% yields (Scheme 3.46).

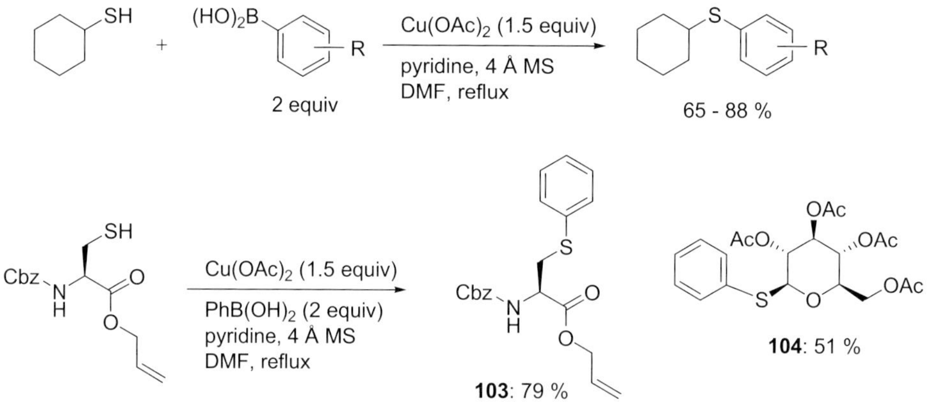

Scheme 3.46 Copper-mediated thioether synthesis.

A new synthesis of thioethers involving copper-catalyzed cross-coupling of aryl boronic acids with N-aryl/heteroaryl thiosuccinmides has been described by Liebeskind and coworkers [68]. The reaction proceeds in the absence of base under mild conditions utilizing various aryl boronic acids. S-arylation is now possible for the first time under nonbasic conditions (Scheme 3.47).

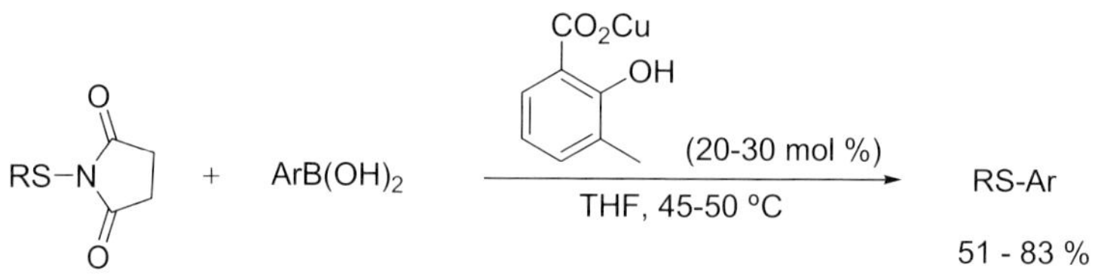

Scheme 3.47 Copper-catalyzed cross-coupling of organo-thiolimides with boronic acids.

3.2.7
Palladium-catalyzed Acylation of Functionalized Aryl Boronic Acids

Carbonylative cross-coupling reactions of aryl boronic acids with aryl electrophiles like aryl bromides, iodides and triflates proceeds smoothly under an atmospheric pressure of carbon monoxide in the presence of $PdCl_2(PPh_3)_2/K_2CO_3$ (for aryl iodides) or $PdCl_2(dppf)/K_2CO_3$, KI (for aryl bromides and triflates) in anisole at 80 °C to give the diaryl ketones in good to excellent yields. The carbonylation of the *o*-substituted phenylboronic acid **105** with benzyl bromide gives the *o*-substituted phenylbenzyl ketone **106** that is readily converted to isoflavone (**107**) (Scheme 3.48) [69].

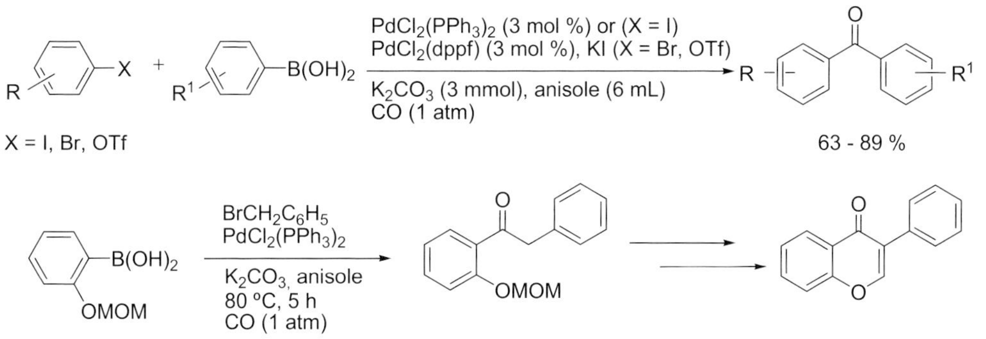

Scheme 3.48 Pd-catalyzed carbonylative cross-coupling of aryl boronic acids.

It was reported by Bumagin that a ligandless palladium-catalyzed reaction of aryl boronic acids with benzoyl chloride gives unsymmetrical diaryl ketones in high yields (Eq. 1, Scheme 3.49) [70]. Goößen has developed a one-pot high yielding synthesis of unsymmetrical aryl (or alkyl) aryl ketones directly from a variety of functionalized alkyl and aryl carboxylic acids by their cross-coupling with a number of substituted aryl boronic acids catalyzed by a Pd/phosphine complex [71]. A small amount (2 equiv) of water was shown to be essential for this reaction and boronic acids with many functional groups were tolerated (Scheme 3.49).

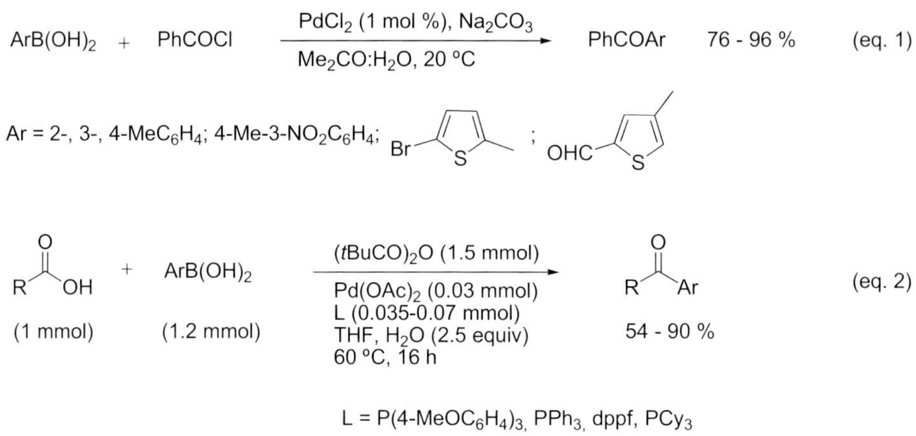

$$ArB(OH)_2 \ + \ PhCOCl \ \xrightarrow[\text{Me}_2\text{CO:H}_2\text{O, 20 °C}]{\text{PdCl}_2 \ (1 \ \text{mol \%}), \ \text{Na}_2\text{CO}_3} \ PhCOAr \quad 76 - 96 \ \% \qquad \text{(eq. 1)}$$

Ar = 2-, 3-, 4-MeC$_6$H$_4$; 4-Me-3-NO$_2$C$_6$H$_4$;

$$R\underset{OH}{\overset{O}{\Vert}} \ + \ ArB(OH)_2 \ \xrightarrow[\substack{\text{Pd(OAc)}_2 \ (0.03 \ \text{mmol}) \\ \text{L} \ (0.035-0.07 \ \text{mmol}) \\ \text{THF, H}_2\text{O} \ (2.5 \ \text{equiv}) \\ 60 \ °C, \ 16 \ h}]{(t\text{BuCO})_2\text{O} \ (1.5 \ \text{mmol})} \ R\underset{Ar}{\overset{O}{\Vert}} \qquad \text{(eq. 2)}$$

(1 mmol)   (1.2 mmol)

54 - 90 %

L = P(4-MeOC$_6$H$_4$)$_3$, PPh$_3$, dppf, PCy$_3$

Scheme 3.49 Pd-catalyzed acylation reactions with boronic acids.

3.2.8
Miscellaneous C–C-bond Formations of Functionalized Aryl Organoboranes

Gooßen has also reported a palladium-catalyzed cross-coupling reaction between aryl boronic acids or esters and α-bromoacetic acid derivatives which allows the synthesis of various substituted aryl acetic acid derivatives in good to excellent yields under mild conditions [72]. Aryl boronic acids with a range of electron-withdrawing and -donating substituents are tolerated in this reaction (Scheme 3.50).

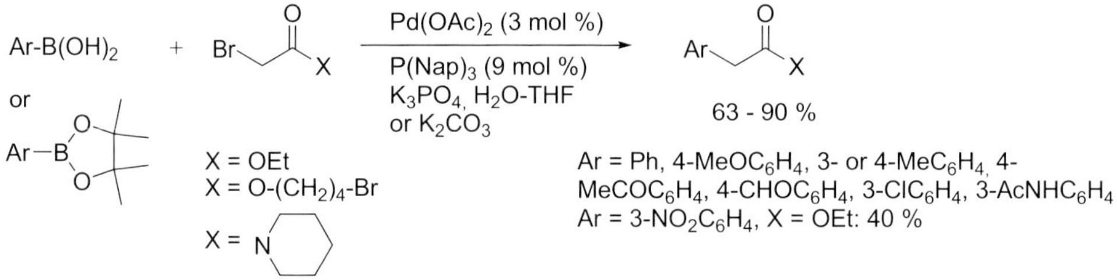

Ar-B(OH)$_2$
or
Ar—B

X = OEt
X = O-(CH$_2$)$_4$-Br

X = N

Pd(OAc)$_2$ (3 mol %)
P(Nap)$_3$ (9 mol %)
K$_3$PO$_4$, H$_2$O-THF
or K$_2$CO$_3$

63 - 90 %

Ar = Ph, 4-MeOC$_6$H$_4$, 3- or 4-MeC$_6$H$_4$, 4-MeCOC$_6$H$_4$, 4-CHOC$_6$H$_4$, 3-ClC$_6$H$_4$, 3-AcNHC$_6$H$_4$
Ar = 3-NO$_2$C$_6$H$_4$, X = OEt: 40 %

Scheme 3.50 Palladium-catalyzed cross-coupling of aryl boronic acids and boronates with α-bromoacetic acid derivatives.

Rhodium(I)-complexes are known to be excellent catalysts for the conjugate additions of aryl- and alkenyl boronic acids to α,β-unsaturated ketones, esters and amides [73,74]. Batey has recently prepared aryl and alkenyl trifluoroborates as air- and moisture-stable reagents for the nucleophilic addition to enones in the presence of a Rh(I)-catalyst to give β-functionalized ketones in good yields [75]. A number of aryl fluoroborates with electron-withdrawing substituents were also tolerated in this reaction. An unprecedented Rh-catalyzed 1,4-addition of aryl boronic acids for stereoselective synthesis of C-glycosides has also been reported by Maddaford[76]. The reaction is stereoselective for the α-isomer and a variety of aryl boronic acids can be used including electron donating or withdrawing, alkenyl and sterically congested groups (Scheme 3.51).

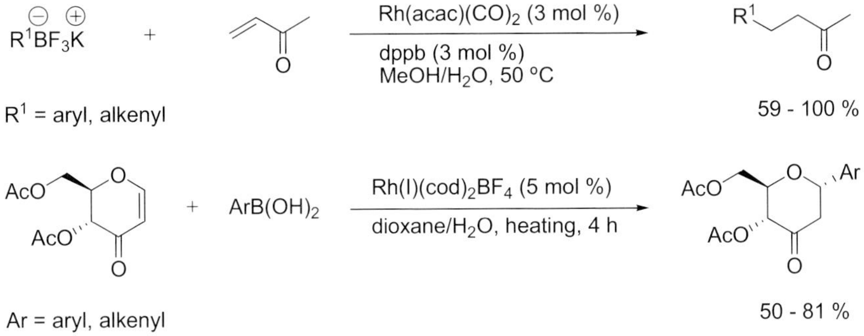

Scheme 3.51 Rhodium-catalyzed 1,4-conjugate additions.

A first example of catalytic asymmetric synthesis of 4-arylpiperidiones (Ar = 4-FC$_6$H$_4$, 4-ClC$_6$H$_4$) has been described using a rhodium-catalyzed 1,4-addition of 4-fluoro- or 4-chlorophenylboroxine and one equivalent of water with the chiral BINAP ligand[77]. The (R)-4-(4-fluorophenyl)-2-piperidione obtained in this reaction is a key intermediate for the synthesis of pharmacologically important (–)-paroxetine (**108**).

The enantioselective Rh-catalyzed addition of aryl boronic acids to dehydroalanine has also been performed in the presence of C_2-symmetric aryl diphosphite ligands allowing the synthesis of unnatural amino acid esters such as **109** in moderate enantioselectivity (Scheme 3.52) [78].

Petasis has demonstrated that aryl and heteroaryl boronic acids participate in one step three component Mannich reactions with glyoxylic acid and diarylamines to give the corresponding α-aryl/heteroarylglycine derivatives in good yields after deprotection of the diarylamine[79]. Several examples of the reaction with *para*-substituted aryl boronic acids as well as 2- and 3-thienyl, -furyl or -benzo[*b*]thienyl boronic acids are described (Scheme 3.53) [80].

Petasis has also reported an efficient one-step and highly versatile three component reaction of boronic acids with amines and α-hydroxyaldehydes to give *anti-β*-amino-alcohols in a highly diastereo-controlled manner[81]. A variety of boronic acids including alkenyl, 2-bromoalkenyl, aryl and heteroaryl derivates participate readily in this process. The experimental procedure is very simple and does not require anhydrous or oxygen free conditions and can be adapted to parallel synthesis for the construction of combinatorial libraries. Since aldehyde racemization does not occur under the reaction conditions, enantiomerically pure (> 99%) amino-alcohols can be obtained by using chiral α-hydroxy aldehydes. Thus, use of enantiomerically pure glyceraldehydes and aminodiphenylmethane followed by hydrogenolysis of the resulting 3-amino-1,2-diol derivative and its subsequent

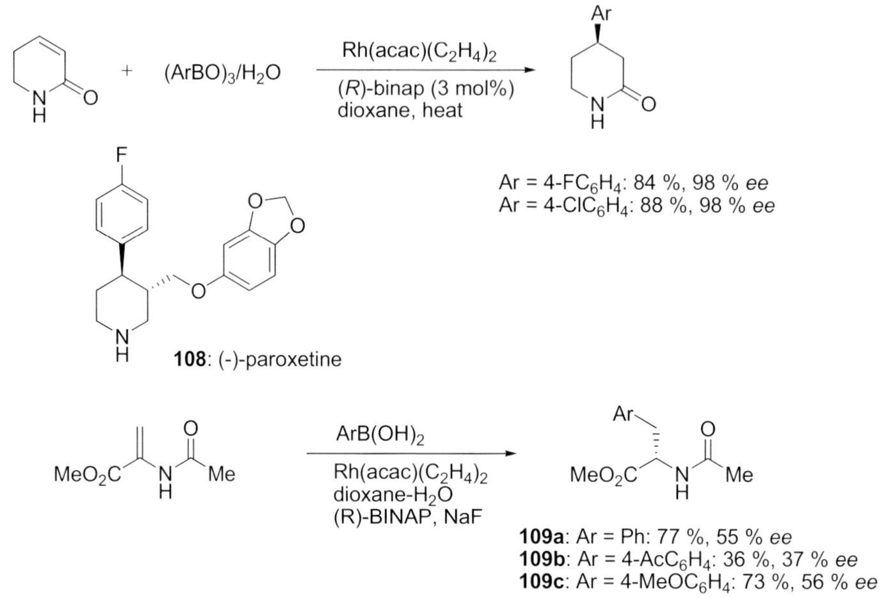
(ArBO)3/H2O
Rh(acac)(C2H4)2
(R)-binap (3 mol%)
dioxane, heat
Ar = 4-FC6H4: 84 %, 98 % ee
Ar = 4-ClC6H4: 88 %, 98 % ee
108: (-)-paroxetine
ArB(OH)2
Rh(acac)(C2H4)2
dioxane-H2O
(R)-BINAP, NaF
109a: Ar = Ph: 77 %, 55 % ee
109b: Ar = 4-AcC6H4: 36 %, 37 % ee
109c: Ar = 4-MeOC6H4: 73 %, 56 % ee

Scheme 3.52 Enantioselective Rh-catalyzed 1,4-conjugate additions.

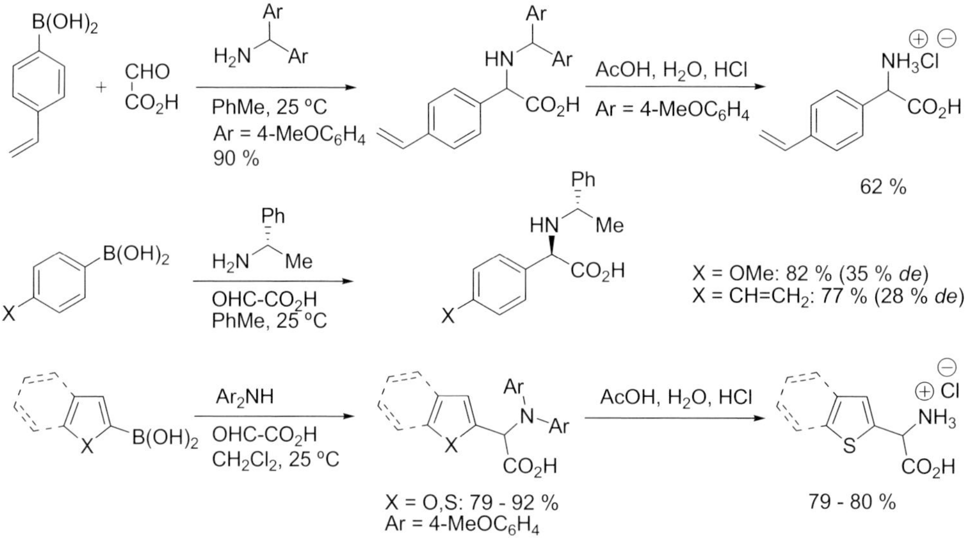
PhMe, 25 °C
Ar = 4-MeOC6H4
90 %
AcOH, H2O, HCl
Ar = 4-MeOC6H4
62 %
OHC-CO2H
PhMe, 25 °C
X = OMe: 82 % (35 % de)
X = CH=CH2: 77 % (28 % de)
Ar2NH
OHC-CO2H
CH2Cl2, 25 °C
AcOH, H2O, HCl
X = O,S: 79 - 92 %
Ar = 4-MeOC6H4
79 - 80 %

Scheme 3.53 Synthesis of amino-acids using the Petasis reaction.

conversion to the NBoc derivative gave N-protected β-amino alcohol in > 99% *ee*. Prakash, Petasis and Olah have extended this reaction for the synthesis of anti-α-(trifluoromethyl)-β-amino alcohols and anti-α-(difluoromethyl)-β-amino alcohols ($R^4 = CF_3$ and CF_2H), in a highly stereoselective fashion [82]. Further, the coupling of enantiomerically pure (*R*)- and (*S*)-α-(trifluoromethyl)-α-hydroxy aldehydes with bromostyryl boronic acid and dibenzylamine gave the expected *anti*-amino alcohols with high enantioselectivity [83]. Also, it was shown that unprotected fluoroalkylamino alcohols can be prepared by this method using diallylamine as the amine component followed by catalytic deallylation of the *bis*(diallyl)amino alcohol **110** in the presence of Pd(PPh₃)₄ and dimethylbarbituric acid as the allyl group scavenger (Scheme 3.54).

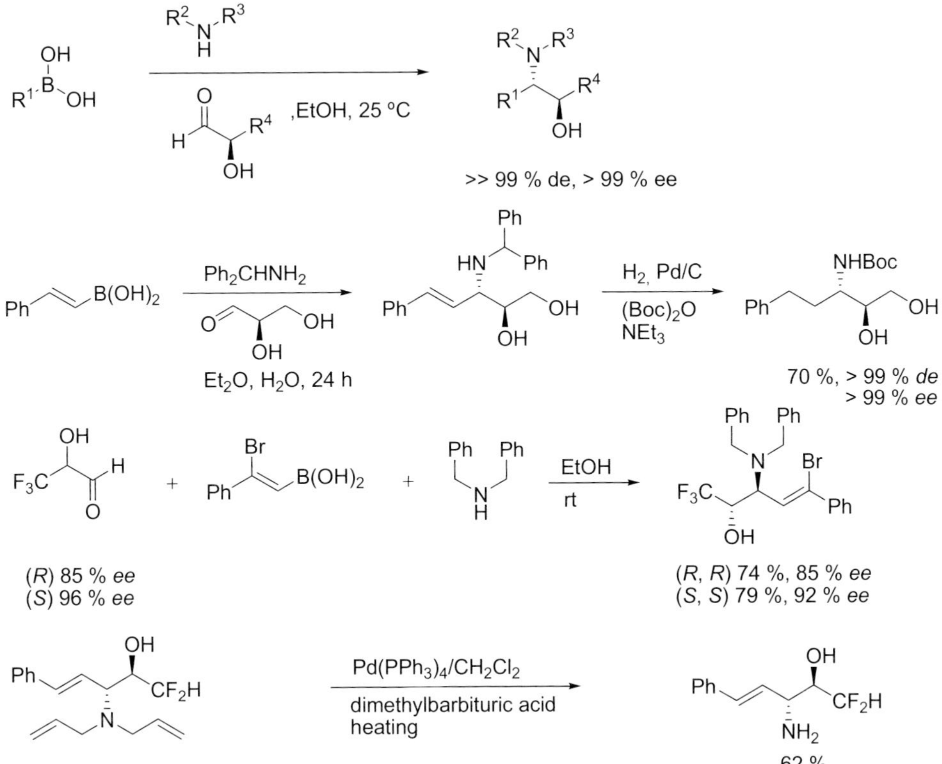

Scheme 3.54 Stereoselective synthesis of β-amino-alcohols using the Petasis-reaction.

A practical synthesis of several aryl bromides and chlorides has been recently described via halodeboronation of a range of arylboronic acids especially with electron-withdrawing groups using DBDMH or DCDMH (1,3-dibromo or 1,3-dichloro-5,5-dimethylhydantoin). [84,85]. This methodology could be also extended for the synthesis of 2-bromo-3-fluorobenzonitrile. Addition of a catalytic amount of NaOMe has beneficial effects on the rate and yields of these reactions.

3.2.9
Miscellaneous Reactions of Functionalized Alkenyl Boronic Acids

In previous studies, Petasis had reported the synthesis of geometrically pure *E*- or *Z*-alkenyl halides by reaction of alkenyl boronic acids with N-halosuccinimides (NIS, NBS or NCS) (Scheme 3.55) [86].

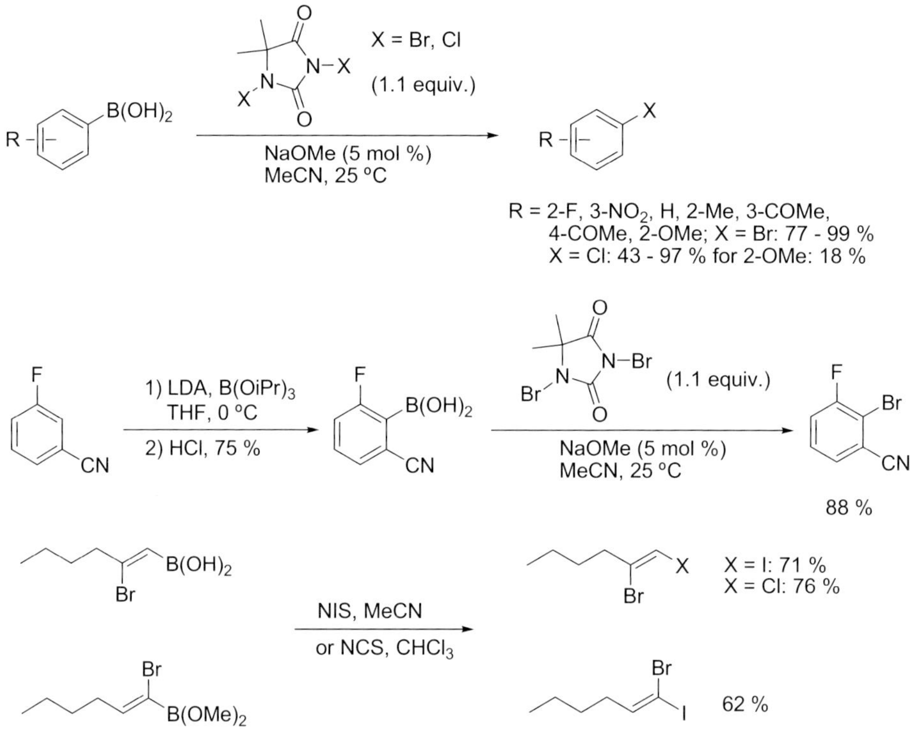

R = 2-F, 3-NO$_2$, H, 2-Me, 3-COMe,
4-COMe, 2-OMe; X = Br: 77 - 99 %
X = Cl: 43 - 97 % for 2-OMe: 18 %

Scheme 3.55 Synthesis of aryl bromides and chlorides by halodeboronation of arylboronic acids.

An efficient conversion of functionalized aryl boronic acids to the correspond-ing phenols has been accomplished by Prakash, Petasis and Olah by simply add-ing aqueous hydrogen peroxide solution [87,88]. They have further developed a one-pot coupling sequence for the preparation of symmetrical diaryl ethers via partial conversion of aryl boronic acids to phenols as precursors for the coupling sequence. The H$_2$O$_2$/arylboronic acid ratio is critical for the outcome of this reac-tion sequence, whereas 0.25 equivalents of H$_2$O$_2$ furnished the highest overall yields. A variety of substituents are tolerated in the aryl boronic acids (Scheme 3.56).

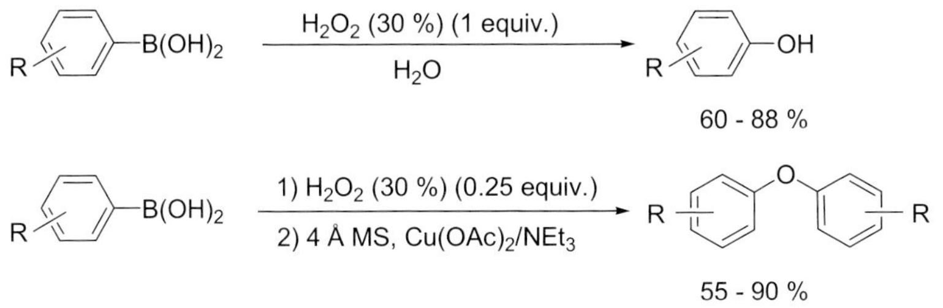

60 - 88 %

55 - 90 %

Scheme 3.56 Oxidation of boronic esters.

3.3
Preparation and Reactions of Functionalized Alkenyl Boranes

3.3.1
Synthesis of Alkenyl Boronic Acids by Transmetallation of Alkenyl Grignard Reagents with Boronate Esters

Jiang has described the successful preparation of α-(trifluoromethyl)ethenyl boronic acid by reaction of readily available 2-bromotrifluoropropene, magnesium and an alkyl borate in an one-pot process [89]. The trifluoromethyl boronic acid **111**, thus obtained, was found to be stable for several months even in the presence of air and moisture (Scheme 3.57).

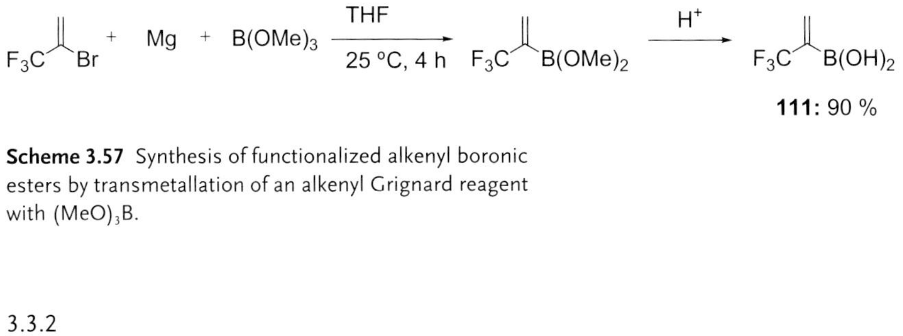

111: 90 %

Scheme 3.57 Synthesis of functionalized alkenyl boronic esters by transmetallation of an alkenyl Grignard reagent with (MeO)$_3$B.

3.3.2
Synthesis of Functionalized Alkenyl Boronic Acids by Hydroboration of Functionalized Alkynes and their Suzuki Cross-coupling Reactions

Witulski has reported the first hydroboration of 1-alkynylamides. Thus, the hydroboration of ynamide **112** with catechol borane in THF proceeded chemo- and regioselectively yielding only the monohydroboration product, alkenyl boronic ester **113** [90]. However, the isolation of the boronic ester **113** was complicated due to its instability and difficulties of storage and purification. Therefore, it was directly subjected to Suzuki–Miyaura cross-coupling yielding (*E*)-β-arylenamide and 3-(2′-amidovinyl)indoles such as **114** (Scheme 3.58).

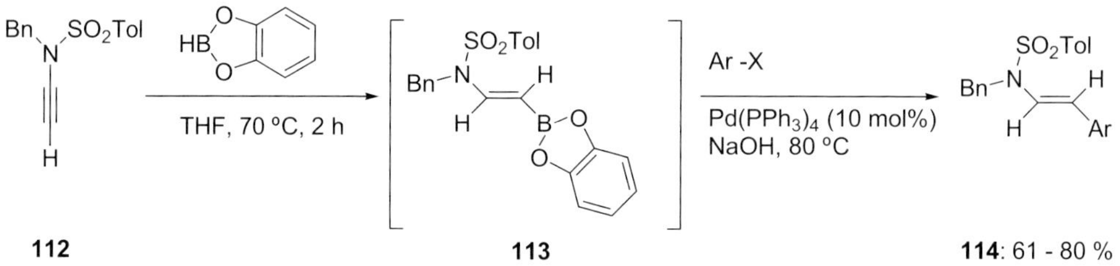

Scheme 3.58 Alkenyl boronic esters by hydroboration of functionalized acetylenes.

Konno has investigated the hydroboration of fluoroalkylated internal alkynes that proceeds in a highly regio- and stereoselective manner to give the corresponding fluoroalkylated alkenyl boranes in excellent yields [91]. These alkenyl boranes were reacted with a range of aryl halides without isolation under Suzuki–Miyaura cross-coupling conditions providing a practical one-pot synthesis of fluoroalkylated trisubstituted olefins **115** in high yields with complete retention of the olefinic geometry (Scheme 3.59).

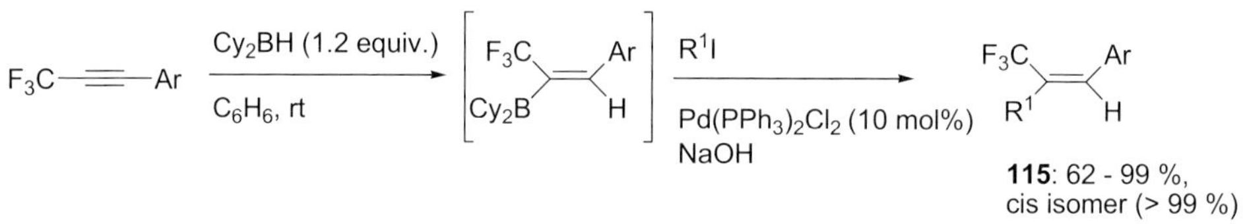

Ar = C_6H_5; 4-ClC_6H_4; 4-$MeOC_6H_4$; 4-$CO_2EtC_6H_4$; 4-$NO_2C_6H_4$; 4-MeC_6H_4.
R^1 = C_6H_5, 4-$MeOC_6H_4$; 4-$CO_2EtC_6H_4$; 4-$NO_2C_6H_4$; 2-ClC_6H_4; 3-ClC_6H_4; 4-ClC_6H_4.

Scheme 3.59 Synthesis of fluoroalkylated alkenyl boranes by hydroboration.

Srebnik has reported the synthesis of phosphono boronates by hydroboration with pinacol borane (**30**) [92]. The reaction proceeds well with terminal alkenyl phosphonates whereas internal alkenyl phosphonates gave complex mixtures. Hydroboration of the corresponding alkynyl phosphonates under identical conditions gave alkenyl phosphonates that were difficult to isolate and were *in situ* subjected to Suzuki coupling with phenyl iodide to give trisubstituted phosphonates providing a new one-pot synthesis of this class of compounds (Scheme 3.60).

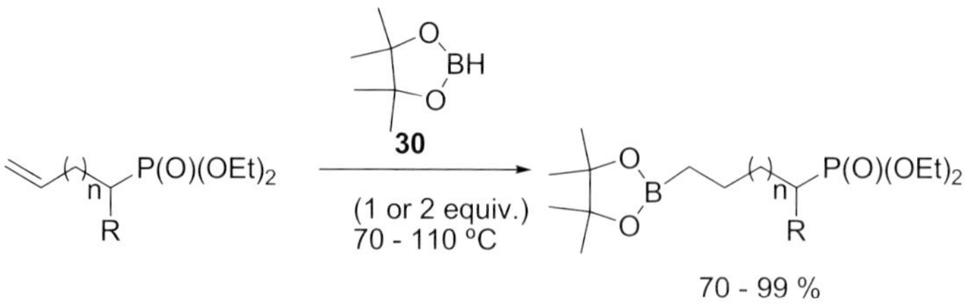

Scheme 3.60 Synthesis of boronic-esters-substituted phosphonates.

3.3.3
Synthesis of Functionalized Alkenyl Boronic Esters by Cross-metathesis

Grubbs has reported the synthesis of functionalized alkenyl pinacol boronates suitable for cross-coupling reactions using ruthenium-catalyzed olefin cross-metathesis of 1-propenyl pinacol boronate (**116**) and various functionalized alkenes [93]. The 1-propenyl pinacol boronate (**116**) is readily synthesized from the commercially available reagents in significantly higher yields than its vinyl analog that led primarily to use **116** along with the carbene-substituted catalyst **117** in these studies [94]. The resultant boronate cross-products are stereoselectively converted into *Z*-alkenyl bromides and *E*-alkenyl iodides, respectively, by bromination or iodination (Scheme 3.61).

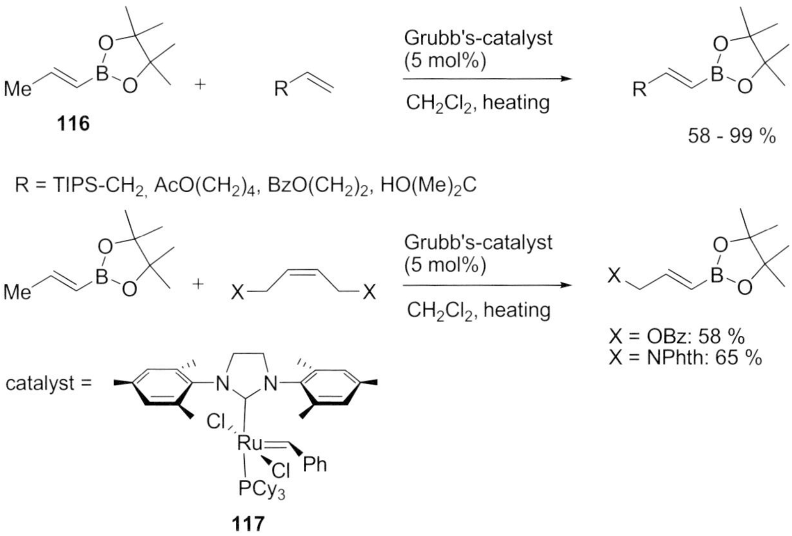

R = TIPS-CH₂, AcO(CH₂)₄, BzO(CH₂)₂, HO(Me)₂C

Scheme 3.61 Synthesis of functionalized alkenyl boranes by cross-metathesis.

Danishefsky has applied the cross-metathesis reaction for the synthesis of highly functionalized alkenyl boronate precursor **118** from the terminal olefin **119** and vinyl pinacol boronate in the presence of first 'generation' Grubbs-catalyst **121** [95]. The metathesis reaction was driven to completion by increasing the amount of vinyl boronate ester **120** affording the polyfunctional boronic ester **118** in 93% yield and with exclusive *trans*-stereochemistry. The alkenyl boronic ester **118** was subsequently used in the synthesis of epothilone 490 via intramolecular Suzuki macrocyclization (Scheme 3.62).

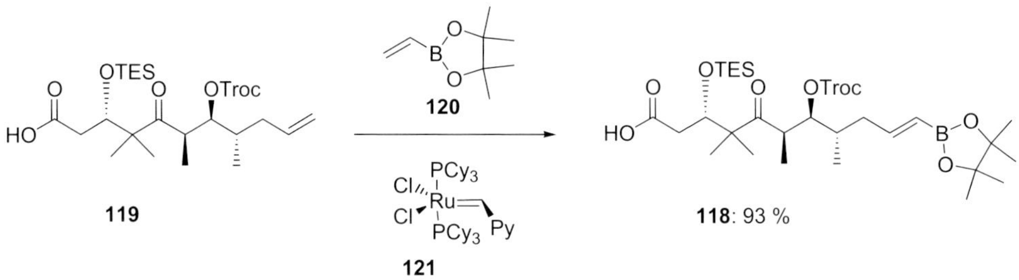

Scheme 3.62 Synthesis of functionalized alkenyl boronates by cross-metathesis.

Pietruszka has reported the synthesis of styryl and cyclopropyl boronic esters with an acrylate functionality by applying a cross-metathesis reaction [96]. Thus, the reaction of cyclopropyl boronate **122** with methyl acrylate gave the E-enoate **123** in 52% yield (Scheme 3.63).

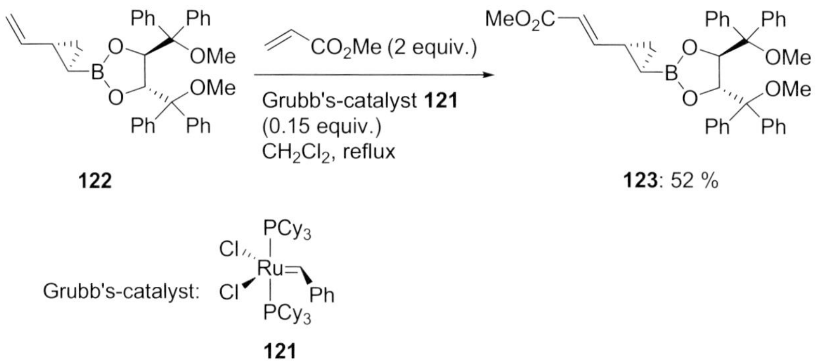

Scheme 3.63 Synthesis of functionalized cyclopropylalkenyl boronic esters.

3.3.4
Synthesis and Palladium-catalyzed Cross-coupling Reactions of Functionalized Alkenyl Trifluoroborates

Cross-coupling reactions of functionalized alkenyl trifluoroborates **124** with 4-bromobenzonitrile in the presence of catalytic amounts of $PdCl_2(dppf)$ has been studied by Molander [97,98]. Herein functionalized 4-cyanostyrenes of type **125** were obtained in satisfactory yields (Scheme 3.64).

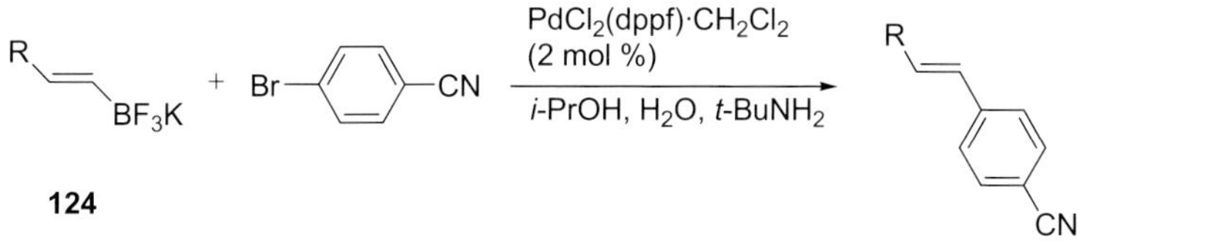

124

125a: R = Cl-(CH$_2$)$_3$: 52 %
125b: R = MeO$_2$C(CH$_2$)$_3$: 33 %

Scheme 3.64 Preparation and cross-coupling reactions of functionalized alkenyl trifluoroborates.

3.3.5
Palladium-catalyzed Cross-coupling of Functionalized Alkenyl Boronates with Cyclopropyl Iodides

Charette has reported the first cross-coupling reaction of aryl/alkenyl boronic esters with cyclopropyl iodides [99]. The alkenyl boronates with the protected hydroxy group gave coupled products in moderate to good yields. Similarly the Pd-catalyzed cross-coupling of benzyloxymethylcyclopropyl boronate with *trans*-benzyloxy-methyliodocyclopropane gave the symmetrical benzyloxy substituted *bis*-cyclopropane in 71% yield (Scheme 3.65) [100].

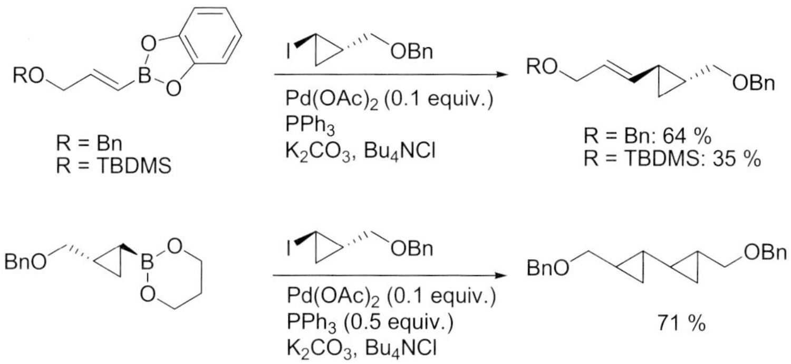

Scheme 3.65 Cross-coupling of boronic esters with cyclopropyl iodides.

3.3.6
Intermolecular Suzuki Cross-coupling Reactions of Functionalized Alkenylborane Derivates: Application in Natural Product Synthesis (Alkenyl B-Alkenyl Coupling)

Roush has demonstrated that thallium(I) ethoxide promotes Suzuki-coupling for a range of functionalized alkenyl boronic acids and functionalized aryl or alkenyl halides in good to excellent yields [101]. This reagent offers distinct advantage over thallium(I) hydroxide in terms of its commercial availability, stability and ease of use (Scheme 3.66).

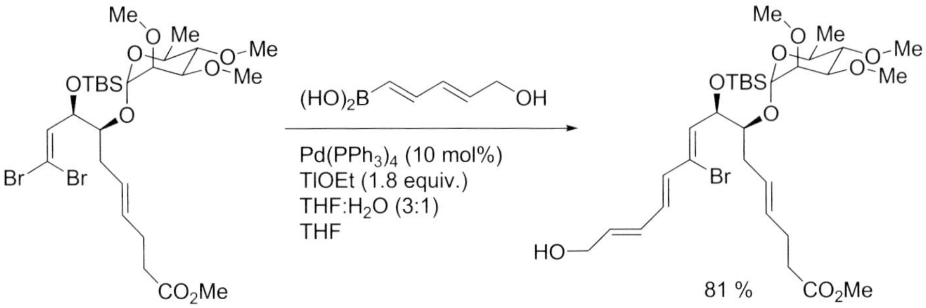

Scheme 3.66 Suzuki cross-coupling of functionalized alkenyl boronic acids.

A highly convergent synthesis of bafilomycin A, a macrolide antibiotic has been recently reported by Roush. A Suzuki cross-coupling reaction was performed between the functionalized alkenyl boronic acid **126** and the alkenyl iodide **127** generating appropriately protected macrocyclic precursor **128** in 65% yield (Scheme 3.67) [102].

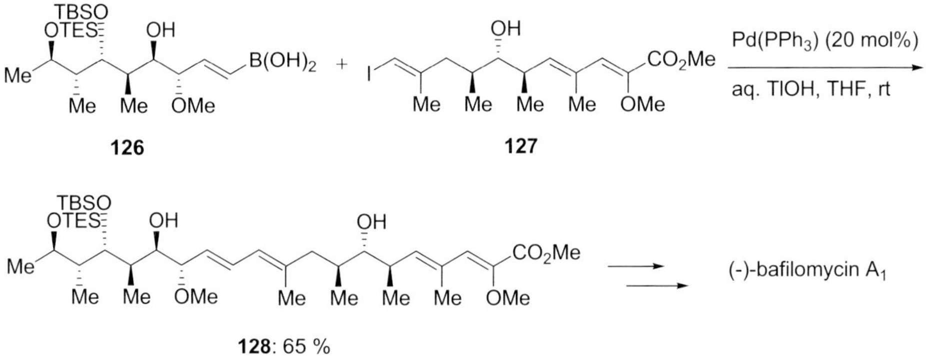

Scheme 3.67 Palladium-catalyzed cross-coupling of functionalized alkenyl boronic acids: synthesis of natural products (by an alkenyl B-alkenyl coupling).

3.3.7
Intramolecular Macrocyclization via Suzuki Cross-coupling of Functionalized Alkenyl Boronic Esters (Alkenyl B-Alkenyl Coupling)

A highly convergent synthesis of rutamycin B, a 26-membered lactone macrolide antibiotic has been achieved by intramolecular macrocyclization through Suzuki coupling of a linear alkenyl boronate with a terminal alkenyl iodide as the key step in high yield as reported by White et al. (Scheme 3.68) [103].

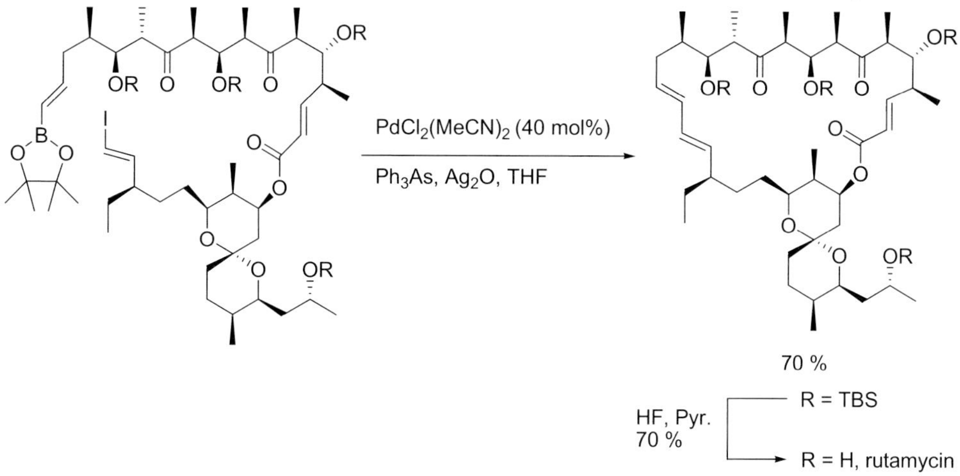

Scheme 3.68 Intramolecular macrocyclization via Suzuki–Miyaura coupling.

Danishefsky has reported the total synthesis of epothilone 490 (**129**) via an intramolecular Suzuki macrocyclization of alkenyl boronic ester **130** bearing a terminal alkenyl iodide group [104]. The corresponding alkenyl boronate fragment was prepared by alkenyl boronate cross-metathesis as reported by Grubbs (Scheme 3.69) [105,106].

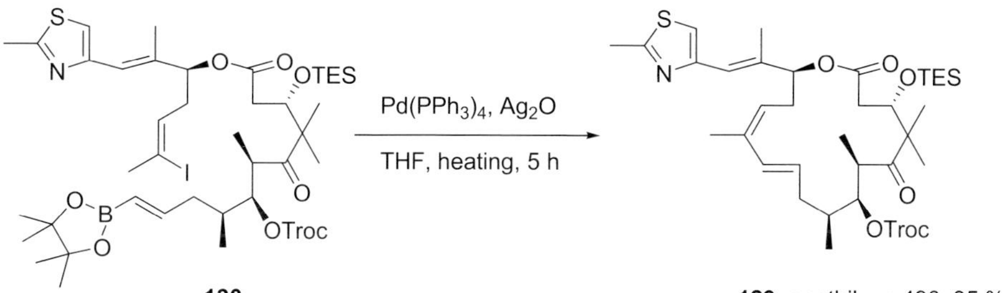

Scheme 3.69 Synthesis of epothilone 490 via Suzuki–Miyaura cross-coupling.

3.3.8
Three-component Mannich Reaction of Functionalized Alkenyl Boronic Acids (Petasis Reaction): Synthesis of β,γ-Unsaturated α-Amino Acids

A new general synthesis of β,γ-unsaturated α-amino acids involving a three-component variant of the Mannich reaction with a number of alkenyl boronic acids, primary and secondary aliphatic or aromatic amines and α-keto acids has been reported by Petasis [107,108]. A remarkable feature of this reaction is that it is triply convergent and gives products with multiple sites for introducing molecular diversity. By using readily cleavable amines, i.e. *bis*(4-methoxyphenyl)methylamine, it is possible to prepare free amino-acids. Use of (*S*)-2-phenylglycinol gave alkenyl amino acid **131** as a single diastereomer (> 99% *de*). Subsequent hydrogenation of **131** gave *R*-homophenylalanine hydrochloride (**132**) with > 99% *ee* (Scheme 3.70).

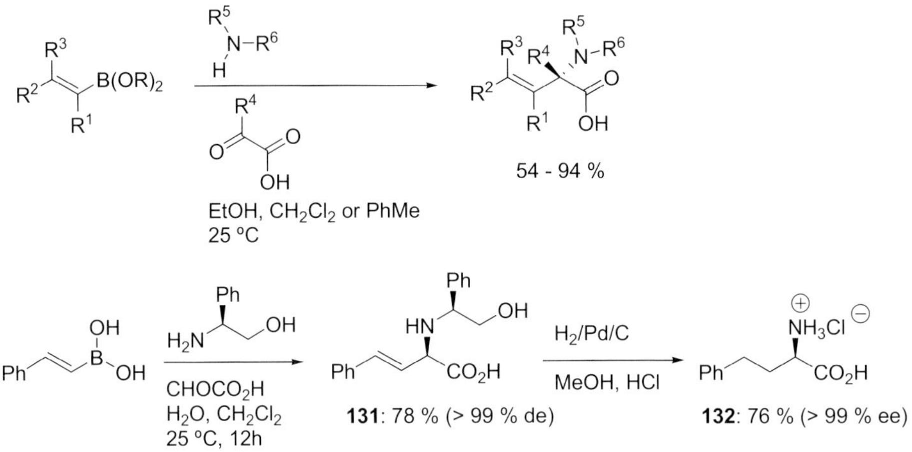

Scheme 3.70 Synthesis of β,γ-unsaturated α-amino acids using the Petasis reaction.

3.3.9
Oxidation of Functionalized Alkenyl Boronic Esters to Aldehydes with Trimethylamine Oxide

Danishefsky has developed a mild oxidative procedure for the conversion of highly functionalized alkenyl boronates (obtained by alkenyl boronate cross-metathesis) to the corresponding aldehydes with trimethylamine N-oxide that was found to be compatible with hydroxyl, ketone and acid functionalities present in alkenyl boronates without protection [109]. For example, by using this methodology, it was possible to oxidize the iodoalkenyl boronate intermediate **133** to the aldehyde **134** that was subjected to intramolecular Nozaki–Kishi macrocyclization yielding 11-hydroxy deoxyepothilone precursor **135** in 40% yield (Scheme 3.71).

3.3.10
Lewis-acid-catalyzed Nucleophilic Addition of Functionalized Alkenyl Boronic Esters to Activated N-acyliminium Ions

Batey has reported the first example of the reaction of alkenyl boronic acids and esters with activated N-acyliminium ion precursors under Lewis-acid catalysis giving 2-functionalized heterocycles in good yields [110]. This methodology has been further extended for the synthesis of fungal metabolite (1*R**,-8a*R**)-1-hydroxyindolizidine (**136**) by concomitant deprotection and tosylation of the adduct followed by hydrogenation and cyclization in the presence of a palladium catalyst (Scheme 3.72).

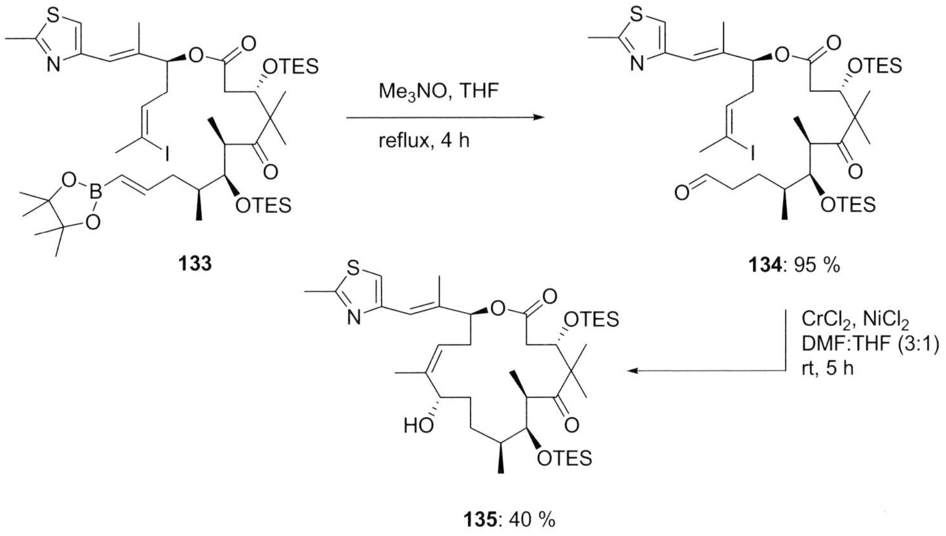
Me3NO, THF
reflux, 4 h
133
134: 95 %
CrCl2, NiCl2
DMF:THF (3:1)
rt, 5 h
135: 40 %

Scheme 3.71 Oxidation of functionalized vinyl boronates to aldehydes.

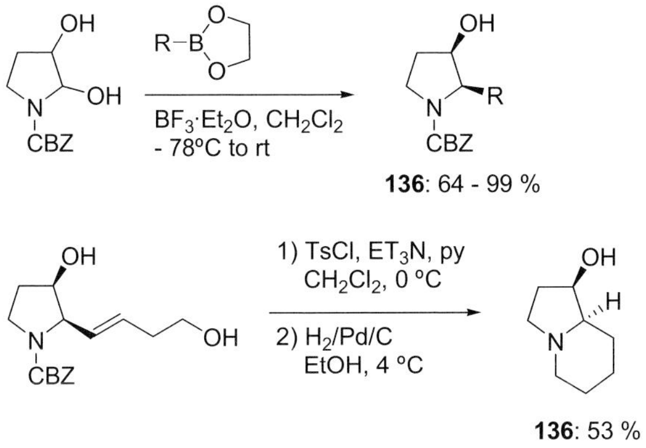
BF3·Et2O, CH2Cl2
- 78°C to rt
136: 64 - 99 %
1) TsCl, ET3N, py
CH2Cl2, 0 °C
2) H2/Pd/C
EtOH, 4 °C
136: 53 %

Scheme 3.72 Nucleophilic addition of functionalized alkenyl boronic acids and esters to activated N-acyliminium ion precursors.

3.4
Preparation and Reactions of Functionalized Alkynlboron Derivatives

Molander has reported the synthesis of functionalized alkynyl trifluoroborates by transmetallation of alkynyllithium compounds with boronates followed by *in situ* treatment with KHF_2 [111]. These alkynyl trifluoroborates are crystalline solids possessing excellent air stability and are shown to undergo facile cross-coupling for example with 4-bromobenzonitrile to give functionalized arylacetylenes in high yields. Thus, the reaction is tolerant to a variety of sensitive functional groups as shown by the reaction of potassium (4-*t*-butyldimethylsiloxy-1-butyn-1-yl) trifluoroborate or potassium (trimethylsilylethynyl) trifluoroborate that afforded coupled products in, respectively, 88% and 60% yields. Interestingly both TBDMS and TMS groups survived cross-coupling reactions despite the presence of fluoride ions (Scheme 3.73).

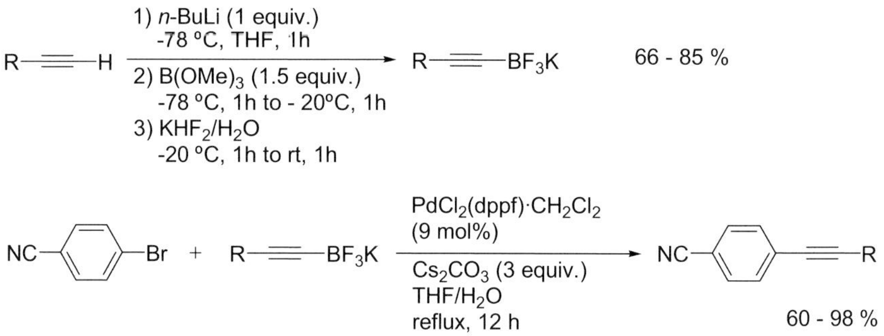

Scheme 3.73 Synthesis and reactions of functionalized alkynyl boronic derivatives.

Alkynyl boronic acid derivatives were not used earlier in Suzuki couplings. An effective Suzuki–Miyaura reaction between alkynyl 'ate' complexes (alkynyltrialkoxy borate complexes) has been reported by Colobert [112] Oh [113]. 1-Alkynyl(triisopropoxy) borates (**137**) were prepared by borylation of the corresponding alkynyl lithium species. These stable borate complexes were subsequently used in Suzuki coupling leading to products of type **138** (Scheme 3.74).

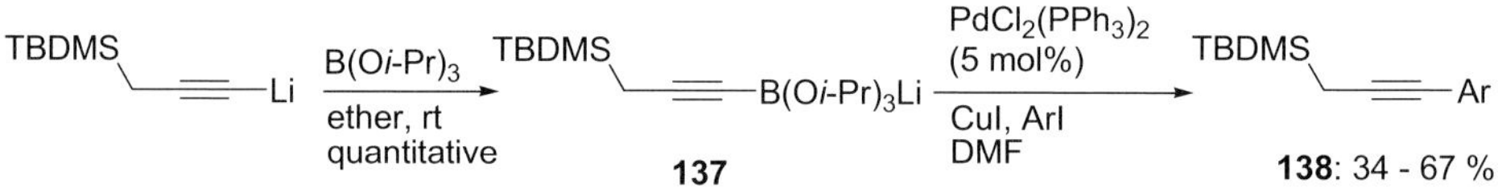

Scheme 3.74 Synthesis and reactions of functionalized alkynyl boronates.

3.5
Synthesis and Reactions of Functionalized Allylic Boronates

Grubbs has reported the synthesis of functionalized allyl boronates via olefin metathesis and developed a one-pot, three-component cross-metathesis/allylboration protocol for the synthesis of highly functionalized homoallylic alcohols by *in situ* reaction of these functionalized allylic boronates with benzaldehyde in high *anti*- diastereoselectivity [114]. Of particular importance is the direct incorporation of a halomethyl side-chain (R = CH$_2$Br and CH$_2$Cl) through previously unknown γ-haloallyl boronate (**139**, R = CH$_2$Br, CH$_2$Cl) thus, highlighting the utility of olefin metathesis for the synthesis of highly functionalized and reactive reagents not available through traditional methods leading here to the *anti*-homoallylic alcohols **140** (Scheme 3.75).

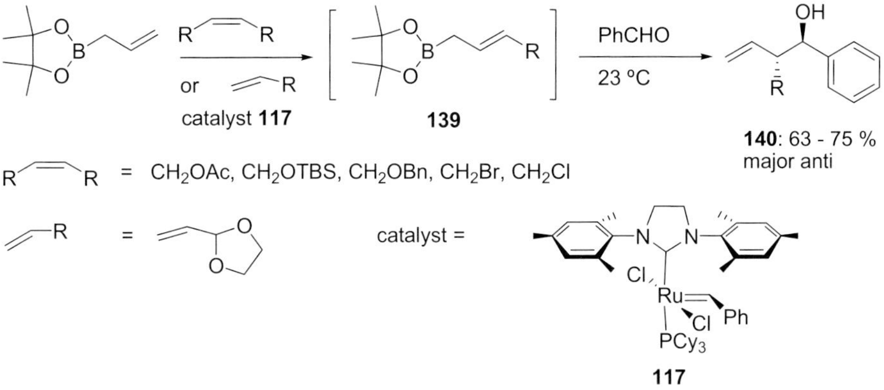

Scheme 3.75 Synthesis of functionalized allylic boronates by cross-metathesis and their one-pot allylboration with aldehydes.

Synthesis of a series of novel functionalized achiral and chiral allyl boronates has been recently reported by Ramachandran via nucleophilic SN$_2$'-type addition of copper boronate species (generated from the boronates **37, 141, 142** under Miyaura conditions) [115,116] to various functionalized allyl acetates that were prepared either via vinylalumination or by Baylis–Hillman reaction with various aldehydes [117]. The resulting allylic boronates bearing an ester moiety (X = OR) were subsequently used for the synthesis of α-alkylidene-β-substituted-γ-butyrolactones by allylboration of aldehydes (Scheme 3.76).

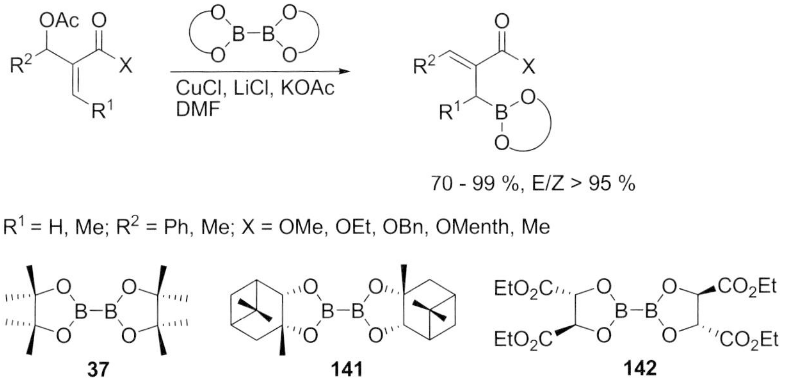

70 - 99 %, E/Z > 95 %

R^1 = H, Me; R^2 = Ph, Me; X = OMe, OEt, OBn, OMenth, Me

Scheme 3.76 Allylic boronates via Hosomi–Miyaura borylation.

Palladium-catalyzed cross-coupling of alkenyl stannanes with pinanediol bromomethyl boronate (**143**) has been reported to give homologous allylic boronates in good yields [118]. Cross-coupling proved compatible with a wide range of functionalities such as methyl ester, nitrile, benzyl ether and even an unprotected hydroxy group (Scheme 3.77).

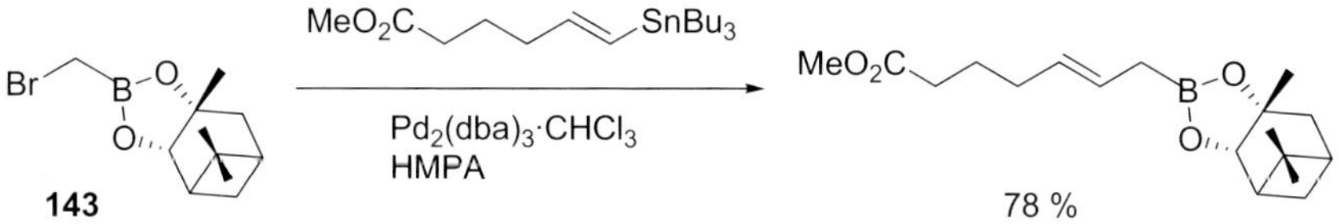

Scheme 3.77 Synthesis of functionalized allylic and benzylic boronates.

3.6
Synthesis and Reactions of Functionalized Cyclopropyl Boronic Esters

Pietruszka has reported the synthesis of stable enantiomerically pure functionalized cyclopropyl boronic esters via highly diastereoselective cyclopropanation of the respective alkenyl boronic esters with diazomethane catalyzed by Pd(OAc)$_2$ [119]. The enantiomerically pure alkenyl boronic esters were prepared by direct hydroboration of the respective alkynes with the chiral 1,3,2-dioxaborolane (**144**). The *ter*-butyldimethylsilyl protecting group in the boronic ester could be selectively deprotected and the resulting hydroxymethyl alkenyl boronate was also cyclopropanated to give hydroxymethylcyclopropyl boronic esters with good diastereoselectivity (Scheme 3.78).

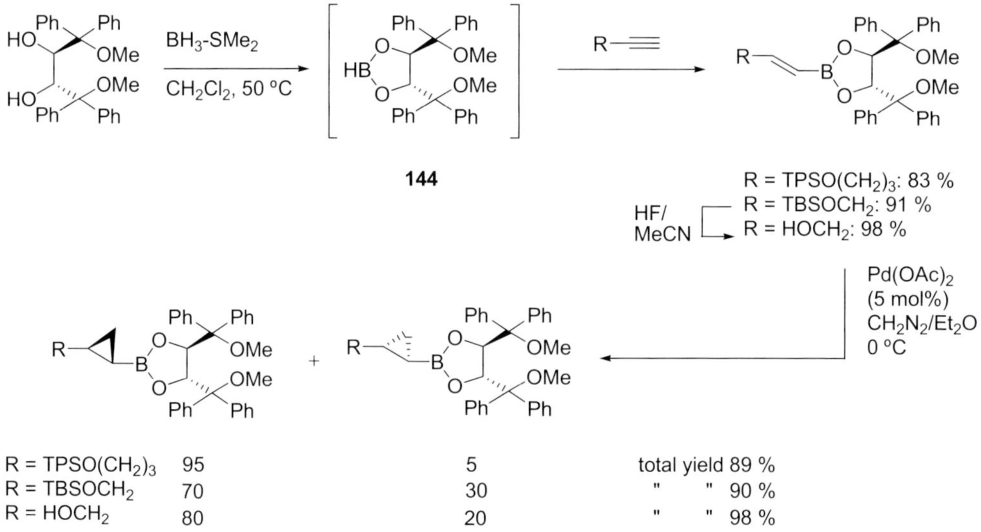

Scheme 3.78 Synthesis of cyclopropyl boronates.

This cyclopropyl boronic ester was converted into *bis*-cyclopropanes through a series of transformations shown in Scheme 3.79 [120]. The iodo *bis*-cyclopropyl boronic acid ester **145**, however gave a complex product mixture on attempted Suzuki coupling with phenylboronic acid. On the other hand, the less bulky hydroxy protected *bis*-cyclopropyldioxaborinane **146** obtained by transesterification underwent smooth cross-coupling with iodobenzene giving the phenyl substituted *bis*-cyclopropane product **147** in 79% yield (Scheme 3.79).

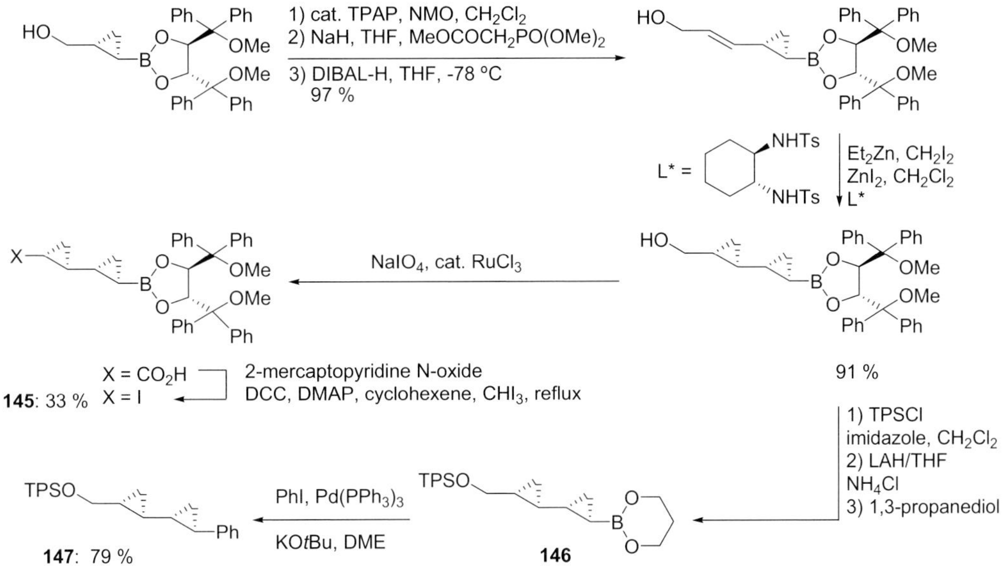

Scheme 3.79 Synthesis of functionalized *bis*-(cyclopropyl)boronic esters.

3.7
Synthesis and Reactions of Functionalized Alkyl Boron Derivates

3.7.1
Synthesis of Aminoalkyl Boranes by Hydroboration and their Suzuki Cross-coupling Reaction

A highly convenient method for introducing alkoxycarbonyl-protected β-aminoethyl groups into arenes and alkenes has been reported by Overman [121]. This one-pot reaction involves hydroboration of benzyl vinylcarbamate to give β-carbobenzyloxyborane that is *in situ* coupled with various aryl and alkenyl halides or triflates to give β-aminoethyl substituted arenes and alkenes in high yields. Overman has subsequently utilized these intermediates **148** for asymmetric synthesis of *trans*-hydroisoquinolones **149** (Scheme 3.80) [122].

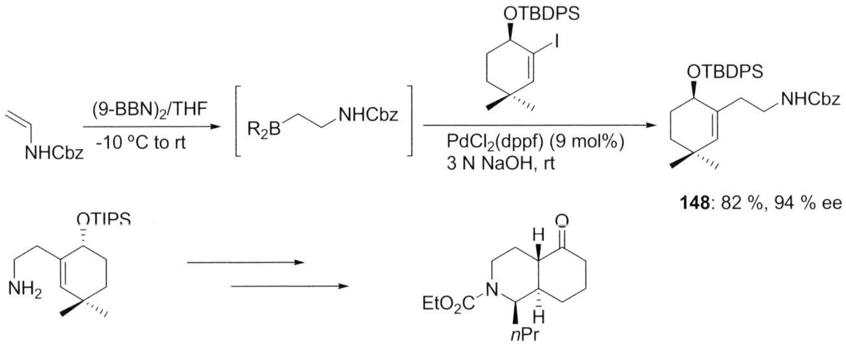

Scheme 3.80 Synthesis of β-amino derivates via Suzuki–Miyaura cross-coupling.

3.7.2
Synthesis of Functionalized Alkyl Boronates by Nucleophilic 1,4-Conjugate Addition of Borylcopper Species to α,β-Unsaturated Carbonyl Compounds

A copper-catalyzed nucleophilic borylation of α,β-unsaturated carbonyl compounds yielding β-borylated carbonyl compounds in good yields has been simultaneously reported by Hosomi and coworkers [123] and Miyaura and coworkers

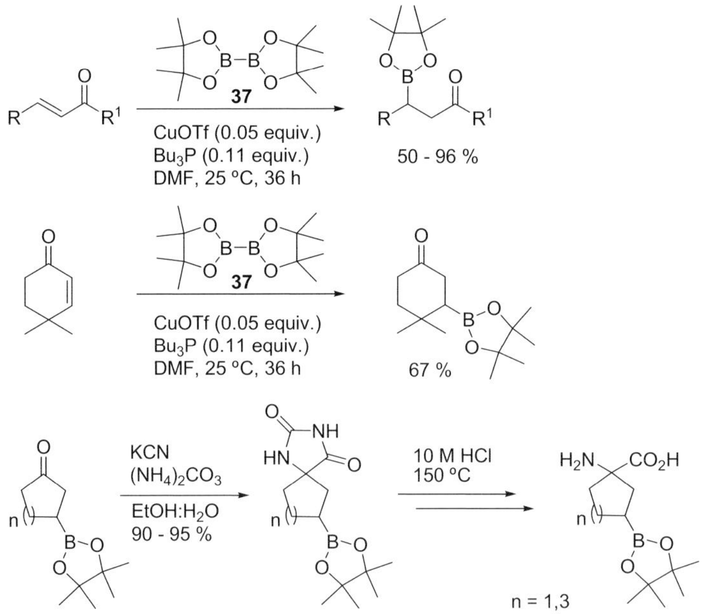

Scheme 3.81 Synthesis of functionalized alkyl boronates by nucleophilic addition of borylcopper species to α,β-unsaturated carbonyl compounds.

[124,125]. The transmetallation between the diboron derivate **37** and Cu(I)-salt generating a boryl copper species has been proposed as the key step in this reaction. Kabalka et al. used this copper-mediated 1,4-borylation reaction for the synthesis of boron containing unnatural amino acids as potential therapeutic agents in boron neutron capture therapy (BNCT) (Scheme 3.81) [126].

3.7.3
Preparation and B-alkyl-Suzuki–Miyaura Cross-coupling Reactions of Functionalized Alkyl Trifluoroborates

Molander has recently reported the preparation of various 3-functionalized propyl trifluoroborates from the corresponding allyl derivatives following various literature protocols for hydroboration of alkenes [127]. The 3-heteropropyl trifluoroborates were obtained as crystalline air-stable solids by treatment of hydroboronated products with KHF_2. Palladium-catalyzed cross-coupling of these 3-heteropropyl trifluoroborates with 4-acetylphenyl triflate gave functionalized alkylarenes in good yields (Scheme 3.82).

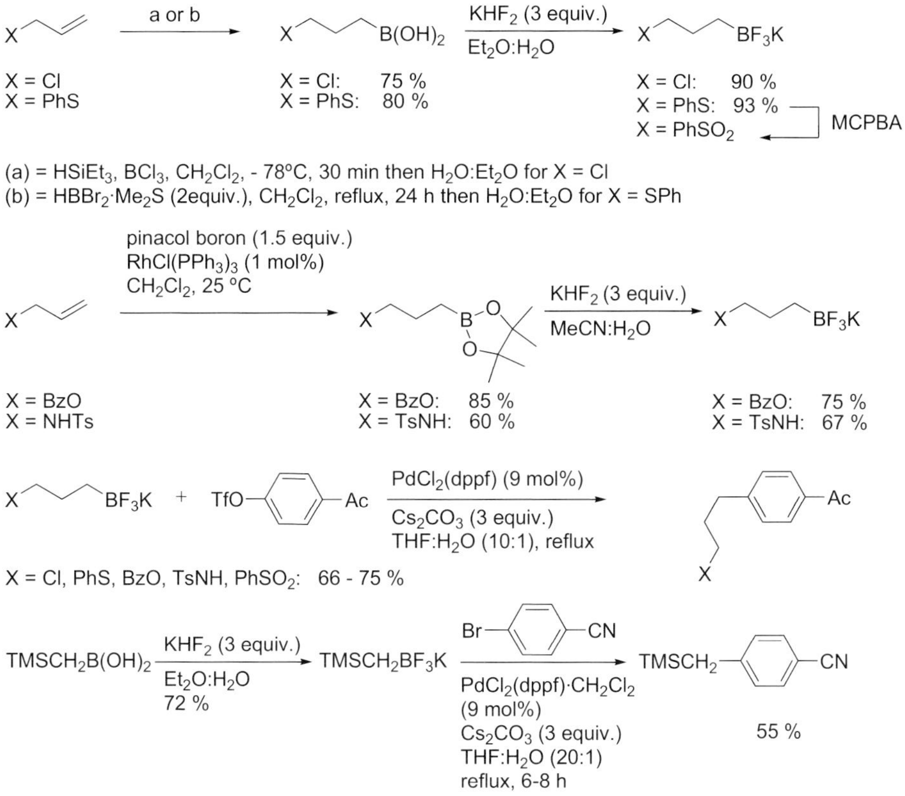

Scheme 3.82 Preparation and Pd-catalyzed Suzuki cross-coupling of functionalized alkyl trifluoroborates.

Molander has also described the preparation and cross-coupling of potassium (tri-methlsilyl)methyl trifluoroborate with various functionalized aryl bromides and tri-flates leading to (trimethylsilyl)methyl substituted arenes in moderate to good yields (Scheme 3.82) [127]. Likewise, potassium alkyl trifluoroborates bearing cyano, keto, bromo and ester groups participate smoothly in cross-coupling reactions with both functionalized aryl and alkenyl triflates in good yields [128]. These functionalized al-kyl trifluoroborates are readily available from Grignard reagents as well as by hydro-boration of the appropriate alkenes employing several different hydroborating proto cols. The inclusion of water in the cross-coupling reaction was found to be essential and Cs_2CO_3 was found to be the most effective base (Scheme 3.83).

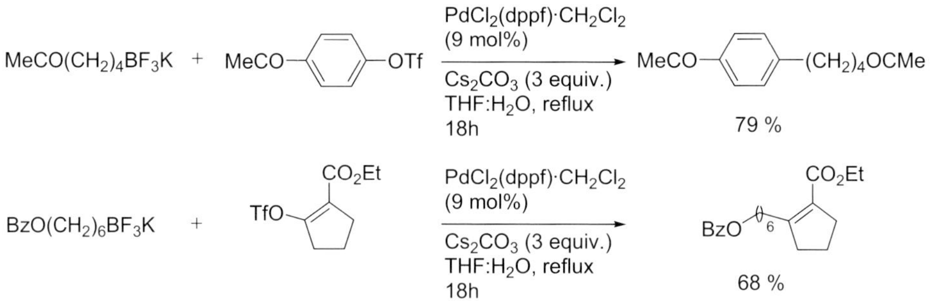

Scheme 3.83 Suzuki cross-coupling reactions of functiona-lized alkyl trifluoroborates with triflates.

3.7.4
Silver(I)-promoted Suzuki Cross-coupling of Functionalized *n*-Alkyl Boronic Acids

n-Alkyl boronic acids are shown to be less reactive in Suzuki–Miyaura cross-cou-pling reactions resulting in poor yields even under forcing conditions. Falck has shown that Ag(I)-salts significantly enhance Suzuki–Miyaura coupling of *n*-alkyl boronic acids and a variety of functional groups are tolerated on the boronic acid moiety (Scheme 3.84) [129].

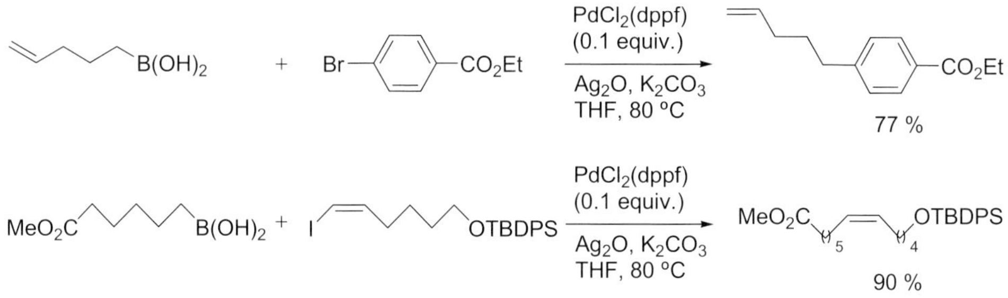

Scheme 3.84 Ag(I)-promoted Suzuki–Miyaura cross-coupling of functionalized *n*-alkylboronic acids.

3.7.5
Alkyl-Alkyl Suzuki Cross-coupling of Functionalized Alkyl Boranes with Alkyl Bromides, Chlorides and Tosylates

Fu has developed an efficient cross-coupling of primary alkyl bromides (Eq. 1), [130] chlorides (Eq. 2) [131] and tosylates (Eq. 3) [132] with functionalized alkyl boranes. The catalytic species are generated from $Pd(OAc)_2$ or $[Pd_2(dba)_3]$ and PCy_3 (for coupling of bromo and chloro alkanes). In the case of alkyl tosylates, $Pd(OAc)_2$ and $Pt\text{-}Bu_2Me$ gives the best results. This process takes advantage of the high tolerance of organoboron derivatives with most functional groups that makes this reaction a powerful synthetic tool (Scheme 3.85) [133].

$$NC\text{-}(CH_2)_6\text{-}Br \quad + \quad MeO_2C(CH_2)_{10}(9\text{-}BBN) \quad \xrightarrow[\substack{K_3PO_4 \cdot H_2O \ (1.2 \ \text{equiv.}) \\ THF, \ 25\,°C}]{\substack{Pd(OAc)_2 \ (4 \ \text{mol\%}) \\ PCy_3 \ (8 \ \text{mol\%})}} \quad MeO_2C(CH_2)_{16}CN$$

(1.2 equiv.)   81 %   (eq.1)

$$t\text{-}BuCO_2(CH_2)_6\text{-}Cl \ + \quad BnO(CH_2)_5(9\text{-}BBN) \quad \xrightarrow[\substack{CsOH \cdot H_2O \ (1.1 \ \text{equiv.}) \\ dioxane, \ 90\,°C}]{\substack{Pd_2(dba)_3 \ (5 \ \text{mol\%}) \\ PCy_3 \ (20 \ \text{mol\%})}} \quad BnO(CH_2)_{11}t\text{-}BuCO_2$$

65 %   (eq.2)

$$MeCO(CH_2)_6OTs \ + \quad TESO(CH_2)_{11}(9\text{-}BBN) \quad \xrightarrow[\substack{NaOH \ (1.2 \ \text{equiv.}) \\ dioxane, \ 50\,°C}]{\substack{Pd(OAc)_2 \ (4 \ \text{mol\%}) \\ PtBu_2Me \ (16 \ \text{mol\%})}} \quad TESO(CH_2)_{17}COMe$$

55 %   (eq.3)

Scheme 3.85 Alkyl-alkyl Suzuki cross-coupling of functionalized alkylboranes with functionalized alkyl bromides, chlorides and tosylates.

3.7.6
Synthesis of Natural and Unnatural Amino Acids via B-alkyl Suzuki Coupling of Functionalized Alkyl Boranes

Functionalized organoboron reagents of type **150**, which are readily available from serine via hydroboration of alkenes **151**, undergo efficient Suzuki-coupling reactions under mild conditions with a range of aryl and alkenyl halides [134]. These adducts **152** are easily transformed into a variety of known and novel nonproteinogenic N-protected amino acids via a one-pot hydrolysis-oxidation procedure leading to amino acid derivatives such as **153**. The methodology has been extended for the synthesis of *R,R*-diaminopimelic acid (DAP) and *R,R*-2,7-diaminosuberic acid (DAS) in enantiopure form (Scheme 3.86).

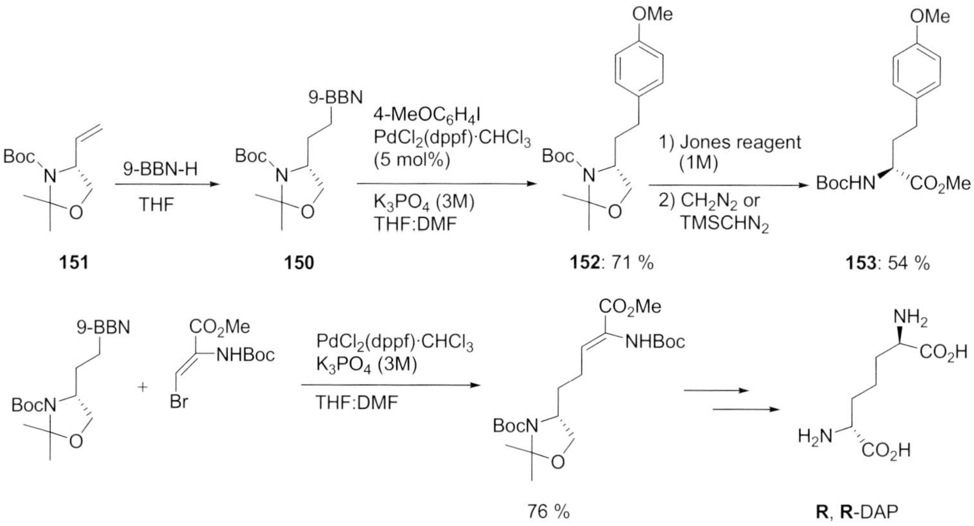

Scheme 3.86 Application of B-alkyl Suzuki cross-coupling for the synthesis of natural and unnatural amino acids.

3.7.7
Application of Intermolecular B-alkyl Suzuki Cross-coupling of Functionalized Alkyl Boranes in Natural Product Synthesis

Glycals are versatile readily available synthetic intermediates with a wide range of application in the synthesis of carbohydrate analogs, C-aryl glycosides and a variety of other natural products. A flexible efficient method for converting glycals to C1-alkyl glycals using a B-alkyl Suzuki–Miyaura cross-coupling has been reported [135]. This method provides access to a range of C1-substituted glycals that are not available by direct alkylation of C1-lithio glycals by utilizing a number of functionalized olefins and glycal coupling partners. It has further been shown that commonly observed side reactions involving reduction of halide coupling partners during B-alkyl Suzuki couplings can be suppressed by preincubation of the borane coupling partner with aqueous NaOH prior to addition to the C1-iodo glycals such as **151** and to the Pd-catalyst leading to polyfunctional products such as **152** (Scheme 3.87).

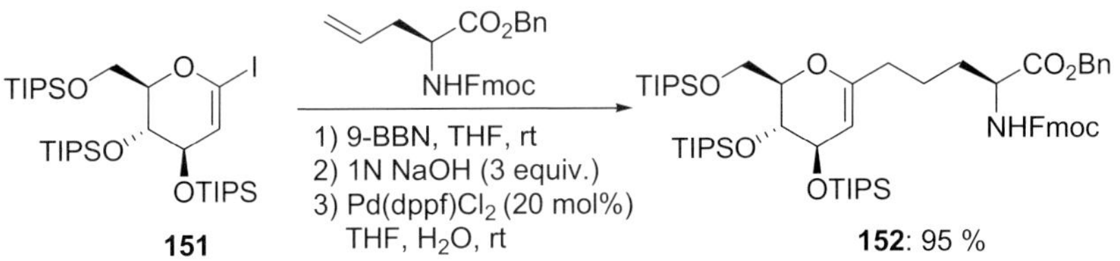

151 **152**: 95 %

Scheme 3.87 Synthesis of polyfunctional glycals.

Polyhydroxy piperidines and related azasugars are known to be potent inhibitors of oligosaccharide-processing enzymes called glycosidases and glycosyl transferases, therefore these class of compounds has received considerable attention. Johnson has developed the synthesis of linked azasugars, a novel class of glycomimetic compounds, involving B-alkyl Suzuki coupling as the key step [136,137]. Thus, Suzuki coupling of the cycloalkenyl bromide intermediate **153** with alkyl boranes derived via hydroboration from olefinated carbohydrate precursors such as **154** was used to form the C-glycoside bond. The azasugar ring of **155** was obtained by subsequent ozonolysis of the coupled products **156** and selective reduction of the resultant carbonyl function. The fully deprotected azasugars were obtained upon acid deprotection (Scheme 3.88).

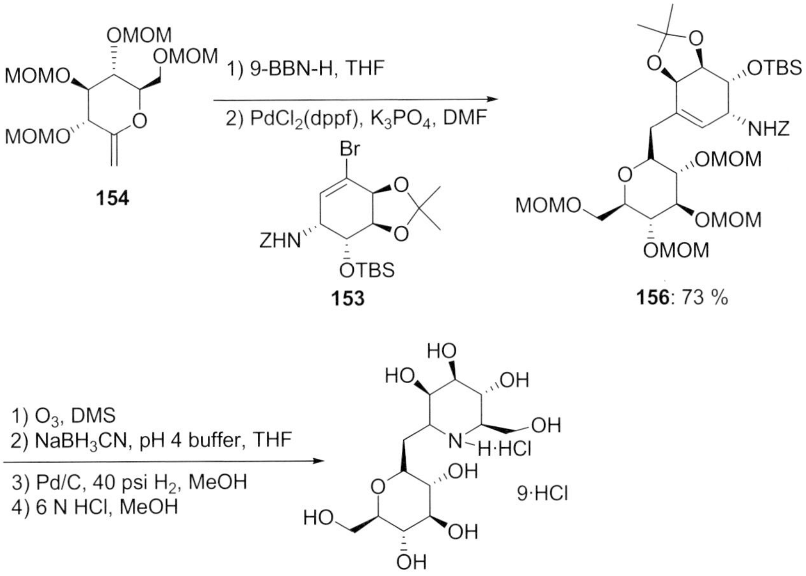

154 **153** **156**: 73 % **155**: 68 %

Scheme 3.88 Synthesis of azasugars.

B-alkyl Suzuki-coupling strategy for elongation of unsaturated side chains has been used as the key step in the synthesis of marine alkaloid (+)-halichlorine [138,139] and the related marine natural product pinnaic acid, [140,141] an inhibitor of cytosolic phospholipase A_2. Thus, the hydroboration of the common protected aminoalkene precursor **157** followed by Pd-mediated Suzuki coupling with iododienoic acid ester (**158**) afforded compound **159** that was converted into pinnaic acid in several steps (Scheme 3.89).

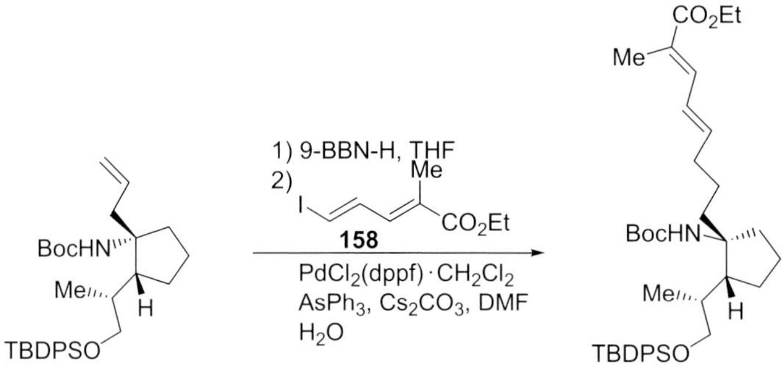

Scheme 3.89 Synthesis of advanced intermediates of pinnaic acid.

The B-alkyl Suzuki coupling reactions are used extensively in the synthesis of the anticancer agents epothilone A, B, and F [142]. In the earlier approaches, [143] alkenyl iodide **160** was coupled with organoborane derived from the alkene **161a**. Zhu [144] has applied a closely related reaction using intermediate **161b** in the preparation of epothilone A. In the subsequent synthesis, Danishefsky has used a more elaborate and sensitive substrate, i.e. β-ketoester **162** as the olefinic coupling partner [145]. Thus, the coupling of iodide **160** in the presence of a Pd-catalyst afforded **163** in 65% yield after acidic work-up while the ester functionality and the two carbonyl groups remains unaffected under these conditions. Recently, Danishefsky and coworkers [146,147] and other groups [148] have utilized the C1–C11 olefinic coupling partner **164** with a protected OH group in 3-position of desoxyepothilone F and B, the corresponding 26-(1,3-dioxanyl) derivative in the synthesis of epothilone A (Scheme 3.90).

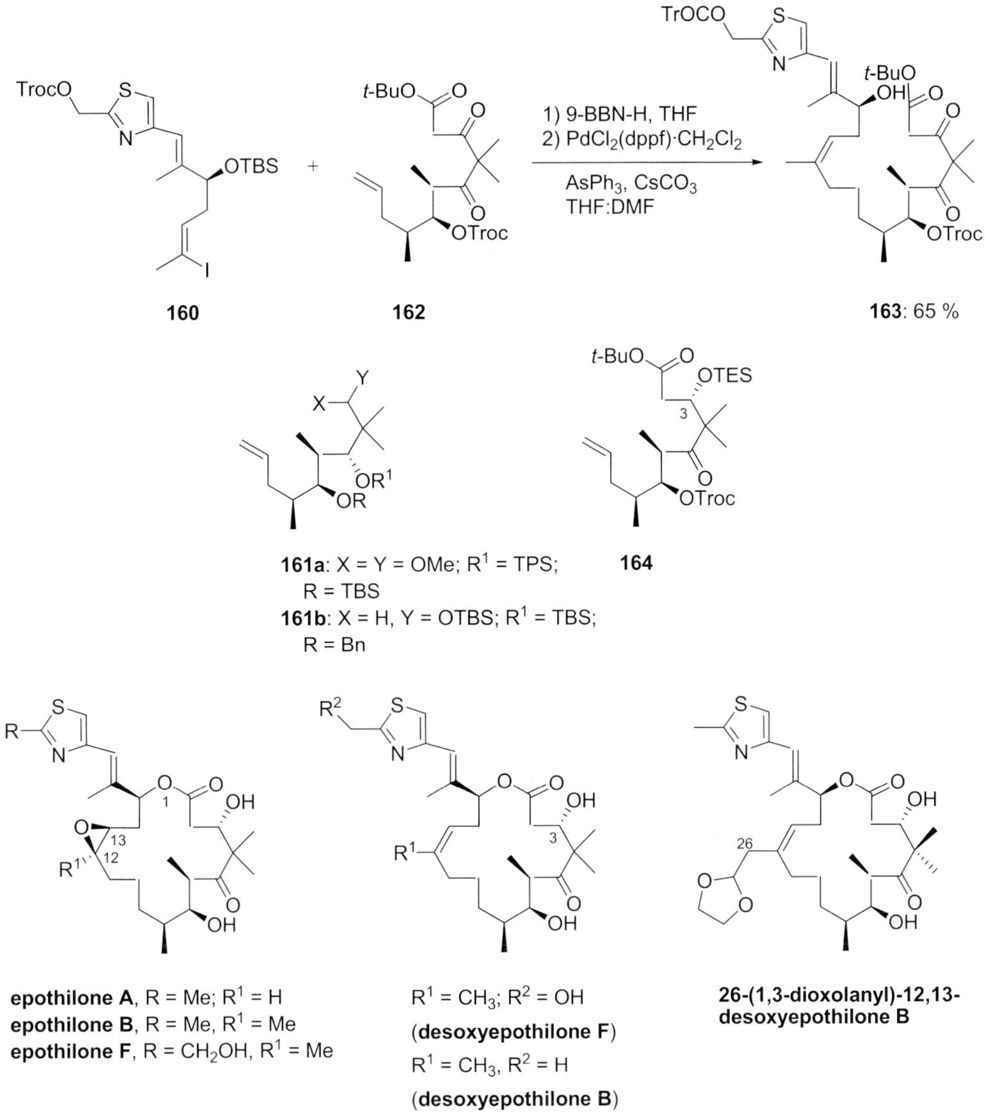

160 + **162**

1) 9-BBN-H, THF
2) PdCl$_2$(dppf)·CH$_2$Cl$_2$

AsPh$_3$, CsCO$_3$
THF:DMF

→ **163**: 65 %

161a: X = Y = OMe; R^1 = TPS; R = TBS
161b: X = H, Y = OTBS; R^1 = TBS; R = Bn

164

epothilone A, R = Me; R^1 = H
epothilone B, R = Me, R^1 = Me
epothilone F, R = CH$_2$OH, R^1 = Me

R^1 = CH$_3$; R^2 = OH
(desoxyepothilone F)
R^1 = CH$_3$, R^2 = H
(desoxyepothilone B)

26-(1,3-dioxolanyl)-12,13-desoxyepothilone B

Scheme 3.90 Synthesis of epithilones using Suzuki–Miyaura cross-couplings.

Sasaki has developed an efficient and practical methodology for the polycyclic ether framework present in marine natural products such as ciguatoxins, brevetoxins, etc., based on palladium-catalyzed B-alkyl Suzuki cross-coupling reactions of alkyl boranes with cyclic ketene acetal triflates or phosphates [149,150,151]. He reported the first total synthesis of (–)-gambierol [152], a marine polyether toxin isolated from *Gambierdiscus toxicus* involving convergent union of ABC and EFGH ring fragments **165** and **166** via a B-alkyl Suzuki coupling as the key step leading to endocyclic enol ether **167** in high yield that was converted to (–)-gambierol through a series of transformations. The same group has reported [153] the synthesis of the FGH ring system of gambierol through PdCl$_2$(dppf)-promoted room-temperature B-alkyl Suzuki coupling of the lactone derived enol phosphate **168** with **169** as the main step leading to the cross-coupling product **170** in 97% yield (Scheme 3.91).

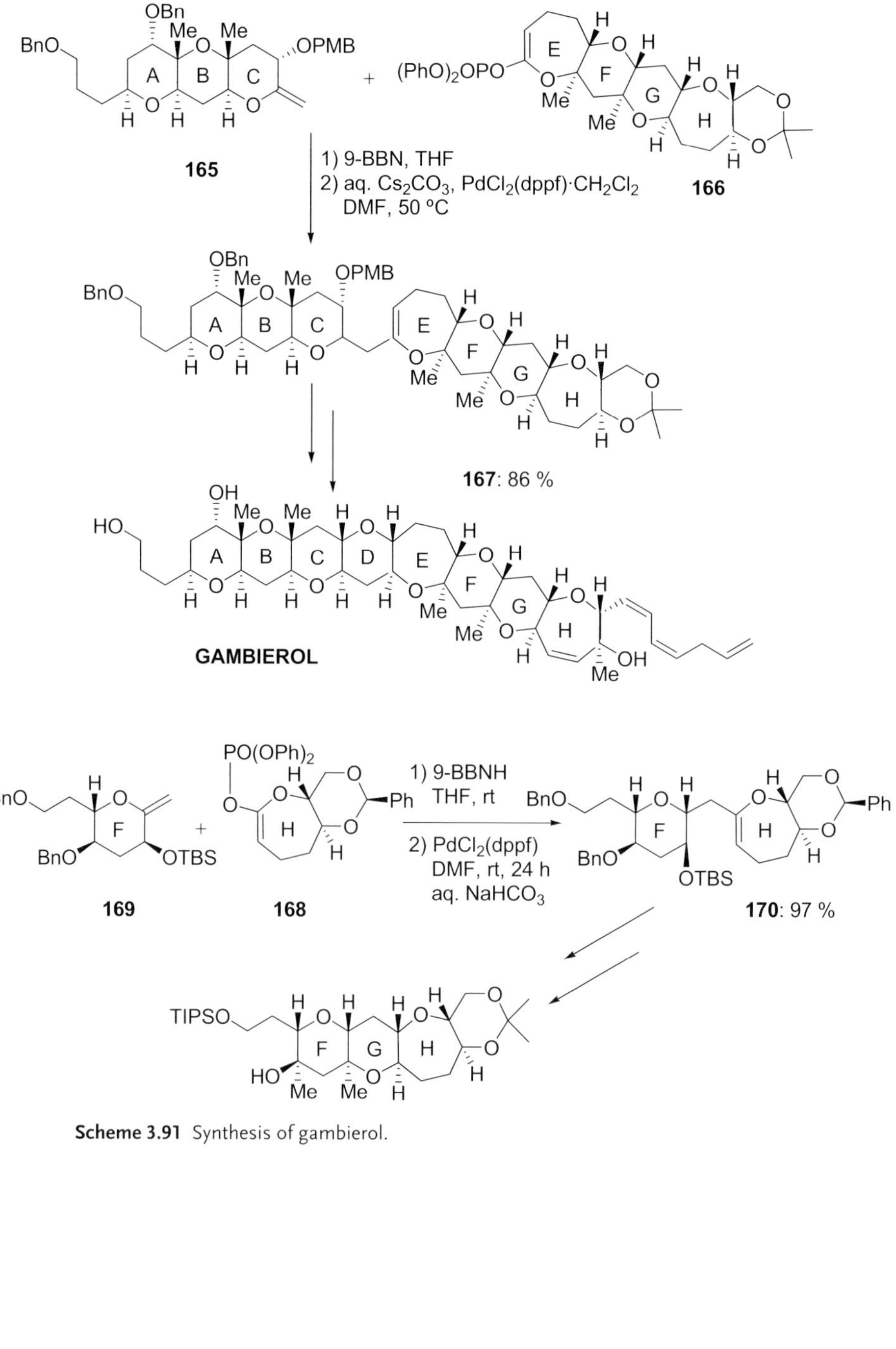

Scheme 3.91 Synthesis of gambierol.

Sasaki has also applied a B-alkyl Suzuki coupling for the preparation of fused polyethers in the convergent synthesis of ABCD ring fragment **171** of cigua-toxin [154,155], the causative toxin for ciguatera fish poisoning, via convergent union of the olefin **172** and the seven-membered lactone phosphate **173** through hydroboration and Pd-mediated C–C coupling leading to the cross-coupling prod-uct **174** in 97% yield. They further elaborated this methodology for the construc-tion of the FGHIJKLM ring fragment of ciguatoxin on the basis of extensive use of B-alkyl Suzuki–Miyaura reaction (Scheme 3.92) [156].

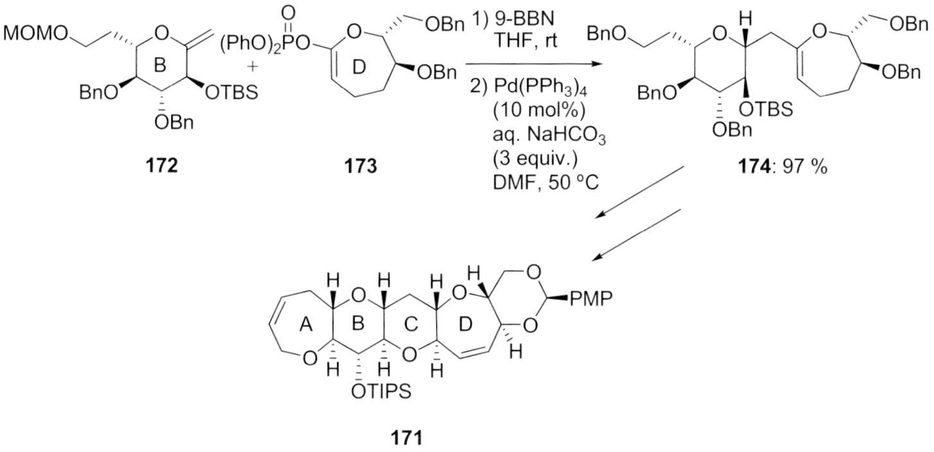

Scheme 3.92 Synthesis of ciguatoxin.

Trost has utilized B-alkyl Suzuki coupling in an enantioselective total synthesis of sphingofungin E (**175**), an antifungal agent that blocks the biosynthesis of sphingolipids leading to apoptosis in both yeast and mammalian cells [157,158]. The strategy involves coupling of the polar head **176** and lipid tail unit **177**. Thus, under palladium catalysis, the alkenyl iodide **176** smoothly reacted with B-alkyl-borane **177** to give the polyfunctional alkene **178** in excellent yield that was subse-quently converted to sphingofungin E in several steps (Scheme 3.93).

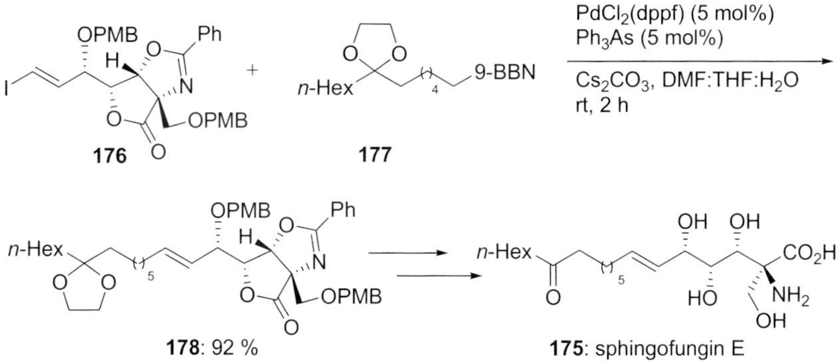

Scheme 3.93 Synthesis of sphingofungin E.

Although formation of five- and six-membered rings via intramolecular B-alkyl Suzuki coupling was reported to occur readily [159,160], the feasibility of this reaction for transannular macrocyclization as an effective method was demonstrated by Chemler and Danishefsky [161]. Regioselective terminal olefin hydroboration with 9-BBN-H followed by palladium-catalyzed intramolecular Suzuki reaction in the presence of base such as thallium ethoxide at high dilution generates macrocycles with high degree of olefin geometry control. Thus, isomerically pure *E* or *Z* alkenyl iodides of type **179** afford the macrocycles **E-180** and **Z-180** in high stereoselectivity and good yields. These reactions are complimentary to ring-closing metathesis macrocyclization and may prove superior in cases where control of olefin geometry is required (Scheme 3.94).

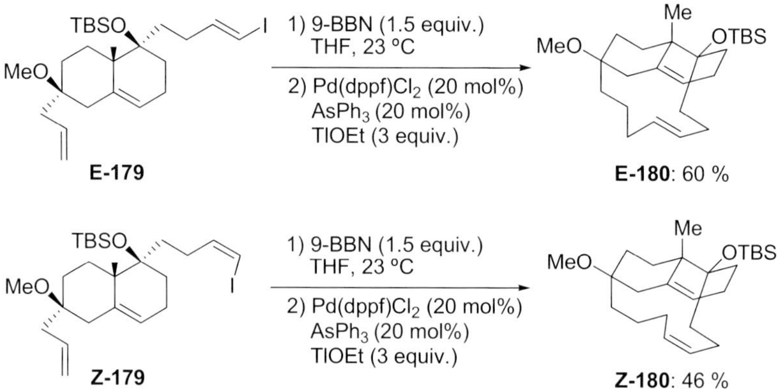

Scheme 3.94 Macrocyclization via B-alkyl Suzuki-coupling.

An enantioselective total synthesis of (+)-phomactin A (**181**) has been recently reported by Halcomb using intramolecular Suzuki coupling of a B-alkyl-9-BBN derivative to prepare the macrocycle in the final step [162,163]. Thus, a regioselective hydroboration of the terminal olefin in the precursor **182** gave an internal alkylborane that was cyclized using modification of Johnson's conditions [164]. The reaction illustrates the mildness of the Suzuki reaction since the coupling was carried out in the presence of the sensitive dihydrofuran ring (Scheme 3.95).

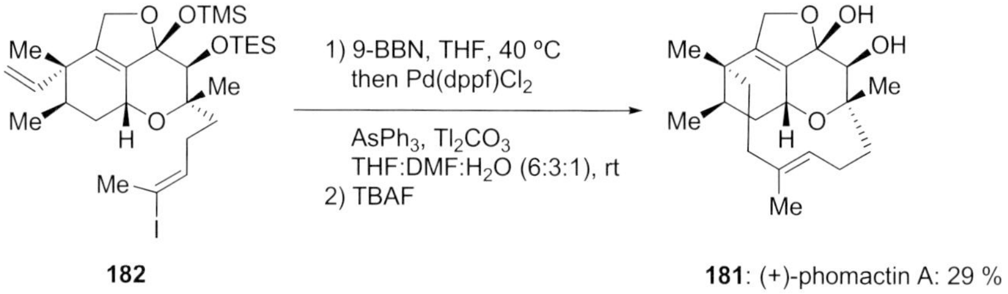

Scheme 3.95 Synthesis of (+)-phomactin A.

An intramolecular B-alkyl Suzuki coupling has been used to construct the core macrocyclic structure of benzolactone enamide salicylhalamide A [165]. Thus, diastereoselective hydroboration of acyclic alkenyl iodide **183** with (dppf)PdCl₂ catalyst/NaOH gave the macrocyclic lactone **184** in 48% yield. Because of the high dilution, a large amount of catalyst (20 mol%) was used. The steric hindrance at the ester group probably retards the cleavage of the lactone under basic conditions (Scheme 3.96).

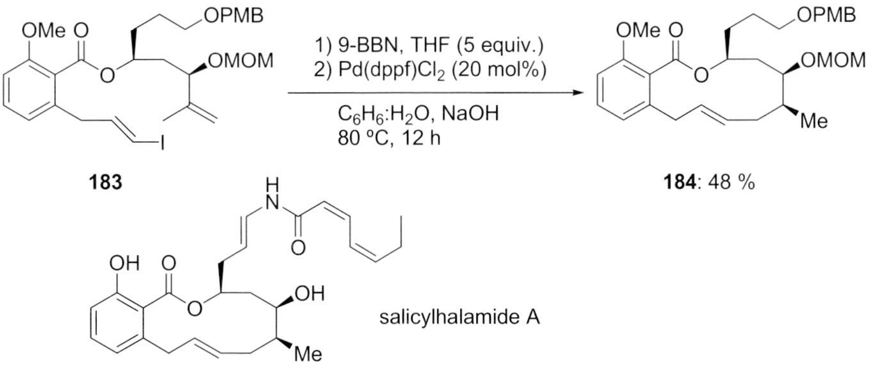

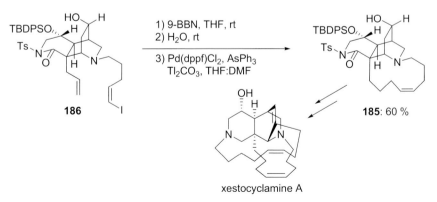

Scheme 3.96 Synthesis of an advanced intermediate for the salicylhalamide A synthesis.

Xestocyclamine A, a polycyclic alkaloid isolated from *xestospongia* sp. is an inhibitor of PKC*β* and is of biomedical interest. During the course of the total synthesis of xestocyclamine A, Danishefsky has reported the elaboration of an *ansa* bridge using intramolecular B-alkyl Suzuki coupling as the key step. Thus, the intermediate **185** was obtained in 60% yield by regioselective hydroboration of **186** at the terminal alkenyl group followed by subsequent treatment with Pd(dppf)Cl₂ catalyst in the presence of triphenylarsine and thallium carbonate as a base (Scheme 3.97) [166].

Scheme 3.97 Macrocyclisation for the xestocyclamine A synthesis.

3.8
Conclusion

The chemistry of polyfunctional boronic esters has greatly been developed because of the multiple synthetic approaches to these molecules and their ability to participate in a variety of cross-coupling reactions. The high functional group compatibility and development of efficient cross-coupling procedures will certainly lead to further spectacular applications of boronic esters.

References

1 (a) A. Pelter, K. Smith, H. C. Brown, *Borane Reagents*, Academic Press, New York, **1988**; (b) D. S. Matteson, *Stereodirected Synthesis with Organoboranes*, Springer, New York, **1995**; (c) V. Snieckus, *Chem. Rev.* **1990**, *90*, 879; (d) T. Ishiyama, N. Miyaura, *J. Organomet. Chem.* **2000**, *611*, 392; (e) E. Tyrell, P. Brookes, *Synthesis* **2003**, 469; (f) T. Ishiyama, N. Miyaura, *J. Organomet. Chem.* **2003**, *680*, 3.

2 (a) A. Suzuki in *Metal-Catalyzed Cross-Coupling Reactions* (Eds.: F. Diederich, P. J. Stang), Wiley-VCH, Weinheim, **1998**, pp. 49–97; (b) N. Miyaura, A. Suzuki, *Chem. Rev.* **1995**, *95*, 2457; (c) S. P. Stanforth, *Tetrahedron* **1998**, *54*, 263; (d) A. Suzuki, *J. Organomet. Chem.* **1999**, *576*, 147; (e) S. R. Chemler, D. Trauner, S. J. Danishefsky, *Angew. Chem.* **2001**, *113*, 4676; *Angew. Chem. Int. Ed.* **2001**, *40*, 4544; (f) A. F. Littke, G. C. Fu, *Angew. Chem.* **2002**, *114*, 4350; *Angew. Chem. Int. Ed.* **2002**, *41*, 4176; (g) S. Kotha, K. Lahiri, D. Kashinath, *Tetrahedron* **2002**, *58*, 9633; (h) A. Suzuki, *J. Organomet. Chem.* **2002**, *653*, 83; (i) S. V. Ley, A. W. Thomas, *Angew. Chem.* **2003**, *115*, 5558; *Angew. Chem. Int. Ed.* **2003**, *42*, 5400.

3 S. Caron, J. M. Hawkins, *J. Org. Chem.* **1998**, *63*, 2054.

4 J. Kristensen, M. Lysén, P. Vedsø, M. Begtrup, *Org. Lett.* **2001**, *3*, 1435.

5 A. Brikh, C. Morin, *J. Organomet. Chem.* **1999**, *581*, 82.

6 J. Clayden, *Organolithiums: Selectivity for Synthesis*, Pergamon Press, Oxford **2002**.

7 M. C. Whisler, S. MacNeil, V. Snieckus, P. Beak, *Angew. Chem.* **2004**, *116*, 2256; *Angew. Chem. Int. Ed.* **2004**, *43*, 2206.

8 For a review, see: P. Knochel, W. Dohle, N. Gommermann, F. F. Kneisel, F. Kopp, T. Korn, I. Sapountzis, V. A. Vuh, *Angew. Chem.* **2003**, *115*, 4438; *Angew.Chem. Int. Ed.* **2003**, *42*, 4302.

9 A. Krasovskiy, P. Knochel, *Angew. Chem.* **2004**, *116*, 3396; *Angew. Chem. Int. Ed.* **2004**, *43*, 3333.

10 O. Baron, A. Krasovskiy, T. Korn, P. Knochel, manuscript in preparation.

11 O. Baron, P. Knochel, *Angew. Chem. Int. Ed.* **2005**, *44*, 3133.

12 M. Murata, S. Watanabe, Y. Masuda, *J. Org. Chem.* **1997**, *62*, 6458.

13 M. Murata, T. Oyama, S. Watanabe, Y. Masuda, *J. Org. Chem.* **2000**, *65*, 164.

14 O. Baudoin, D. Guénard, F. Guéritte, *J. Org. Chem.* **2000**, *65*, 9268.

15 C. Christophersen, M. Begtrup, S. Ebdrup, H. Petersen, P. Vedsø, *J. Org. Chem.* **2003**, *68*, 9513.

16 T. Ishiyama, M. Murata, N. Miyaura, *J. Org. Chem.* **1995**, *60*, 7508.

17 T. Ishiyama, Y. Itoh, T. Kitano, N. Miyaura, *Tetrahedron Lett.* **1997**, *38*, 3447.

18 T. Ishiyama, K. Ishida, N. Miyaura, *Tetrahedron* **2001**, *57*, 9813.

19 A. Fürstner, G. Seidel, *Org. Lett.* **2002**, *4*, 541.

20 L. Zhu, J. Duquette, M. Zhang, *J. Org. Chem.* **2003**, *68*, 3729.

21 W. Ni, H. Fang, G. Springsteen, B. Wang, *J. Org. Chem.* **2004**, *69*, 1999.

22 H. Nakamura, M. Fujiwara, Y. Yamamoto, *J. Org. Chem.* **1998**, *63*, 7529.

23 C. Malan, C. Morin, *J. Org. Chem.* **1998**, *63*, 8019.

24 For a review, see: T. Ishiyama, N. Miyaura, *J. Organomet.Chem.* **2003**, *680*, 3.

25 H. Chen, J. F. Hartwig, *Angew. Chem.* **1999**, *111*, 3597; *Angew. Chem. Int. Ed.* **1999**, *38*, 3391.

26 H. Chen, S. Schlecht, T. C. Semple, J. F. Hartwig, *Science* **2000**, *287*, 1995.

27 J.-Y. Cho, M. K. Tse, D. Holmes, R. E. Maleczka, Jr., M. R. Smith, III, *Science* **2002**, *295*, 305.

28 S. Shimada, A. S. Batsanov, J. A. K. Howard, T. B. Marder, *Angew. Chem.* **2001**, *113*, 2226; *Angew. Chem Int. Ed.* **2001**, *40*, 2168.

29 J.-Y. Cho, C. N. Iverson, M. R. Smith, III, *J. Am. Chem. Soc.* **2000**, *122*, 12868.

30 T. Ishiyama, J. Takagi, K. Ishida, N. Miyaura, N. R. Anastasi, J. F. Hartwig, *J. Am. Chem. Soc.* **2002**, *124*, 390.

31 T. Ishiyama, J. Takagi, J. F. Hartwig, N. Miyaura, *Angew. Chem.* **2002**, *114*, 3182; *Angew. Chem. Int. Ed.* **2002**, *41*, 3056.
For a related reference with bipyridyl (bpy) ligand see: T. Ishiyama, J. Takagi, K. Ishida, N. Miyaura, N. R. Anastasi, J. F. Hartwig, *J. Am. Chem. Soc.* **2002**, *124*, 390.

32 J. Takagi, K. Sato, J. F. Hartwig, T. Ishiyama, N. Miyaura, *Tetrahedron Lett.* **2002**, *43*, 5649.

33 G. A. Molander, B. Biolatto, *J. Org. Chem.* **2003**, *68*, 4302.

34 G. A. Molander, B. Biolatto, *Org. Lett.* **2002**, *4*, 1867.

35 R. A. Batey, T. D. Quach, *Tetrahedron Lett.* **2001**, *42*, 9099.

36 H.-J. Frohn, N. Yu. Adonin, V. V. Bardin, V. F. Starichenko, *Tetrahedron Lett.* **2002**, *43*, 8111.

37 J. P. Wolfe, R. A. Singer, B. H. Yang, S. L. Buchwald, *J. Am. Chem. Soc.* **1999**, *121*, 9550.

37 A. F. Littke, G. C. Fu, *Angew. Chem.* **1998**, *110*, 3586; *Angew. Chem. Int. Ed.* **1998**, *37*, 3387.

39 A. F. Littke, C. Dai, G. C. Fu, *J. Am. Chem. Soc.* **2000**, *122*, 4020.

40 A. N. Cammidge, K. V. L. Crépy, *J. Org. Chem.* **2003**, *68*, 6832.

41 A. Nadin, J. M. S. López, A. P. Owens, D. M. Howells, A. C. Talbot, T. Harrison, *J. Org. Chem.* **2003**, *68*, 2844.

42 C.-y. Chen, P. Dageneau, E. J. J. Grabowski, R. Oballa, P. O'Shea, P. Prasit, J. Robichaud, R. Tillyer, X. Wang, *J. Org. Chem.* **2003**, *68*, 2633.

43 E. C. Western, J. R. Daft, E. M. Johnson, II, P. M. Gannett, K. H. Shaughnessy, *J. Org. Chem.* **2003**, *68*, 6767.

44 Y. Satoh, C. Gude, K. Chan, F. Firooznia, *Tetrahedron Lett.* **1997**, *38*, 7645.

45 F. Firooznia, C. Gude, K Chan, Y. Satoh, *Tetrahedron Lett.* **1998**, *39*, 3985.

46 F. Firooznia, C. Gude, K. Chan, N. Marcopulos, Y. Satoh, *Tetrahedron Lett.* **1999**, *40*, 213.

47 E. C. Gravett, P. J. Hilton, K. Jones, F. Romero, *Tetrahedron Lett.* **2001**, *42*, 9081.

48 A. Bouillon, A. S. Voisin, A. Robic, J.-C. Lancelot, V. Collot, S. Rault, *J. Org. Chem.* **2003**, *68*, 10178.

49 B. K. Albrecht, R. M. Williams, *Org. Lett.* **2003**, *5*, 197.

50 D. Ma, Q. Wu, *Tetrahedron Lett.* **2001**, *42*, 5279.

51 D. Ma, Q. Wu, *Tetrahedron Lett.* **2000**, *41*, 9089.

52 E. Vedejs, M. A. Zajac, *Org. Lett.* **2001**, *3*, 2451.

53 N. Yasuda, M. A. Huffmann, G.-J. Ho, L. C. Xavier, C. Yang, K. M. Emerson, F.-R. Tsay, Y. Li, M. H. Kress, D. L. Rieger, S. Karady, P. Sohar, N. L. Abramson, A. E. DeCamp, D. J. Mathre, A. W. Douglas, U.-H. Dolling, E. J. J. Grabowski, P. J. Reider, *J. Org. Chem.* **1998**, *63*, 5438.

54 S. Sakamuri, C. George, J. Flippen-Anderson, A. P. Kozikowski, *Tetrahedron Lett.* **2000**, *41*, 2055.

55 N. K. Garg, R. Sarpong, B. M. Stoltz, *J. Am. Chem. Soc.* **2002**, *124*, 13179.

56 A.-C. Carbonnelle, J. Zhu, *Org. Lett.* **2000**, *2*, 3477.

57 V. Lobrégat, G. Alcaraz, H. Bienaymé, M. Vaultier, *Chem. Commun.* **2001**, 817.

58 M. Smet, B. Metten, W. Dehaen, *Tetrahedron Lett.* **2001**, *42*, 6527.

59 S. V. Ley, A. W. Thomas, *Angew. Chem. Int. Ed.* **2003**, *42*, 5400.

60 S. Yu, J. Saenz, J. K. Srirangam, *J. Org. Chem.* **2002**, *67*, 1699.

61 J. C. Antilla, S. L. Buchwald, *Org. Lett.* **2001**, *3*, 2077.

62 M. Sasaki, S. Dalili, A. K. Yudin, *J. Org. Chem.* **2003**, *68*, 2045.

63 H. M. Petrassi, K. B. Sharpless, J. W. Kelly, *Org. Lett.* **2001**, *3*, 139.

64 D.A. Evans, J. L. Katz, T. R. West, *Tetrahedron Lett.* **1998**, *39*, 2937.

65 C. P. Decicco, Y. Song, D. A. Evans, *Org. Lett.* **2001**, *3*, 1029.

66 D. A. Evans, J. L. Katz, G. S. Peterson, T. Hintermann, *J. Am. Chem. Soc.* **2001**, *123*, 12411.

67 P. S. Herradura, K. A. Pendola, R. K. Guy, *Org. Lett.* **2000**, *2*, 2019.

68 C. Savarin, J. Srogl, L. S. Liebeskind, *Org. Lett.* **2002**, *4*, 4309.

69 T. Ishiyama, H. Kizaki, T. Hayashi, A. Suzuki, N. Miyaura, *J. Org. Chem.* **1998**, *63*, 4726.

70 N. A. Bumagin, D. N. Korolev, *Tetrahedron Lett.* **1999**, *40*, 3057.

71 L. J. Gooßen, K. Ghosh, *Angew. Chem.* **2001**, *113*, 3566; *Angew. Chem. Int. Ed.* **2001**, *40*, 3458.

72 L. J. Gooßen, *Chem. Commun.* **2001**, 669.

73 R. Itooka, Y. Iguchi, N. Miyaura, *J. Org. Chem.* **2003**, *68*, 6000.

74 Y. Takaya, M. Ogasawara, T. Hayashi, M. Sakai, N. Miyaura, *J. Am. Chem. Soc.* **1998**, *120*, 3379.

75 R. A. Batey, A. N. Thadani, D. V. Smil, *Org. Lett.* **1999**, *1*, 1683.

76 J. Ramnauth, O. Poulin, S. S. Bratovanov, S. Rakhit, S. P. Maddaford, *Org. Lett.* **2001**, *3*, 2571.

77 T. Senda, M. Ogasawara, T. Hayashi, *J. Org. Chem.* **2001**, *66*, 6852.

78 C. J. Chapman, K. J. Wadsworth, C. G. Frost, *J. Organomet. Chem.* **2003**, *680*, 206.

79 N. A. Petasis, A. Goodman, I. A. Zavialov, *Tetrahedron Lett.* **1997**, *53*, 16463.

80 For reaction of arylboronic acids with salicylaldehyde/amine: N. A. Petasis, S. Boral, *Tetrahedron Lett.* **2001**, *42*, 539.

81 N. A. Petasis, I. A. Zavialov, *J. Am. Chem. Soc.* **1998**, *120*, 11798.

82 G. K. S. Prakash, M. Mandal, S. Schweizer, N. A. Petasis, G. A. Olah, *Org. Lett.* **2000**, *2*, 3173.

83 G. K. S. Prakash, M. Mandal, S. Schweizer, N. A. Petasis, G. A. Olah, *J. Org. Chem.* **2002**, *67*, 3718.

84 R. H. Szumigala, P. N. Devine, D. R. Gauthier, R. P. Volante, *J. Org. Chem.* **2004**, *69*, 566.

85 C. Theibes, G. K. S. Prakash, N. A. Petasis, G. A. Olah, *Synlett* **1998**, 141.

86 N. A. Petasis, I. A. Zavialov, *Tetrahedron Lett.* **1996**, *37*, 567.

87 J. Simon, S. Salzbrunn, G. K. S. Prakash, N. A; Petasis, G. A. Olah, *J. Org. Chem.* **2001**, *66*, 633.

88 For *ipso*-nitration of arylboronic acids, see: S. Salzbrunn, J. Simon, G. K. S. Prakash, N. A. Petasis, G. A. Olah, *Synlett* **2000**, 1485.

89 B. Jiang, Q.-F. Wang, C.-G. Yang, M. Xu, *Tetrahedron Lett.* **2001**, *42*, 4083.

90 B. Witulski, N. Buschmann, U. Bergsträßer, *Tetrahedron* **2000**, *56*, 8473.

91 T. Konno, J. Chae, T. Tanaka, T. Ishihara, H. Yamanaka, *Chem. Commun.* **2004**, 690.

92 I. Pergament, M. Srebnik, *Org. Lett.* **2001**, *3*, 217.

93 C. Morrill, R. H. Grubbs, *J. Org. Chem.* **2003**, *68*, 6031.

94 H. E. Blackwell, D. J. O'Leary, A. K. Chatterjee, R. A; Washenfeldes, D. A. Bussmann, R. H. Grubbs, *J. Am. Chem. Soc.* **2000**, *122*, 58.

95 J. T. Njardarson, K. Biswas, S. J. Danishefsky, *Chem. Commun.* **2002**, 2759.

96 P. G. Garcia, E. Hohn, J. Pietruszka, *J. Organomet. Chem.* **2003**, *680*, 281.

97 G. A. Molander, C. R. Bernardi, *J. Org. Chem.* **2002**, *67*, 8424.

98 G. A. Molander, M. R. Rivero, *Org. Lett.* **2002**, *4*, 107.

99 A. B. Charette, A. Giroux, *J. Org. Chem.* **1996**, *61*, 8718.

100 A. B. Charette, R. P. De Freitas-Gil, *Tetrahedron Lett.* **1997**, *38*, 2809.

101 S. A. Frank, H. Chen, R. K. Kunz, J. Schnaderbeck, W. R. Roush, *Org. Lett.* **2000**, *2*, 2691.

102 K. A. Scheidt, A. Tasaka, T. D. Bannister, M. D. Wendt, W. R. Roush, *Angew. Chem. Int. Ed.* **1999**, *38*, 1652.

103 J. D. White, R. Hanselmann, R. W. Jackson, W. J. Porter, Y. Ohba, T. Tiller, S. Wang, *J. Org. Chem.* **2001**, *66*, 5217.

104 J. T. Njardarson, K. Biswas, S. J. Danishefsky, *Chem. Commun.* **2002**, 2759.

105 A. K. Chatterjee, R. H. Grubbs, *Angew. Chem. Int. Ed.* **2002**, *41*, 3172.

106 H. E. Blackwell, D. J. O'Leary, A. K. Chatterjee, R. A. Washenfelder, D. A. Bussmann, R. H. Grubbs, *J. Am. Chem. Soc.* **2000**, *122*, 58.

107 N. A. Petasis, I. A. Zavialov, *J. Am. Chem. Soc.* **1997**, *119*, 445.

108 N. A. Petasis, Z. D. Patel, *Tetrahedron Lett.* **2000**, 41, 9607.

109 J. T. Njardarson, K. Biswas, S. J. Danishefsky, *Chem. Commun.* **2002**, 2759.

110 R. A. Batey, D. B. Mackay, V. Santhakumar, *J. Am. Chem. Soc.* **1999**, 121, 5075.

111 G. A. Molander, B. W. Katona, F. Machrouhi, *J. Org. Chem.* **2002**, *67*, 8416.

112 A.-S. Castanet, F. Colobert, T. Schlama, *Org. Lett.* **2000**, *2*, 3559.

113 C. H. Oh, S. H. Jung, *Tetrahedron Lett.* **2000**, *41*, 8513.

114 S. D. Goldberg, R. H. Grubbs, *Angew. Chem. Int. Ed.* **2002**, *41*, 807.

115 K. Takahashi, T. Ishiyama, N. Miyaura, *Chem. Lett.* **2000**, 982.

116 K. Takahashi, T. Ishiyama, N. Miyaura, *J. Organomet. Chem.* **2001**, *625*, 47.

117 P. V. Ramachandran D. Pratihar, D. Biswas, A. Srivastava, M. V. R. Reddy, *Org. Lett.* **2004**, *6*, 481.

118 J. R. Falck, M. Bondlela, J. Ye, S.-D. Cho, *Tetrahedron Lett.* **1999**, *40*, 5647.

119 J. E. Luithle, J. Pietruszka, *J. Org. Chem* **1999**, *64*, 8287.

120 J. E. A. Luithle, J. Pietruszka, *J. Org. Chem.* **2000**, *65*, 9194.

121 A. Kamatani, L. E. Overman, *J. Org. Chem.* **1999**, *64*, 8743.

122 A. Kamatani, L. E. Overman, *Org. Lett.* **2001**, *3*, 1229.

123 H. Ito, H. Yamanaka, J.-I. Tateiwa, A. Hosomi, *Tetrahedron Lett.* **2000**, *41*, 6821.

124 K. Takahashi, T. Ishiyama, N. Miyaura, *Chem. Lett.* **2000**, 982.

125 K. Takahashi, T. Ishiyama, N. Miyaura, *J. Organomet. Chem.* **2001**, *625*, 47.

126 G. W. Kabalka, B. C. Das, S. Das, *Tetrahedron Lett.* **2001**, *42*, 7145.

127 G. A. Molander, C-S. Yun, M. Ribagorda, B. Biolatto, *J. Org. Chem.* **2003**, *68*, 5534.

128 G. A. Molander, T. Ito, *Org. Lett.* **2001**, *3*, 393.

129 G. Zou. Y. K. Reddy, J. R. Falck, *Tetrahedron Lett.* **2001**, *42*, 7213.

130 M. R. Netherton, C. Dai, K. Neuschütz, G.C. Fu, *J. Am. Chem. Soc.* **2001**, *123*, 10099.

131 J. H. Kirchhoff, C. Dai, G. C. Fu, *Angew. Chem. Int. Ed.* **2002**, *41*, 1945.

132 M. R. Netherton, G. C. Fu, *Angew. Chem. Int. Ed.* **2002**, *41*, 3910.

133 D. J. Cardenas, *Angew. Chem. Int. Ed.* **2003**, *42*, 384.

134 P. N. Collier, A. D. Campbell, I. Patel, T. M. Raynham, R. J. K. Taylor, *J. Org. Chem.* **2002**, *67*, 1802; For similar synthesis of amino acids, see also: M. Sabat, C. R. Johnson, *Org. Lett.* **2000**, *2*, 1089.

135 J. S. Potuzak, D. S. Tan, *Tetrahedron Lett.* **2004**, *45*, 1797.

136 B. A. Johns, Y. T. Pan, A. D. Elbein, C. R. Johnson, *J. Am. Chem. Soc.* **1997**, *119*, 4856.

137 B. A. Johns, C. R. Johnson, *Tetrahedron Lett.* **1998**, *39*, 749.

138 D. Trauner, J. B. Schwarz, S. J. Danishefsky, *Angew. Chem. Int. Ed.* **1999**, *38*, 3542.

139 D. Trauner, S. J. Danishefsky, *Tetrahedron Lett.* **1999**,*40*, 6513.

140 M. W. Carson, G. Kim, M. F. Hentemann, D. Trauner, S. J. Danishefsky, *Angew Chem. Int. Ed.* **2001**, *40*, 4450.

141 M. W. Carson, G. Kim, S. J. Danishefsky, *Angew. Chem. Int. Ed.* **2001**, *40*, 4453.

142 A. Balog, C. Harris, K. Savin, X.-G. Zhang, T.-C. Chou, S. J. Danishefsky, *Angew. Chem. Int. Ed.* **1998**, *37*, 2675.

143 D. Meng, P. Bertinato, A. Balog, D.-S. Su, T. Kamenecka, E. J. Sorensen, S. J. Danishefsky, *J. Am. Chem. Soc.* **1997**, *119*, 10073.

144 B. Zhu, J. S. Panek, *Org. Lett.* **2000**, *2*, 2575.

145 C. B. Lee, T. C. Chou, X.-G. Zhang, Z.-G. Wang, S. D. Kuduk, M. D. Chappell, S. J. Stachel, S. J. Danishefsky, *J. Org. Chem.* **2000**, *65*, 6525.

146 C. B. Lee, Z. Wu, F. Zhang, M. D. Chappell, S. J. Stachel, T. C. Chou, Y. Guan, S. J. Danishefsky, *J. Am. Chem. Soc.* **2001**, *123*, 5249.

147 M. D. Chappell, C. R. Harris, S. D. Kuduk, A. Balog, Z. Wu, F. Zhang, C. B. Lee, S. J. Stachel, S. J. Danishefsky, T.-C. Chou, Y. Guan, *J. Org. Chem.* **2002**, *67*, 7730.

148 K.-H. Altmann, G. Bold, G. Caravatti, D. Denni, A. Flörsheimer, A. Schmidt, G. Rihs, M. Wartmann, *Helv. Chim. Acta* **2002**, *85*, 4086.

149 M. Sasaki, H. Fuwa, M. Inoue, K. Tachibana, *Tetrahedron Lett.* **1998**, *39*, 9027.

150 M. Sasaki, K. Noguchi, H. Fuwa, K. Tachibana, *Tetrahedron Lett.* **2000**, *41*, 1425.

151 M. Sasaki, C. Tsukano, K. Tachibana, *Org. Lett.* **2002**, *4*, 1747.

152 H. Fuwa, N. Kainuma, K. Tachibana, M. Sasaki, *J. Am. Chem. Soc.* **2002**, *124*, 14983.

153 H. Fuwa, M. Sasaki, K. Tachibana, *Tetrahedron Lett.* **2000**, *41*, 8371.

154 M. Sasaki, M. Ishikawa, H. Fuwa, K. Tachibana, *Tetrahedron* **2002**, *58*, 1889.

155 M. Sasaki, H. Fuwa, M. Ishikawa, K. Tachibana, *Org. Lett.* **1999**, *1*, 1075.

156 H. Takakura, M. Sasaki, S. Honda, K. Tachibana, *Org. Lett.* **2002**, *4*, 2771.

157 B. M. Trost, C. Lee, *J. Am. Chem. Soc.* **2001**, *123*, 12191.

158 B. M. Trost, C. H. Lee, *J. Am. Chem. Soc.* **1998**, *120*, 6818.

159 N. Miyaura, M. Ishikawa, A. Suzuki, *Tetrahedron Lett.* **1992**, *33*, 2571.

160 J. A. Soderquist, G. Leon, J. C. Coleberg, I. Martinez, *Tetrahedron Lett.* **1995**, *36*, 3119.

161 S. R. Chemler, S. J. Danishefsky, *Org. Lett.* **2000**, *2*, 2695.

162 P. J. Mohr, R. L. Halcomb, *J. Am. Chem. Soc.* **2003**, *125*, 1712.

163 N. C. Kallan, R. L. Halcomb, *Org. Lett.* **2000**, *2*, 2687.

164 C. R. Johnson, M. P. Braun, *J. Am. Chem. Soc.* **1993**, *115*, 11014.

165 M. Bauer, M. E. Maier, *Org. Lett.* **2002**, *4*, 2205.

166 A. Gagnon, S. J. Danishefsky, *J. Angew. Chem. Int. Ed.* **2002**, *41*, 1581.

4
Polyfunctional Magnesium Organometallics for Organic Synthesis

Paul Knochel, Arkady Krasovskiy, and Ioannis Sapountzis

4.1
Introduction

Since their discovery at the beginning of the last century by Victor Grignard, [1] organomagnesium reagents have played a pivotal role in synthetic organic and organometallic chemistry. This pioneering work was honored 1912 with the Nobel Price – *"for the discovery of the so-called Grignard reagent, which in recent years has greatly advanced the progress of organic chemistry"* – and this statement still holds true today. Their easy synthesis, good stability and their excellent reactivity towards a wide range of different electrophiles made Grignard reagents one of the most powerful means for carbon–carbon bond formation used by generations of chemists. Furthermore, organomagnesium reagents have found numerous applications in industrial processes and a variety of these organometallics have become commercially available.

Besides their use in nucleophilic addition or substitution reactions, Grignard reagents can act as a base, transfer an electron in a single-electron transfer processes (SET) to other organic molecules or generate another Grignard reagent in a halogen–magnesium exchange reaction. The capability of organomagnesium reagents to undergo transmetallation reactions with a variety of main group- and transition-metal salts, particularly to organocopper reagents, [2] opened new synthetic avenues in organic chemistry. In addition, the work of Kharasch [3] and later, the nickel-catalysis of Kumada [4] as well as Corriu [5] are often regarded as the first reactions of modern cross-coupling chemistry [6,7,8]. Several comprehensive reviews and books have been published, encompassing the preparation and use of Grignard reagents [9] as well as the chemical and physical properties, [10] mechanistic investigations of the formation [11] and studies of the structures in solution and in solid state [10d,12].

In this chapter, we will focus on the formation of *functionalized* alkyl-, alkenyl-, aryl- and heteroaryl-magnesium halides and describe the scope and limitations of their applications in carbon–carbon and carbon–heteroatom bond forming reactions [13].

Where possible we will try to point out the advantages of Grignard reagents and compare them to other organometallics. A separate section will deal with recent developments in transition-metal-catalyzed reactions, where Grignard reagents again occupy a central position, especially in the interesting field of iron-catalyzed reactions.

4.2
Methods of Preparation of Grignard Reagents and their Uncatalyzed Reactions

4.2.1
Direct Oxidative Addition of Magnesium to Organic Halides

Grignard reagents are sensitive to air and moisture. An inert atmosphere is therefore always advantageous for their preparation and further reactions. The usual method used for the preparation of organomagnesium reagents is the reaction of organic halides with magnesium metal in a polar, aprotic solvent like THF or diethyl ether (Scheme 4.1, Eq. 1). For large-scale industrial process, [14] these volatile and highly flammable ethers represent safety hazards and can be substituted by "butyl diglyme" ($C_4H_9OC_2H_4OC_4H_9$) that possesses a high flashpoint (118 °C) and a low water solubility.

$$RX \xrightarrow[\text{THF or Et}_2\text{O}]{\text{Mg}} RMgX \qquad (1)$$

$$2\ RMgX \rightleftharpoons R_2Mg + MgX_2 \quad (2)$$

Scheme 4.1 Synthesis of Grignard reagents by oxidative addition (Eq. 1) and Schlenk equilibrium (Eq. 2).

Magnesium turnings or powder are usually covered with a small amount of solvent and to this suspension a solution of the organic halide is added. This reaction is exothermic and cooling is often necessary after the induction period. The magnesium metal is, as received or after exposure to air, covered with an "oxide" layer (mainly $Mg(OH)_2$), [11e] that passivates the metal. This "oxide" layer is responsible for the induction period that is normally observed in the synthesis of Grignard reagents. An activation with the promoter 1,2-dibromoethane can help to reduce this induction time. 1,2-Dibromoethane reacts with magnesium to ethene and $MgBr_2$. It also accelerates the formation of the Grignard reagent, leading to an activated magnesium metal surface [15]. The mechanism of this reaction is not yet fully clarified, but a radical mechanism is generally accepted [11,16].

In solution, a Grignard reagent (RMgX) is in equilibrium (so-called Schlenk equilibrium, Scheme 4.1, Eq. 2) with R_2Mg and MgX_2, depending on temperature, solvent and the anion X. This equilibrium can be shifted to the right side, by pouring a solution of RMgX into dioxane, which does not dissolve MgX_2, leaving R_2Mg in solution. Most Grignard reagents (RMgX) or diorganomagnesium compounds crystallize with four-coordinated Mg in a distorted tetrahedron, but if the ligands can fit, five- (CH_3MgBr in THF) and six- ($MgBr_2$ in THF) coordinated structures can be observed [17]. All experimental evidence indicates similar coordination numbers in solution, emphasizing the role of coordinating etheral solvents in Grignard reagents.

The presence of sensitive functional groups makes the preparation of Grignard reagents more complicated and many functional groups are not tolerated with this insertion method. If, however, the direct oxidative addition is conducted at low temperatures with activated metals, such as Rieke magnesium (Mg^*), the preparation of functionalized Grignard reagents is possible, but generally still shows limitations for the functional-group tolerance (Scheme 4.2) [18].

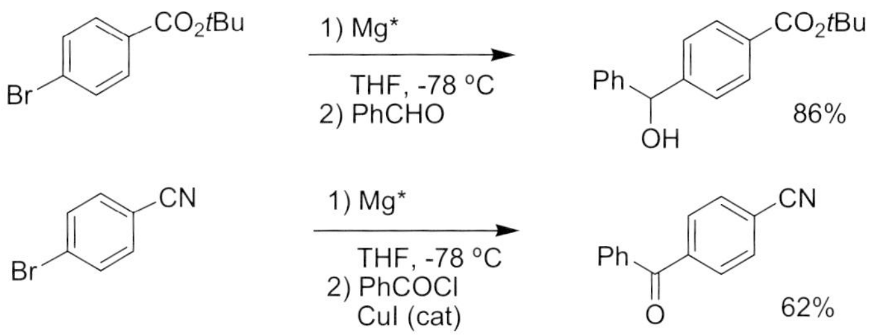

Scheme 4.2 Preparation of functionalized Grignard reagents using Rieke-magnesium (Mg^*).

These Grignard reagents could be trapped directly with benzaldehyde or in the presence of a catalytic amount of CuI (10 mol%) reacted with benzoyl chloride.

4.2.2
Metalation Reactions with Magnesium Amides

The direct deprotonation of organic molecules with kinetically poor bases, such as organolithium or magnesium compounds is limited. However, the addition of amines or the presence of directing groups, breaking the aggregation of these reagents can lower this barrier. Indeed, various alkyllithiums or lithium dialkyla-mides have been used for the directed *ortho*-metallation and metallation of aromatic and heteroaromatic compounds [19]. The major drawback of these lithium reagents is the high reactivity towards electrophilic groups excluding the presence of many sensitive functionalities. In comparison to their lithium counterparts, the use of magnesium dialkylamides or Grignard reagents in metallation reactions has received little attention.

The direct metallation of organic substrates with alkylmagnesium halides requires a greater kinetic acidity for the C–H bond than for the conjugated acid of the Grignard reagent. Strongly coordinating solvents like HMPT help to promote these reactions. One of the major applications is the metallation of acetylene derivatives, like the monometallation of acetylene by *n*BuMgCl to form ethynylmagnesium chloride [20].

Unlike to their lithium analogues, Hauser bases (R_2NMgBr) are much more stable in THF (up to reflux conditions). Eaton reported in 1989 the use of magnesium bis(2,2,6,6-tetramethylpiperamide), $(TMP)_2Mg$, as a selective metalating reagent (Scheme 4.3) [21].

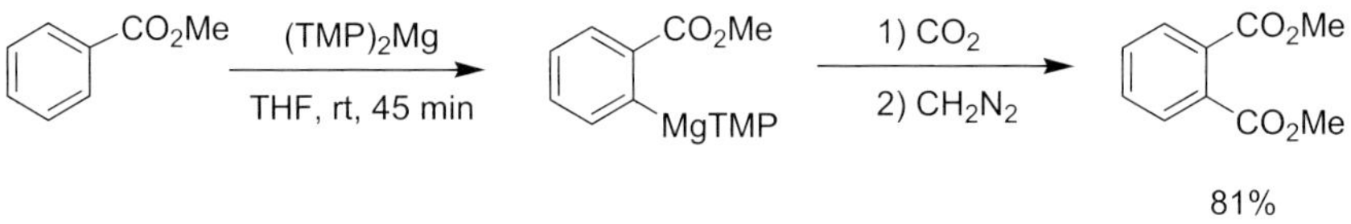

Scheme 4.3 Selective *ortho*-magnesiation of methyl benzoate.

Of special interest is the ease with which *ortho*-magnesiation reactions can be accomplished in the presence of an ester function, which is normally susceptible to nucleophilic attack using conventional Li-based reagents [22]. This methodology was applied to the deprotonation of cyclopropylamides [23] as well as to numerous heterocycles (Scheme 4.4) [24].

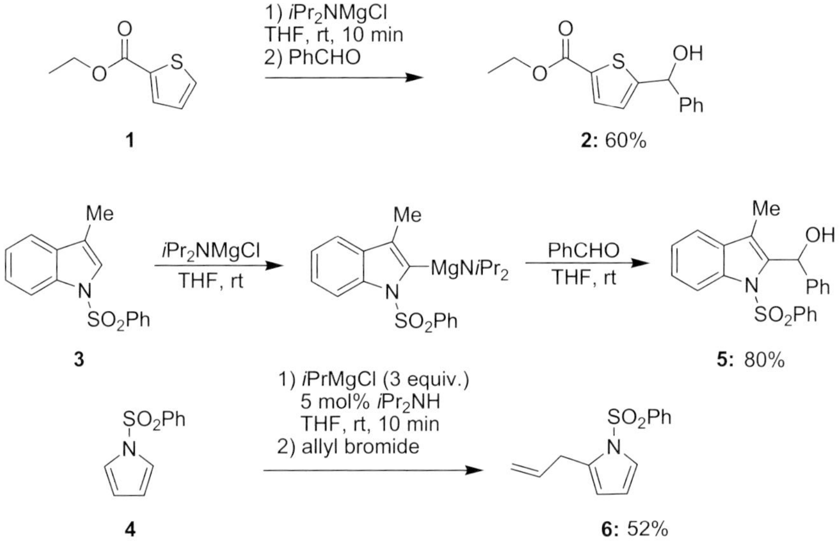

Scheme 4.4 Selective *ortho*-magnesiation of heterocycles 1,3 and 4.

Thus, ethyl thiophene-2-carboxylate (**1**) is selectively metalated in the 5-position using iPr_2NMgCl, readily prepared from $nBuMgCl$ and iPr_2NH, and afterwards reacts with benzaldehyde furnishing **2** in 60% yield (Scheme 4.4). Nitrogen-heterocycles, such as indole **3** and pyrrole **4** undergo a metallation reaction as well, leading, for example, to the magnesiated intermediates, which can react with a variety of electrophiles, leading to the functionalized heterocycles **5** and **6** in 80% and 52% yield, respectively. In the case of the *N*-phenylsulfonylpyrrole (**4**) it turns out that a catalytic amount of iPr_2NH is sufficient to accelerate the deprotonation reaction. In addition, magnesium bisamides are used for the regio- and stereoselective formation of enolates [25] and the use of optically pure magnesium amides opens the field of asymmetric synthesis to this versatile substance class [26].

4.2.3
The Halogen–Magnesium Exchange Reaction

The halogen–lithium exchange reaction discovered by Wittig [27] and Gilman [28] allows the preparation of a broad range of organolithium compounds and has become one of the most important ways for the preparation of aromatic, heteroaromatic and alkenyl lithium compounds, starting from the commercially available alkyllithium reagents and the corresponding organic halides, mainly bromides and iodides [29]. Although this reaction is very fast and normally proceeds at low temperatures (typically $-78\,°C$) the functional-group tolerance is only moderate. In contrast, the halogen–magnesium exchange has been found to be the method of choice for preparing new functionalized organomagnesium reagents of considerable synthetic utility. The major reason for this great functional-group tolerance of Grignard reagents is that the reactivity of carbon–magnesium bonds is strongly dependent on the reaction temperature. Only reactive electrophiles like aldehydes and most ketones react rapidly at temperatures below $0\,°C$. Performing the halogen–magnesium exchange at temperatures below $0\,°C$ has therefore the potential for the preparation of magnesium organometallics bearing reactive and sensitive functional groups.

4.2.3.1   Early Studies
The first example of a bromine–magnesium exchange reaction was briefly reported in 1931 by Prévost [30]. Thus, the reaction of cinnamyl bromide **7** with EtMgBr furnished cinnamylmagnesium bromide **8** although only in modest yield (Scheme 4.5). Similarly, Urion reported the preparation of cyclohexylmagnesium bromide **9** via a Br/Mg-exchange [31].

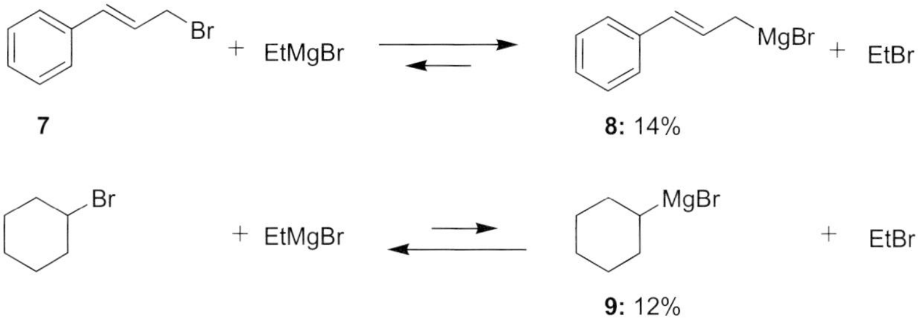

Scheme 4.5 First examples of a bromine–magnesium exchange.

The halogen–magnesium exchange is an equilibrium process favoring the formation of the most stable organomagnesium compound. To shift this equilibrium to the desired side, the resulting organomagnesium species has to be more stable than the Grignard reagent used for the exchange reaction ($sp > sp^2$(vinyl) $> sp^2$(aryl) $> sp^3$(prim.) $> sp^3$(sec.)). Although the mechanism of the exchange reaction is not fully clarified, a halogen-ate complex is believed to be an intermediate, as proposed for the halogen–lithium exchange [32].

One of the first synthetically useful procedures, employing a halogen–magnesium exchange reaction, was reported by McBee and coworkers who were successful in preparing perfluoroalkylmagnesium halides of type **10** starting from the perfluorinated iodide **11** and phenylmagnesium bromide (Scheme 4.6) [33].

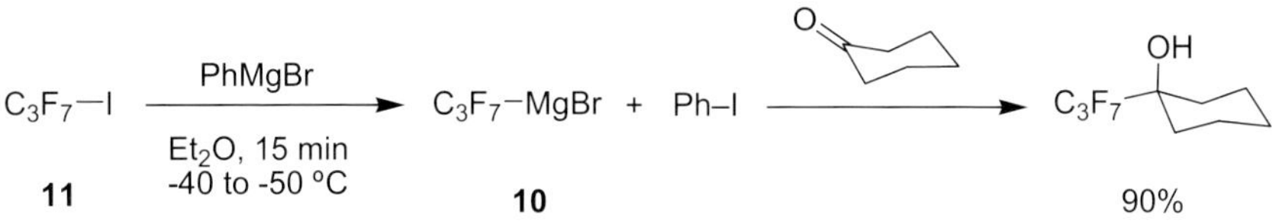

Scheme 4.6 Synthesis and reaction of heptafluoropropylmagnesium bromide (10).

This procedure had significant advantages compared to the oxidative addition, such as higher yields or less side reactions and is still one of the best methods for the synthesis of perfluorinated Grignard reagents [34].

The halogen–magnesium exchange reaction was later used by Villiéras, who developed a general approach to magnesium carbenoids [35]. He showed, that the reaction of iPrMgCl with CHBr$_3$ at –78 °C furnishes the corresponding magnesium carbenoid **12** that could be trapped with chlorotrimethylsilane leading to product **13** in 90% yield (Scheme 4.7). This pioneering work paved the way to the systematic study of magnesium carbenoids [36] demonstrating that the halogen–magnesium exchange rate is enhanced by the presence of electronegative substituents.

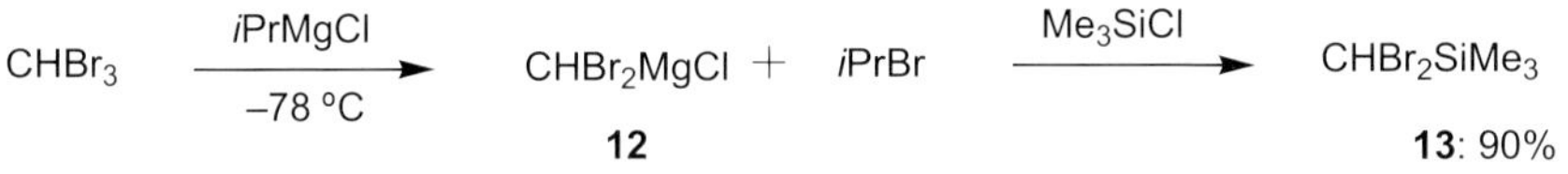

Scheme 4.7 Preparation of magnesium carbenoid 12 *via* bromine–magnesium exchange.

This behavior was confirmed a few years later by the work of Tamborski who showed that the electronic properties of both, the halogen atom and the organic molecule play an important role in the formation rate of the new Grignard reagent [37]. Only for very electron-poor systems, such as the tetra- or pentafluorobenzenes, was the exchange of a chlorine possible, requiring elevated temperatures and longer reaction times then for the corresponding bromines and iodines. The reactivity order (I>Br>Cl>>F) is influenced by the bond-strength, the electronegativity and the ease of polarizability of the halide. For these reasons, aryl iodides are usually used as starting materials, although the use of bromides would be advantageous from an economical point of view. For instance, the exchange reaction of 1-chloro-2,3,4,5,6-tetrafluorobenzene (**14a**) with EtMgBr, requires 1 h at room temperature to reach complete conversion to the Grignard reagent, whereas the corresponding bromo- and iodo-congeners **14b** and **14c** react already at 0 °C within 1 min to compound **15** (Scheme 4.8).

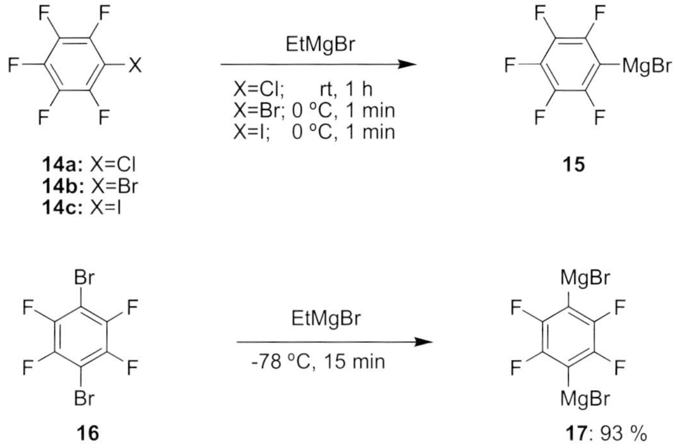

Scheme 4.8 Preparation of polyhalogenated Grignard reagents 15 and 17.

It was shown that 1,4-dibromo-2,3,5,6-tetrafluorobenzene (**16**) is readily converted to the corresponding 1,4-dimagnesium species **17** with EtMgBr (Scheme 4.8). Similarly, Furukawa showed that 2-iodopyridine leads to the corresponding Grignard reagent within 0.5 h by reaction with EtMgBr at 25 °C [38]. These early results indicate the synthetic potential of the halogen–magnesium exchange reaction and recent developments will be discussed in the following chapters [39].

4.2.3.2 The Preparation of Functionalized Arylmagnesium Reagents

Functionalized aryl iodides react readily with *i*PrMgBr or *i*PrMgCl in THF below 0 °C leading to a range of functionalized arylmagnesium iodides [40]. Sensitive carbonyl group derivatives like nitriles, esters or amides are well tolerated. Typically, the treatment of methyl 4-iodobenzoate (**18**) with *i*PrMgBr in THF at –20 °C for 30 min produces the functionalized Grignard reagent **19**, which is stable for several hours below 0 °C and reacts smoothly with aldehydes at –20 °C leading to the expected alcohols **20a** and **b** in 72–83% yield (Scheme 4.9) [41].

Aromatic iodides bearing electron-donating groups, such as compound **21**, undergo the iodine–magnesium exchange as well, but usually higher temperatures (25 °C) and elongated reaction times are necessary [40,42]. The addition of the resulting arylmagnesium species to diethyl *N*-Boc-iminomalonate **22** [43] furnishes adduct **23** in 79% yield. Saponification followed by decarboxylation leads to α-amino-acid **24** in 81% yield (Scheme 4.10) [42].

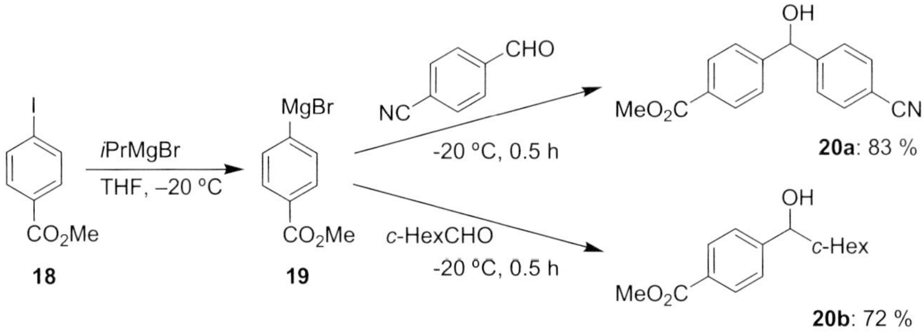

Scheme 4.9 The reaction of ester containing arylmagnesium reagent 19 with aldehydes.

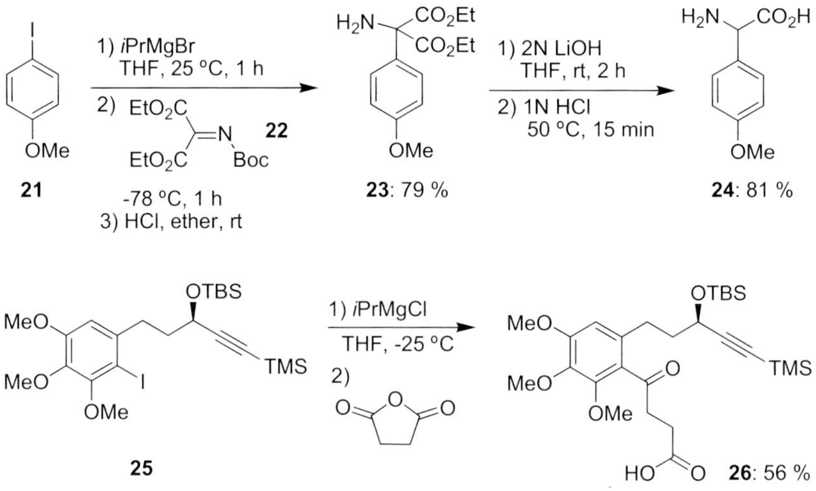

Scheme 4.10 Reactions of electron-rich arylmagnesium reagents.

Schmalz showed in their synthesis of colchicine, [44] that even building block 25, possessing a very electron-rich aromatic system with three methoxy groups and an additional alkyl chain, undergoes an I/Mg exchange reaction under very mild conditions. Thus, the addition of *i*PrMgCl at −25 °C furnishes the magnesiated intermediate that reacts with succinic anhydride leading to compound 26 in 56% yield (Scheme 4.10).

As already mentioned, the use of bromides would be advantageous, but the Br/Mg-exchange reaction is significantly slower than the I/Mg-exchange. Using *i*PrMgCl or *i*PrMgBr, the exchange is sufficiently fast below 0 °C only for systems bearing electron-withdrawing groups [45,46]. A few further examples are reported in the literature, where aryl bromides are used as starting materials.

For example, Leazer from Merck Process Research (USA) found the Br/Mg-exchange reaction the most reliable and safe method for the preparation of Grignard reagent 27, a valuable building block for the synthesis of a new neurokin 1 receptor agonist (Scheme 4.11) [47]. It is known, that trifluoromethylphenyl

Grignard reagents and some polyhalogenated Grignard reagents can detonate at high temperatures like their lithium counterparts [48]. This makes a low-temperature process more eligible for this potent substance class, frequently encountered in pharmaceutical drugs and synthetic intermediates.

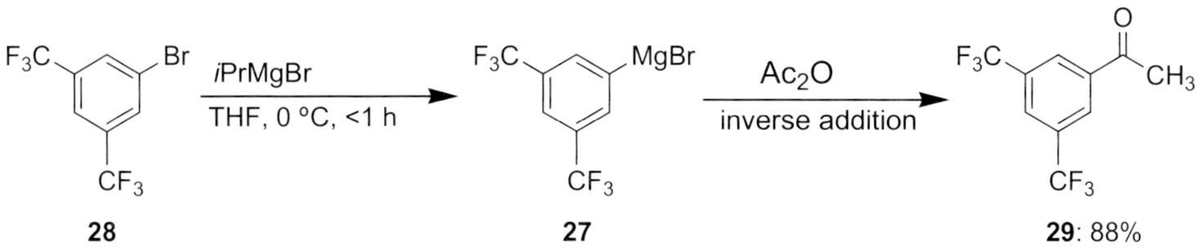

Scheme 4.11 Synthesis of Grignard reagent 27 via a Br/Mg-exchange reaction.

The use of iPrMgBr allows the preparation of Grignard reagent **27** at 0 °C within 30 min starting from the readily available aryl bromide **28**. Inverse addition to a solution of acetic anhydride furnishes substituted acetophenone **29** in 88% yield. This procedure is also suitable for a multikilogram scale-up.

Polyfunctional aromatic bromides, such as **30** [46] and **31** [45] (Scheme 4.12) bearing a chelating group at the *ortho*-position rapidly undergo the Br/Mg-exchange. The chelating group complexes iPrMgBr prior to the Br/Mg-exchange, is facilitating this exchange *via* intramolecular reaction. Thus, the dibromide **30** reacts regioselectively in *ortho*-position to the amidine functionality leading to reagent **32**. After the addition to 2-butylacrolein, the expected alcohol **33** is formed in 68% yield [46]. An oxygen chelating functional group like an ethoxymethyl group in the aryl bromide **31** enhances the Br/Mg-exchange rate allowing the preparation of the magnesium derivative **34** at −30 °C within 2 h. In the presence of a catalytic amount of CuCN·2LiCl (10 mol%), [49] the Grignard reagent **34** undergoes an allylation with allyl bromide leading to the aromatic nitrile **35** in 80% yield [45].

The electron-releasing methoxy function is a less effective chelating group and requires higher reaction temperature, showing the limitations of this method. Thus, 2,4-dibromoanisole **36** is converted to the corresponding arylmagnesium compound **37** by the treatment with iPrMgCl (2 equiv) in THF at 40 °C for 5 h. After the addition of CO_2, the corresponding carboxylic acid **38** is obtained in 90% yield (Scheme 4.12) [39d].

Although several examples are given above allowing a Br/Mg-exchange reaction, a more general, low-temperature process is highly desirable since the higher temperatures used in this exchange reaction preclude the presence of many functional groups. In addition, the elimination of HBr from the alkyl bromide formed during the exchange process is favored.

A promising solution to this problem was recently reported by Knochel and Krasovskiy, who found that lithium salts can accelerate the Br/Mg-exchange reaction considerably [50]. A stoichiometric amount of LiCl was best, resulting in the formation of a stable Grignard reagent iPrMgCl·LiCl. This reagent is superior for the

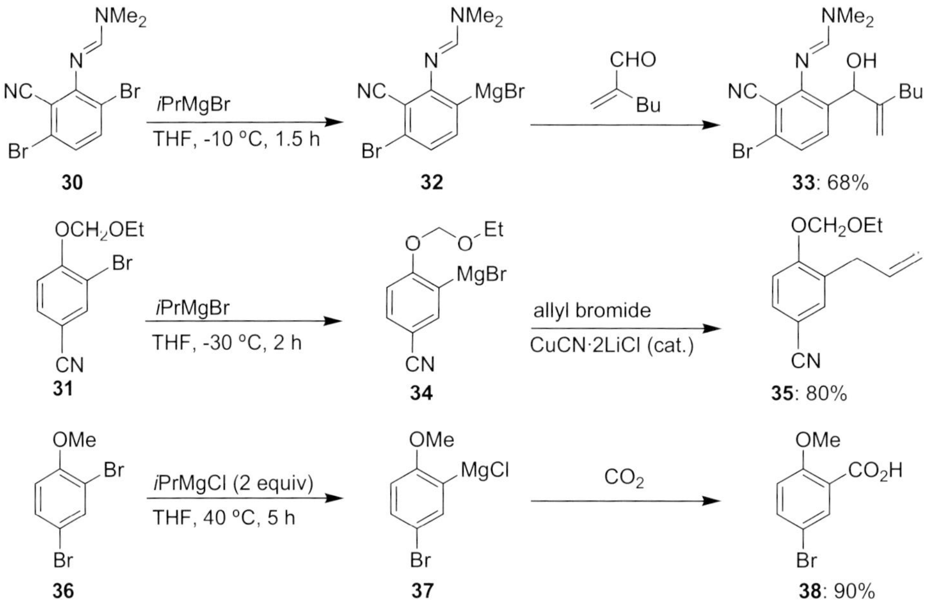

Scheme 4.12 Br/Mg-exchange of functionalized aromatic bromides using *i*PrMgBr or *i*PrMgCl.

exchange reaction and considerably enhances the scope of the Br/Mg-exchange reaction. The addition of LiCl breaks the aggregates of the dimeric *i*PrMgCl producing a more reactive complex. This reagent can be used for the preparation of a variety of substrates bearing different functional groups (Scheme 4.13). Thus, 3-bromobenzonitrile (**39**) undergoes a fast Br/Mg-exchange at −10 °C, leading to the 3-magnesiated species **40** that reacted upon transmetallation to CuCN·2LiCl with benzoyl chloride furnishing compound **41** in 88% yield (Scheme 4.13).

The use of *i*PrMgCl·LiCl also allows for the preparation of Grignard reagent **42** at −50 °C without any formation of the 4-bromo dehydrobenzene side product that is usually observed when performing this reaction with a lithium reagent or at higher temperatures (Scheme 4.13). Reaction with *t*BuCHO furnished product **43** in 89% yield.

New types of Grignard reagents such as 2-bromocyclopentenylmagnesium chloride (**44**) can be readily prepared by the reaction of 1,2-dibromocyclopentene (**45**) with *i*PrMgCl·LiCl (20 °C, 24 h). This reagent has a remarkable stability and displays a good reactivity toward various electrophiles such as an aldehyde leading to the alcohol **46**. Copper-catalyzed reactions require the formation of an intermediate mixed Grignard reagent of the type RMgCH$_2$SiMe$_3$. After the addition of benzoyl chloride in the presence of CuCN·2LiCl, the bromoenone **47** is obtained in 81% yield. In the absence of the addition of TMSCH$_2$Li, but in the presence of *i*PrMgCl (1 equiv.) and catalytic amount of CuCl$_2$·2LiCl (0.5 mol%) a rearrangement occurs providing after a quenching with benzaldehyde, the allylic alcohol **48** in 91% yield (Scheme 4.14) [51].

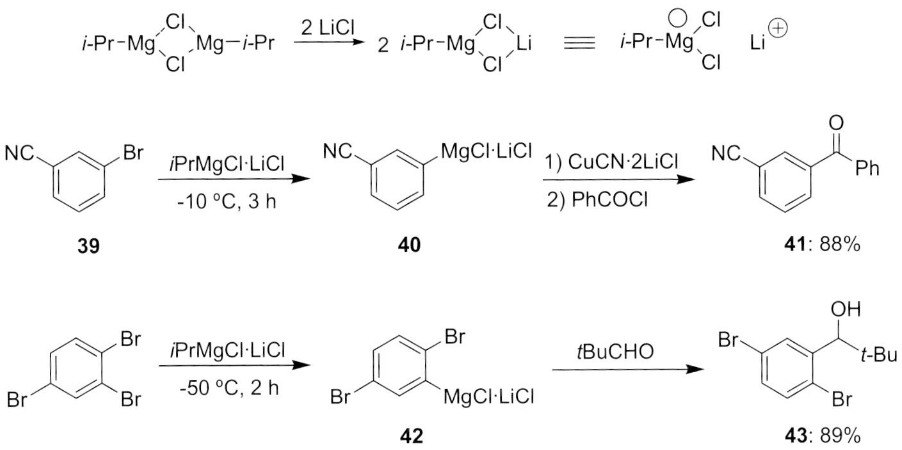

Scheme 4.13 Br/Mg-exchange using iPrMgCl·LiCl.

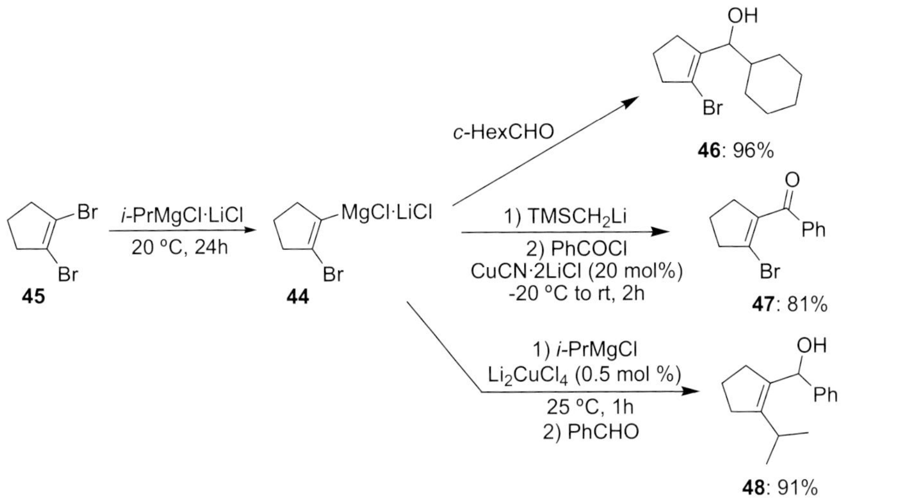

Scheme 4.14 Br/Mg-exchange of 1,2-dibromocyclopentene (45) using iPrMgCl·LiCl.

A wide range of basic nitrogen functionalities are compatible with the iodine–magnesium exchange and many protecting groups are well tolerated. For example, diallylaniline 49 is allylated via the intermediate Grignard reagent 50 leading to the functionalized aniline derivative 51 in 81% yield (Scheme 4.15) [41].

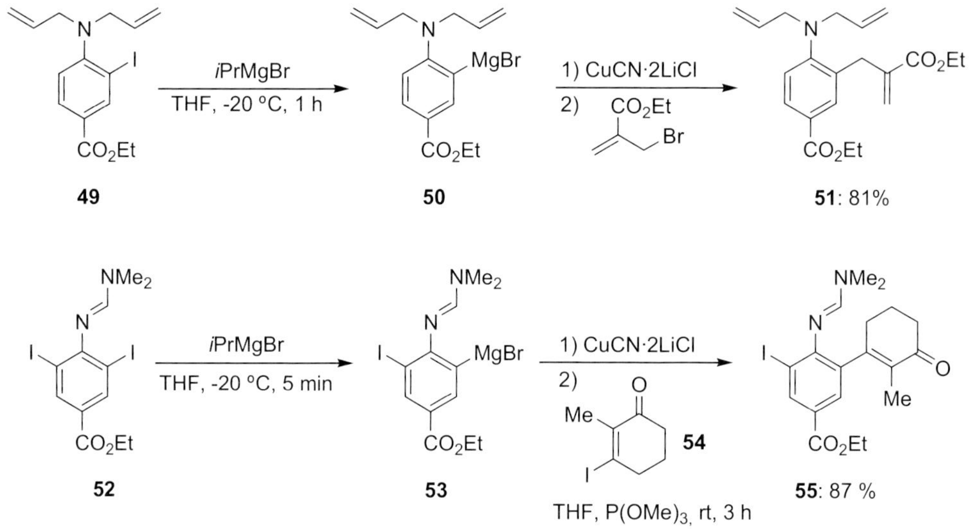

Scheme 4.15 Arylmagnesium compounds containing nitrogen functional groups.

The more labile amidine [52] protecting group is also compatible with a magnesium–halogen exchange and is a convenient mean for introducing primary amines in a molecule. The diiodo-amidine **52** is converted within 5 min at −20 °C into the arylmagnesium species **53**. Remarkably, only one exchange reaction takes place. After the first I/Mg-exchange the electron density of the aromatic ring increases and thus hampers a second exchange. Transmetallation of **53** with CuCN·2LiCl [49] provides the corresponding arylcopper derivative, which readily undergoes an addition-elimination reaction with α,β-unsaturated carbonyl compound **54**, leading to product **55** in 87% yield (Scheme 4.15) [53].

Likewise, an imine is a suitable way to protect both, anilines and aromatic aldehydes (Scheme 4.16). Thus, 2-iodophenylenediamine **56** undergoes an iodine–magnesium exchange with iPrMgBr (2 equiv.) at −10 °C in 3 h leading to Grignard reagent **57**. Transmetallation to the copper derivative by treatment with CuCN·2LiCl [49] and subsequent allylation with allyl bromide gives the diimine **58** in 83% yield [54].

Whereas aryl iodides bearing an aldehyde group preferentially react with the aldehyde function during attempted iodine–magnesium exchange, the corresponding imine (**59**) undergoes a smooth exchange reaction leading to the Grignard reagent **60**. The addition of BiCl₃ followed by a silica gel column chromatographical purification, provides the resulting functionalized triarylbismuthane **61** (Scheme 4.16) [55]. The tedious introduction and removal of a protecting group can in principle be avoided for proton-donating groups through additional equivalents of base. Although halogen–metal exchange reactions on aryl halides bearing acidic protons have been successfully conducted with alkyllithium reagents, the low temperatures (−78 °C) and the considerable amounts of side-products make this methodology less attractive. The formation of unprotected

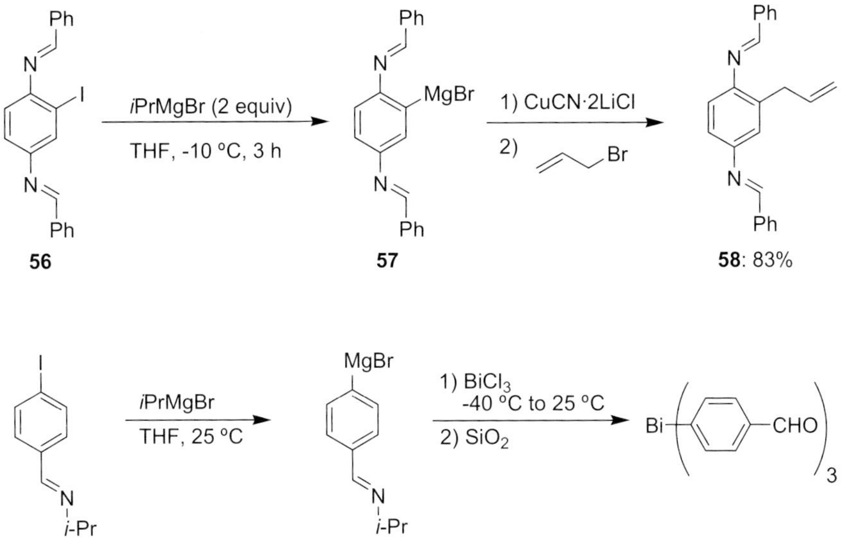

Scheme 4.16 Preparation of imino-arylmagnesium reagents **57** and **60**.

functionalized Grignard reagents can be easily accomplished (Scheme 4.17). Addressing the problem of intermolecular quenching, in the first step, a deprotonation of the acidic amine proton in **62** is accomplished with methyl- or phenyl-magnesium chloride. These two Grignard reagents only reluctantly undergo exchange reactions and lead to an intermediate of type **63**. In a second step, the I/Mg exchange reaction is carried out with *i*PrMgCl, leading to the desired Grignard reagent of type **64**. In particular, the successive addition of PhMgCl (–30 °C, 10 min) and *i*PrMgCl (–25 °C, 10 min) to the diiodoaniline **65** gives the dimagnesium derivative **66** that reacts in good yield with cinnamaldehyde leading to the polyfunctional benzylic alcohol **67** (Scheme 4.17) [56].

Furthermore, after transmetallation with CuCN · 2LiCl [49], Grignard reagent **69** reacts with propargyl bromide affording polyfunctionalized amine **70** in 89% yield. Other proton-donating groups are also compatible with this approach. Thus, Grignard reagents bearing a hydroxy group can be prepared by a deprotonation with MeMgCl · LiCl and subsequent exchange reaction with *i*PrMgCl. A variety of additional functional groups, such as an ester, a cyano group or a bromide are well tolerated. For example, 4-bromo-2,6-diiodophenol (**71**) furnishes Grignard reagent **72** under very mild conditions. Reaction with benzaldehyde leads to diol **73** in 84% yield (Scheme 4.18) [57].

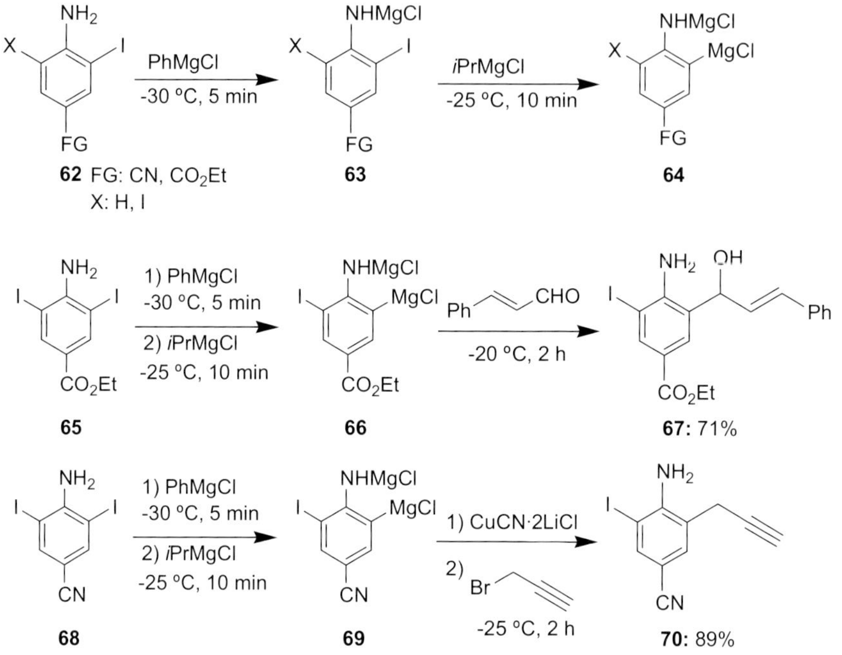

Scheme 4.17 Reactions of unprotected amino-arylmagnesium reagents.

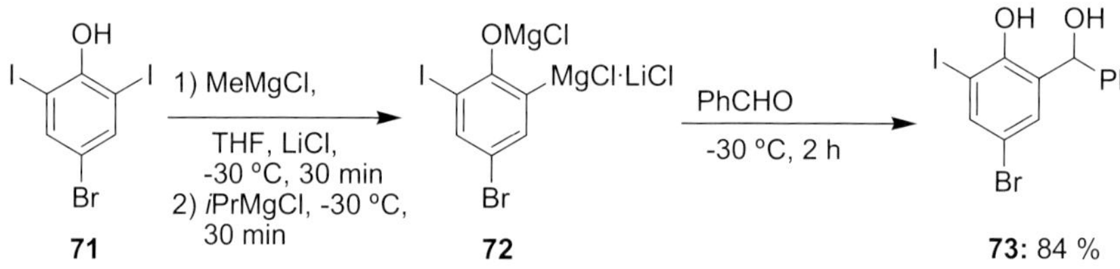

Scheme 4.18 Preparation of unprotected hydroxy-arylmagnesium reagents.

Unprotected acids, amides or benzylic alcohols can be tolerated as well using a combination of organo-magnesium and -lithium reagents [58].

Another very important nitrogen-containing functional group, which is not compatible with the direct oxidative addition of magnesium metal into a carbon–halogen bond, is the nitro group. Nitro groups are found in numerous fine chemicals, dyes, high-energy materials and biologically active substances, and many nitrogen substituents in an aromatic molecule are initially introduced by nitration [59]. Therefore, it can be regarded as a masked amino functionality and the easy transformation to a variety of derivatives allows the application of nitro chemistry to numerous total syntheses [60]. Due to the high electrophilicity of the nitro functionality, organometallics can trigger either nucleophilic attack or electron-transfer reactions. However, it has been shown that *ortho*-lithiated nitrobenzene is stable at very low temperature [61]. Interestingly, the corresponding zinc and copper species obtained by transmetallation with zinc or copper(I) salts, exhibit excellent stability and show, under appropriate reaction conditions, no tendency to undergo electron-transfer reactions [62].

Although *i*PrMgCl is the magnesium reagent of choice for performing an iodine–magnesium exchange, in the case of nitroarenes, the use of a less reactive organomagnesium compound is essential. A broad range of functionalized aryl-magnesium compounds bearing a nitro function in the *ortho*-position, can be prepared by an iodine–magnesium exchange using phenylmagnesium chloride as exchange reagent (Scheme 4.19) [63]. In particular, the nitro-substituted aryl iodide **74** undergoes a smooth I/Mg-exchange with phenylmagnesium chloride only in the *ortho*-position to the nitro group leading to Grignard reagent **75**. This excellent selectivity can be explained by the precoordination of phenylmagnesium chloride with the oxygen. Reaction with hexanal provides the benzylic alcohol **76** in 86% yield (Scheme 4.19).

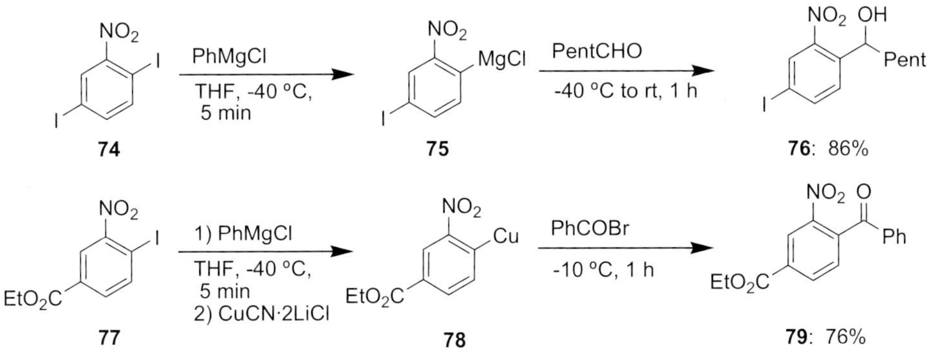

NO$_2$ PhMgCl   THF, -40 °C, 5 min   **74**   →   NO$_2$ MgCl   **75**   PentCHO   -40 °C to rt, 1 h   →   NO$_2$ OH Pent   **76**: 86%

NO$_2$ EtO$_2$C   1) PhMgCl   THF, -40 °C, 5 min   2) CuCN·2LiCl   **77**   →   NO$_2$ Cu EtO$_2$C   **78**   PhCOBr   -10 °C, 1 h   →   NO$_2$ O Ph EtO$_2$C   **79**: 76%

Scheme 4.19 Preparation of polyfunctional arylmagnesium compounds of type 69 bearing a nitro function.

A transmetallation of the Grignard reagent with CuCN·2LiCl [49] furnishes the corresponding copper reagent **78** that can react, for instance with benzoyl bromide, leading to ketone **79** in 76% yield (Scheme 4.19) [63].

The *ortho*-relationship between the carbon–magnesium bond and the nitro function is essential for a clean and fast exchange reaction. *Meta*- and *para*-substituted iodonitroarenes lead to unselective exchange reactions with addition of the organometallic species to the nitro group. The *ortho*-nitro-substitution pattern facilitates the I/Mg exchange by precomplexation of the Grignard reagent to the nitro function prior to I/Mg exchange. This precomplexation can be accomplished by other chelating groups, such as an ester group in **80** as well (Scheme 4.20). They accelerate the exchange reaction and stabilize the newly formed Grignard reagent **81**, allowing the preparation of *meta*- and *para*-magnesiated nitroarenes. Due to the sensitivity of nitro groups, lower reaction temperatures are necessary and the reaction is typically carried out at –78 °C showing a complete exchange reaction within 10 min. Addition of benzaldehyde and subsequent cyclization furnishes lactone **82** in 78% yield (Scheme 4.20) [64].

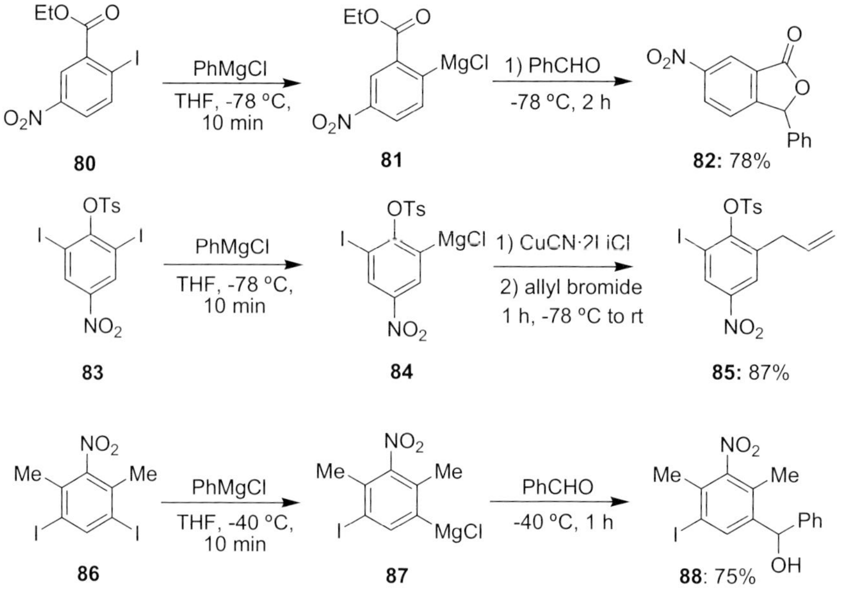

Scheme 4.20 Preparation of *meta* and *para*-nitroarylmagnesium compounds.

Likewise, diiodide **83** can be transferred into Grignard reagent **84** that upon transmetallation to copper reacts smoothly with allyl bromide to compound **85** in 87% yield. An I/Mg-exchange is also possible when the nitro group is sterically hindered. Thus, the reaction of the diiodonitrobenzene derivative **86** with PhMgCl (THF, –40 °C, 10 min) furnishes the corresponding Grignard compound **87** that reacts with benzaldehyde affording the expected benzyl alcohol **88** in 75% yield (Scheme 4.20) [64]. These results indicate that, contrary to general belief, one-electron-transfer reactions between nitro groups and organometallics, especially organomagnesium compounds, are often less favorable than the halogen–magnesium exchange reaction.

The triazene functionality finds more applications in organic synthesis [65]. This versatile functionality reacts with *i*PrMgCl when attempted an I/Mg-exchange on an iodoarene bearing a triazene functionality. However, by using the more reactive *i*PrMgCl·LiCl the exchange reaction can be performed at lower temperature (–40 °C) and is now compatible with the triazene function. Thus, the reaction of the iodotriazene **89** with *i*PrMgCl·LiCl provides the desired Grignard reagent **90** that undergoes a smooth addition-elimination with 3-iodo-2-cyclohexenone in the presence of CuCN·2LiCl providing the enone **91** in 88% yield. Since the triazene is a synthetic equivalent of an iodide functionality [65], the enone **91** is readily converted to the aryl iodide **92** by treatment with CH₃I [66]. Interestingly, the polyfunctional iodide **92** can be converted to the corresponding Grignard reagent after a transient protection with TMSCN in the presence of CsF in CH₃CN leading to the silylated cyanhydrin **93** in 90% yield. The highly active exchange reagent *i*PrMgCl·LiCl converts **93** to the corresponding Grignard reagent. After a copper-catalyzed acylation, a smooth deprotection with Bu_4NF regenerates the enone functionality leading to the diketone **94** in 87% yield (Scheme 4.21) [67].

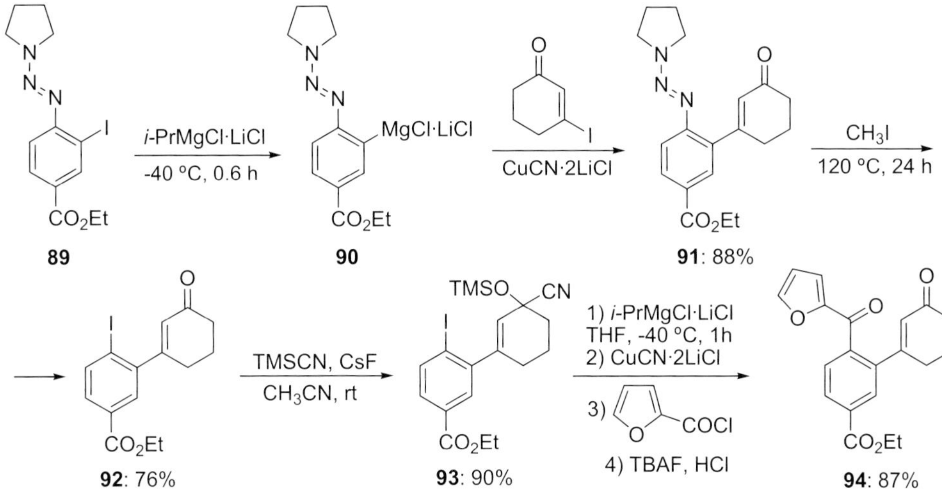

Scheme 4.21 Preparation and reaction of magnesiated triazenes.

Organoboronic esters are valuable reagents capable of undergoing many catalytic carbon–carbon bond formations in organic synthesis [68]. A general method for the preparation and functionalization of organoboron compounds is therefore of common interest. Although a variety of methods are known for the preparation of boronic acids and esters using either main group or transition metals, their further functionalization has received notably less attention. The selective metallation of an organoboron compound would allow the synthesis of bimetallic species that are very useful in multistep sequences. Using iPrMgCl · LiCl [50] as exchange reagent, the reaction becomes fast enough for aromatic iodides of type **95** at low temperatures ($< -20\,°C$) and no attack on the boronic ester is observed (Scheme 4.22).

Thus, the formation of Grignard reagents of type **96** possessing a pinacol borane (PinB) unit is possible and several additional functional groups are tolerated [69]. These bimetallic organomagnesium halides can react directly with an electrophile or be transmetalated to the corresponding organocopper reagents leading to the functionalized arylboronic esters **97a–c** in 70–96% yield (Scheme 4.22). Amazingly, starting from iodide **98**, a one-pot sequence consisting of an exchange reaction, followed by the addition to benzaldehyde of the magnesiated boronic ester and subsequent Suzuki cross-coupling furnishes biphenyl **99** in 73% overall yield (Scheme 4.22).

How far can this functional-group tolerance be extended? A keto group usually reacts with a Grignard reagent even at $-78\,°C$. In fact, iPrMgCl reacts with benzophenone affording the addition product and a large amount of diphenylmethanol resulting from a β-hydrogen reductive transfer. Nevertheless, by tuning the reac-

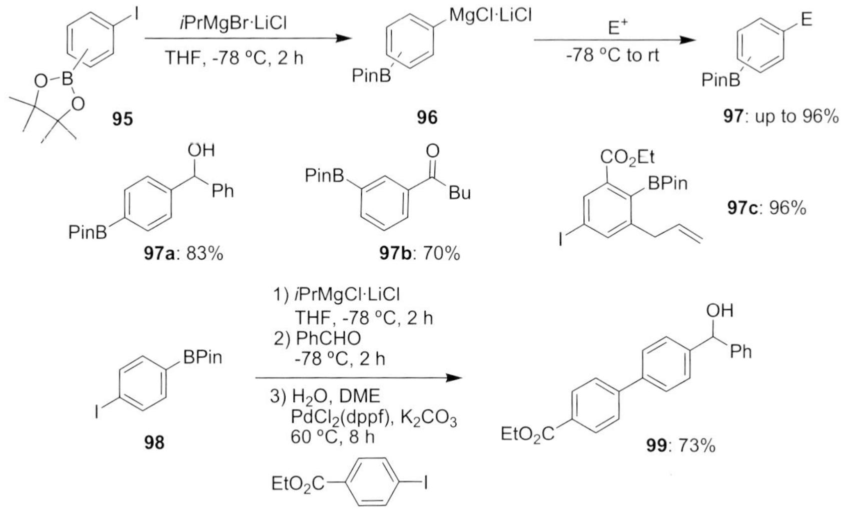

Scheme 4.22 Boron-functionalized arylmagnesium reagents.

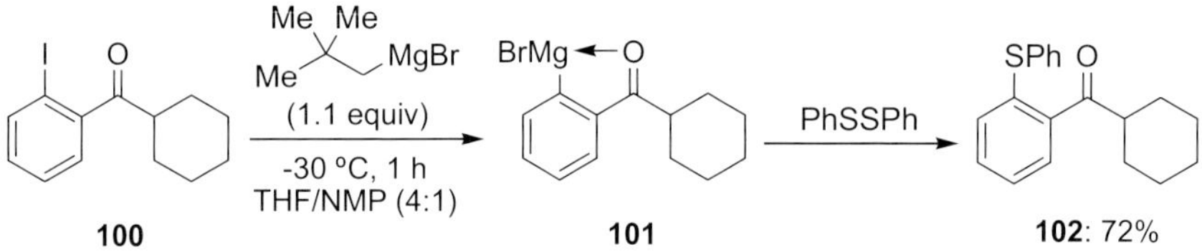

Scheme 4.23 Preparation of an arylmagnesium compound bearing a keto group.

tion conditions, the preparation of ketone-containing arylmagnesium species group can be achieved. To avoid side reactions a sterically hindered but reactive Grignard reagent was chosen: neopentylmagnesium bromide (NpMgBr) [70] in conjunction with *N*-methylpyrrolidinone (NMP) as a polar cosolvent to increase the rate of the iodine–magnesium exchange.

Using these modifications, 2-iodophenyl cyclohexyl ketone (**100**) reacts with NpMgBr (1.1 equiv) at –30 °C within 1 h in a 4:1 mixture of THF and NMP affording the arylmagnesium reagent **101**. The *ortho*-keto function facilitates formation of the Grignard reagent by precoordination of NpMgBr and stabilizes the resulting arylmagnesium species by chelation. Reaction with diphenyl disulfide furnishes the thioether **102** in 72% yield (Scheme 4.23) [71].

Incorporating electrophilic functional groups in the *ortho*-position to the carbon–magnesium bond allows two sequential alkylations leading to ring closure (Scheme 4.24). Reacting benzylic chloride **103** with *i*PrMgBr in THF (–30 °C, 1 h) furnishes the corresponding Grignard reagent **104** that reacts at –10 °C with phenyl isocyanate leading to the functionalized *N*-arylphthalimide derivative **105** in 75% yield [72].

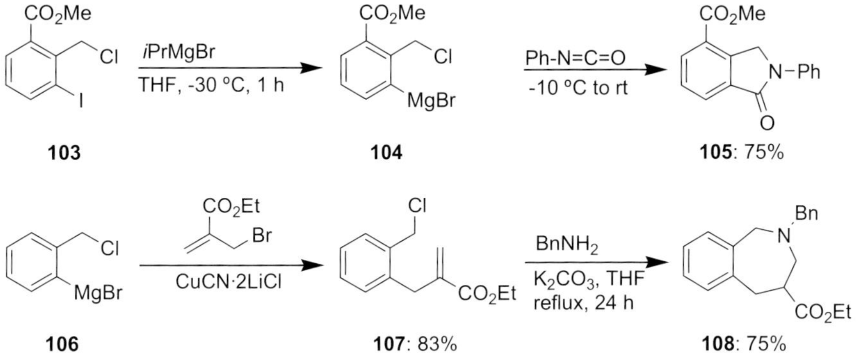

Scheme 4.24 Reaction of chloromethyl substituted arylmagnesium species.

The reaction of the related arylmagnesium species **106** with ethyl (2-bromomethyl)acrylate [73] furnishes the polyfunctionalized product **107** in 83% yield. Subsequent treatment of **107** with benzylamine in the presence of K_2CO_3 in refluxing THF provides the benzoazepine **108** in 75% yield [72]. In strong contrast to the corresponding lithium reagent, which is stable only at −100 °C, [74] the magnesium species **106** is stable for several hours at −30 °C. Recently, the reagent **106** has also been used for the preparation of an oxaphenanthrene [74].

Under slightly acidic conditions a formamidine used as protecting group reacts in an intramolecular addition as an electrophilic function leading to different heterocycles. Hence, the use of *i*PrMgBr allows the selective mono-exchange on protected diiodoaniline derivatives of type **109** and the corresponding Grignard reagents can react with a variety of isocyanates leading to amides such as **110**. Addition of HCl and treatment with silica gel furnishes the desired quinazolinones **111** in good yields (Scheme 4.25) [54].

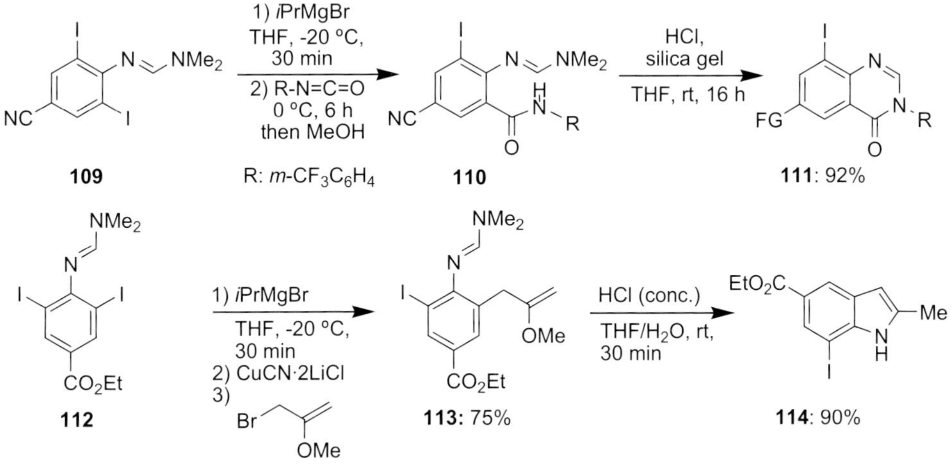

Scheme 4.25 Synthesis of quinazolinones and indoles.

Similarly, reaction of Grignard reagent **112** with 2-methoxyallyl bromide [75] leads to compound **113** that leads after acidic deprotection of the formamidine moiety and the enol ether to the indole cyclization product **114** in 90% yield (Scheme 4.25).

Cyclizations can be achieved with functionalized arylmagnesium reagents bearing a remote leaving group like a tosylate **115** or an allylic acetate **116** as well. In both cases, a stereoselective substitution reaction is observed (Scheme 4.26) [76]. The S_N2 ring closure of **115** is catalyzed by CuCN·2LiCl [49] and proceeds with complete inversion of configuration leading to **117** without eroding the original enantiomeric excess of 60%*ee*.

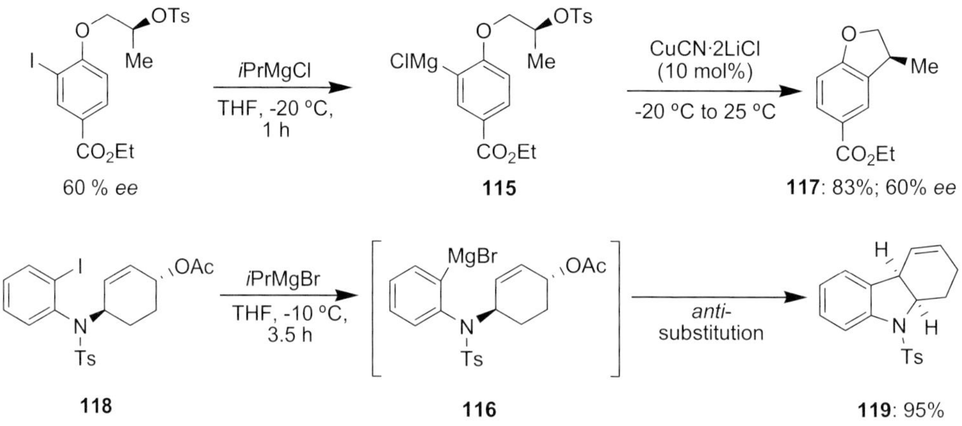

Scheme 4.26 Stereoselective ring closure of arylmagnesium intermediates 115 and 116.

An *anti*-S_N2' substitution is observed with Grignard reagent **116**, readily available from aryl iodide **118**, providing the *cis*-tetrahydrocarbazole **119** in almost quantitative yield. In this case, the Grignard reagent undergoes the ring closure in the absence of a catalyst [76].

4.2.3.3 Halogen–Magnesium Exchange Using Lithium Trialkylmagnesiates

Oshima have shown that besides alkylmagnesium halides, lithium trialkylmagnesiates (R_3MgLi) readily undergo iodine- or bromine–magnesium exchange reactions at low temperatures [77,78]. Lithium trialkylmagnesiates are prepared by the reaction of an organolithium reagent (RLi; 2 equiv) with an alkylmagnesium halide (RMgX; 1 equiv) in THF at 0 °C (30 min). Either 1 equiv or 0.5 equiv of the lithium dibutylmagnesiate (Bu_3MgLi), relative to the aromatic halide, can be used, showing that two of the three butyl groups undergo the exchange reaction. Compared to the halogen–magnesium exchange performed with *i*PrMgBr, lithium trialkylmagnesiates undergo the exchange reaction more readily and are less sensitive to the electronic density of the aromatic ring. Importantly, trialkylmagnesiates react more rapidly with aryl bromides than does *i*PrMgCl. Thus, the reaction of 3-bromobenzonitrile **120** provides lithium diarylbutylmagnesiate **121** that is allylated in the presence of CuCN·2LiCl [49] leading to the nitrile **122** in 85% yield (Scheme 4.27).

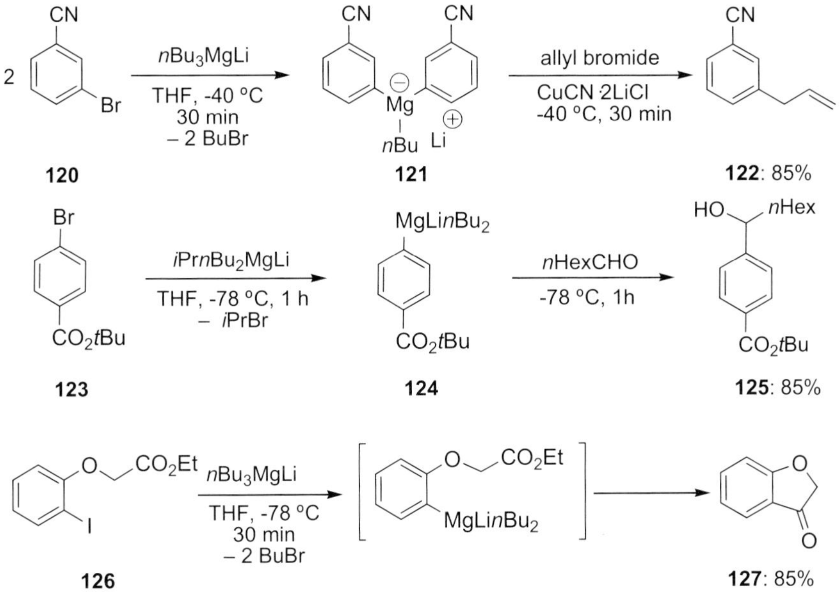

Scheme 4.27 The use of a bromine–magnesiate exchange for the preparation of functionalized arylmagnesium reagents.

However, the resulting lithium triorganomagnesiates are more sensitive to the presence of electrophilic functional groups displaying a reactivity that is intermediate between organolithium and organomagnesium species. Therefore the greater reactivity limits the number of functional groups usually tolerated with these exchange reagents. An ester group is only tolerated when the *t*Bu-halobenzoates are used. Thus, bromide **123** can be converted to the corresponding Grignard reagent **124** at –78 °C with the more reactive *i*Pr(*n*Bu)$_2$MgLi (1.2 equiv). Reaction with heptanal furnishes the benzylic alcohol **125** in 71% yield (Scheme 4.27). On the other hand, the exchange reaction on ethyl (2-iodophenoxy)acetate (**126**) leads after intramolecular addition to the ester group, to the formation of 3-coumaranone **127** in 85% yield (Scheme 4.27). The presence of at least one extra butyl group in the reagents of type **121** or **124** complicates quenching reactions due to competitive reactivity with electrophiles or requires additional amounts of the electrophile.

4.2.3.4 The Preparation of Functionalized Heteroarylmagnesium Reagents
A variety of functionalized heterocyclic Grignard reagents can be prepared using an iodine– or bromine–magnesium exchange reaction [45,79]. The electronic na-

ture of the heterocycle influences the halogen–magnesium exchange rate significantly. Electron-poor heterocycles are reacting faster in the halogen–magnesium exchange reaction. Also electron-withdrawing substituents strongly accelerate the exchange. Thus, 2-chloro-4-iodopyridine (**128**) reacts with *i*PrMgBr at –40 °C within 0.5 h [45,80] furnishing selectively the magnesium species **129** that adds to hexanal leading to the alcohol **130** in 85% yield (Scheme 4.28).

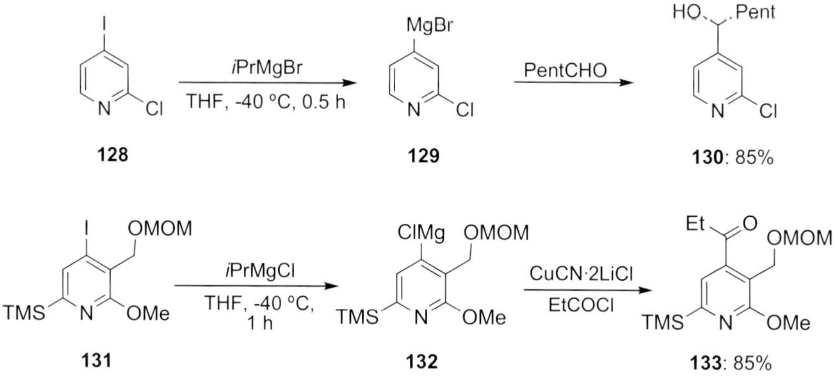

Scheme 4.28 Preparation of functionalized pyridines using an I/Mg exchange.

The highly functionalized pyridine **131** also undergoes a very clean and selective I/Mg-exchange reaction on addition of *i*PrMgCl. The reaction is facilitated by the MOM-protecting group and after transmetallation of Grignard reagent **132** to the corresponding copper species the reaction with propionyl chloride furnishes ketone **133** in 69% yield (Scheme 4.28) [81]. If instead of a pyridine, a pyrimidine derivative such as **134** is used, a selective iodine–magnesium exchange occurs at –80 °C within 10 min providing the organomagnesium compound **135**. Subsequent reaction of **135** with allyl bromide in the presence of CuCN·2LiCl [49] gives the 2-allylpyrimidine **136** in 81% yield (Scheme 4.29) [45].

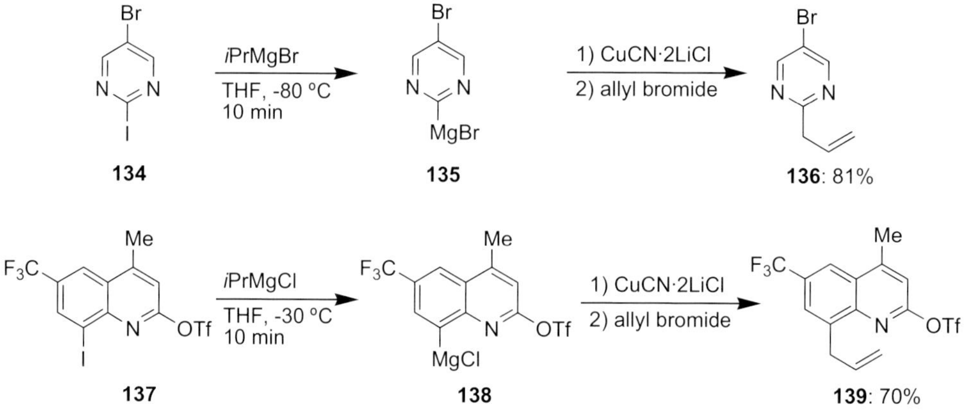

Scheme 4.29 Halogen–magnesium exchange reaction on pyrimidines and quinolines.

The functionalized iodoquinoline **137** is converted to the corresponding magnesium reagent **138** at –30 °C within 10 min. Transmetallation with CuCN · 2LiCl [49] and reaction with allyl bromide furnishes the allylated quinoline **139** in 70% yield (Scheme 4.29). The further functionalization of the triflate group in a cross-coupling reaction is also possible making this heterocycle a versatile building block [82]. Similarly, the I/Mg-exchange reaction turned out to be the best for the regioselective functionalization of imidazo[1,2-*a*]pyridines, a substance class that has demonstrated great potential in the search of new drugs. The preparation of a range of functionalized 6-substituted-2-aminoimidazo[1,2-*a*]pyridines of type **140** has been realized starting from iodide **141**, performing the I/Mg exchange at –40 °C (Scheme 4.30) [83]. The use of lithium reagents gave mixtures of exchange and metallation products in 3- and 5-positions of the imidazo[1,2-*a*]pyridine **141**.

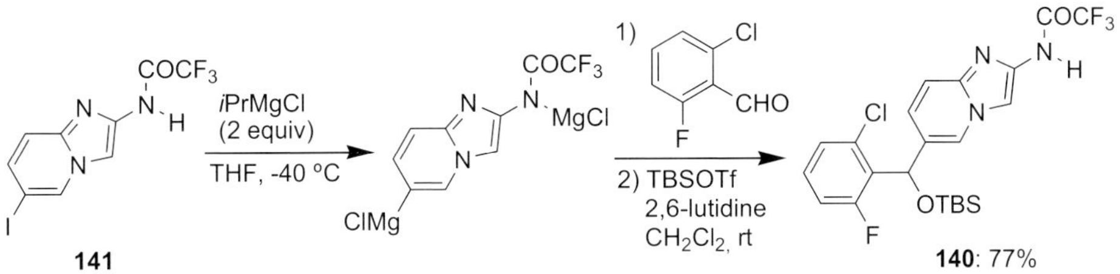

Scheme 4.30 Preparation of imidazo[1,2-*a*]pyridines using a I/Mg exchange.

A range of functionalized iodinated heterocycles have been magnesiated using an iodine–magnesium exchange allowing a rapid synthesis of polyfunctional heterocycles [84,85]. Thus, the protected iodopyrrole **142** undergoes an iodine–magnesium exchange at –40 °C within 1 h, leading to the magnesiated pyrrole **143** that reacts with DMF furnishing the formyl derivative **144** in 75% yield (Scheme 4.31) [86]. Similarly, iodo-isoxazoles [87] and -pyroles[88] are readily converted into the corresponding Grignard reagents.

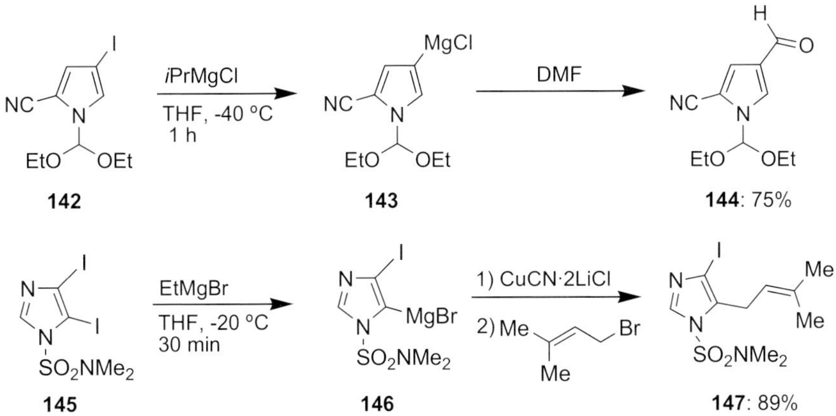

Scheme 4.31 Magnesiation of 5-membered heterocycles.

Polyhalogenated substrates usually undergo a single, selective halogen–magnesium exchange (Scheme 4.31). After a first magnesiation, the electron density of the heterocycle increases to such an extent that a subsequent second exchange is very slow. Selectivity can be gained through chelating groups or by the electronic effects of the heterocycle (Schemes 4.31–4.33). The *ortho*-directing sulfamoyl *N*-protecting group in **145** greatly enhances the stability of the magnesiated species **146** allowing a selective mono-exchange reaction. Here, the use of EtMgBr turned out to be beneficial. Transmetallation to the copper derivative and reaction with 3,3-dimethylallyl bromide furnishes compound **147** in excellent yield (Scheme 4.31) [89]. The selective formation of substituted 2- or 3-bromothiophenes can be achieved by a bromine– or iodine–magnesium exchange reaction. Thus, treatment of 2,3-dibromothiophene (**148**) with EtMgCl at room temperature in THF leads exclusively to the 2-magnesiated heterocycle **149**. On the other hand, 2-bromo-3-iodothiophene (**150**) leads to the expected Grignard reagent **151** under the same conditions (Scheme 4.32) [90]. Reaction with ethyl cyanoformate furnishes the two regioisomeric esters **152** and **153** in 81% and 71% yield, respectively.

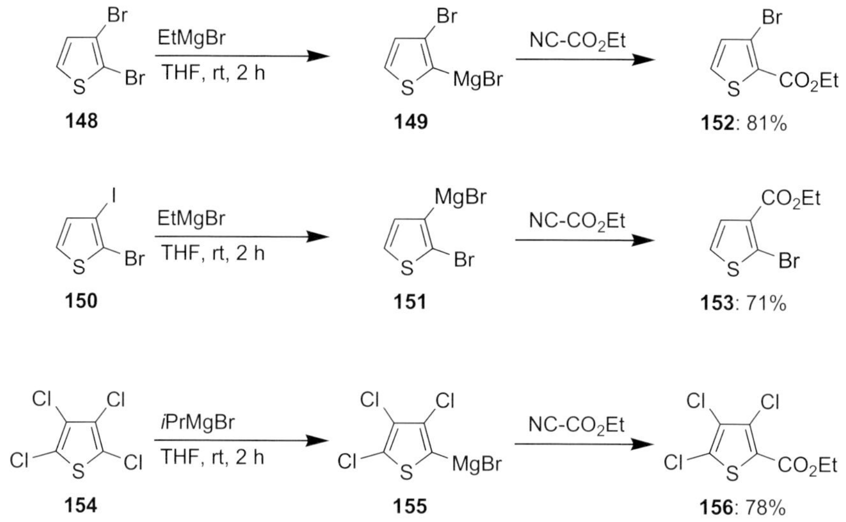

Scheme 4.32 Selective functionalization of halothiophenes.

Although a chlorine–magnesium exchange is a very slow reaction, the presence of four chlorine atoms in tetrachlorothiophene (**154**) accelerates this exchange (25 °C, 2 h) leading to the magnesiated heterocycle **155** that reacts with ethyl cyanoformate providing the thienylester **156** in 78% yield (Scheme 4.32) [45b].

Usually, the bromine–magnesium exchange on heterocyclic substrates is easier compared to the aromatic counterparts. Thus, the extensive functionalization of tribromoimidazole **157** [91] is possible. The first exchange reaction occurs at the 2-position leading to the allylated derivative **158**, after a copper-catalyzed allylation, in 57% yield. Treatment of **158** with a second equivalent of *i*PrMgBr leads to an exchange only in position 5, due to intramolecular chelation of the Grignard reagent. By the reaction with ethyl cyanoformate (–40 °C to 25 °C, 2 h), the corresponding 4-bromo-5-carbethoxyimidazole **159** is obtained in 55% yield (Scheme 4.33) [45].

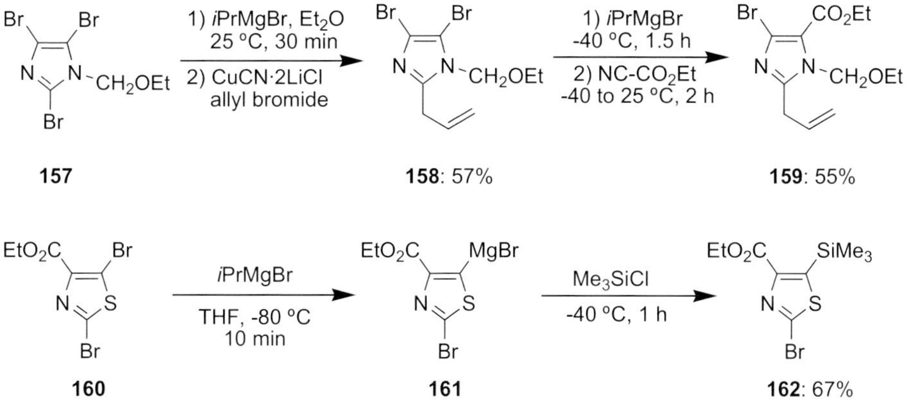

157   →   1) *i*PrMgBr, Et₂O 25 °C, 30 min; 2) CuCN·2LiCl allyl bromide →   **158**: 57%   →   1) *i*PrMgBr –40 °C, 1.5 h; 2) NC-CO₂Et –40 to 25 °C, 2 h →   **159**: 55%

160   →   *i*PrMgBr, THF, –80 °C, 10 min →   **161**   →   Me₃SiCl, –40 °C, 1 h →   **162**: 67%

Scheme 4.33 Regioselective Br/Mg-exchange reactions.

The strong influence of chelating groups on the regioselectivity of the exchange is well demonstrated in the case of dibromothiazole **160**. The Br/Mg-exchange takes place selectively at position 5 due to the chelating effect of the ethoxycarbonyl group leaving the bromide in the electronically favored 2-position unaffected (Scheme 4.33). The reaction of the intermediate Grignard reagent **161** with Me₃SiCl provides the expected product **162** in 67% yield [45b]. Bach used the Br/Mg-exchange for the synthesis of substituted 4-bromothiazoles, starting form the corresponding 2,4-dibromothiazoles [92].

The bromine–magnesium exchange can be accomplished using the more reactive *i*PrMgCl·LiCl as well. Thus, 3,5-dibromopyridine (**163**) undergoes a selective mono-exchange at –10 °C within 10 min and after transmetallation and reaction with allyl bromide leads to the 3-allylated pyridine **164** in 93% yield (Scheme 4.34) [50]. This substrate was converted to the corresponding Grignard reagent using *i*PrMgCl as well, but the yields are significantly lower [93].

Similarly, 2,6-dibromopyridine (**165**) reacts selectively exchanged with *i*PrMgCl·LiCl leading to the mono-magnesium species, whereas the bromine–lithium exchange usually shows extensive decomposition of the starting material [45,94]. In the search for a large-scale synthesis of aldehyde **166**, an intermediate in the synthesis of a muscarinic receptor antagonist, the Merck group found trialkylmagnesiate species *n*-Bu₃MgLi to be advantageous. The use of only 0.35 equivalents of *n*-Bu₃MgLi is sufficient for a selective mono-exchange reaction leading to the ate complex **167** that rapidly reacts with DMF furnishing the aldehyde **166** in 94% yield [95]. This reaction was carried out on up to 25 kg scale. However, the use of magnesiate reagents for preparing various pyridylmagnesium species generally requires one equivalent of *n*-BuMe₂MgLi and the yields are only

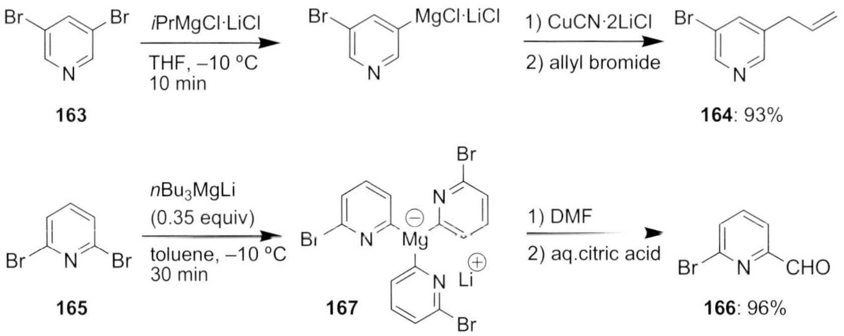

Scheme 4.34 Selective functionalization of dibromopyridines 153 and 155.

moderate [78,96]. Thus, the reagent of choice for the bromine–magnesium exchange should be *i*PrMgCl·LiCl.

The preparation of functionalized uracils and purines is of high interest due to the biological properties of these important classes of heterocycles [97]. Starting from various protected 5-iodouracils such as **168**, the addition of *i*PrMgBr (–40 °C, 45 min) leads to the formation of the corresponding magnesium compound **169** that can be trapped by various aldehydes, ketones and acid chlorides, leading for instance, after transmetallation to copper and reaction with benzoyl chloride to ketone **170** in 73% yield (Scheme 4.35) [98].

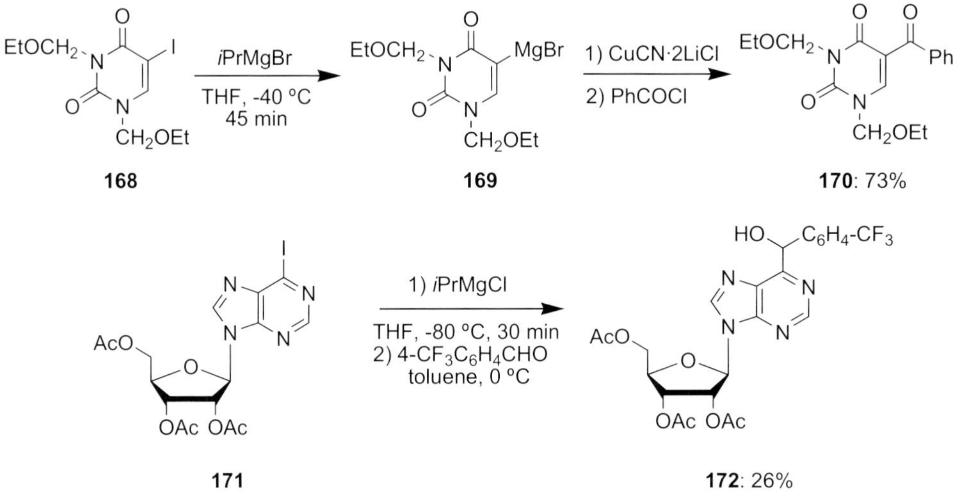

Scheme 4.35 Selective functionalization of uraciles and purines.

Similarly, the triacetylated nucleoside **171** undergoes an I/Mg-exchange reaction at –80 °C within 30 min and subsequent addition of 4-trifluorobenzaldehyde in toluene affords the expected alcohol **172** in 26% yield (Scheme 4.35) [99].

Sensitive "benzylic" heterocyclic magnesium species such as **173** are readily obtained by a Br/Mg-exchange from the bromomethyloxazole **174**. Grignard reagent **173** is generated at –78 °C in the presence of δ-valerolactone in order to minimize self-condensation, leading to the hemi-ketal in 66% yield. This reaction was used by Smith in the course of the total synthesis of (+)-phorboxazole A leading to intermediate **175** in 76% yield (Scheme 4.36) [100].

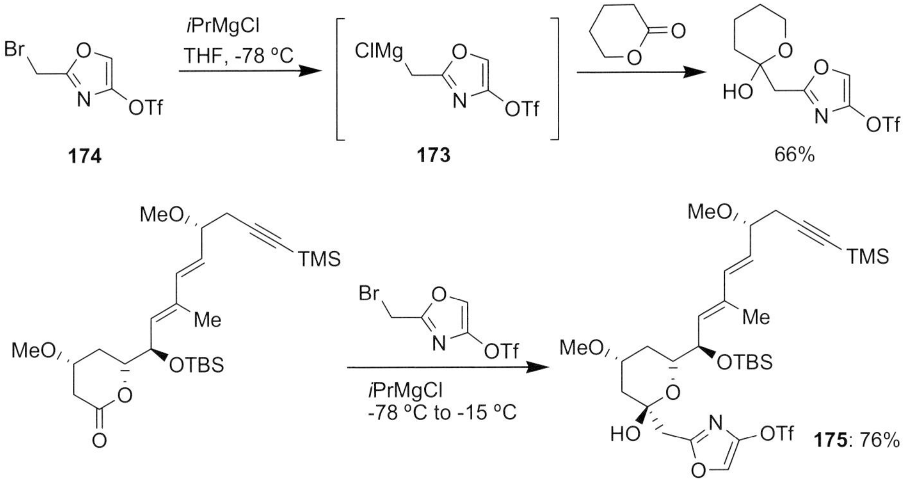

Scheme 4.36 Preparation of (+)-phorboxazole A intermediate 175 using a Br/Mg-exchange.

Finally, a number of these heterocyclic Grignard reagents can be generated with solid-phase reagents and reacted with typical electrophiles in excellent yield [40]. Since numerous heterocyclic bromides are available, this exchange method is anticipated to become a major method for the functionalization of sensitive poly-functional heterocycles. The carbon–magnesium bond possesses a good intrinsic reactivity, which can be enhanced by appropriate transmetallations. The presence of electron-poor substituents attached on the heterocyclic ring somewhat reduces the reactivity of a neighboring carbon–magnesium bond and further improves the functional group compatibility of this carbon–metal bond.

The treatment of heterocyclic iodide bearing acidic protons such as 3-hydroxy-2-iodopyridine **176** with MeMgCl·LiCl followed by *i*PrMgCl furnishes soluble dimagnesiated species **177** that react in satisfactory yields with electrophiles [101]. Similarly, 5-iodouracil **178** produces the trimagnesiated species **179** that reacts smoothly with benzaldehyde affording the uracil derivative **180** without the need of a protecting group. The role of LiCl is clearly to break oligomeric magnesium intermediates having moderate solubility and to produce highly soluble mixed Li, Mg-species [102].

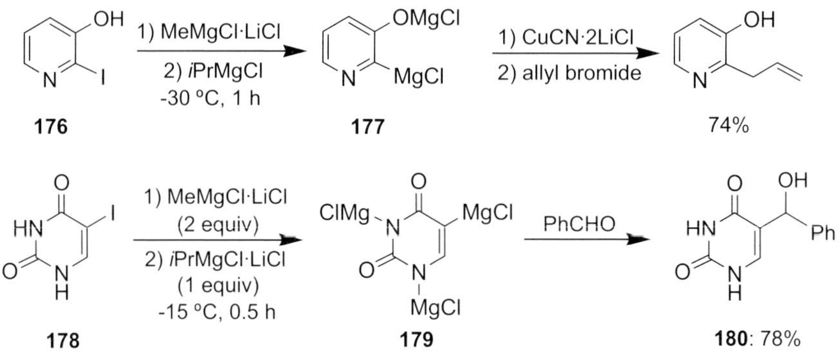

Scheme 4.37 Preparation of heterocyclic Grignard reagents bearing acidic protons.

4.2.4
The Preparation of Functionalized Alkenylmagnesium Reagents

Alkenyl iodides react with iPrMgBr, iPrMgCl, iPr$_2$Mg or iPrMgCl·LiCl leading, after an I/Mg-exchange, to the corresponding alkenylmagnesium halides. This exchange reaction is slower than with aryl iodides and therefore the use of more reactive reagents, such as iPr$_2$Mg and iPrMgCl·LiCl is advantageous. Thus, (*E*)-iodooctene (**181**) undergoes the exchange reaction at 25 °C and the reaction requires 18 h, precluding the presence of sensitive functional groups at a remote position in iodoalkenes when iPr$_2$Mg is used (Scheme 4.38) [103]. However, the reaction of Grignard reagent **182** with tosyl cyanide leads to unsaturated nitrile **183** in 71% yield. The use of iPrMgCl·LiCl allows the conversion of the alkenyl iodide **184**, with ester function in the molecule, into Grignard reagent **185** already at –40 °C within 12 h furnishing after addition of propanal to the corresponding allylic alcohol **186** with excellent stereoselectivity (*E:Z* = 99:1). Similarly, (*Z*)-1-iodoalkene **187** furnishes the corresponding *Z*-alkenylmagnesium chloride **188** that after reaction with diphenyl disulfide leads to the *cis*-product **189** in 81% yield [104].

The silylated cyanhydrin derivative **190**, which is prepared *in situ* from the corresponding ketone [105] is converted to Grignard reagent **191** under very mild conditions. Transmetallation with CuCN·2LiCl, reaction with benzoyl chloride and deprotection furnishes the unsaturated diketone **192** in 74% yield (Scheme 4.38) [104]. The reaction can also be extended to various cyclic dienic systems **193** and **194** with good success. The use of the highly active exchange reagent iPrMgCl·LiCl is essential for the success of the reaction (Scheme 4.39) [106].

The presence of a chelating group greatly enhances the iodine–magnesium exchange reaction. As a result, the functionalized (*Z*)-allylic ether **195** reacts at –78 °C with iPrMgBr providing the corresponding alkenylmagnesium reagent **196**. Reaction of **196** with benzaldehyde furnishes the *Z*-alcohol **197** in 87% yield (Scheme 4.40) [103]. This methodology is also applicable to the resin-attached allylic ether **198** that reacts smoothly with *i*PrMgBr in THF:NMP (40:1) within 1.5 h at −40 °C leading to the desired Grignard reagent. In the absence of NMP, the exchange reaction is considerably slower. Quenching with benzaldehyde and cleavage from the resin with TFA in CH_2Cl_2 provides the dihydrofuran **199** in 86% yield and 97% HPLC-purity [103,107].

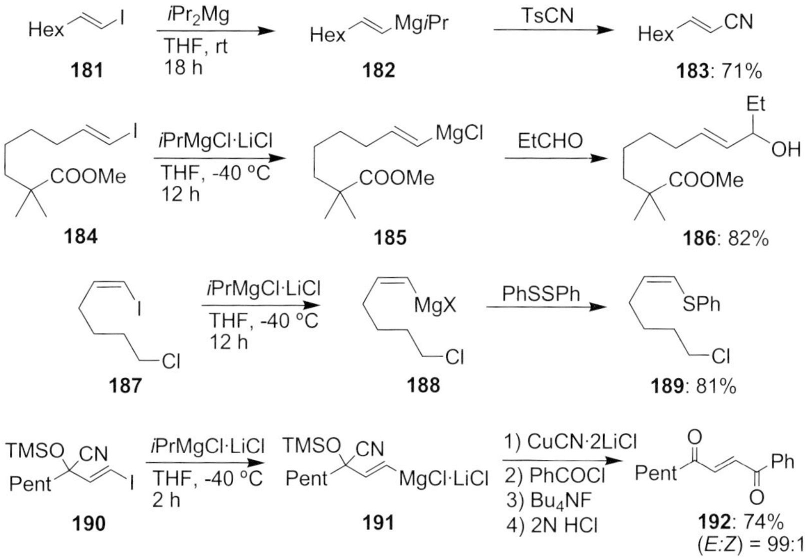

Scheme 4.38 Synthesis of alkenylmagnesium halides via I/Mg exchange.

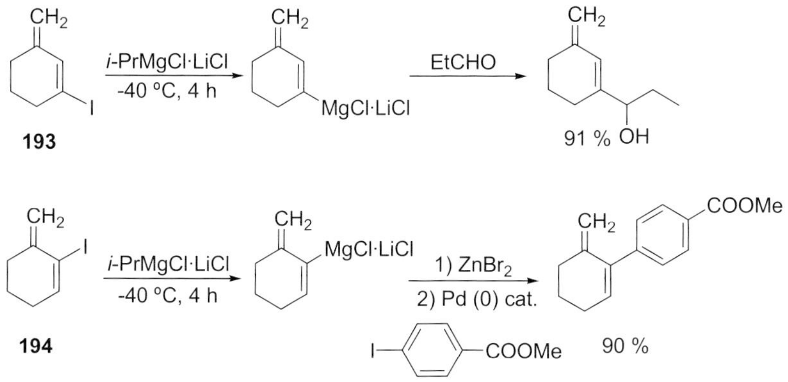

Scheme 4.39 Synthesis of dienyl Grignard reagents.

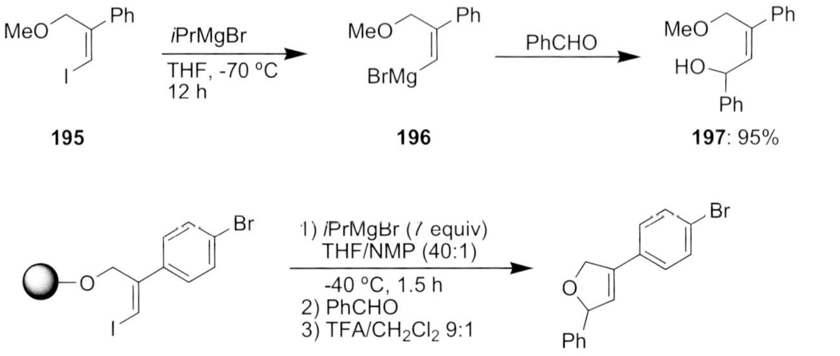

Scheme 4.40 Preparation of functionalized alkenylmagnesium reagents.

Electron-withdrawing groups that are directly attached to the double bond increase its propensity for undergoing the iodine–magnesium exchange reaction. A range of β-iodoenoates like **200** are converted to the corresponding Grignard reagent **201** (–20 °C, 0.5 h to 2 h) leading, after an addition-elimination reaction on 3-iodo-2-methylcyclopentenone in the presence of CuCN · 2LiCl, [49] to the *E*-enoate **202** demonstrating a high configurational stability of the intermediate alkenyl-magnesium species **201** (Scheme 4.41) [108].

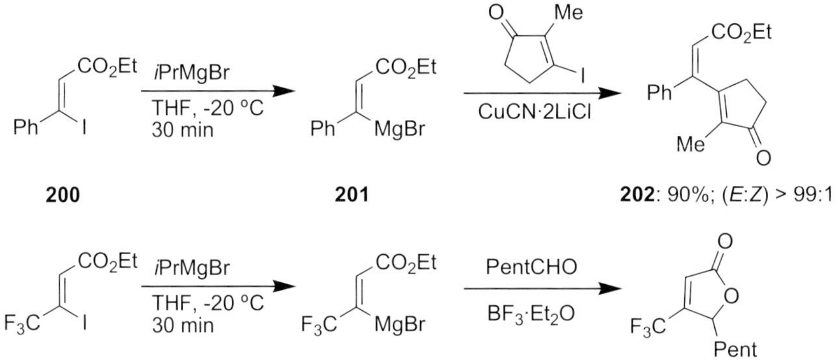

Scheme 4.41 Preparation of carbonyl-containing alkenylmagnesium compounds 201 and 205.

Similarly, Abarbri used this method for the synthesis of 3-perfluoroalyl-buteno-lides, such as **203**. Alkenyl iodide **204** reacts at –78 °C with *i*PrMgBr furnishing Grignard reagent **205** that reacts under Lewis-acid catalysis (BF$_3$ · OEt$_2$) with penta-nal, leading to butenolide **203** in 73% yield [109].

A diphenylphosphanoxide group also accelerates the I/Mg-exchange reaction of alkenyl iodides opening applications in phosphane ligand synthesis. Thus, chiral vinylic iodide **206** undergoes an exchange reaction at –30 °C with *i*PrMgBr leading to Grignard reagent **207** that reacted with CO_2 furnishing compound **208** in an almost quantitative yield (Scheme 4.42) [110].

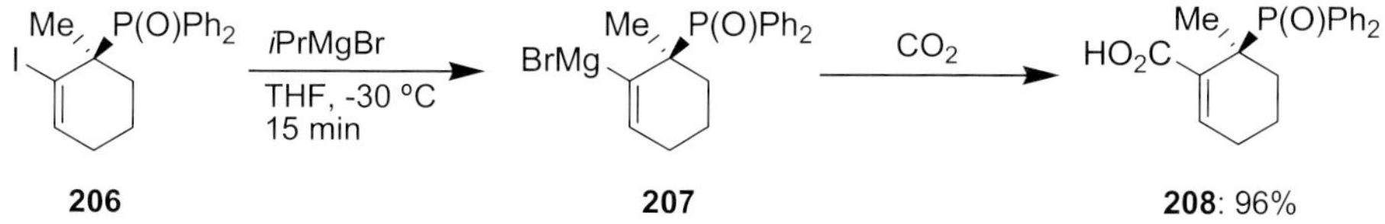

Scheme 4.42 Application of alkenylmagnesium halides in phosphane ligand synthesis.

Whereas alkenylmagnesium compounds bearing a β-leaving group such as a halide or alkoxyde are elusive, [111] the incorporation of the leaving group in a ring system leads to more robust reagents. The reaction of 5-iodo-6-methyl-1,3-dioxin-4-one (**209**) [112] with iPrMgCl at −30 °C furnishes the desired Grignard reagent **210** that proves to have a half-life of ca. 2 h at −30 °C. After transmetallation with $CuCN \cdot 2LiCl$, **210** undergoes a smooth allylation leading to the enone **211** in 77% yield (Scheme 4.43) [113,114].

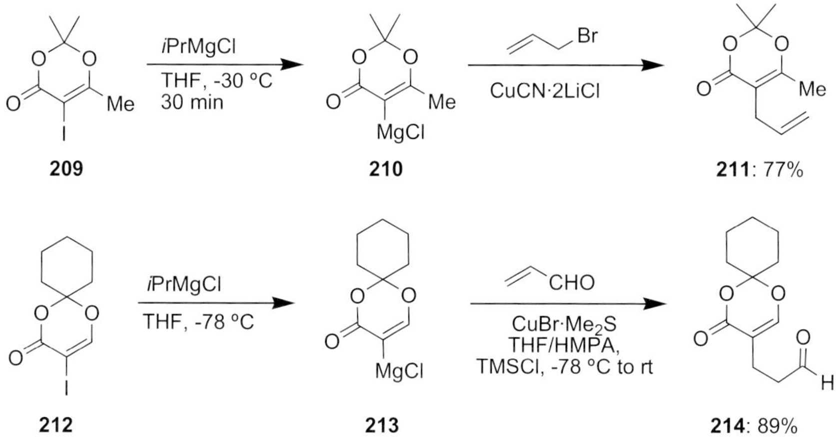

Scheme 4.43 Copper-catalyzed reactions of a functionalized alkenylmagnesium reagent.

The preparation of related carbonyl-containing alkenylmagnesium reagents has been reported by Hiemstra in the course of synthetic studies toward the synthesis of solanoeclepin A [115,116]. The treatment of the cyclic alkenyl iodide **212** with iPrMgCl in THF at −78 °C furnished the desired Grignard reagent **213** that reacts with acrolein and catalytic amounts of $CuBr \cdot Me_2S$ in THF:HMPA in the presence of TMSCl furnishing the Michael adduct **214** in 89% yield (Scheme 4.43) [116].

If the sp^2-carbon atom bears an electron-withdrawing group and a bromine atom, a very fast Br/Mg-exchange reaction is usually observed (−40 °C, 15 min to 1 h). This behavior is very general for alkenyl bromides of type **215** (Y=CN, SO_2Ph, CO_2tBu and $CONEt_2$) that readily react with iPrMgBr affording Grignard reagents of type **216**. Reaction of **216** with electrophiles is not always stereoselective, [117] producing a mixture of diastereoisomers of type **217**, although this method provides an efficient synthesis of tri- and tetra-substituted alkenes **217a–d** [118, 119] (Scheme 4.44).

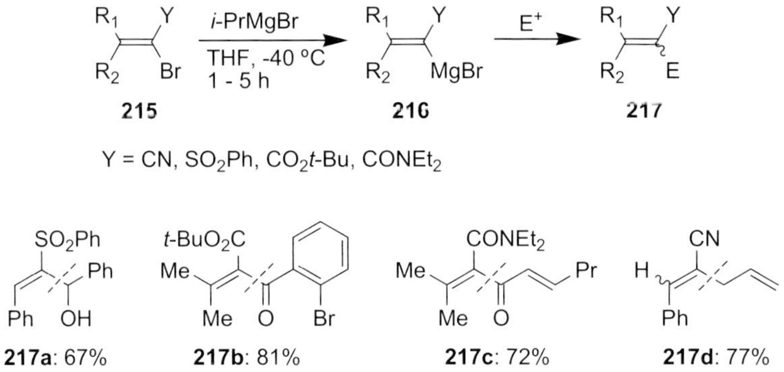

Y = CN, SO$_2$Ph, CO$_2$t-Bu, CONEt$_2$

217a: 67%   **217b**: 81%   **217c**: 72%   **217d**: 77%

Scheme 4.44 Functionalized alkenylmagnesium compounds bearing an electron-withdrawing group in α-position. The dotted lines indicate the new carbon–carbon bond formed.

Similarly, the exchange reaction on readily available dibromoester **218** [120] furnishes, after reaction with *i*PrMgCl (1 equiv) Grignard reagent **219** that reacts with cyclopentanone to butenolide **220** in 71% yield (Scheme 4.45) [121].

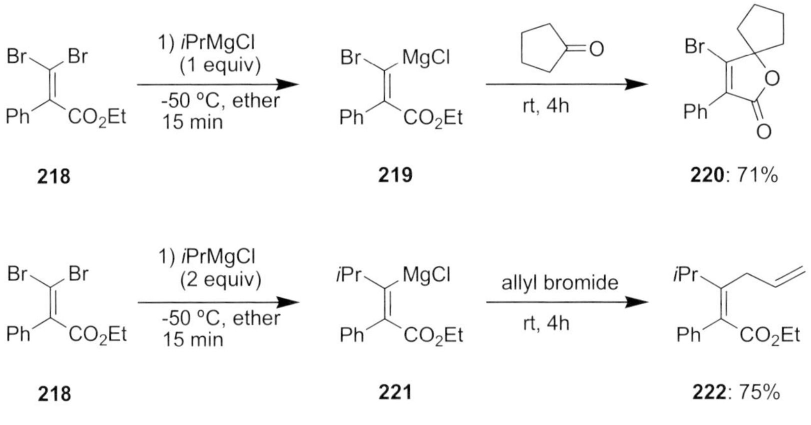

218   **219**   **220**: 71%

218   **221**   **222**: 75%

Scheme 4.45 Selective functionalization of geminal dibromoalkenyl compounds.

The addition of 2 equivalents of *i*PrMgCl leads, after exchange of one bromine and 1,2-migration at the carbenoid center, to the new substituted Grignard reagent **221**. This can further react with an electrophile, furnishing, for example, product **222** in 75% yield (Scheme 4.45) [121]. Recently, Satoh introduced a sulfoxide–magnesium exchange reaction for the synthesis of magnesium carbenoids [122]. The 1-chlorovinylidene **223**, reacts readily with EtMgCl leading to the carbenoid **224** that furnishes after addition of benzaldehyde to the alcohol **225** in 66% yield (Scheme 4.46) [122a].

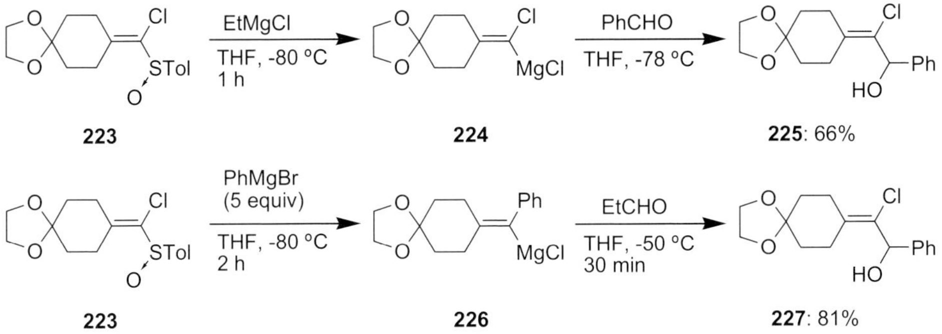

Scheme 4.46 Sulfoxide–magnesium exchange reaction.

Perfoming this ligand exchange reaction with a large excess (5 equiv) of the less reactive PhMgCl, Grignard reagent **226** is generated. Further reaction with an electrophile leads to tetrasubstituted olefin **227** in 81% yield (Scheme 4.46) [122a].

Remarkably, the conjugate addition of various Grignard reagents to the alkynyl-nitrile **228** generates the stabilized and unreactive cyclic magnesium chelate **229** that, after protonation, furnishes the polyfunctionalized nitrile **230**. Fleming has shown that the reactivity of the cyclic organomagnesium reagent of type **229** can be dramatically enhanced by generating an intermediate magnesiate species **231**. This magnesiate species now reacts with PhCHO leading to the allylic diol **232**, with complete retention of the double-bond stereochemistry in 60% yield (Scheme 4.47) [123].

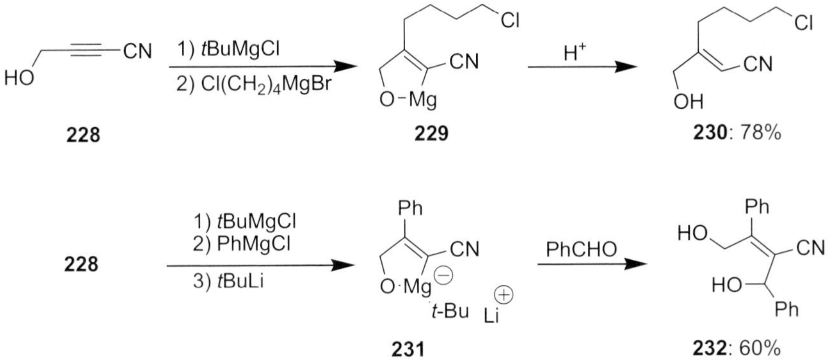

Scheme 4.47 Functionalized alkenylmagnesium compounds obtained by carbomagnesiation.

The I/Mg-exchange on 2-iodo-5-chloro-1-pentene (**233**) provides a functional-ized alkenylmagnesium species **234** that reacts with high diastereoselectivity with the magnesiated unsaturated nitrile **235**. After conjugate addition-alkylation of the *o*-chloroalkyl Grignard reagent **234** only the *cis*-decalin **236** is obtained in 62% yield (Scheme 4.48) [123,124].

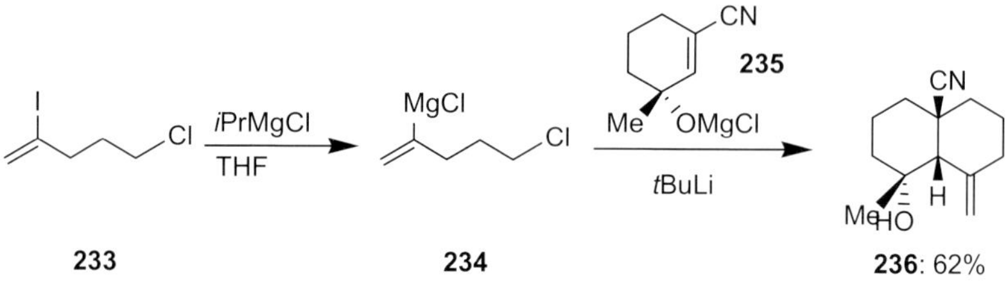

Scheme 4.48 Functionalized alkenylmagnesium compound obtained by I/Mg exchange.

4.2.5
Preparation of Functionalized Alkylmagnesium Reagents

Although the preparation of polyfunctional alkylmagnesium reagents may be envisioned, only a few examples have been reported [125]. The difficulties arise from the higher reactivity of the resulting alkylmagnesium compounds compared to alkenyl-, aryl- or heteroaryl-magnesium species. Also, the rate of the I/Mg-exchange is considerably slower with alkyl iodides. However, a range of polyfunctional cyclopropylmagnesium compounds can be prepared using the iodine–magnesium exchange [126]. Thus, the *cis*-cyclopropyliodoester (*cis*-237) and the corresponding *trans*-isomer (*trans*-237) are readily converted to the corresponding Grignard reagents (*cis*-238 and *trans*-238). The formation of the magnesium organometallics 238 is stereoselective and their reaction with benzoyl chloride furnishes, after a transmetallation of 238 with CuCN · 2LiCl, [49] the expected *cis*- and *trans*-1,2-ketoester 239 with retention of configuration [127, 128] in 73% and 65% yield, respectively (Scheme 4.49) [126].

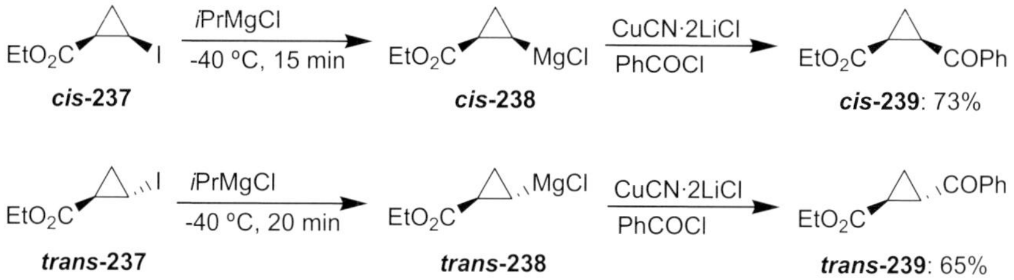

Scheme 4.49 Stereoselective preparation of functionalized cyclopropylmagnesium compounds.

Interestingly, the radical cyclization of allylic β-iodoacetals of type 240 has been shown by Oshima to provide the corresponding organomagnesium compound 241 in DME, which leads, after iodolysis, to the primary alkyl iodide 242 (Scheme 4.50) [125].

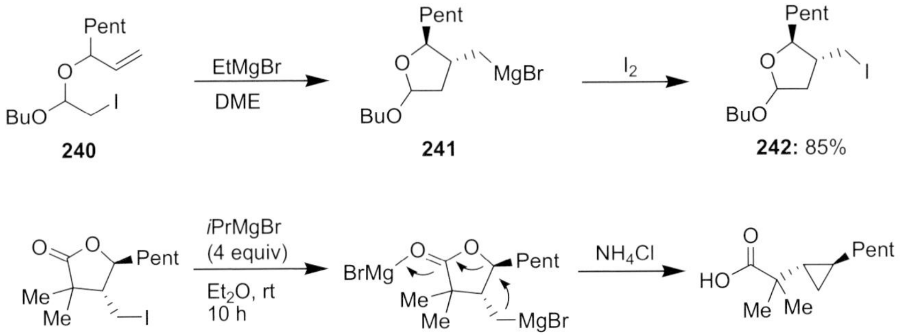

Scheme 4.50 Reactions of functionalized alkylmagnesium compounds.

The I/Mg exchange of 3-iodomethyl-1-oxacyclopentanone derivative **243** is followed by an intramolecular nucleophilic substitution and opening of the oxacyclopentane ring leads after aqueous work-up to cyclopropane derivative **244** (Scheme 4.50) [129].

Metalated nitriles are versatile synthetic intermediates, because of their high nucleophilicity and the easy transformation of the nitrile moiety into a plethora of functional groups. They are typically generated by deprotonation with metal amide base or *t*BuOK [130]. Another possibility to access this substance class is the metallation of α-halonitriles, such as **245** with *i*PrMgBr. This reaction is very fast, furnishing the corresponding Grignard reagent **246** almost instantaneously at –78 °C. Quenching of the metalated nitrile **246** with cyclohexanone leads to the sterically hindered hydroxy-nitrile **247** in 73% yield (Scheme 4.51) [131].

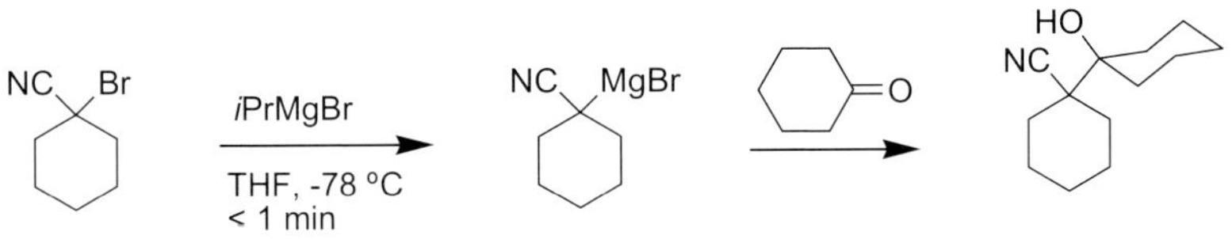

Scheme 4.51 Metalation of α-halonitrile **245**.

4.2.6
Preparation of Functionalized Alkylmagnesium Carbenoids

The pioneering work of Villieras allows, through a Br/Mg-exchange, a general preparation of magnesium carbenoids [35,36,132]. This very fast reaction enables the preparation of sensitive cyclopropylmagnesium carbenoids such as **248** and **249** starting from the corresponding 1,1-dihalocyclopropanes **250** and **251**. By performing the halogen–magnesium exchange in ether, a highly stereoselective exchange reaction is observed, due to the assistance of the ester group. The oxygen precomplexes the Grignard reagent and breaks the aggregates. Quenching of the magnesium carbenoids proceeds with retention of configuration providing the two diastereomeric products **252** and **253** in 80–85% yield (Scheme 4.52) [126].

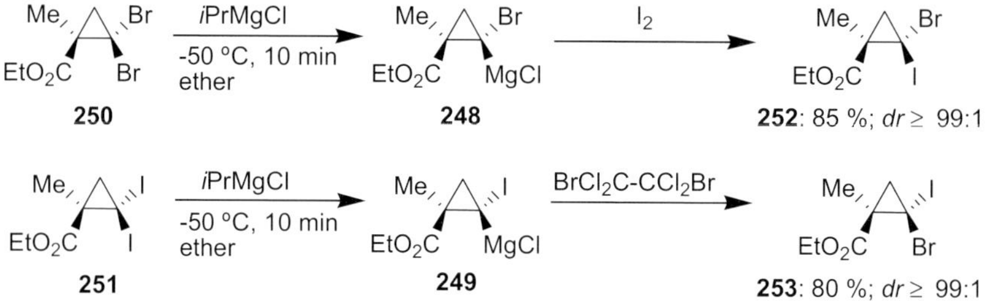

Scheme 4.52 Stereoselective preparation of cyclopropylmagnesium carbenoids.

Functionalized acyclic magnesium carbenoids can be prepared in THF/*N*-butyl-pyrrolidinone (NBP) mixtures at low temperatures. Thus, the reaction of the *bis*-iodomethylcarboxylate **254** with *i*PrMgCl in THF:NBP is complete within 15 min at −78 °C. [133]. The resulting chiral *bis*-carbenoid **255** is quenched with PhSSPh giving the *bis*-adduct **256** in 70% yield (Scheme 4.53) [133].

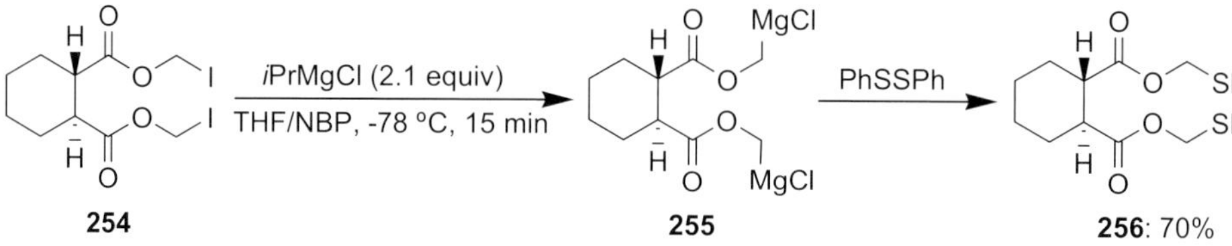

Scheme 4.53 Preparation of functionalized acyclic magnesium carbenoids.

The recently introduced sulfoxide/magnesium-exchange gives rise to the formation of alkylmagnesium halides as well. Thus, the reaction of the sulfoxide **257** with *i*PrMgBr at −78 °C furnished the desired magnesium carbenoid **258** that reacts with PhCHO with an excellent diastereoselectivity providing the mono-protected 1,2-diol **259** in 61% yield (*dr* = 93:7) (Scheme 4.54) [133].

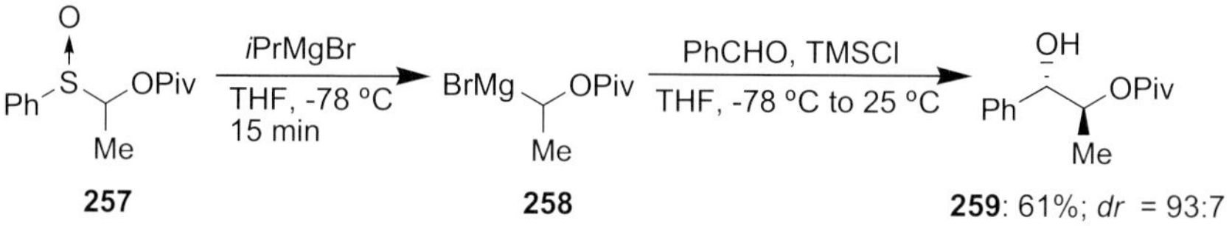

Scheme 4.54 Sulfoxide–magnesium exchange reaction.

Acyclic geminal diiodo-alkanes, such as **260** have been extensively studied by Hoffmann, in order to get a better understanding of the mechanism of I/Mg exchange and in the quest for a chiral Grignard reagent [134]. Although, these systems proved to undergo a fast exchange reaction leading to Grignard reagent **261**, no enantioselective I/Mg-exchange reaction could be observed for the enantiotopic iodides of **260** by using chiral ligands on the exchange reagent, such as **262** (Scheme 4.55). Thus, alcohol **263** is obtained in good yield, but the enantiomeric excess is only moderate (83%, 53%*ee*).

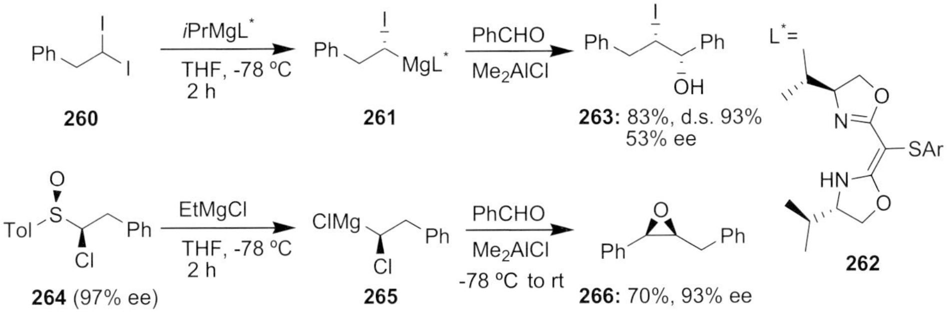

Scheme 4.55 Synthesis of chiral Grignard reagents.

A solution to this problem was found by using the sulfoxide–magnesium exchange. Starting with the chiral sulfoxide **264**, the reaction with EtMgCl leads to Grignard reagent **265**, which reacts with benzaldehyde and leads after cyclization to epoxide **266** in 70% yield, without almost any loss of enantiomeric excess (93% *ee*) (Scheme 4.55). A very interesting application of carbenoid chemistry is found in the synthesis of substituted pyrimidines **267**, developed by Oshima. Using a geminal dibromo-oxime ether **268** and *n*BuMgBr as exchange reagent, magnesium carbenoid **269** is easily accessible and provides an interesting way to pyrimidines (Scheme 4.56).

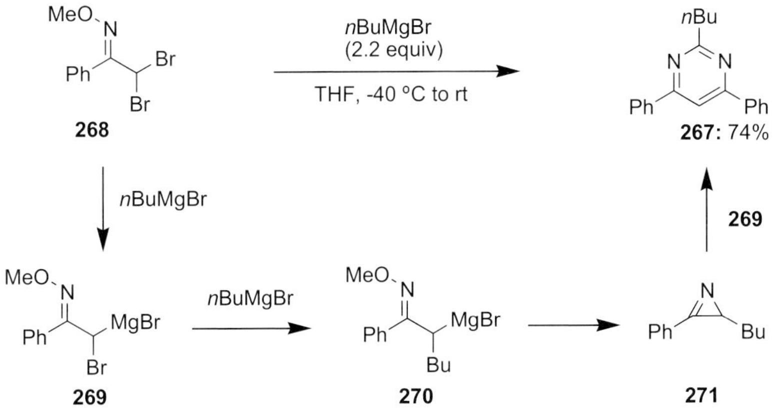

Scheme 4.56 Pyrimidine synthesis using magnesium carbenoids.

Thus, the reaction with another equivalent of *n*BuMgCl provides substituted Grignard **270** that cyclizes to aziridine **271**. Further reaction with another molecule of carbenoid **269**, followed by a rearrangement and elimination of methanol leads to pyrimidine **267** in 74% yield (Scheme 4.56).

4.3
Further Applications of Functionalized Grignard Reagents

Along with the standard 1,2-addition reactions to carbonyl groups or copper-catalyzed substitution reactions that were shown in the previous sections, functionalized Grignard reagents have found several other applications in synthesis. Thus, the addition to imminium salts, for example, is a very potent method for the preparation of benzylic amines. The Grignard reagent **272** adds to the imminium trifluoroacetate **273** at –40 °C within 30 min in a THF/CH$_2$Cl$_2$ mixture providing the *bis*-allylamine **274** in 76% yield [135]. Similarly, heterocyclic Grignard reagents undergo this aminomethylation. For instance, magnesiated uracil **275** reacts within 1 h with imminium salt **273** and leads to compound **276** in 85% yield (Scheme 4.57) [98,136].

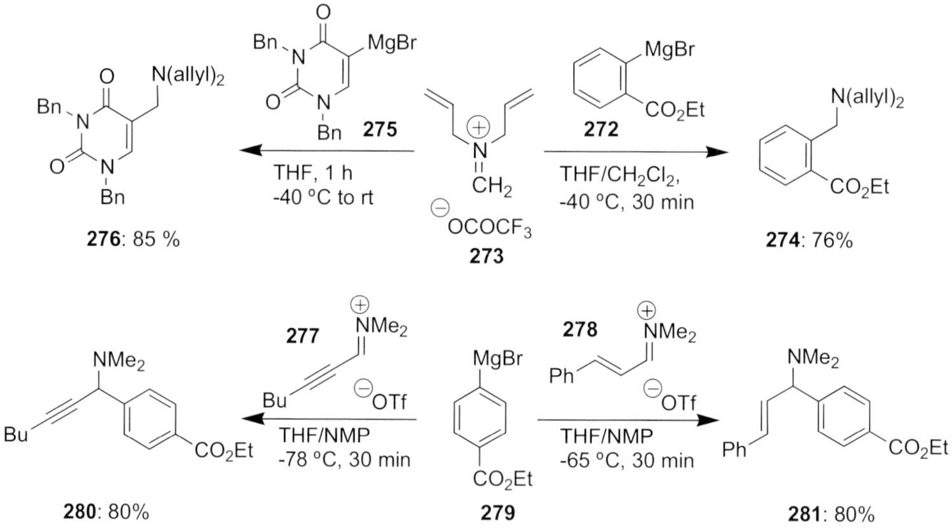

Scheme 4.57 Reaction of functionalized arylmagnesium compounds with imminium salts.

Various unsaturated imminium salts like **277** and **278** react with functionalized arylmagnesium halides, such as **279** furnishing the expected benzylic amines **280** and **281** in 80% yield (Scheme 4.57) [137]. Aryl sulfonamides such as **282** are of considerable medicinal importance [138]. In order to provide a generally useful, mild methodology for the synthesis of sulfonamides, allowing the variation at both nitrogen and sulfur, Barrett and coworkers developed a procedure starting from functionalized Grignard reagents (Scheme 4.58) [139]. They found, that sulfinylation of aryl Grignard reagent **283** using sulfur dioxide afforded sulfinate **284**, which upon direct addition of neat sulfuryl chloride gave the corresponding arylsulfonyl chloride. Subsequent addition of secondary amines, such as diethylamine at room temperature gave the desired sulfonamide **282** in 67% overall yield (Scheme 4.58). This method is applicable to heterocyclic Grignard reagents as well.

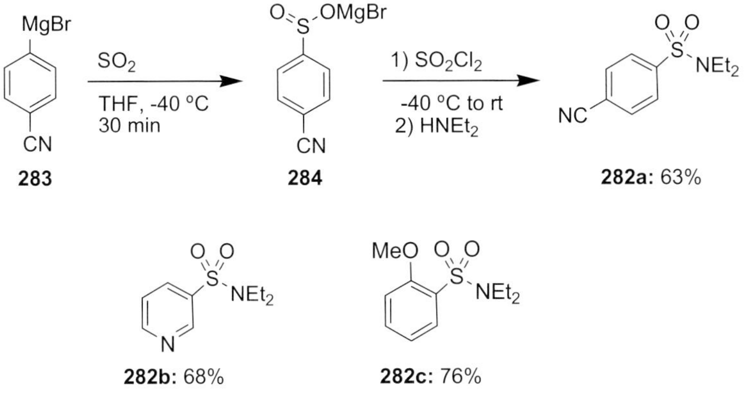

Scheme 4.58 General one-pot synthesis of sulfonamides 282.

In the presence of catalytic amounts of $CuI \cdot 2LiCl$ and Me_3SiCl (1 equiv), [140] functionalized arylmagnesium compounds, such as **285** can be added to various cyclic and acyclic enones providing Michael-addition products of type **286** in good yields (Scheme 4.59) [141].

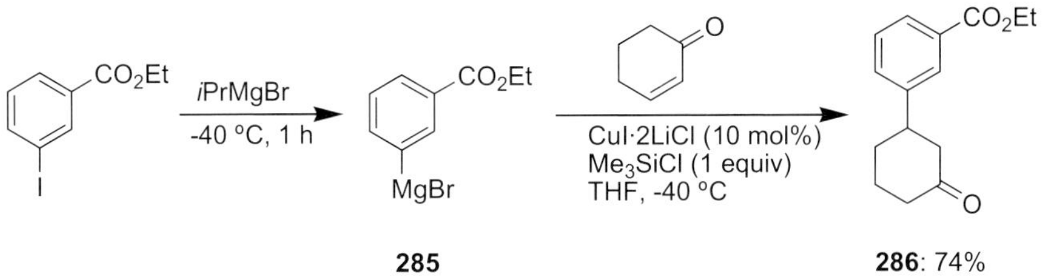

Scheme 4.59 Cu(I)-catalyzed addition of functionalized arylmagnesium compounds to enones.

Grignard reagents are useful intermediates for the synthesis of boronic acids and esters, but alkoxyborates $B(MeO)_3$ and $B(iPrO)_3$ have a reduced reactivity towards arylmagnesium halides at low temperatures and therefore lack of general applicability. In contrast, the more reactive methoxyboron pinacolate (MOBPin) **287** is an excellent reagent for the introduction of a boronic ester moiety, allowing the synthesis of various functionalized heteroaryl- and arylboronic esters such as **288** or **289** starting form the corresponding Grignard reagents (Scheme 4.60) [69].

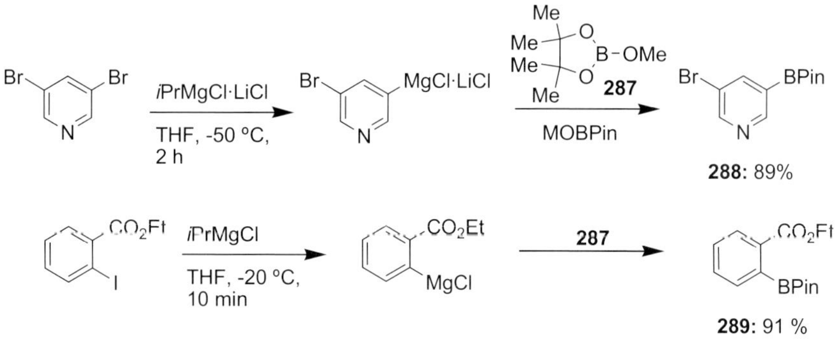

Scheme 4.60 Synthesis of functionalized boronic esters.

A similar reaction was applied to the total synthesis of the antibiotic vancomycin by Nicolaou. The selective conversion of aryl iodide **290** (X=I) to the corresponding phenol **291** (X=OH) was needed in the final steps of the synthesis (Scheme 4.61) [142]. Aryl iodide **290** was converted to the corresponding Grignard reagent **292** by first deprotonating all acidic hydrogens with an excess of MeMgBr and subsequent addition of *i*PrMgBr at −40 °C. Reaction of reagent **292** with B(OMe)₃ leads to the boronic ester **293** that is readily oxidized to the corresponding phenol **291** with an alkaline solution of H₂O₂ in ca. 50% overall yield.

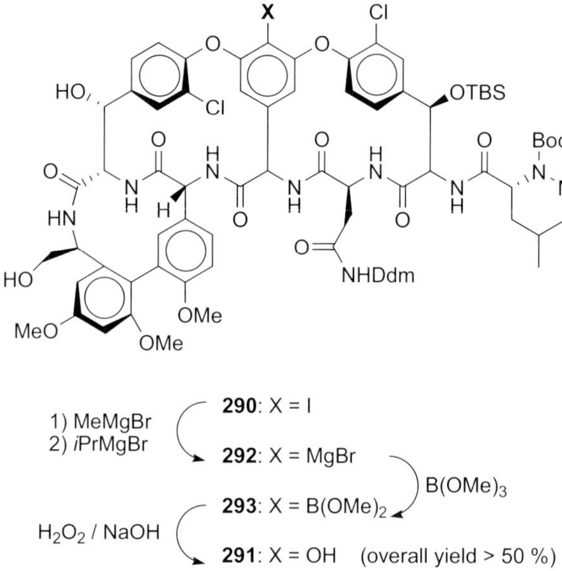

1) MeMgBr
2) *i*PrMgBr

290: X = I

292: X = MgBr

293: X = B(OMe)₂) B(OMe)₃

H₂O₂ / NaOH

291: X = OH (overall yield > 50 %)

Scheme 4.61 Formation of a functionalized arylmagnesium during the synthesis of vancomycin.

This two-step oxidation of a Grignard reagent to the corresponding phenol is a very important reaction. Ricci developed a methodology that allows the direct conversion of a Grignard reagent to either the corresponding phenol or the aniline derivative using *N,O*-bis(trimethylsilyl)hydroxylamine **294** as reagent [143]. For example, the direct reaction of magnesiated aryl derivative **295** with **294** provides exclusively the corresponding aminophenol **296** in 64% yield (Scheme 4.62; see also Scheme 4.17) [56].

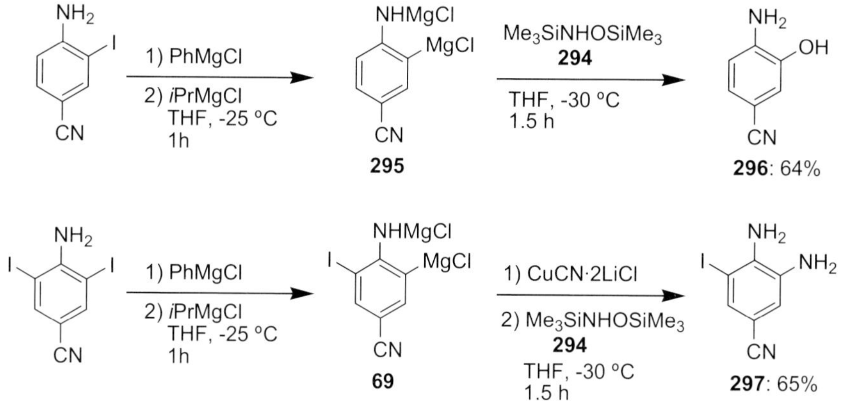

Scheme 4.62 Oxidation of Grignard reagents using *N,O*-bis(trimethylsilyl)hydroxylamine **294**.

On the other hand, a transmetallation of Grignard reagent **69** to the corresponding copper derivative with CuCN·2LiCl [49] leads to the formation of the diamino derivative **297** in 65% yield (Scheme 4.62). Besides this, several other reagents are known that allow the formal oxidation of a Grignard reagent to an amino function and their use has been excellently reviewed elsewhere [144].

Diarylamines, often found in pharmaceuticals, are usually accessed by the reaction of a nitrogen nucleophile with an aromatic halide following a S$_N$Ar mechanism. Generally, activating groups are required together with a good leaving group on the aromatic ring (pathway a, Scheme 4.63) [145]. More recently, various arylamines were prepared by palladium-catalyzed cross-coupling reactions of amines with aryl halides [146,147,148]. Other transition metals such as copper [149,150] and nickel [151] have also allowed the performance of C(aryl)-N bond formation reactions. Oxidative coupling procedures between arylboronic acids and aromatic or heterocyclic amines mediated by Cu(II) salts proved to be effective as well [149]. In all these approaches, aromatic amines were used as precursors following pathway a (Scheme 4.63).

On the other hand, one can envision the reaction of an electrophilic nitrogen synthon with a carbon nucleophile such as a Grignard reagent. In this case, nitrogen will act as an electrophile, resulting in an "umpolung" of the reactivity (pathway b, Scheme 4.63) [152]. The polarization of the nitro group would, in principle, permit such a retrosynthetic analysis. Indeed, the reaction of nitroarenes with Grignard reagents was first investigated in pioneering work by Wieland in 1903 [153]. Gilman and McCracken observed the formation of diarylamine, when reacting a Grignard reagent with nitrosobenzene [154] and later Köbrich sug-

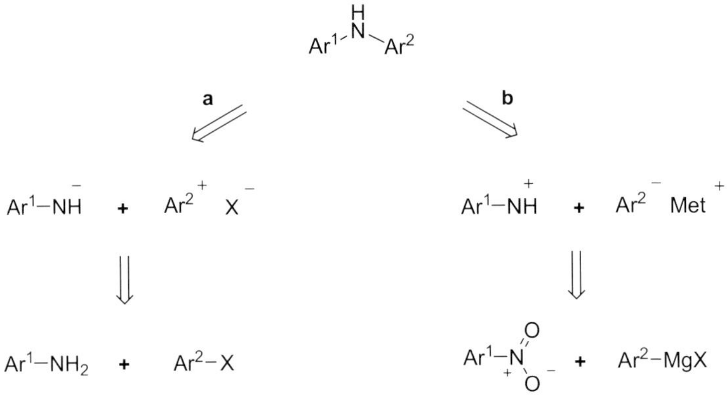

$$Ar^1\text{-}\overset{\displaystyle H}{\underset{}{N}}\text{-}Ar^2$$

a   **b**

$$Ar^1\text{-}\overset{-}{NH} \quad + \quad Ar^{2\,+}X^- \qquad\qquad Ar^1\text{-}\overset{-}{NH} \quad + \quad Ar^{2\,-}Met^+$$

$$Ar^1\text{-}NH_2 \quad + \quad Ar^2\text{-}X \qquad\qquad Ar^1\text{-}\overset{+}{N}\overset{\displaystyle O}{\underset{O^-}{\big\Vert}} \quad + \quad Ar^2\text{-}MgX$$

Scheme 4.63 Synthesis of diarylamines using pathway a or b.

gested a mechanism for this reaction that was later confirmed by Knochel (Scheme 4.64) [61a,155]. However, no synthetically valuable procedure was reported, although Bartoli carefully studied the reactions between nitro aromatics and Grignard reagents [156].

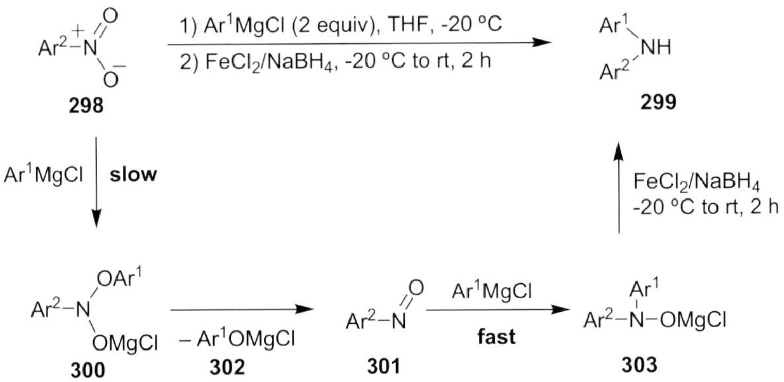

$$
\begin{array}{c}
Ar^2\text{-}\overset{+}{N}\overset{O}{\underset{O^-}{\big\Vert}} \\
\mathbf{298}
\end{array}
\quad
\begin{array}{c}
\text{1) } Ar^1MgCl \ (2 \ \text{equiv}), \ THF, \ \text{-20 °C} \\
\hline
\text{2) } FeCl_2/NaBH_4, \ \text{-20 °C to rt, 2 h}
\end{array}
\quad
\begin{array}{c}
Ar^1\diagdown \\
\qquad NH \\
Ar^2\diagup \\
\mathbf{299}
\end{array}
$$

Ar^1MgCl | **slow**

$FeCl_2/NaBH_4$
-20 °C to rt, 2 h

$$
\begin{array}{c}
Ar^2\text{-}N\overset{\displaystyle OAr^1}{\underset{\displaystyle OMgCl}{\diagup\diagdown}} \\
\mathbf{300}
\end{array}
\xrightarrow{\ -\,Ar^1OMgCl\ }
\begin{array}{c}
\mathbf{302} \\
Ar^2\text{-}N\overset{O}{\big\Vert} \\
\mathbf{301}
\end{array}
\xrightarrow[\mathbf{fast}]{\ Ar^1MgCl\ }
\begin{array}{c}
Ar^1 \\
Ar^2\text{-}\overset{|}{N}\text{-}OMgCl \\
\mathbf{303}
\end{array}
$$

Scheme 4.64 Proposed mechanism for the reaction of aryl magnesium compounds with nitroarenes **298**, leading to diarylamines **299**.

Starting from a nitroarene **298** and a arylmagnesium halide, the first aryl group is transferred to the oxygen of the nitro group, furnishing **300** that can produce as an intermediate arylnitroso derivative **301** after elimination of magnesium phenolate **302**. Reaction of this intermediate nitroso species **302** with the second equivalent of Grignard reagent leads to the formation of the C–N bond and produces the air-sensitive magnesium diarylhydroxylamide **303**. In order to turn this reaction into a preparative useful tool, a subsequent reduction of **303** with FeCl$_2$/NaBH$_4$ [157] is required providing the diarylamine **299** (Scheme 4.64). This method allows the arylation of a variety of nitrobenzene derivatives and therefore is ideal for the synthesis of a range of functionalized diarylamines such as **299a–c** with high yields (Scheme 4.65) [158].

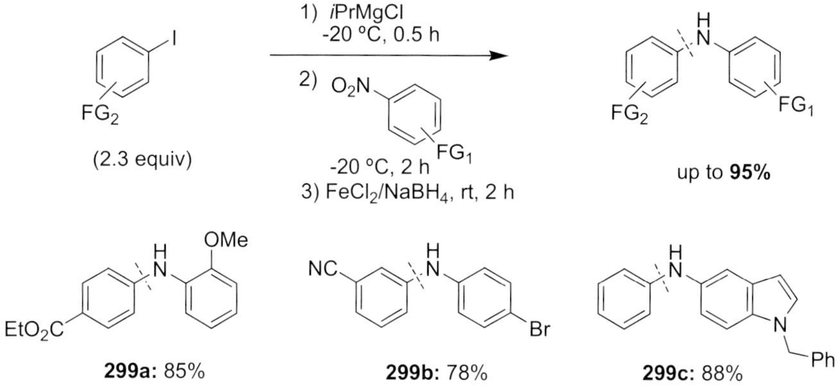

Scheme 4.65 Polyfunctional diarylamines 299 obtained by the reaction of a functionalized arylmagnesium compound with a nitroarene. The dotted lines indicate the newly formed C–N bond.

The Grignard reagent may bear electron-withdrawing groups or electron-donating groups. This is also the case for nitroarene and even heterocyclic nitroarenes can be used in this reaction. Interestingly, sensitive functions like an iodine, bromine or triflate [158] group can be present in either reaction partner. This is difficult to realize for transition-metal catalyzed amination procedures [159,160,161].

This method shows an excellent functional-group tolerance and is almost indifferent to the electronic properties of the reaction partners. However one equivalent of Grignard reagent is wasted in the first reduction step. This can be avoided by using a nitrosoarene instead of a nitroarene as the electrophilic reagent. The reaction of 4-dimethylaminonitrosobenzene (**304**) with PhMgCl (1.2 equiv) provides the expected diarylhydroxylamine **305** that, after reductive treatment (FeCl$_2$/NaBH$_4$), gives the diarylamine **299d** in 73% isolated yield (Scheme 4.66) [162,163].

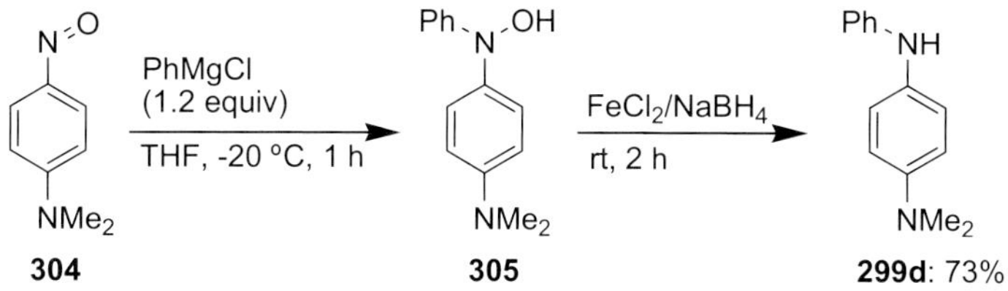

Scheme 4.66 Synthesis of diarylamines using nitrosoarenes and Grignard reagents.

The difficult access to nitrosoarenes and their tendency to dimerize makes these reagents less attractive. The substitution of the nitroso group with a toluenesulfonamide leads to aryl 4-tolylazo sulfones of type **306**, which can be used as amination reagents as well. The formation of these building blocks is easily accomplished in two steps, starting from the corresponding anilines **307** by the reaction of diazonium tetrafluoroborate **308** with sodium toluenesulfinic acid, (Scheme 4.67).

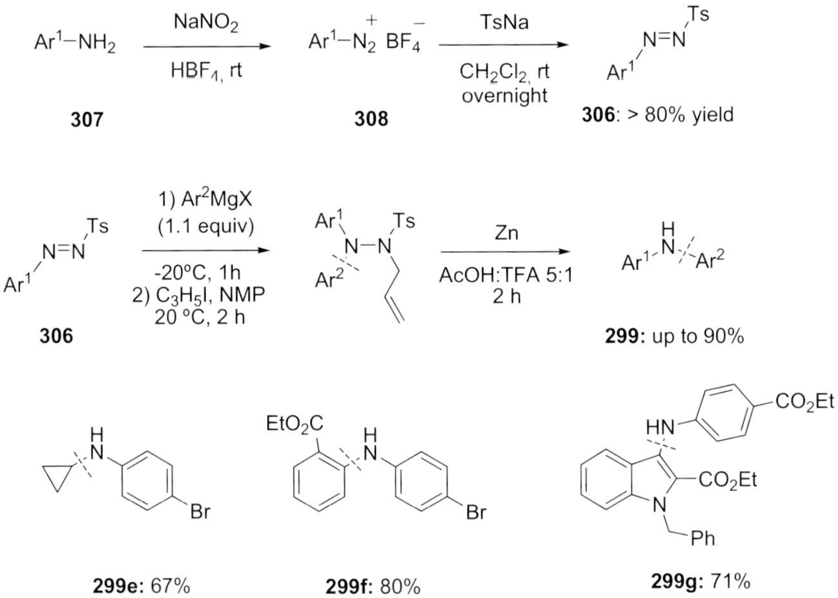

299e: 67%   **299f**: 80%   **299g**: 71%

Scheme 4.67 Synthesis of aryl 4-tolylazo sulfones 306 and diarylamines 299e–g.

The addition of a Grignard reagent leads to a hydrazine derivative, which after allylation and subsequent reduction of the N–N bond using zinc in glacial acetic acid and TFA provides diarylamines **299** in good yields (Scheme 4.67) [164]. A wide range of functional groups is tolerated either on the azo sulfones or on the Grignard reagents. Furthermore, this methodology can by applied to alkyl- and heteroaryl-magnesium halides.

These reactions complement recently developed palladium(0)-catalyzed amination reactions [146,147,148] and related procedures using a copper(I) [149] – or nickel(0) [151] – catalysis. As indicated above, the mild reaction conditions are compatible with a range of functional groups. Functionalized arylmagnesium chlorides such as **309** prepared by an I/Mg-exchange readily undergo addition reactions to aryl oxazolines. The addition-elimination of **309** to the β-methoxy aryloxazoline followed by an *ortho*-lithiation and substitution with ethylene oxide leads to a polyfunctionalized aromatic intermediate **310** for alkaloid synthesis (Scheme 4.68) [165].

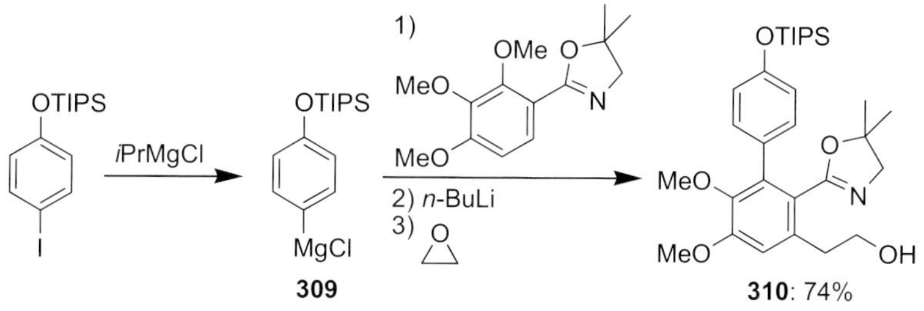

Scheme 4.68 Formation of a functionalized arylmagnesium in the course of an alkaloid synthesis.

Aromatic iodides and bromides bearing a good leaving group in the *ortho*-position can generate 1,2-dehydrobenzene and related substrates (arynes) through an elimination reaction [166]. This approach has been used successfully for a variety of organometallic reagents, but the harsh reaction conditions or the high reactivity of the organometallic reagent (especially the lithio-derivatives) made this method incompatible with a variety of functional groups [167,168]. Performing the I/Mg-exchange on *ortho*-iodotosylates **311** leads to stable Grignard reagents of type **312** that react readily with various electrophiles at low temperatures, leading to products of type **313** such as **313a** in 95% yield (Scheme 4.69) [169].

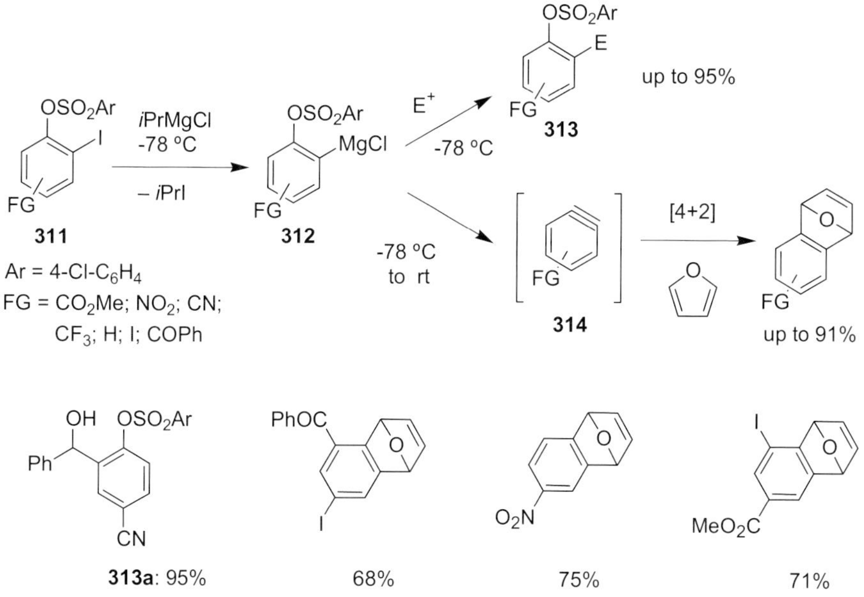

Scheme 4.69 Preparation of functionalized arynes.

At higher temperatures, the elimination reaction is favored, and thus functionalized benzynes of type **314** are formed, which react with furan in [4+2]-cycloaddition (Scheme 4.69). Remarkably, benzyne **314** reacts with magnesium thiolates and amides as nucleophiles in an addition reaction, leading to a new carbon–magnesium bond, which can further react with a variety of electrophiles (Scheme 4.70) [170].

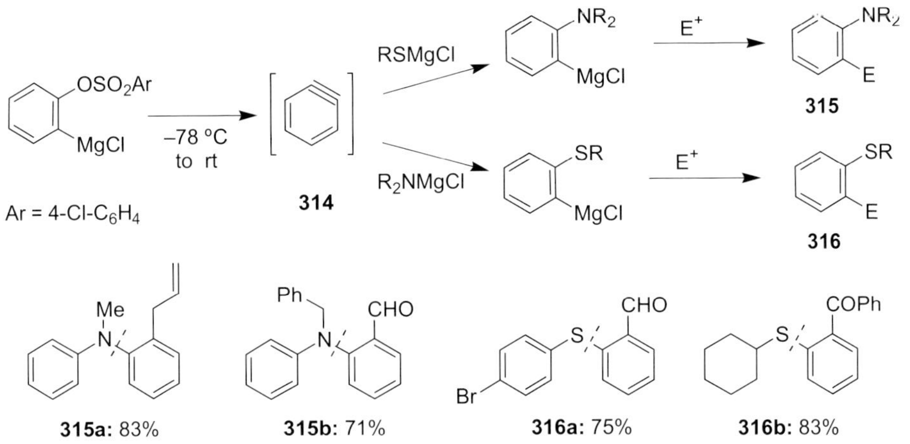

Scheme 4.70 Preparation of arylamines and aryl thioethers by addition reactions to benzyne.

This multicomponent reaction gives rise to a variety of *ortho*-functionalized arylamines **315** and thioethers **316** in good overall yields. The reaction can be extended to functionalized benzynes. Thus, the sulfonate **317** readily provides the expected aryne that adds regioselectively magnesium thiolate **318** providing the magnesium derivative **319**. Its copper(I)-catalyzed trapping furnishes the tetrasubstituted benzene **320** in 72% yield (Scheme 4.71) [171].

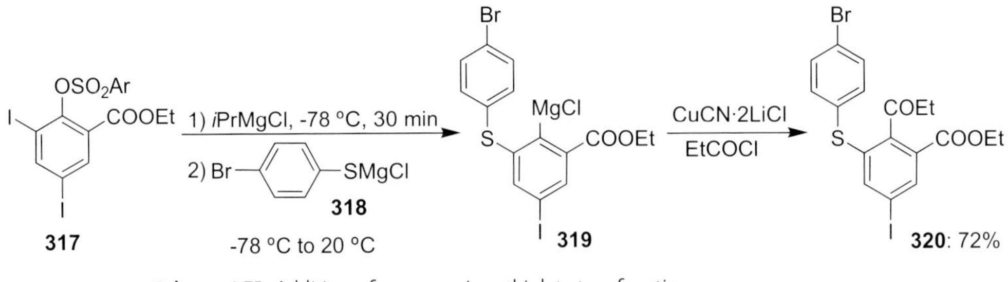

Scheme 4.71 Addition of a magnesium thiolate to a functionalized aryne leading to a tetrasubstituted benzene derivative.

4.4
Application of Functionalized Magnesium Reagents in Cross-coupling Reactions

The availability of functionalized Grignard reagents considerably enhances the scope of these reagents for performing cross-coupling reactions with various transition metals.

4.4.1
Palladium-catalyzed Cross-coupling Reactions

The direct palladium-catalyzed cross-coupling reaction of Grignard reagents (Kumada cross-coupling) has been extensively studied [172]. However, the high reactivity of Grignard reagents limits the functional-group tolerance at the higher temperatures that are normally required for these reactions. Thus, most reactions have to be conducted at low temperatures and therefore most often activated heterocycles are used as substrates. Functionalized Grignard reagents such as **321** participate in cross-coupling reactions with various 2-halopyridines of type **322** in the presence of Pd(0)-catalysts. These remarkably fast cross-coupling reactions require the presence of a Pd(0)-catalyst and are therefore not addition-elimination reactions of the Grignard reagent. In the absence of a Pd(0)-complex, no reaction is observed. These reactions may proceed via the formation of an organo-palladate [173] of the type $ArPdL_2^{(-)} MgX^{(+)}$ that would undergo a fast addition-elimination reaction with the 2-chloropyridine derivative **322** leading to the functionalized pyridine **323** in 87% yield (Scheme 4.72) [174].

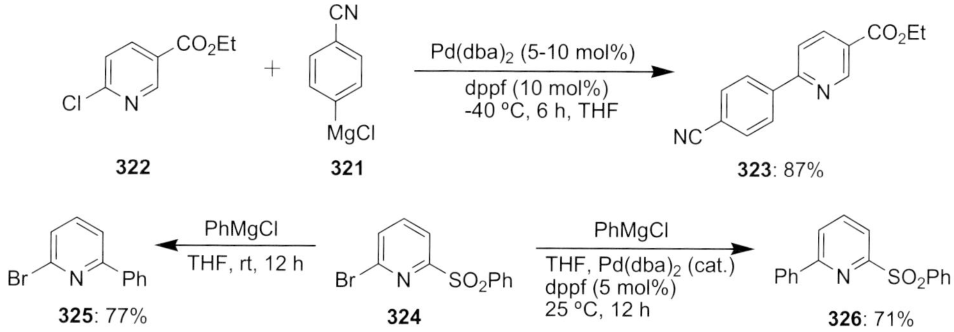

Scheme 4.72 Pd-catalyzed cross-coupling with 2-halopyridines.

This reaction can be extended to several haloquinolines and Quéguiner found an interesting selectivity in the cross-coupling of bromosulfone **324** [175]. Thus, PhMgCl reacts with the disubstituted pyridine **324** by direct substitution of the phenylsulfonyl group leading to the bromopyridine **325** in 77% yield. Using a Pd(0)-catalyst, the highly functionalized biaryl **326** is obtained in 71% yield (Scheme 4.72).

Most often a Grignard reagent is transmetalated *in situ* to the corresponding zinc reagent, which shows a lower reactivity towards nucleophilic additions. Thus, Negishi cross-coupling reactions have found significantly more applications [176]. Especially interesting are arylmagnesium reagents bearing amino groups [41,177]. A range of 2-arylated-1,4-phenylenediamines of type **327** can be prepared starting from the *bis*-imine **56** with the I/Mg-exchange being complete within 3 h at −10 °C. After transmetallation to the zinc reagent with ZnBr$_2$, *bis*(dibenzylideneacetone)palladium (Pd(dba)$_2$; 5 mol%), *tris-o*-furylphosphane (tfp; 10 mol%) [178] and 4-iodoanisole are added. The Negishi cross-coupling reaction is usually complete after 16 h at 25 °C leading to the 1,4-phenylenediamine **327** in 79% yield (Scheme 4.73) [177].

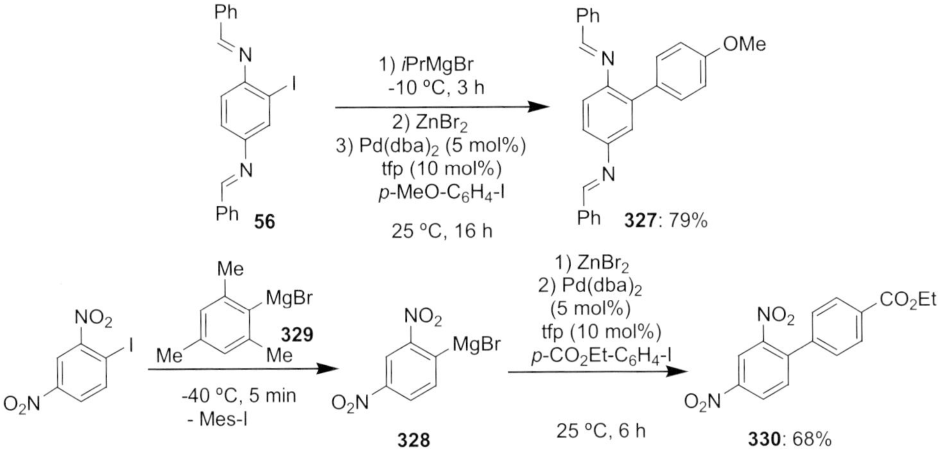

Scheme 4.73 Cross-coupling with nitrogen-functionalized Grignard reagents.

The nitro-substituted arylmagnesium species **328** is best prepared using sterically hindered mesitylmagnesium bromide **329** [179]. Thus, the reaction of the zinc derivative of **328** with ethyl 4-iodobenzoate (THF, −40 °C to rt, 3 h) in the presence of Pd(dba)$_2$ (5 mol%) and tfp (10 mol%) provides the biaryl **330** in 68% yield (Scheme 4.73) [179].

The cross-coupling of heterocycles is also possible and thus, polyfunctional zinc reagent **331** obtained from the iodide **332** via a I/Mg exchange, followed by a transmetallation, reacts readily in the presence of the highly active palladium catalyst Pd(*t*-Bu$_3$P)$_2$ [180] under mild conditions, furnishing the biaryl **333** in 87% yield (Scheme 4.74) [181].

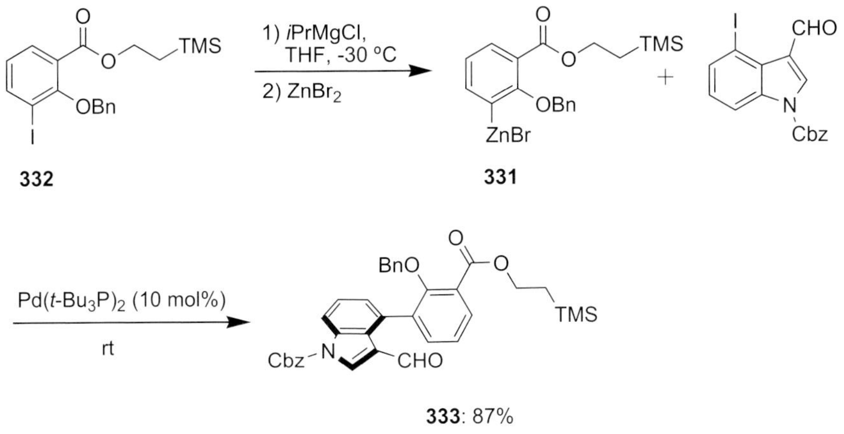

Scheme 4.74 Pd-catalyzed cross-coupling of highly functionalized arylzinc reagents.

4.4.2
Nickel-catalyzed Cross-coupling Reactions

As already mentioned, Kumada and Corriu simultaneously developed the first cross-coupling reactions, between aromatic halides and Grignard reagents catalyzed by nickel salts [4,5,172b]. A variety of nickel salts and ligands have been studied [182] since these early reports and even enantioselective versions have been reported. Herrmann reported a cross-coupling between inactivated aryl chlorides and aryl Grignard reagents in the presence of Ni(acac)$_2$ and *tris*-(*tert*-butyl)-phosphane as ligand [183]. Thus, chlorobenzene (**334**) reacts with the magnesium derivative **335** at room temperature within 18 h and leads to the desired biphenyl **336** in 99% yield (Scheme 4.75).

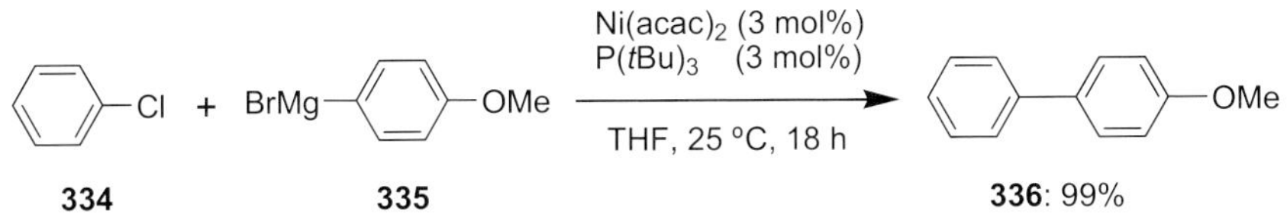

Scheme 4.75 Ni-catalyzed cross-coupling using aryl chlorides and aryl Grignard reagents

The functional-group tolerance is again moderate and therefore a transmetallation to the corresponding zinc reagent is advantageous. Functionalized aryl-zinc compounds allow the performance of sp^2-sp^3 cross-coupling reactions using Ni(acac)$_2$ (10 mol%) as catalyst in the presence of 4-trifluoromethylstyrene or 4-fluorostyrene as promoter of the reductive elimination step. Under these conditions, the Grignard reagent **337** reacts with the iodothioketal **338** providing the desired cross-coupling product **339** in 72% yield (Scheme 4.76) [184].

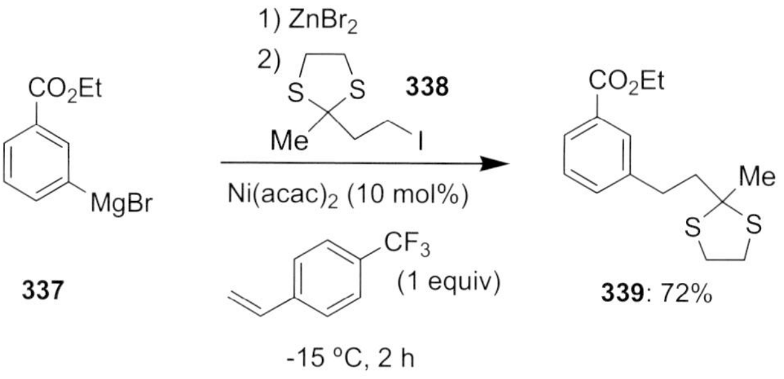

Scheme 4.76 Ni-catalyzed cross-coupling between functionalized Grignard reagents and functionalized alkyl iodides.

An alternative to this Ni-catalyzed reaction is the corresponding copper-mediated reaction. In this case, the functionalized arylmagnesium species is transmetalated to the corresponding arylcopper reagent with $CuCN \cdot 2LiCl$ [49] in the presence of trimethylphosphite (1.9 equiv) (Scheme 4.77). This last additive confers an
excellent stability to the copper reagent that can be handled at room temperature
under these conditions. Thus, the reaction of the magnesium species **19** with
$CuCN \cdot 2LiCl$ [49] and $P(OMe)_3$ furnishes the stable arylcopper **340** that undergoes
a smooth cross-coupling reaction with functionalized alkyl iodides such as the
iodopivalate **342**, leading to the substitution product **341** in 89% yield [184].

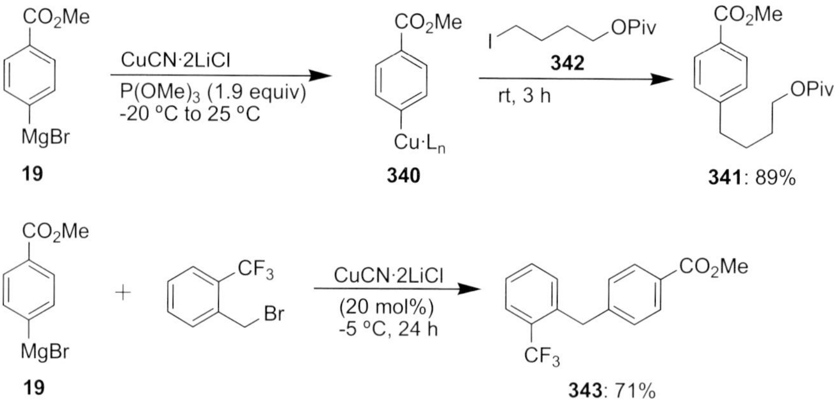

Scheme 4.77 Cu-mediated cross-coupling reactions of functionalized arylmagnesium compounds.

Interestingly, reactive benzylic halides undergo the cross-coupling reaction in
the presence of a catalytic amount of $CuCN \cdot 2LiCl$ [49] leading to diphenyl-
methane derivatives such as **343** (Scheme 4.77) [185].

4.4.3
Iron-catalyzed Cross-coupling Reactions

Despite the seminal contribution of Kochi and coworkers, [186] iron-catalyzed cross-coupling reactions received less attention then the corresponding nickel- and palladium-mediated ones. The low cost and low toxicity of iron(III) salts, initiated a renaissance of this cross-coupling procedures in the search for new environmentally benign carbon–carbon bond-forming processes. In addition, the iron-catalyzed cross-coupling reactions most often can be carried out under ligand-free conditions. Although the mechanisms of these reactions are far from clear, it is speculated that highly reduced iron–magnesium clusters of the formal composition $[Fe(MgX)_2]_n$ generated in situ may play a decisive role in the catalytic cycle [187].

Cahiez and Knochel demonstrated the generality of iron-catalyzed alkenylation and recognized the advantageous use of cosolvents, such as NMP [188,189]. For example, the reaction of bromide **344** with BuMgCl furnishes in the presence of only 1 mol% Fe(acac)$_3$ product **345** in 79% yield (Scheme 4.78) [188].

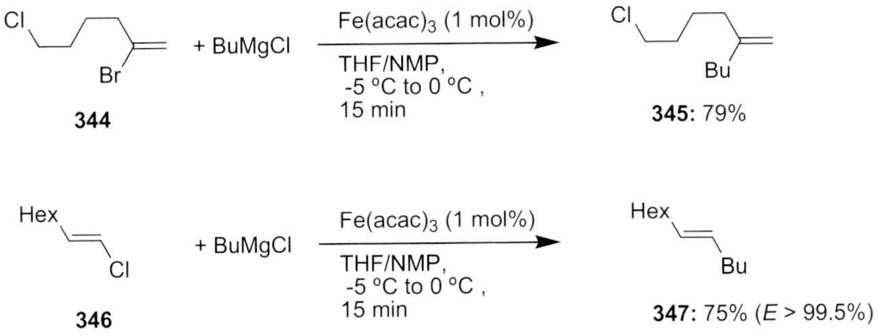

Scheme 4.78 Fe(iii)-catalyzed cross-coupling reactions of alkylmagnesium halides with bromo- and chloroolefins.

Similarly, (*E*)-1-chlorooctene (**346**) reacts with BuMgCl and leads exclusively to the *E*-olefin **347** without isomerization of the double bond (Scheme 4.78).

This methodology can be extended to aromatic Grignard reagents as well, leading to sp^2-sp^2 cross-coupling reactions [189]. Thus, functionalized arylmagnesium reagent **19** undergoes efficient cross-coupling reactions with polyfunctionalized alkenyl iodides such as **348** in the presence of Fe(acac)$_3$ (5 mol%) leading to the styrene derivative **349** in 69% yield.

Remarkably, the cross-coupling reaction is complete at –20 °C within 15–30 min (Scheme 4.79). The arylmagnesium compound can bear various electrophilic functions like a nonaflate [190] (see Grignard reagent **350**). The iron(III)-cross-coupling reaction still proceeds with a good yield leading to the highly functionalized nonaflate **351** in 73% yield (Scheme 4.79) [189].

Fürstner showed that the iron-catalyzed cross-coupling is a very powerful tool for the performance of sp^2-sp^3 cross-couplings. The reaction proceeds best when

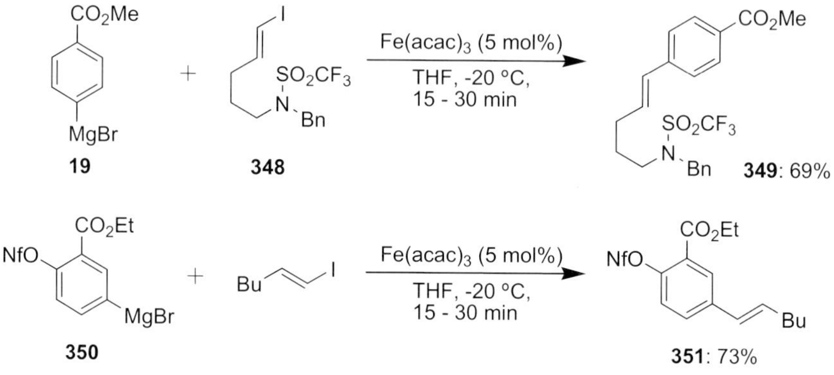

Scheme 4.79 Fe(iii)-catalyzed cross-coupling reactions with functionalized arylmagnesium species.

aryl- or heteroaryl chlorides, -tosylates or -triflates are used. Aryl bromides and iodides are less effective, leading mostly to the undesired dehalogentated side products [191]. Thus, the cross-coupling product **352** is obtained in good yields (81–91%), when reacting *n*OctMgBr with various benzoates **353a–c** (Scheme 4.80).

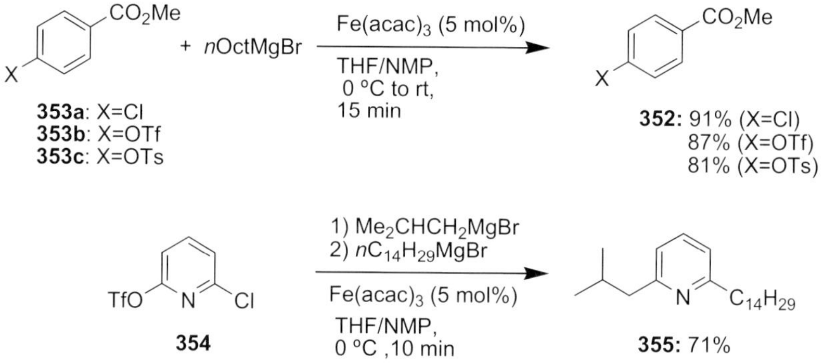

Scheme 4.80 Cross-coupling of aryl- and heteroaryl derivatives with alkylmagnesium halides.

The slightly lower reactivity of chlorides compared to triflates was used in a one-pot synthesis of compound **355** starting from **354**, to where alkyl chains were subsequently introduced (Scheme 4.80) [192]. Fürstner also showed that alkenyl triflates undergo an iron-catalyzed cross-coupling reaction with various Grignard reagents [193]. For instance, alkenyl triflate **356**, bearing an ester moiety, can be selectively cross-coupled with Grignard reagent **357** in the presence of Fe(acac)$_3$. leading to **358** in 97% yield (Scheme 4.81).

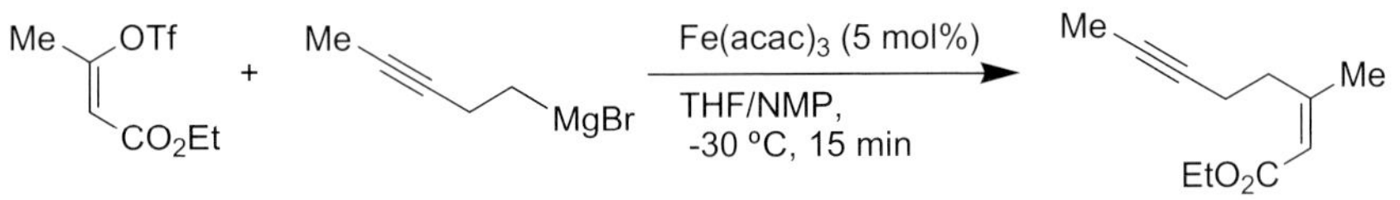

356 **357** **358**: 97%

Scheme 4.81 Iron-catalyzed cross-coupling of alkenyl triflates.

More recently, several groups independently studied the cross-coupling of arylmagnesium halides with various alkyl halides [194]. Nakamura showed that by using TMEDA as additive it was possible to suppress undesired side reactions such as olefin formation by the loss of hydrogen halide from the halide substrate or dehalogenation [194a]. Using $FeCl_3$ (5 mol%) as a catalyst, 4-methoxyphenylmagensium bromide is reacted with iodoester **359** furnishing product **360** in 88% yield (Scheme 4.82). Hayashi improved the reaction conditions with $Fe(acac)_3$ as catalyst in refluxing diethyl ether [194b]. Thus, the reaction of **361** with 4-tolylmagnesium bromide gave 69% of the cross-coupling product **362** exclusively with the alkyl bromide, leaving the triflate group unreacted (Scheme 4.82).

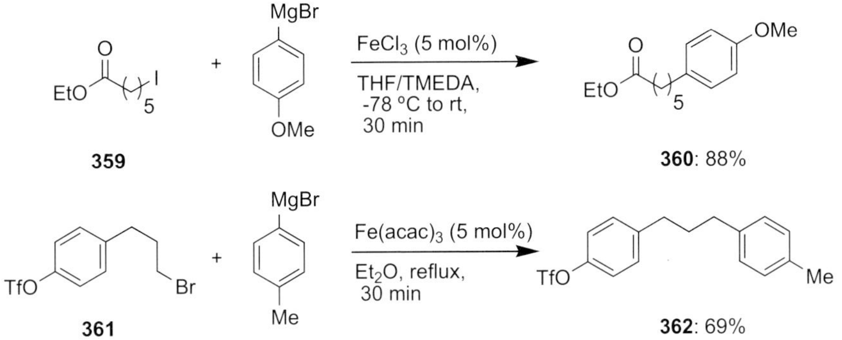

359 **360**: 88%

361 **362**: 69%

Scheme 4.82 Cross-coupling reactions of alkyl halides with arylmagnesium derivatives.

Finally, Fürstner showed that the low-valent tetrakis(ethylene)ferrate complex $[Li(tmeda)]_2[Fe(C_2H_4)_4]$ is an excellent catalyst for the cross-coupling reaction between alkyl halides and aryl Grignard reagents (Scheme 4.83) [194c].

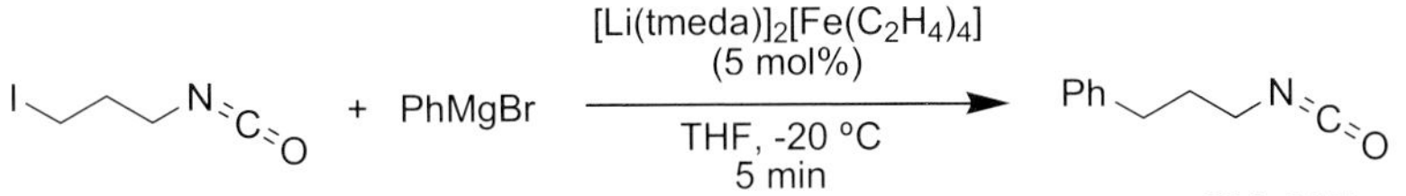

363 **364**: 90%

Scheme 4.83 Cross-coupling reactions of alkyl halides with arylmagnesium derivatives.

The reaction of iodide **363** affords the cross-coupling product **364** without the addition of any further additives. These reactions clearly demonstrate that iron-catalyzed cross-coupling reactions have an excellent reaction scope especially for the formation of Csp^3–Csp^2 bond-formation type of cross-coupling where palladium catalysis usually gives moderate results. Recently, Knochel found that the coupling of two aryl moieties is possible by transmetalating the Grignard reagents to the corresponding copper derivatives **365**. Thus, biaryls of type **366** are easily prepared in the presence of catalytic amounts of $Fe(acac)_3$ with good yields (Scheme 4.84) [195].

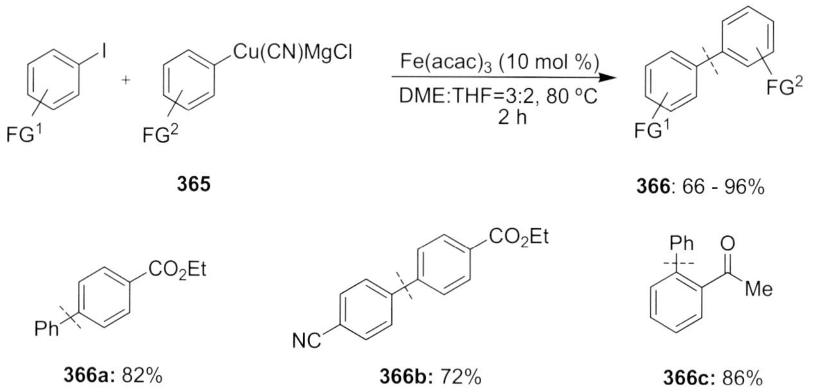

365

366: 66 - 96%

366a: 82%

366b: 72%

366c: 86%

Scheme 4.84 Iron-catalyzed aryl-aryl cross-coupling.

The transmetallation of the Grignard reagent to the corresponding copper reagent has two advantages. First, the dehalogenation, resulting tentatively from a halogen–magnesium exchange reaction is suppressed and secondly, the competitive reductive homo-coupling of two arylcopper reagents is significantly reduced. This reaction has a broad scope and can efficiently be applied to heterocyclic systems (Scheme 4.85) [195].

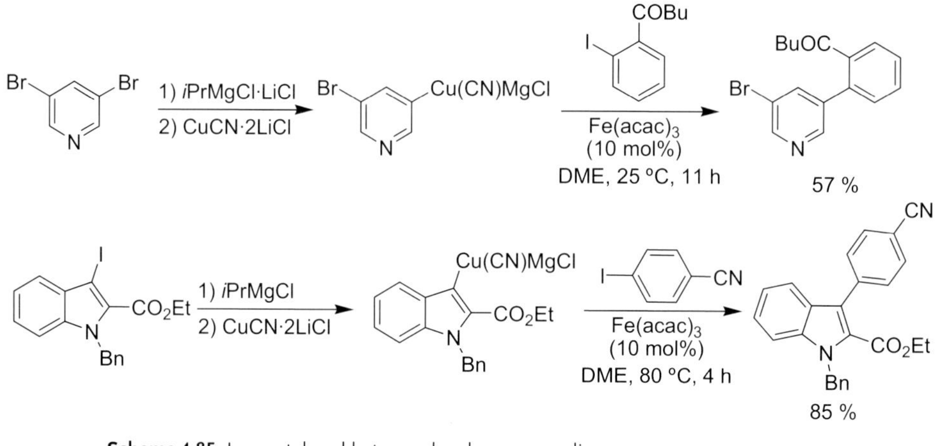

57 %

85 %

Scheme 4.85 Iron-catalyzed heteroaryl-aryl cross-coupling.

The catalytic activity of iron salts in the cross-coupling reaction of Grignard reagents with acyl chlorides and thiolesters was already discovered in the 1980s, but the relevance of this method was for a long time not fully explored [196]. Due to the very good functional-group tolerance of Grignard reagents, very powerful ketone syntheses were reported by Knochel and Fürstner, who extended this method far beyond the scope of the initial reports. Diarylketones (**367**) are easily accessible in high yields, by the reaction of aroyl cyanides of type **368** with functionalized Grignard reagents (Scheme 4.86) [197].

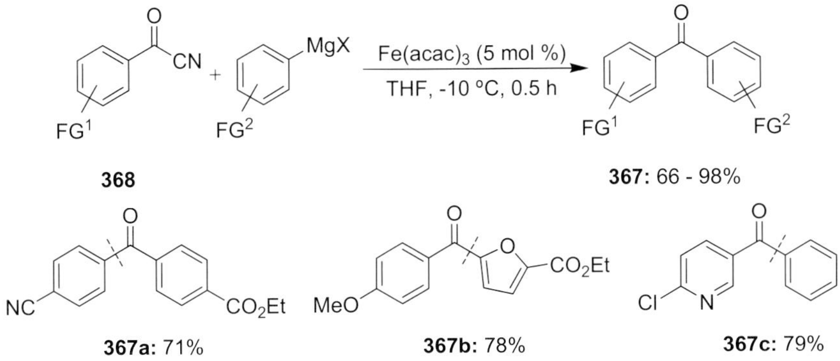

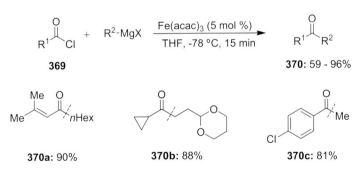

Scheme 4.86 Synthesis of diarylketones **367**. The dotted lines indicate the new C–C bond.

The more reactive alkyl Grignard reagents require lower temperatures for the cross-coupling reaction, but the yields are generally high (Scheme 4.87) [193]. The reaction of acyl chlorides (**369**) with a variety of Grignard reagents furnishes at –78 °C the desired ketones **370a–c** in good yields.

Scheme 4.87 Synthesis of ketones **370**. The dotted lines indicate the new C–C bond.

4.5
Summary and Outlook

The halogen–magnesium exchange reaction has opened new perspectives in organic synthesis. Practical questions may arise for the optimum choice of the best reaction conditions. Which are the best reagents for performing this exchange reaction? *i*PrMgCl is usually the best reagent for performing an I/Mg-exchange. Whereas the more active exchange reagent *i*PrMgCl·LiCl is well suited for performing a Br/Mg-exchange and allows an I/Mg-exchange to be performed under exceedingly mild reaction conditions. Sensitive aromatic iodides bearing a nitro group react unselectively with *i*PrMgCl but can also be converted to the corresponding Grignard reagent with the selective reagent PhMgCl. Thus, these general rules give an access to numerous new functionalized Grignard reagents. Many more functional groups than previously thought are compatible with magnesium organometallics. The mild conditions required for performing a halogen–magnesium exchange are the key for assuring a high functional-group tolerance. This again places Grignard reagents in a central position for organic chemistry and opens fascinating new perspectives. The high functional-group tolerance shows that organic chemistry have only partially mastered the reactivity of organometallic reagents for the elaboration of complex organic molecules and much progress will be done in future years.

References and Notes

1 V. Grignard, *Compt. Rend. Acad. Sci. Paris*, **1900**, *130*, 1322.
2 B. H. Lipshutz, S. Sengupta, *Org. Reactions.* **1992**, *41*, 135.
3 M. S. Kharasch, C. F. Fuchs *J. Am. Chem. Soc.* **1943**, *65*, 504.
4 K. Tamao, K. Sumitani, M. Kumada, *J. Am. Chem. Soc.* **1972**, *94*, 4374.
5 R. J. P. Corriu, J. P. Masse, *J. Chem. Soc. Chem. Commun.* **1972**, 144.
6 a) H. Urabe, F. Sato, in *Handbook of Grignard Reagents* Eds.: G. S. Silverman, P. E. Rakita; Marcel Dekker, New York, 1996, p. 577; b) B. J. Wakefield, *Organomagnesium Methods in Organic Synthesis*; Academic Press, London **1995**.
7 K. Tamao, M. Kumada, *The Chemistry of the Metal-Carbon Bond*, Vol. 4 Ed.: F. R. Hartley; Wiley, Chichester, 1987, p. 819.
8 J. Tsuji, *Transition Metal Reagents and Catalysts*, Wiley, Chichester 2000.
9 a) H. G. Richey, Jr., *Grignard Reagents* Wiley, New York, **2000**; b)

M. S. Kharasch, O. Reinmuth, *Grignard Reactions of Nonmetallic Substances*, Prentice-Hall, New York, **1954**, c) G. S. Silverman, P. E. Rakita, *Handbook of Grignard-Reagents*, Marcel Dekker: New York, **1996**;
10 W. E. Lindsell, *Comprehensive Organometallic Chemistry II*, Vol. 1, Pergamon Press, Oxford, **1995**, Chapter 3, pp. 72–78 and references therein.
11 a) C. Hamdouchi, H. M. Walborsky, in: G. S. Silverman, P. E. Rakita (Eds.), *Handbook of Grignard-Reagents*, Marcel Dekker: New York, **1996**, pp.145; b) J. F. Garst, F. Ungvary, in H. G. Richey, Jr. (Ed.), *Grignard Reagents* Wiley, Chichester, **2000**, pp. 185; c) M. S. Kharasch, O. Reinmuth, *Grignard Reactions of Nonmetallic Substances*, Prentice-Hall, New York, **1954**; d) J. F. Garst, F, Ungváry, J. T. Baxter, *J. Am. Chem. Soc.* **1997**, *119*, 253; e) J. F. Garst, M. P. Soriaga, *Coordination Chemistry Reviews* **2004**, *248*, 623.

12 a) F. Bickelhaupt in H. G. Richey, Jr. (Ed.), *Grignard Reagents* Wiley, Chichester, **2000**, pp. 299; b) H. L. Uhm, in G. S. Silverman, P. E. Rakita (Eds.), *Handbook of Grignard-Reagents*, Marcel Dekker: New York, **1996**, pp.117.

13 a) A. Boudier, L. O. Bromm, M. Lotz, P. Knochel, *Angew. Chem.* **2000**, *112*, 4584; *Angew. Chem. Int. Ed.* **2000**, *39*, 4415; b) C. Najera, M. Yus, *Recent Res. Dev. Org. Chem.* **1997**, *1*, 67; c) A. Inoue, K. Oshima in, *Main Group Metals in Organic Synthesis*, H. Yamamoto, K. Oshima, Eds., Wiley-VCH, Weinheim, **2004**.

14 P. E. Rakita, J. F. Aultman, L. Stapleton, *Chem. Eng.* **1990**, *97*, 110.

15 For factors controlling Grignard reagent formation, see: W. E. Lindsell, *Comprehensive Organometallic Chemistry I*, Vol. 1, Pergamon Press, Oxford, **1982**, Chapter 3, pp. 155–252 and references therein.

16 a) H. M. Walborsky, *Acc. Chem. Res.* **1990**, *23*, 286; b) J. F. Garst, *Acc. Chem. Res.* **1991**, *24*, 95; c) C. Walling, *Acc. Chem. Res.* **1991**, *24*, 255; c) H. R. Rogers, C. L. Hill, Y. Fujiwara, R. J. Rogers, H. L. Mitchell und G. M. Whitesides, *J. Am. Chem. Soc.* **1980**, *102*, 217; d) K. S. Root, C. L. Hill, L. M. Lawrence, G. M. Whitesides, *J. Am. Chem. Soc.* **1989**, *111*, 5404; e) E. C. Ashby, J. Oswald, *J. Org. Chem.* **1988**, *53*, 6068.

17 M. Vallino, *J. Organomet. Chem.* **1969**, *20*, 1.

18 a) R. D. Rieke, *Science* **1989**, *246*, 1260; b) T. P. Burns, R. D. Rieke, *J. Org. Chem.* **1987**, *52*, 3674; c) J. Lee, R. Velarde-Ortiz, A. Guijarro, J. R. Wurst, R. D. Rieke, *J. Org. Chem.* **2000**, *65*, 5428; d) R. D. Rieke, T.-J. Li, T. P. Burns, S. T. Uhm, *J. Org. Chem.* **1981**, *54*, 4323; e) R. D. Rieke, M. S. Sell, W. R. Klein, T. Chen, J. D. Brown, M. V. Hansan, in *Active Metals*, A. Fuerstner, Ed., Wiley-VCH, Weinheim, **1995**.

19 a) V. Snieckus, *Chem. Rev.* **1990**, *90*, 879; b) C. G. Hartung, V. Snieckus, *Modern Arene Chemistry*, **2002**, 330.

20 a) A. B. Holmes, C. N. Sporikou, *Org. Synth.* **1987**, *65*, 61; b) H. J. Bestman, T. Brosche, K. H. Koschatzky, K. Michaelis, H. Platz, K. Roth, J. Suess, O. Vostrowsky, W. Knauf, *Tetrahedron Lett.* **1982**, *23*, 4007; c) L. Poncini, *Bulletin des Societes Chimiques Belges* **1983**, *92*, 215.

21 P. E. Eaton, C. H. Lee, Y. Xiong, *J. Am. Chem. Soc.* **1989**, *111*, 8016.

22 P. Beak, C. J. Upton, *J. Org. Chem.* **1975**, *40*, 1094.

23 a) P. E Eaton, K. A. Lukin, *J. Am. Chem. Soc.* **1993**, *115*, 11370; b) M.-X. Zhang, P. E. Eaton, *Angew. Chem.* **2002**, *114*, 2273; *Angew. Chem. Int. Ed.* **2002**, *41*, 2169.

24 a) M. Shilai, Y. Kondo, T. Sakamoto, *J. Chem. Soc., Perkin Trans. 1* **2001**, 442; b) W. Schlecker, A. Huth, E. Ottow, *J. Org. Chem.* **1995**, *60*, 8414; c) W. Schlecker, A. Huth, E. Ottow, J. Mulzer, *Liebigs Ann.* **1995**, 1441; d) Y. Kondo, A. Yoshida, T. Sakamoto, *J. Chem. Soc. Perkin Trans.1* **1996**, 2331; e) A. Dinsmore, D. G. Billing, K. Mandy, J. P. Michael, D. Mogano, S. Patil, *Org. Lett.* **2004**, *6*, 293.

25 a) D. Bonafoux, M. Bordeau, C. Biran, J. Dunogues, *J. Organomet. Chem.* **1995**, *493*, 27; b) G. Lessene, R. Tripoli, P. Cazeau, C. Biran, M. Bordeau, *Tetrahedron Lett.* **1999**, *40*, 4037. for a review see: K. W. Henderson, W. J. Kerr, *Chem. Eur. J.* **2001**, *7*, 3430

26 a) D. A. Evans, S. G. Nelson, *J. Am. Chem. Soc.* **1997**, *119*, 6452; b) K. W. Henderson, W. J. Kerr, J. H. Moir, *Tetrahedron* **2002**, *58*, 4573.

27 G. Wittig, U. Pockels, H. Dröge, *Chem. Ber.* **1938**, *71*, 1903.

28 a) R. G. Jones, H. Gilman, *Org. Reactions* **1951**, *6*, 339; b) H. Gilman, W. Langham, A. L. Jacoby, *J. Am. Chem. Soc.* **1939**, *61*, 106.

29 a) W. E. Parham, L. D. Jones, *J. Org. Chem.* **1976**, *41*, 1187; b) W. E. Parham, L. D. Jones, Y. Sayed, *J. Org. Chem.* **1975**, *40*, 2394; c) W. E. Parham, L. D. Jones, *J. Org. Chem.* **1976**, *41*, 2704; d) W. E. Parham, D. W. Boykin, *J. Org. Chem.* **1977**, *42*, 260; e) W. E. Parham, R. M. Piccirilli, *J. Org. Chem.* **1977**, *42*, 257; f) C. E. Tucker, T. N. Majid, P. Knochel, *J. Am. Chem. Soc.* **1992**, *114*, 3983.

30 C. Prévost, *Bull. Soc. Chim. Fr.* **1931**, *49*, 1372.

31 E. Urion, *Comp. Rend. Acad. Sci. Paris* **1934**, *198*, 1244.

32 a) W. F. Bailey, J. J. Patricia, *J. Organomet. Chem.* **1988**, *352*, 1; b) H. J. Reich, N. H. Phillips, I. L. Reich, *J. Am. Chem. Soc.* **1985**, *107*, 4101; c) W. B. Farnham, J. C. Calabrese, *J. Am. Chem. Soc.* **1986**, *108*, 2449.

33 a) O. R. Pierce, A. F. Meiners, E. T. McBee, *J. Am. Chem. Soc.* **1953**, *75*, 2516; b) E. T. McBee, C. W. Roberts, A. F. Meiners, *J. Am. Chem. Soc.* **1957**, *79*, 335; c) P. Moreau, R. Albachi, A. Commeyras, *Nouv. J. Chim.* **1977**, *1*, 497.

34 a) R. D. Chambers, W. K. R. Musgrave, J. Savory, *J. Chem. Soc.* **1962**, 1993; b) for a review on fluorinated organometallics, see: D. J. Burton, Z. Y. Yang, *Tetrahedron* **1992**, *48*, 189.

35 a) J. Villiéras, *Bull. Soc. Chim. Fr.* **1967**, *5*, 1520; b) J. Villiéras, B. Kirschleger, R. Tarhouni, M. Rambaud, *Bull. Soc. Chim. Fr.* **1986**, 470.

36 For recent examples, see a) A. Müller, M. Marsch, K. Harms, J. C. W. Lohrenz, G. Boche, *Angew. Chem.* **1996**, *108*, 1639; *Angew. Chem. Int. Ed. Engl.* **1996**, *35*, 1518; R. W. Hoffmann, M. Julius, F. Chemla, T. Ruhland, G. Frenzen, *Tetrahedron* **1994**, *50*, 6049.

37 C. Tamborski, G. J. Moore, *J. Organomet. Chem.* **1971**, *26*, 153.

38 N. Furukawa, T. Shibutani, H. Fujihara, *Tetrahedron Lett.* **1987**, *28*, 5845.

39 For other examples of halogen–magnesium exchange reactions, see a) H. H. Paradies, M. Görbing, *Angew. Chem.* **1969**, *81*, 293; *Angew. Chem. Int. Ed. Engl.* **1969**, *8*, 279; b) G. Cahiez, D. Bernard, J. F. Normant, *J. Organomet. Chem.* **1976**, *113*, 107; c) D. Seyferth, R. L. Lambert, *J. Organomet. Chem.* **1973**, *54*, 123; d) H. Nishiyama, K. Isaka, K. Itoh, K. Ohno, H. Nagase, K. Matsumoto, H. Yoshiwara, *J. Org. Chem.* **1992**, *57*, 407; e) C. Bolm, D. Pupowicz, *Tetrahedron Lett.* **1997**, *38*, 7349.

40 L. Boymond, M. Rottländer, G. Cahiez, P. Knochel, *Angew. Chem.* **1998**, *110*, 1801; *Angew. Chem. Int. Ed. Engl.* **1998**, *37*, 1701.

41 A. E. Jensen, W. Dohle, I. Sapountzis, D. M. Lindsay, V. A. Vu, P. Knochel, *Synthesis* **2002**, 565.

42 P. Cali, M. Begtrup, *Synthesis* **2002**, 63.

43 R. Kober, W. Hammes, W. Steglich, *Angew. Chem.* **1982**, *94*, 213; *Angew. Chem. Int. Ed. Engl.* **1982**, *21*, 203; b) D. von der Brück, R. Bühler, H. Plieninger, *Tetrahedron* **1972**, *28*, 791.

44 T. Graening, W. Friedrichsen, J. Lex, H.-G. Schmalz, *Angew. Chem. Int. Ed.* **2002**, *41*, 1524–1526.

45 a) M. Abarbri, F. Dehmel, P. Knochel, *Tetrahedron Lett.* **1999**, *40*, 7449; b) M. Abarbri, J. Thibonnet, L. Bérillon, F. Dehmel, M. Rottländer, P. Knochel, *J. Org. Chem.* **2000**, *65*, 4618.

46 G. Varchi, A. E. Jensen, W. Dohle, A. Ricci, G. Cahiez, P. Knochel, *Synlett* **2001**, 477.

47 J. L. Leazer Jr., R. Cvetovich, F.-R. Tsay, U. Dolling, T. Vickery, D. Bachert, *J. Org. Chem.* **2003**, *68*, 3695–3698.

48 a) E. C. Ashby, D. M. Al-Ferki, *J. Organomet. Chem.* **1990**, *390*, 275; b) M. C. Jones, *Plant/Oper. Prog.* **1989**, *8*, 200; c) J. Broeke, B.-J. Deelman, G. van Koten, *Tetrahedron Lett.* **2001**, *42*, 8085; d) P. Pinho, D. Guijarro, P. Andersson, *Tetrahedron* **1998**, *54*, 7897; e) F. A. R. Kaul, G. T. Puchta, H. Schneider, M. Grosche, D. Mihalios, W. A. Herrmann, *J. Organomet. Chem.* **2001**, *621*, 184.

49 P. Knochel, M. C. P. Yeh, S. C. Berk, J. Talbert, *J. Org. Chem.* **1988**, *53*, 2390.

50 A. Krasovskiy, P. Knochel, *Angew. Chem.* **2004**, *116*, 3369; *Angew. Chem. Int. Ed.* **2004**, *43*, 3333.

51 A. Krasovskiy, P. Knochel, manuscript in preparation.

52 a) L. Gottlieb, A. I. Meyers, *Tetrahedron Lett.* **1990**, *31*, 4723; b) A. I. Meyers, T. R. Elsworthy, *J. Org. Chem.* **1992**, *57*, 4732; c) A. I. Meyers, G. Milot, *J. Am. Chem. Soc.* **1993**, *115*, 6652.

53 W. Dohle, *Ph.D. thesis*, LMU Munich (Germany) **2002**.

54 D. M. Lindsay, W. Dohle, A. E. Jensen, F. Kopp, P. Knochel, *Org. Lett.* **2002**, *4*, 1819.

55 T. Murafuji, K. Nishio, M. Nagasue, A. Tanabe, M. Aono, Y. Sugihara, *Synthesis* **2000**, 1208.

56 G. Varchi, C. Kofink, D. M. Lindsay, A. Ricci, P. Knochel, *Chem. Commun.* **2003**, 396.

57 F. Kopp, A. Krasovskiy, P. Knochel, *Chem. Commun.* **2004**, in press.

58 S. Kato, N. Nonoyama, K. Tomimoto, T. Mase, *Tetrahedron Lett.* **2002**, *43*, 7315.

59 N. Ono, *The Nitro Group in Organic Synthesis*, Ed.: H. Feuer; Wiley, 2001.

60 a) J. Bondoch, D. Sole, S. G. Rubio, J. Bosch, *J. Am. Chem. Soc.* **1997**, *119*, 7230; b) D. Sole, J. Bondoch, S. G. Rubio, E. Peidro, J. Bosch, *Angew, Chem.* **1999**, *111*, 408; *Angew, Chem. Int. Ed.* **1999**, *38*, 395; c) D. Sole, J. Bonjoch, S. G. Rubio, E. Peidro, J. Bosch, *Chem. Eur. J.* **2000**, *6*, 655; d) C. Szantay, Z. K. Balogh, I. Moldvai, C. Szantay, E. T. Major, G. Blasko, *Tetrahedron*, **1996**, *52*, 11053.

61 a) G. Köbrich, P. Buck, *Chem. Ber.* **1970**, *103*, 1412; b) P. Buck, R. Gleiter, G. Köbrich, *Chem. Ber.* **1970**, *103*, 1431; c) P. Wiriyachitra, S. J. Falcone, M. P. Cava, *J. Org. Chem.* **1979**, *44*, 3957; d) J. F. Cameron, J. M. J. Fréchet, *J. Am. Chem. Soc.* **1991**, *113*,4303.

62 C. E. Tucker, T. N. Majid, P. Knochel, *J. Am. Chem. Soc.* **1992**, *114*, 3983.

63 I. Sapountzis, P. Knochel, *Angew. Chem.* **2002**, *114*, 1680; *Angew. Chem. Int. Ed.* **2002**, *41*, 1610.

64 I. Sapountzis, H. Dube. R. Lewis, P. Knochel, manuscript in preparation.

65 a) S. Bräse, *Acc. Chem. Res*, **2004**, *37*,804; b) D. B. Kimball, M. M. Haley, *Angew. Chem. Int. Ed.* **2002**, *41*, 3338; c) A. de Meijere, P. von Zezeschwitz, H. Nuske, B. Stulgies, *J. Organomet. Chem.* **2002**, *653*, 129.

66 W. B. Wan, R. C. Chiechi, T. J. R. Weakley, M. M. Haley, *Eur. J. Org. Chem.* **2001**, *18*, 3485.

67 C.-Y. Liu, P. Knochel, manuscript in preparation.

68 a) *Metal-catalyzed Cross-Coupling Reactions*, F. Diederich, P.J. Stang, Eds. Wiley-VCH, Weinheim, **1998**; b) *Cross-Coupling Reactions. A practical Guide*, N. Miyaura, Ed. Springer-Verlag, Berlin, **2002**; c) *Organometallics in Organic Synthesis*, E. Negishi, Wiley, New York, **1980**.

69 O. Baron, P. Knochel, *Angew. Chem. Int. Ed.* **2005**, *44*, 3133.

70 For the utility of neopentyl organometallics in zinc and copper organometallic chemistry, see a) P. Jones, C. K. Reddy, P. Knochel, *Tetrahedron* **1998**, *54*, 1471; b) P. Jones, P. Knochel, *J. Chem. Soc., Perkin Trans 1*, **1997**, 3117.

71 F. F. Kneisel, P.Knochel, *Synlett* **2002**, *11*, 1799.

72 T. Delacroix, L. Bérillon, G. Cahiez, P. Knochel, *J. Org. Chem.* **2000**, *65*, 8108.

73 J. Villiéras, M. Rambaud, *Synthesis* **1982**, 924.

74 A. Y. Fedorov, F. Carrara, J.-P. Finet, *Tetrahedron Lett.* **2001**, *42*, 5875.

75 R. M. Jacobsen, R. A. Raths, J. H. McDonald, *J. Org. Chem.* **1977**, *42*, 2545.

76 F. F. Kneisel, Y. Monguchi, K. M. Knapp, H. Zipse, P. Knochel, *Tetrahedron Lett.* **2002**, *43*, 4875.

77 K. Oshima, *J. Organomet. Chem.* **1999**, *575*, 1.

78 a) K. Kitagawa, A. Inoue, H. Shinokubo, K. Oshima, *Angew. Chem.* **2000**, *112*, 2594; *Angew. Chem. Int. Ed. Engl.* **2000**, *39*, 2481; b) A. Inoue, K. Kitagawa, H. Shinokubo, K. Oshima, *J. Org. Chem.* **2001**, *66*, 4333; c) A. Inoue, K. Kitagawa, H. Shinokubo, K. Oshima, *Tetrahedron* **2000**, *56*, 9601; see also d) R. I. Yousef, T. Rüffer, H. Schmidt, D. Steinborn, *J. Organomet. Chem.* **2002**, *655*, 111.

79 M. Rottländer, L. Boymond, L. Bérillon, A. Leprêtre, G. Varchi, S. Avolio, H. Laaziri, G. Quéguiner, A. Ricci, G. Cahiez, P. Knochel, *Chem. Eur. J.* **2000**, *6*, 767.

80 L. Bérillon, A. Leprêtre, A. Turck, N. Plé, G. Quéguiner, G. Cahiez, P. Knochel, *Synlett*, **1998**, 1359.

81 a) A. E. Gabarda, W. Du, T. Isarno, R. S. Tangirala, D. P. Curran, *Tetrahedron* **2002**, 6329–6341; b) A. E. Gabarda, D. P. Curran, *J. Comb. Chem.* **2003**, *5*, 617–624; c) K. Yabu, S. Masumoto, S. Yamasaki, Y. Hamashima, M. Kanai, W. Du, D. P. Curran, M. Shibasaki, *J. Am. Chem. Soc.* **2001**, *123*, 9908.

82 A. Staubitz, W. Dohle, P. Knochel, *Synthesis* **2003**, *2*, 233.

83 C. Jaramillo, J. C. Carretero, J. E. de Diego, M. del Prado, C. Hamdouchi, J. L. Roldán and C. Sánchez-Martínez, *Tetrahedron Lett.* **2002**, *43*, 9051.

84 I. Collins, *J. Chem. Soc., Perkin Trans. 1*, **2000**, 2845.

85 G. Quéguiner, F. Marsais, V. Snieckus, J. Epsztajin, *Adv. in Heterocyclic Chem.* **1991**, *52*, 187.

86 M. Bergauer, P. Gmeiner, *Synthesis* **2001**, 2281.

87 H. Kromann, F. A. Slok, T. N. Johansen, P. Krogsgaard-Larsen, *Tetrahedron* **2001**, *57*, 2195.

88 J. Felding, J. Kristensen, T. Bjerregaard, L. Sander, P. Vedso, M. Begtrup, *J. Org. Chem.* **1999**, *64*, 4196.

89 a) M. R. Dobler, *Tetrahedron Lett.* **2003**, *44*, 7115; b) Y. Chen, H. V. Rasika Dias, C. J. Lovely, *Tetrahedron Lett.* **2003**, *44*, 1379–1382; c) C. J. Lovely, H. Du, H. R. Dias, *Org. Lett.* **2001**, *3*, 1319; see also d) R. S. Loewe, S. M. Khersonsky, R. D. McCullough; *Adv. Mater.* **1999**, *11*, 250.

90 C. Christophersen, M. Begtrup, S. Ebdrup, H. Petersen, P. Vedso, *J. Org. Chem.* **2003**, *68*, 9513–9516.

91 B. H. Lipshutz, W. Hagen *Tetrahedron Lett.* **1992**, *33*, 5865.

92 A. Spiess, G. Heckmann, T. Bach, *Synlett* **2004**, 131.

93 W. Cai, D. H. Brown Ripin, *Synlett* **2002**, 273–274.

94 a) F. Trécourt, G. Breton, F. Mongin, F. Marsais, G. Quéguiner, *Tetrahedron Lett.* **1999**, *40*, 4339; b) A. Leprêtre, A. Turck, N. Plé, P. Knochel, G. Quéguiner, *Tetrahedron* **2000**, *56*, 265.

95 a) T. Mase, I. N. Houpis, A. Akao, I. Dorziotis, K. Emerson, T. Hoang, T. Iida, T. Itoh, K. Kamei, S. Kato, Y. Kato, M. Kawasaki, F. Lang, J. Lee, J. Lynch, P. Maligres, A. Molina, T. Nemoto, S. Okada, R. Reamer, J. Z. Song, D. Tschaen, T. Wada, D. Zewge, R. P. Volante, P. J. Reider, K. Tomimoto, *J. Org. Chem.* **2001**, *66*, 6775; b) T. Ida, T. Wada, K. Tomimoto, T. Mase, *Tetrahedron Lett.* **2001**, *42*, 4841.

96 for the metalation of quinolines and pyrimidines using trialkymagnesium-ate complexes, see: a) F. D. Therkelsen, M. Rottlaender, N. Thorup, E. Bjerregaard Pedersen, *Org. Lett.* **2004**, *6*, 1991; b) S. Dumouchel, F. Mongin, F. Trecourt, G. Queguiner, *Tetrahedron Lett.* **2003**, *44*, 2033.

97 G. R. Newkome, W. W. Pandler, *Contemporary Heterocyclic Chemistry*, Wiley, New York, **1982**.

98 M. Abarbri, P. Knochel, *Synlett* **1999**, 1577.

99 T. Tobrman, D. Dvorak, *Org. Lett.* **2003**, *5*, 4289.

100 A. B. Smith III, K. P. Minbiole, P. R. Verhoest, M. Schelhaas, *J. Am. Chem. Soc.* **2001**, *123*, 10942.

101 F. Kopp, A. Krasovskiy, P. Knochel, *Chem. Commun.* **2004**, *20*, 2288.

102 F. Kopp, P. Knochel, manuscript in preparation

103 M. Rottländer, L. Boymond, G. Cahiez, P. Knochel, *J. Org. Chem.* **1999**, *64*, 1080.

104 H. Ren, A. Krasovskiy, P. Knochel, *Org. Lett.* **2004**, *6*, submitted.

105 S. S. Kim, G. Rajagopal, D. H. Song, *J. Organomet. Chem.* **2004**, *689*, 1734.

106 H. Ren, A. Krasovskiy, P. Knochel, *Chem. Commun.* Submitted.

107 M. Rottlaender, P. Knochel, *J. Comb. Chem.* **1999**, *1*, 181.

108 I. Sapountzis, W. Dohle, P. Knochel, *Chem. Commun.* **2001**, 2068.

109 J. Thibonnet, A. Duchene, J.-L. Parrain, M. Abarbri, *J. Org. Chem.* **2004**, *69*, 4262.

110 Dissertation, Matthias Lotz LMU Munich (Germany), 2002.

111 a) H. Gurien, *J. Org. Chem.* **1963**, *28*, 878; b) J. Ficini, J. C. Depezay, *Bull. Soc. Chim. Fr.* **1966**, 3878; c) F. G. Mann, F. H. Stewart, *J. Chem. Soc.* **1954**, 2826; d) T. Reichstein, J. Baud, *Helv. Chim. Acta* **1937**, *20*, 892; see also e) M. I. Calaza, M. R. Paleo, F. J. Sardina, *J. Am. Chem. Soc.* **2001**, *123*, 2095; f) F. Foubelo, A. Gutierrez, M. Yus, *Synthesis* **1999**, 503; g) F. F. Fleming, B. C. Shook, *Tetrahedron Lett.* **2000**, *41*, 8847.

112 T. Iwaoka, T. Murohashi, N. Katagiri, M. Sato, C. Kaneko, *J. Chem. Soc., Per-*

kin Trans. 1 **1992**, 1393; b) M. Sato, H. Ogasawara, K. Oi, T. Kato, *Chem. Pharm. Bull.* **1983**, *31*, 1896.

113 V. A. Vu, L. Bérillon, P. Knochel, *Tetrahedron Lett.* **2001**, *42*, 6847.

114 J. Thibonnet, V. A. Vu, L. Bérillon, P. Knochel, *Tetrahedron*, **2002**, *58*, 4787.

115 R. H. Blaauw, J. C. J. Benningshof, A. E. Van Ginkel, J. H. van Maarseveen, H. Hiemstra, *J. Chem. Soc., Perkin Trans. 1*, **2001**, 2250.

116 J.-F. Briére, R. H. Blaauw, J. C. J. Benningshof, A. E. van Ginkel, J. H. van Maarseveen, H. Hiemstra, *Eur. J. Org. Chem.* **2001**, *12*, 2371.

117 J. Thibonnet, P. Knochel, *Tetrahedron Lett.* **2000**, *41*, 3319.

118 a) N. Krause, *Tetrahedron Lett.* **1989**, *30*, 5219; b) J. W. J. Kennedy, D. G. Hall, *J. Am. Chem. Soc.* **2002**, *124*, 898.

119 F. F. Fleming, V. Gudipati, O. W. Steward, *Org. Lett.* **2002**, *4*, 659.

120 P. J. Colson, L. S. Hegedus, *J. Org. Chem.* **1993**, *58*, 5918.

121 V. A. Vu, I. Marek, P. Knochel, *Synthesis* **2003**, 1797.

122 a) Satoh, K. Takano, H. Ota, H. Someya, K. Matsuda, M. Koyama, *Tetrahedron* **1998**, *54*, 5557; b) T. Satoh, T. Sakamoto, M. Watanabe, K. Takano, *Chem. Pharm. Bull.* **2003**, *51*, 966; c) T. Satoh, A. Kondo, J. Musashi, *Tetrahedron* **2004**, *60*, 5453; d) T. SAtoh, T. Kurihara, K. Fujita, *Tetrahedron* **2001**, *57*, 5369; e) T. Satoh, S. Saito, *Tetrahedron Lett.* **2004**, *45*, 347.

123 F. F. Fleming, Z. Zhang, Q. Wang, O. W. Steward, *Org. Lett.*, **2002**, *4*, 2493.

124 F. F. Fleming, Z. Zhang, Q. Wang, O. W. Steward, *J. Org. Chem.* **2003**, *68*, 7646.

125 A. Inoue, H. Shinokubo, K. Oshima, *Org. Lett.* **2000**, *2*, 651.

126 V. A. Vu, I. Marek, K. Polborn, P. Knochel, *Angew. Chem.* **2002**, *114*,361; *Angew. Chem. Int. Ed. Engl.* **2002**, *41*, 351.

127 a) C. Hamdouchi, C. Topolski, M. Goedken, H. M. Walborsky, *J. Org. Chem.* **1993**, *58*, 3148; b) G. Boche, D. R. Schneider, *Tetrahedron Lett.* **1978**, *19*, 2327; c) G. Boche, D. R. Schneider, H. Wintermayr, *J. Am. Chem. Soc.* **1980**, *102*, 5697.

128 A. de Meijere, S. I. Kozhushkov, *Chem. Rev.* **2000**, *100*, 93.

129 T. Tsuji, T. Nakamura, H. Yorimitsu, H. Shinokubo, K. Oshima, *Tetrahedron* **2004**, *60*, 973.

130 a) F. F. Fleming, B. C. Shook, *Tetrahedron* **2002**, *58*, 1; b) S. Arseniyadis, K. S. Kyler, D. S. Watt, *Org. React.* **1984**, *31*, 1.

131 F. F. Fleming, Z. Zhang, P. Knochel, *Org. Lett.* **2004**, *6*, 501–503.

132 H. Hart, T. Ghosh, *Tetrahedron Lett.* **1988**, *29*, 881.

133 S. Avolio, C. Malan, I. Marek, P. Knochel, *Synlett* **1999**, 1820.

134 a) V. Schulze, M. Broenstrup, V. P. W. Boehm, P. Schwerdtfeger, M. Schimeczek, R. W. Hoffmann, *Angew. Chem.* **1998**, *110*, 869; *Angew. Chem. Int. Ed.* **1998**, *37*, 824; b) V. Schulze, R. W. Hoffmann, *Chem. Eur. J.* **1999**, *5*, 337; R. W. Hoffmann, *Chem. Soc. Rev.* **2003**, *32*, 225; c) R. W. Hoffmann, P. G. Nell, *Angew. Chem.* **1999**, *111*, 354; *Angew. Chem. Int. Ed. Engl.* **1999**, *38*, 338.

135 N. Millot, C. Piazza, S. Avolio, P. Knochel, *Synthesis* **2000**, 941.

136 F. Dehmel, M. Abarbri, P. Knochel, *Synlett* **2000**, 345.

137 N. Gommermann, C. Koradin, P. Knochel, *Synthesis* **2002**, 2143.

138 *Pharmaceutical Substances. Syntheses. Patents. Applications*; A. Kleemann, J. Engel, B. Kutshcer, D. Reichert, Eds., Thieme, Stuttgart, **1999**.

139 R. Pandya, T. Murashima, L. Tedeschi, A. G. M. Barrett, *J. Org. Chem.* **2003**, *68*, 8274–8276.

140 a) M. T. Reetz, A. Kindler, *J. Chem. Soc. Chem. Commun.* **1994**, 2509; b) E. Nakamura, I. Kuwajima, *J. Am. Chem. Soc.* **1984**, *106*, 3368; c) E. J. Corey, N. W. Boaz, *Tetrahedron Lett.* **1985**, *26*, 6019; d) A. Alexakis, J. Berlan, Y. Besace, *Tetrahedron Lett.* **1986**, *27*, 1047.

141 G. Varchi, A. Ricci, G. Cahiez, P. Knochel, *Tetrahedron* **2000**, *56*, 2727.

142 a) K. C. Nicolaou, M. Takayanagi, N. F. Jain, S. Natarajan, A. E. Koumbis, T. Bando, J. M. Ramanjulu, *Angew. Chem.* **1998**, *110*, 2881; *Angew. Chem. Int. Ed. Engl.* **1998**, *37*, 2717; b) K. C. Nicolaou, A. E. Koumbis,

M. Takayanagi, S. Natarajan, N. F. Jain, T. Bando, H. Li, R. Hughes, *Chem. Eur. J.* **1999**, *5*, 2622.

143 a) A. Casarini, P. Dembech, D. Lazzari, E. Marini, G. Reginato, A. Ricci, G. Seconi, *J. Org. Chem.* **1993**, *58*, 5620; b) A. Alberti, F. Cane, P. Dembech, D. Lazzari, A. Ricci, G. Seconi, *J. Org. Chem.* **1996**, *61*, 1677; c) F. I. Knight, J. M. Brown, D. Lazzari, A. Ricci, A. J. Blacker, *Tetrahedron* **1997**, *53*, 11411; d) P. Dembach, G. Seconi, A. Ricci, *Chem. Eur. J.* **2000**, *6*, 1281

144 for some general reviews on electrophilic aminations of organometallic reagents, see: a) E. Erdik, M. Ay, *Chem. Rev.* **1989**, *89*, 1947; b) G. Boche, in *Houben-Weyl, Methods of Organic Chemistry*, G. Helmchen, R. W. Hoffmann, J. Mulzer, E. Schaumann Eds., Thieme, Stuttgart, **1995**.

145 a) F. Terrier, *Nucleophilic Aromatic Displacement: The Influence of the Nitro Group*, VCH, New York, **1991**; b) J. F. Bunnett, E. W. Garbisch, K. M. Pruitt, *J. Am. Chem. Soc.* **1957**, *79*, 385.

146 a) N. Kataoka, Q. Shelby, J. P. Stambuli, J. F. Hartwig, *J. Org. Chem.* **2002**, *67*, 5533; b) S. L. Buchwald, *J. Am. Chem. Soc.* **2003**, *125*, 6653; c) S. L. Buchwald, *J. Am. Chem. Soc.* **2002**, *124*, 11684; d) S. L. Buchwald, *Org. Lett.* **2002**, *4*, 2885.

147 a) J. P. Wolfe, S. Wagaw, J.-F. Marcoux, S. L. Buchwald, *Acc. Chem. Res.* **1998**, *31*, 805; b) J. F. Hartwig, *Angew. Chem.* **1998**, *110*, 2154; *Angew. Chem. Int. Ed. Engl.* **1998**, *37*, 2046; c) L. M. Alcazar-Roman, J. F. Hartwig, A. L. Rheingold, L. M. Liable-Sands, I. A. Guzei, *J. Am. Chem. Soc.* **2000**, *122*, 4618; d) J. F. Hartwig in *Handbook of Organopalladium Chemistry for Organic Synthesis, Vol. 1*, (Eds. E.-I. Negishi, A. de Meijere), John Wiley & Sons, New York, pp. 1051; e) A. R. Muci, S. L. Buchwald, *Top. Curr. Chem.* **2002**, *219*, 131.

148 U. K. Singh, E. R. Strieter, D. G. Blackmond, S. L. Buchwald, *J. Am. Chem. Soc.* **2002**, *124*, 14104.

149 for an excellent recent review, see: S. V. Ley, A. W. Thomas, *Angew. Chem.* **2004**, *116*, 1061; *Angew. Chem. Int. Ed.* **2003**, *42*, 5400.

150 a) J. C. Antilla, S. L. Buchwald, *Org. Lett.* **2001**, *3*, 2077; b) J. P. Collman, M. Zhong, *Org. Lett.* **2000**, *2*, 1233; c) P. Y. S. Lam, G. Vincent, D. Bonne, C. G. Clark, *Tetrahedron Lett.* **2003**, *44*, 4927; d) P. Y. S. Lam, S. Deudon, K. M. Averill, R. Li, M. Y. He, P. DeShong, C. G. Clark, *J. Am. Chem. Soc.* **2000**, *122*, 7600. e) J. Zanon, A. Klapars, S. L. Buchwald, *J. Am. Chem. Soc.* **2003**, *125*, 2890; f) F. Y. Kwong, S. L. Buchwald, *Org. Lett.* **2003**, *5*, 793; g) D. Zim, S. L. Buchwald, *Org. Lett.* **2003**, *5*, 2413.

151 a) B. H. Lipshutz, H. Ueda, *Angew. Chem.* **2000**, *112*, 4666; *Angew. Chem. Int. Ed.* **2000**, *39*, 4492; b) C. Desmartes, R. Schneider, Y. Fort, *Tetrahedron Lett.* **2001**, *42*, 247.

152 D. Seeebach, *Angew. Chem.* **1979**, *91*, 259; *Angew. Chem. Int. Ed. Engl.* **1979**, *18*, 239.

153 a) H. Wieland, *Chem. Ber.* **1903**, *36*, 2315; b) T. Severin, R. Schmitz, *Chem. Ber.* **1963**, *96*, 3081; c) T. Severin, M. Adam, *Chem. Ber.* **1964**, *97*, 186.

154 H. Gilman, R. McCracken, *J. Am. Chem. Soc.* **1927**, *49*, 1052.

155 a) G. Bartoli, M. Bosco, G. Cantagalli, R. Dalpozzo, F. Ciminale, *J. Chem. Soc., Perkin Trans. 2* **1985**, 773; b) G. Bartoli, M. Bosco, R. Dalpozzo, G. Calmieri, E. Marcantoni, *J. Chem. Soc. Perkin Trans. 1*, **1991**, 2757.

156 a) G. Bartoli, *Acc. Chem. Res.* **1984**, *17*, 109; b) M. Bosco, R. Dalpozzo, G. Batoli, G. Calmieri, M. Petrini, *J. Chem. Soc., Perkin Trans. 2* **1991**, 657.

157 A. Ono, H. Sasaki, F. Yaginuma, *Chem. Ind. (London)* **1983**, 480.

158 a) I. Sapountzis, P. Knochel, *J. Am. Chem. Soc.* **2002**, *124*, 9390; b) I. Sapountzis, P. Knochel, manuscript in preparation; c) I. Sapountzis, N. Gommermann, P. Knochel, manuscript in preparation.

159 a) B. H. Yang, S. L. Buchwald, *J. Organomet. Chem.* **1999**, *576*, 125; b) J. P. Wolfe, S. Wagan, J.-F. Marcoux, S. L. Buchwald, *Acc. Chem. Res.* **1998**,

31, 805; J. F. Hartwig, *Angew. Chem.* **1998**, *110*, 2155; *Angew. Chem. Int. Ed. Engl.* **1998**, *37*, 2046; c) L. M. Alcazar-Roman, J. F. Hartwig, A. L. Rheingold, L. M. Liable-Sands, I. A. Guzei, *J. Am. Chem. Soc.* **2000**, *122*, 4618.

160 a) A. Klapaus, J. C. Antilla, X. Huang, S. L. Buchwald, *J. Am. Chem. Soc.* **2001**, *123*, 7727. b) M. Wolter, A. Klapaus, S. L. Buchwald, *Org. Lett.* **2001**, *3*, 3803; c) R. Shen, J. A. Porco Jun., *Org. Lett.* **2000**, *2*, 1333; d) A. V. Kalinin, J. F. Bower, P. Riebel, V. Snieckus, *J. Org. Chem.* **1999**, *64*, 2986.

161 a) B. H. Lipshutz, H. Ueda, *Angew. Chem.* **2000**, *112*, 4666; *Angew. Chem. Int. Ed. Engl.* **2000**, *39*, 4492; b) C. Desmarets, R. Schneider, Y. Fort, *Tetrahedron Lett.* **2001**, *42*, 247.

162 F. Kopp, I. Sapountzis, P. Knochel, *Synlett*, **2003**, *6*, 885.

163 a) N. Momiyama, H. Yamamoto, *Org. Lett.* **2002**, *4*, 3579; b) N. Momiyama, H. Yamamoto, *Angew. Chem.* **2002**, *114*, 4666; *Angew. Chem. Int. Ed. Engl.* **2002**, *41*, 2986.

164 I. Sapountzis, P. Knochel, Angew. Chem. **2004**, *116*, 915; *Angew. Chem. Int. Ed. Engl.* **2004**, *43*, 897.

165 K. S. Feldman, T. D. Cutarelli, *J. Am. Chem. Soc.* **2002**, *124*, 11600.

166 For some reviews on the synthesis of 1,2-dehydrobenzyne, see: a) R. W. Hoffmann *Dehydrobenzene and Cycloalkenes*, Academic Press, New York, **1967**; b) S. V. Kessar *Nucleophilic Coupling of Arynes* in *Comprehensive Organic Synthesis*, Eds. B. M. Trost, I. Fleming, Pergamon Press, Oxford, **1991**; c) for a recent review see: H. Pellisier, M. Santinelli, *Tetrahedron* **2003**, *59*, 701; d) W. Oppolzer *Intermolecular Diels–Alder Reactions* in *Comprehensive Organic Synthesis*, Eds. B. M. Trost, I. Fleming, Pergamon Press, Oxford, **1991**.

167 a) M. Schlosser, E. Castagnetti, Eur. J. Org. Chem. 2001, 3991; b) K. C. Caster, C. G. Keck, R. D. Walls, *J. Org. Chem.* 2001, 66, 2932; c) S. E. Whitney, M. Winters, B. Rickborn, *J. Org. Chem.* **1990**, *55*, 929; d) K. Dachriyanus, M. V. Sargent, B. W. Skelton, A. H. White, *Aust. J. Chem.* **2000**, *53*, 267; see also: a) Z. Liu, R. C. Larock,

Org. Lett. **2003**, *5*, 4673; b) P. P. Wickham, K. H. Hazen, H. Guo, G. Jones, K. H. Reuter, W. J. Scott, *J. Org. Chem.* **1991**, *56*, 2045; c) K. H. Reuter, W. J. Scott, *J. Org. Chem.* **1993**, *58*, 4722; d) S. Triphaty, R. LeBlanc, T. Durst, *Org. Lett.* **1999**, *1*, 1973.

168 a) T. Hamura, T. Hosoya, H. Yamaguchi, Y. Kuriyama, M. Tanabe, M. Miyamoto, Y. Yasui, T. Matsumoto, K. Suzuki *Helv. Chim. Acta* **2002**, *85*, 3589; b) T. Hosoya, T. Hamura, Y. Kuriyama, M. Miyamoto, T. Matsumoto, K. Suzuki *Synlett* **2000**, *4*, 520; c) T. Matsumoto, T. Sohma, H. Yamaguchi, S. Kurata, K. Suzuki *Synlett* **1995**, 263; d) T. Hamura, Y. Ibusuki, K. Sato, T. Matsumoto, Y. Osamura, K. Suzuki, *Org. Lett.* **2003**, *5*, 3551.

169 I. Sapountzis, W. Lin, M. Fischer, P. Knochel, Angew. Chem. **2004**, *116*, 4464; *Angew. Chem. Int. Ed. Engl.* **2004**, *43*, 4364.

170 W. Lin, I. Sapountzis, P. Knochel, manuscript in preparation

171 W. Lin, P. Knochel, manuscript in preparation

172 a) A. Minato, K. Tamao, T. Hayashi, K. Suzuki, M. Kumada, *Tetrahedron* **1981**, *22*, 5319; b) for an excellent review, see: J. Hassan, M. Sevignon, C. Gozzi, E. Schulz, M. Lemaire, *Chem. Rev.* **2002**, *102*, 1359.

173 a) C. Amatore, A. Jutand, *J. Organomet. Chem.* **1999**, *576*, 254; b) J. F. Fauvarque, F. Pflüger, M. Troupel, *J. Organomet. Chem.* **1981**, *208*, 419.

174 V. Bonnet, F. Mongin, F. Trécourt, G. Quéguiner, P. Knochel, *Tetrahedron Lett.* **2001**, *42*, 5717.

175 V. Bonnet, F. Mongin, F. Trécourt, G. Quéguiner, P. Knochel, *Tetrahedron* **2002**, *58*, 4429.

176 a) E. Negishi, *Acc. Chem. Res.* **1982**, *15*, 340; b) E. Negishi, H. Matsushita, M. Kobayashi, C. L. Rand, *Tetraherdon Lett.* **1983**, *24*, 3823; c) E. Negishi, T. Takahashi, S. Baba, D. E. Van Horn, N. Okukado, *J. Am. Chem. Soc.* **1987**, *109*, 2393; d) E. Negishi, Z. Owczarczyk, *Tetrahedron Lett.* **1991**, *32*, 6683.

177 A. E. Jensen, P. Knochel, *J. Organomet. Chem.* **2002**, *653*, 122.

178 a) V. Farina, B. Krishnan, *J. Am. Chem. Soc.* **1991**, *113*, 9585; b) V. Farina, S. Kapadia, B. Krishnan, C. Wang, L. S. Liebeskind, *J. Org. Chem.* **1994**, *59*, 5905.

179 I. Sapountzis, H. Dube, P. Knochel, *Adv. Synth. Catal.* **2004**, *346*, 709.

180 C. Dai, C. G. Fu, *J. Am. Chem. Soc.* **2001**, *123*, 2719.

181 K. S. Feldman, K. J. Eastman, G. Lessene, *Org. Lett.* **2002**, *4*, 3525.

182 a) A.-S. Rebstock, F. Mongin, F. Trécourt, G. Quéguiner, *Tetrahedron* **2003**, *59*, 4973–4977 b) S. Sengupta, M. Leite, D. S. Raslan, C. Quesnelle, V. Snieckus, *J. Org. Chem.* **1992**, *57*, 4066.

183 V. P. W. Boehm, T. Weskamp, C. W. K. Gstoettmayr, W. A. Herrmann, *Angew. Chem Int. Ed.* **2000**, *39*, 1602.

184 a) R. Giovannini, P. Knochel, *J. Am. Chem. Soc.* **1998**, *120*, 11186; b) R. Giovannini, T. Stuedemann, A. Devesagayaraj, G. Dussin, P. Knochel, *J. Org. Chem.* **1999**, *64*, 3544.

185 W. Dohle, D. M. Lindsay, P. Knochel, *Org. Lett.* **2001**, *3*, 2871.

186 a) M. Tamaru, J. K. Kochi, *J. Am. Chem. Soc.* **1971**, *93*, 1487; b) M. Tamaru, J. K. Kochi, *Synthesis* **1971**, *93*, 303; c) M. Tamaru, J. K. Kochi, *J. Organomet. Chem.* **1971**, *31*, 289; d) M. Tamaru, J. K. Kochi, *Bull. Chem. Soc. Jpn.* **1971**, *44*, 3063; e) J. K. Kochi, *Acc. Chem. Res.* **1974**, *7*, 351; f) S. Neumann, J. K. Kochi, *J. Org. Chem.* **1975**, *40*, 599; g) R. S. Smith, J. K. Kochi, *J. Org. Chem.* **1976**, *41*, 502.

187 B. Bogdanovic, M. Schwickardi, *Angew. Chem.* **2000**, *112*, 4788; *Angew. Chem. Int. Ed.* **2000**, *39*, 4610.

188 a) G. Cahiez, S. Marquais, *Pure Appl. Chem.* **1996**, *68*, 53; b) G. Cahiez,

S. Marquais, *Tetrahedron Lett.* **1996**, *37*, 1773; c) G. Cahiez, H. Advedissian, *Synthesis* **1998**, 1199.

189 W. Dohle, F. Kopp, G. Cahiez, P. Knochel, *Synlett* **2001**, 1901.

190 M. Rottländer, P. Knochel, *J. Org. Chem.* **1998**, *63*, 203.

191 a) A. Fürstner, A. Leitner, M. Mendez, H. Krause, *J. Am. Chem. Soc.* **2002**, *124*, 13856; b) A. Fürstner, A. Leitner, *Angew. Chem.* **2002**, *114*, 632; *Angew. Chem. Int. Ed.* **2002**, *41*, 609.

192 for application of this methodology to the synthesis of natural product muscopyridine, see: A. Fuerstner, A. Leitner, *Angew. Chem. Int. Ed.* **2003**, *42*, 308.

193 B. Scheiper, M. Bonnekessel, H. Krause, A. Fürstner, *J. Org. Chem.* **2004**, *69*, 3943

194 a) M. Nakamura, K. Matsuo, S. Ito, E. Nakamura, *J. Am. Chem. Soc.* **2004**, *126*, 3686; b) T. Nagano, T. Hayashi, *Org. Lett.* **2004**, *6*, 1297; c) R. Martin, A. Fürstner, *Angew. Chem.* **2004**, *116*, 4045; *Angew. Chem. Int. Ed.* **2004**, *43*, 3955.

195 I. Sapountzis, C. Kofink, P. Knochel, manuscript in preparation.

196 a) W. C. Percival, R. B. Wagner, N. C. Cook, *J. Am. Chem. Soc.* **1953**, *75*, 3731; b) C. Cardellicchio, V. Fiandanese, G. Marchese, L. Ronzini, *Tetrahedron Lett.* **1987**, *28*, 2053; c) V. Fiandanese, G. Marchese, V. Martina, L. Ronzini, *Tetrahedron Lett.* **1984**, *25*, 4805; d) V. Fiandanese, G. Marchese, L. Ronzini, *Tetrahedron Lett.* **1983**, *24*, 3677; e) K. Reddy, P. Knochel, *Angew. Chem.* **1996**, *108*, 1812; *Angew. Chem. Int. Ed.* **1996**, *35*, 1700.

197 C. Duplais, F. Bures, I. Sapountzis, T. J. Korn, G. Cahiez, P. Knochel, *Angew. Chem.* **2004**, *116*, 2984; *Angew. Chem. Int. Ed.* **2004**, *43*, 2968.

5
Polyfunctional Silicon Organometallics for Organic Synthesis

Masaki Shimizu and Tamejiro Hiyama

5.1
Introduction

Organosilicon compounds are, in general, stable enough to be employed for a variety of uses as functional materials. The carbon–silicon bond is akin to carbon–carbon bond and thus much less reactive than other carbon–metal bonds due to low polarization of the bond [electronegativity (Allred): C, 2.50; Si, 1.74] [1]. Consequently, silicon-based compounds are easily prepared and handled, inert to a wide range of functional groups, and tolerate the conditions employed for various synthetic manipulations, whereas nucleophilic activation of a C–Si bond or electrophilic activation of substrates makes organosilicon compounds extremely versatile as nucleophilic reagents in organic synthesis [2]. Namely, such moderate reactivity allows one to incorporate diverse functional groups into organosilicon compounds at any place in any manner. Moreover, it is possible for both the reaction partners to be present during the activation or for substrate to contain the reagent moiety, so that a tandem reaction and/or an intramolecular reaction leading to cyclic molecules is conceivable to achieve rapid synthesis of complex structures that are hardly accessible with other organometallic reagents. Thus, organosilicon compounds are a prodigious class of polyfunctional organometallics. Actually, polyfunctional organosilicon compounds are widely used in total synthesis of natural products [3]. In addition, low toxicity and wide availability of silicon-containing compounds make their synthetic potential even greater.

In this chapter, we describe preparation of polyfunctional organosilicon reagents and demonstrate their high versatility by selecting some recent examples of allylic silanes, alkenylsilanes, alkylsilanes, and miscellaneous types of silanes, in which C–Si bonds are utilized as C–Metal ones to be converted into C–C bonds.

Organometallics. Paul Knochel
Copyright © 2005 WILEY-VCH Verlag GmbH & Co. KGaA, Weinheim
ISBN: 3-527-31131-9

5.2
Allylic Silanes

5.2.1
Intermolecular Reactions of Polyfunctional Allylic Silanes

Silyl-substituted allyl methoxyacetate **1** undergoes the Claisen–Ireland rearrangement to diastereoselectively give optically active allylic silane **2** as exemplified by Scheme 5.1. The resulting silane is employed for asymmetric addition to both aliphatic and aromatic acetals **3a** and **3b** in the presence of Me₃SiOTf, giving rise to homoallylic ethers **4a** and **4b**, respectively, with high diastereo- and enantioselectivities [4]. The whole sequence of reactions is an example of 1,4- and 1,5-remote

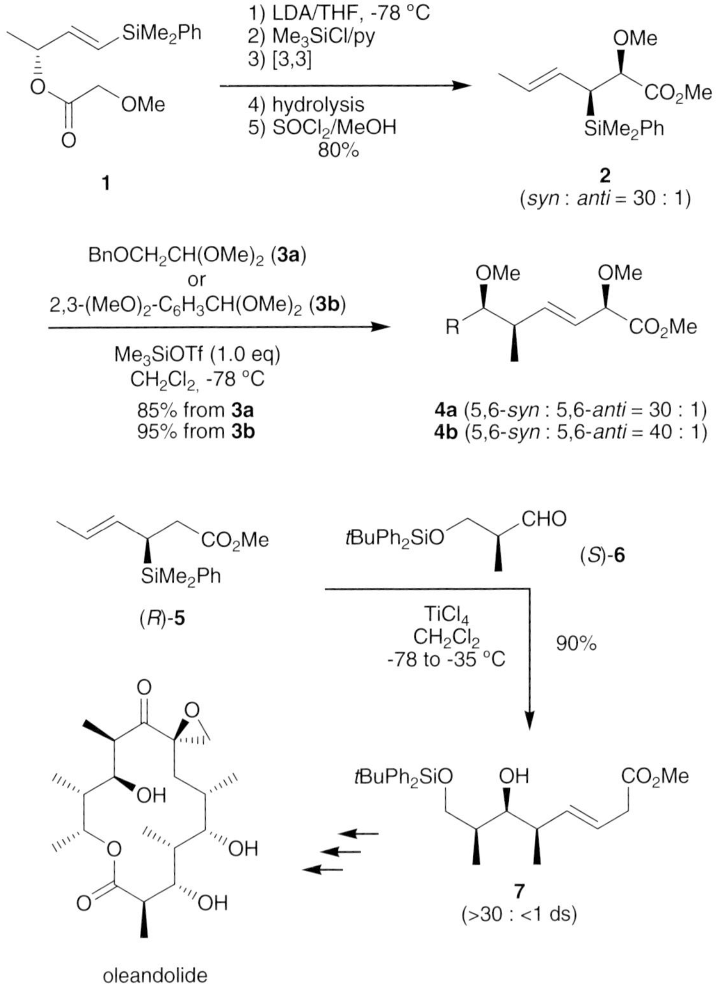

Scheme 5.1 Preparation and allylation to aldehydes of allylic silane **2**.

stereocontrol [5]. Similar chiral crotyl-type silane **5** undergoes double stereodifferentiating crotylation reaction of chiral aldehyde (*S*)-**6** in the presence of TiCl$_4$ as the Lewis acid promoter [6]. The utility of the reaction is demonstrated by total synthesis of oleandolide [7].

Similarly, *α*-substituted crotylsilane (*R*)-**5** is shown to add smoothly to imine **8**, *in situ* prepared from benzaldehyde acetal and methyl carbamate in the presence of BF$_3$·O·Et$_2$ at −78 °C, giving *syn*-adduct **9** stereoselectively (Scheme 5.2) [8].

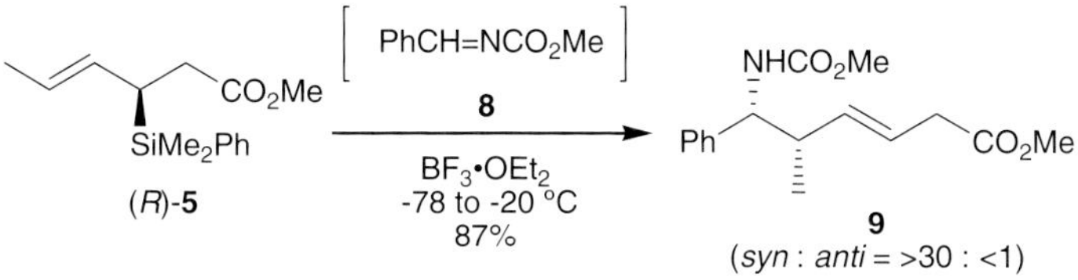

Scheme 5.2 Allylation to imine with allylic silane **5**.

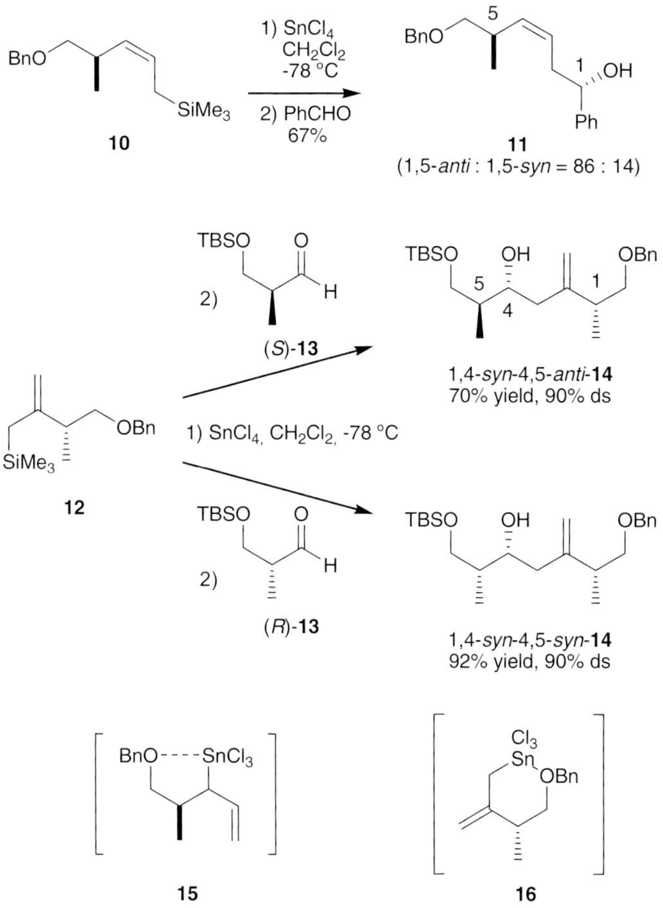

Scheme 5.3 *α*-Selective allylation to aldehydes with chiral allylic silanes **10** and **12**.

The above three examples demonstrate that allylic silanes react with acetals and imines at the γ-carbon. However, with SnCl$_4$ as the Lewis acid catalyst, the addition takes place at the α-position as shown by the examples in Scheme 5.3. Transmetallation is considered to take place first at the γ-carbon to give respectively **15** and **16** as intermediates, which then add to aldehydes in an S$_E$2' manner with high 1,4- and 1,5-remote asymmetric induction [9]. Because such high stereocontrol is best achieved by pretreatment of **10** and **12** with tin(IV) chloride before the reaction with aldehydes, intermediacy of **15** and **16** is confirmed.

Tris(trimethylsilyl)silylmethacrylate **17** undergoes carbosilylation under radical conditions (Scheme 5.4) [10]. Radical acceptors suitable for the reaction are electron-deficient alkenes, terminal alkynes, and aromatic aldehydes.

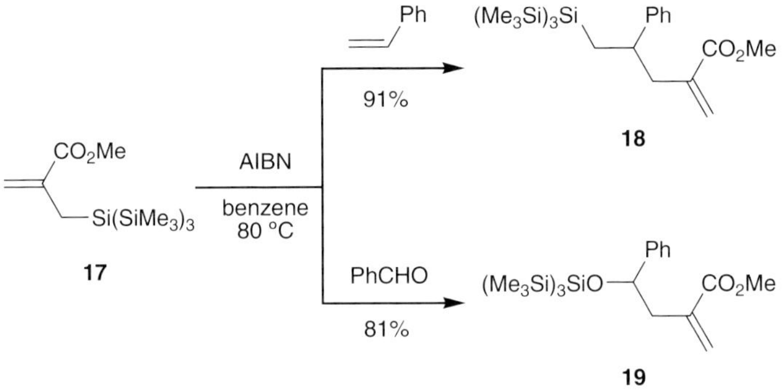

Scheme 5.4 Allylsilylation of unsaturated bonds under radical conditions.

5.2.2
Intramolecular Reactions of Polyfunctional Allylic Silanes

Allylic silanes **20** containing an alkynyl moiety undergo intramolecular allylsilylation in the presence of a HfCl$_4$/Me$_3$SiCl catalyst system to produce five-, six-, and

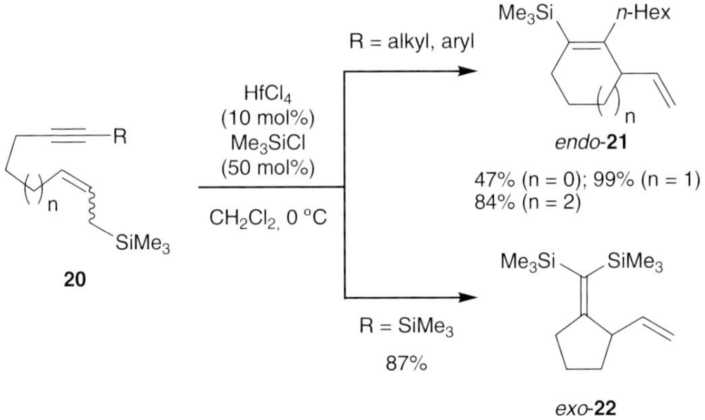

Scheme 5.5 Lewis acid-catalyzed intramolecular allylsilylation of alkynes.

seven-membered carbocycles **21** in an *endo*-fashion (Scheme 5.5) [11], in sharp contrast to *exo*-selective transition metal-catalyzed or -mediated carbocyclization (*vide infra*) [12]. On the other hand, a trimethylsilyl group on the terminal alkyne carbon switches the cyclization mode to 5-*exo* to give *gem*-bis(trimethylsilyl)methylenecyclopentane **22**.

In the presence of an electrophilic catalyst like PtCl$_2$, terminal alkyne **23** containing an allylic silane functionality undergoes *exo*-carbocyclization to give **24** possibly through intramolecular trapping of a Pt(II)-coordinated triple bond or a vinyl cation intermediate with the allylic silane moiety (Scheme 5.6) [13]. Pd(II), Ru(II), and Ag(I) salts also serve as a catalyst of this cyclization.

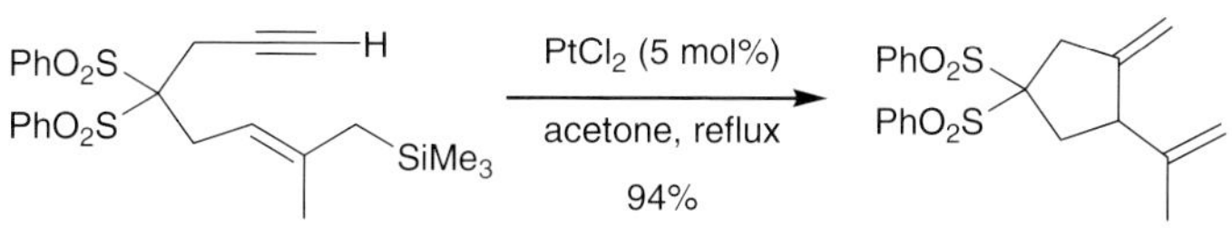

Scheme 5.6 Transition metal-catalyzed carbocyclization of alkyne with allylic silane.

Cyclic conjugate diene **25** is also activated by co-use of Li$_2$PdCl$_4$, benzoquinone, and LiCl, and the resulting complex (**26**) is trapped intramolecularly by an allylic silane moiety, giving rise to *syn*-bicyclic allylic chlorides **28** and **29** (Scheme 5.7)

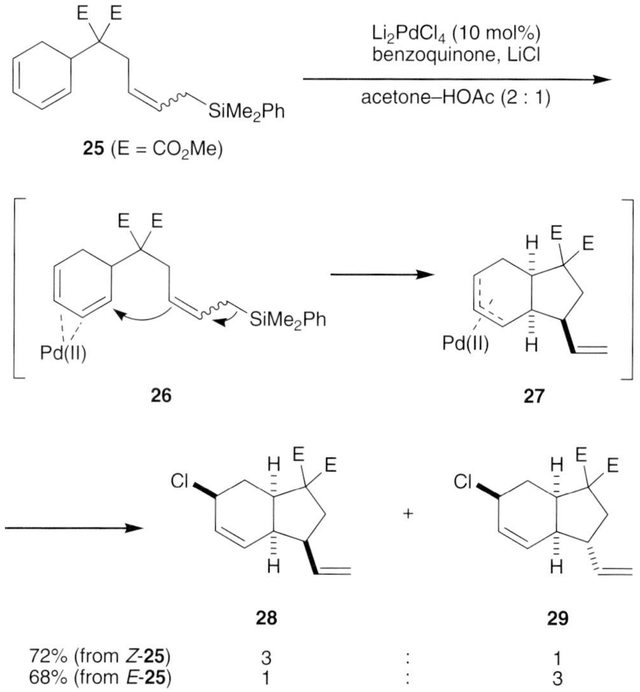

Scheme 5.7 Intramolecular Pd-catalyzed 1,4-addition to 1,3-dienes.

[14]. The whole transformation is an oxidative intramolecular 1,4-addition. Stereochemistry of the 1,4-addition is explained by an intramolecular *anti* attack of the allylsilane moiety to a Pd(II)-coordinated diene functionality in **26** to generate (π-allyl)palladium complex **27** followed by an intermolecular *anti* attack of a chloride ion. The substrate allylsilane is easily prepared by the reaction of the corresponding allylic acetate with PhMe₂SiLi.

The intramolecular cyclization strategy is applied to efficient synthesis of oxacycles starting with trimethylsiloxy-containing allylic silanes (Scheme 5.8). Treatment of **30** with benzaldehyde in the presence of a catalytic amount of Me₃SiOTf and PrOSiMe₃ gives tetrahydropyran **32**. An oxonium ion intermediate (**31**) is considered to be generated first and then undergo intramolecular nucleophilic attack by an allylsilane part [15]. When ortholactones are employed in lieu of aldehydes, spiroketals are readily prepared.

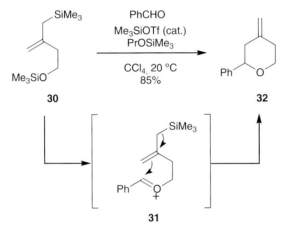

Scheme 5.8 Lewis acid-catalyzed formation of an oxonium ion and intramolecular trapping with an allylic silane moiety.

Similar tetrahydropyran synthesis is performed with 2-(trimethylsiloxymethyl)propen-3-ylsilane **33** and two molecules of aldehydes (Scheme 5.9) [16]. For example, **33** reacts with propanal in the presence of BF₃·OEt₂ to give *exo*-methylene tetrahydropyran **36** as a single diastereomer without any [3+2] formation of tetrahydrofuran derivatives. A proposed mechanism involves an ene-type reaction leading to silyl enol ether **34** followed by formation of oxonium ion **35** and intramolecular cyclization.

Diastereo- and enantioselective synthesis of trisubstituted dihydropyrans is realized by Me₃SiOTf-catalyzed condensation of optically active allylic silanes with aldehydes (Scheme 5.10) [17]. Configuration of a homoallylic position in silanes **37** and **39** effectively controls the stereochemistry of the reaction at 2- and 6-positions of pyrans **38** and **40**. The stereochemical outcome is explained by a boat-like six-membered transition state **41**, which is derived from **37** and prefers the silyl group at a pseudoaxial position to optimize σ–p overlap, in preference to chair-like transition state **42**.

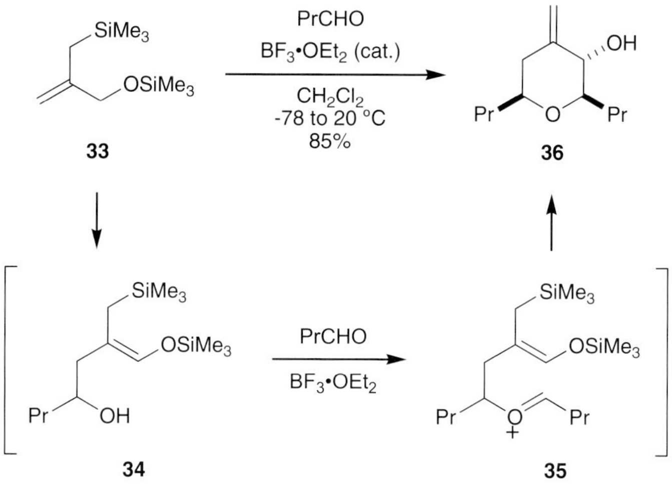

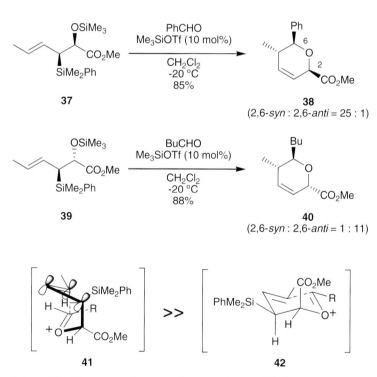

Scheme 5.9 Stereoselective synthesis of tetrasubstituted tetrahydropyran.

Scheme 5.10 Diastereo- and enantioselective synthesis of dihydropyrans.

5.2.3
Tandem Reactions of Polyfunctional Allylic Silanes

Allylic silanes having a functional group at the β-carbon are extremely versatile synthetic reagents. For example, (2-trimethylsiloxy)allylsilane **43** behaves as an acetone α,α'-dianion equivalent as is evidenced by the reaction with acetals or aldehydes in the presence of titanium tetrachloride. Double C–C bond formation readily takes place to give rise to β,β'-dioxygenated ketone **44** (Scheme 5.11) [18]. From a mechanistic point of view, the reagent is working as a double silyl enol ether rather than an allylsilane.

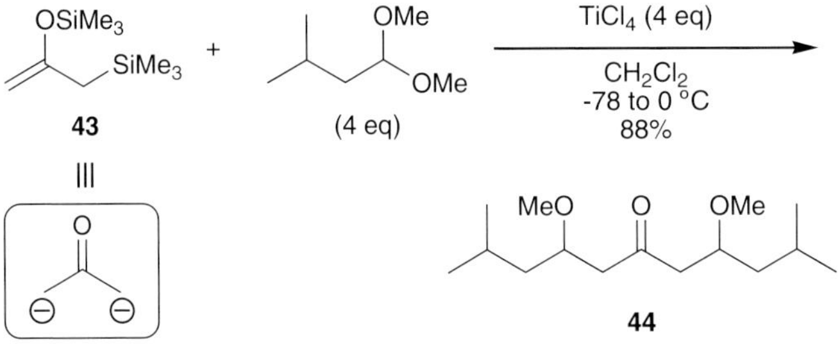

Scheme 5.11 Double condensation reaction of **43**.

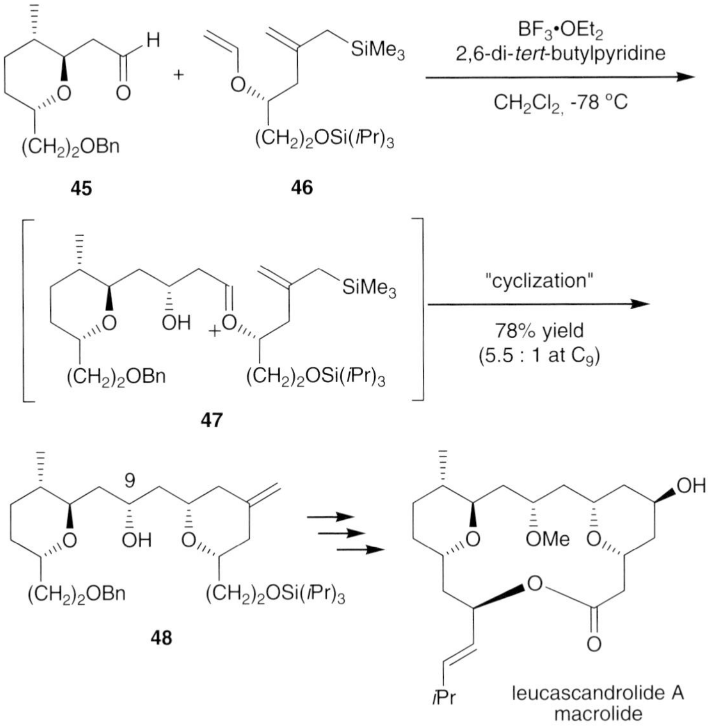

Scheme 5.12 Tandem aldol–Prins cyclization.

Although alkyl enol ethers often undergo oligomerization under electrophilic conditions due to extremely reactive oxocarbenium ion intermediates, intramolecular trapping by an allylsilane moiety of such reactive species leads to an extremely versatile strategy. Aldol-type reaction of aldehyde **45** with allylic silane enol ether **46** is catalyzed by BF$_3$•OEt$_2$ to generate oxonium ion **47**, which stereoselectively undergoes cyclization through intramolecular allylation to produce *cis*-2,6-disubstituted tetrahydropyran **48** (Scheme 5.12) [19]. The synthetic utility of this method is demonstrated by a formal total synthesis of leucascandrolide A.

The tandem aldol–allylation strategy is also applicable to stereocontrolled polyketide/macrolide synthesis. (*E*)- and (*Z*)-Crotyl(enol)(pinacolato)silanes **49** and **51** react stereoselectively with cyclohexanecarbaldehyde to produce 1,3-diols **50** and **52**, respectively, with high diastereoselectivities (Scheme 5.13) [20]. It is noteworthy that the reaction of (*E,E*)-crotyl(enol)silane **53** is capable of constructing of

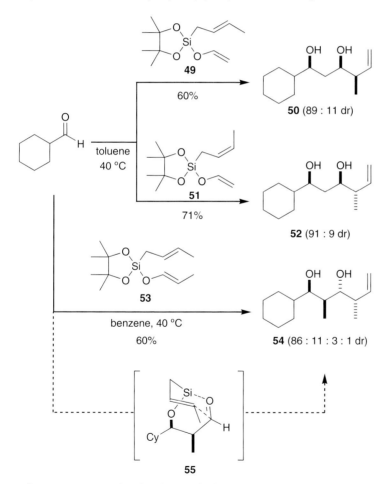

Scheme 5.13 Strain-induced tandem aldol–allylation.

four contiguous chiral centers with fairly high diastereoselectivity. The strain on silicon induced by the pinacol ligand is essential for the stereoselective aldol reaction; no reaction takes place when pinacol is replaced by 2,4-dimethyl-2,4-pentanediol. The stereochemical outcome of the reaction is explained in terms of a chairlike six-membered transition state **55** for intramolecular crotylsilylation of β-siloxy aldehydes.

Bis(allyl)homoallyloxysilanes **56a** and **56b** are designed for a tandem intramolecular silylformylation–allylsilylation reaction, which has turned out to be an efficient approach to construct polyol and polyketide frameworks [21]. For example, heating a solution of **56** in benzene at 60 °C in the presence of Rh(acac)(CO)$_2$ under CO atmosphere followed by the Tamao oxidation gives *syn,syn*-triols **59** stereoselectively via oxasilacyclopentanes **57** and **58** (Scheme 5.14). Bis(*cis*-crotyl)silane **56b** is readily prepared by double Pd-catalyzed 1,4-hydrosilylation of 1,3-butadiene with dichlorosilane followed by reduction with LiAlH$_4$ and alcoholysis with the corresponding homoallylic alcohol.

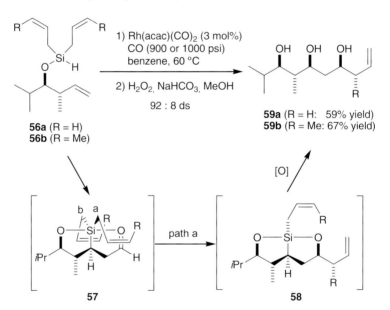

Scheme 5.14 Tandem silylformylation–allylation of alkenes.

The tandem silylformylation–allylation methodology is extended to remote 1,5-stereocontrol [22]. Thus, treatment of homopropargylic hydrosilyl ethers **60**, produces, in a manner similar to homoallylic hydrosilyl ethers **56**, 3-silyl-1,5-diol **61** whose oxidation or protodesilylation/acetylation gives, respectively, 1,5-*anti* diol **62** or diacetate **63** with high diastereoselectivity (Scheme 5.15). The 1,5-*anti* selectivity contrasts sharply to the one obtained in the tandem reaction of homoallylic silyl ethers **56**.

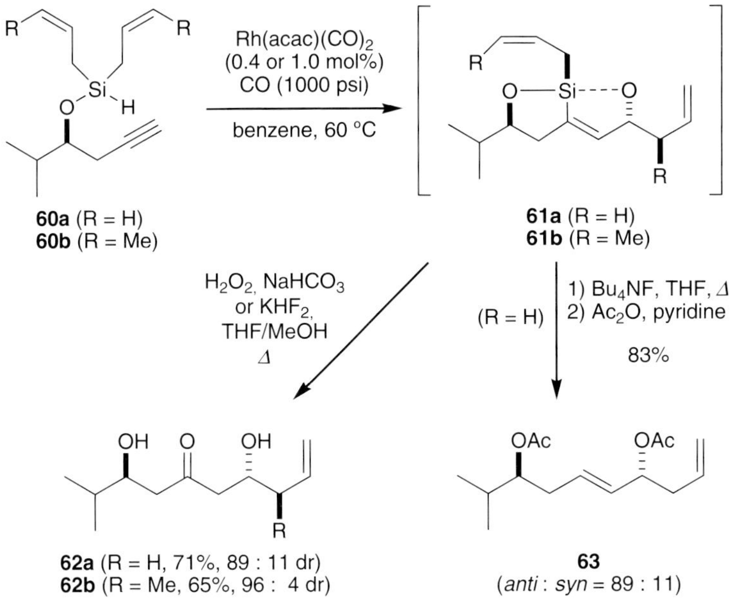
R
R
Si
O
H
Rh(acac)(CO)2
(0.4 or 1.0 mol%)
CO (1000 psi)
benzene, 60 °C
R
O
Si
O
R
60a (R = H)
60b (R = Me)
61a (R = H)
61b (R = Me)
H2O2, NaHCO3
or KHF2,
THF/MeOH
Δ
(R = H)
1) Bu4NF, THF, Δ
2) Ac2O, pyridine
83%
OH
O
OH
R
OAc
OAc
62a (R = H, 71%, 89 : 11 dr)
62b (R = Me, 65%, 96 : 4 dr)
63
(anti : syn = 89 : 11)

Scheme 5.15 Tandem silylformylation–allylation of alkynes.

Diallyl(diisopropyl)silane **64** delivers its two allyl groups on silicon sequentially to methyl vinyl ketone in the presence of BF_3•OEt_2 (Scheme 5.16) [23]. β-Silyl cation **65** is first generated by 1,4-addition and then intramolecularly allylated by the remaining allyl group to afford double allylation product **66**.

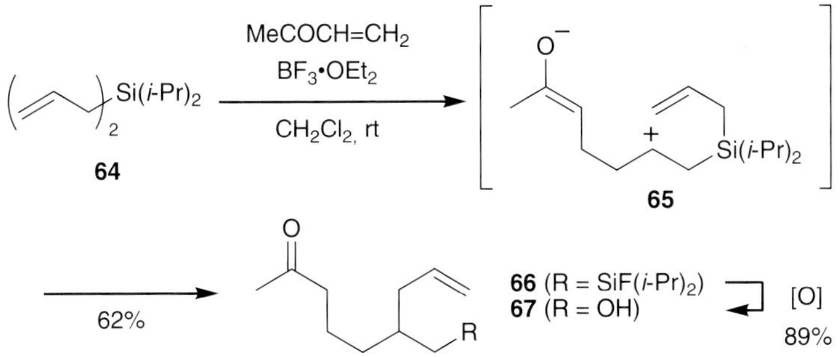
Si(i-Pr)2
2
64
MeCOCH=CH2
BF3•OEt2
CH2Cl2, rt
O−
+
Si(i-Pr)2
65
62%
O
R
66 (R = SiF(i-Pr)2)
67 (R = OH)
[O]
89%

Scheme 5.16 Tandem double allylation of α,β-unsaturated ketone.

5.2.4
Sequential Synthetic Reactions of Metal-containing Allylic Silanes

Metal-containing organosilanes are versatile reagents from two synthetic viewpoints. Synthetic reactions of the silicon functionality provide an efficient method for the preparation of polyfunctional organometallic reagents, whereas synthetic transformations based on the metal part provides polyfunctional organosilicon reagents.

Typical examples are α-boryl allylic silanes **69** that are stereoselectively prepared by *gem*-silylborylation of allylic chlorides **68** (Scheme 5.17) [24]. Treatment of **68** with LDA at –98 °C generates the corresponding lithium carbenoids, which smoothly react with coexisting (dimethylphenylsilyl)(pinacolato)borane to give **69** with complete retention of configuration. With benzaldehyde as an electrophile, **69** reacts as an allylic silane in the presence of Me₃SiOBn and Me₃SiOTf in CH₂Cl₂ at –78 °C to afford (*E*)-alkenylboranes **70** with high stereospecificity. On the other hand, allylation as an allylic borane takes place with benzaldehyde under thermal conditions, giving rise to (Z)-alkenylsilane **71** with opposite stereospecificity. Silane **71** is converted into disubstituted dihydropyran **72** through the Overman's procedure (see Section 5.3.2).

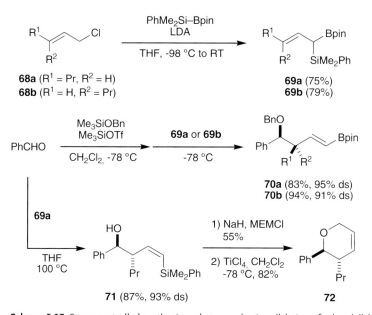

Scheme 5.17 Stereocontrolled synthesis and stereoselective allylation of α-borylallylsilanes.

Silylborylation of allenes readily produces β-borylallylsilanes, which are useful for stereoselective preparation of functional alkenylboranes and *trans*-1,2-benzoxadecalines (Scheme 5.18) [25]. TiCl₄-promoted allylation of propanal diethyl acetal with **73** gives (*E*)-alkenylborane **74** stereoselectively. Meanwhile, treatment of **73** with propanal in the presence of Me₃SiOTf produces tricyclic compound **75**, in which two propanal molecules are incorporated, as a single diastereomer. The proposed mechanism involves allylation of propanal followed by acetal formation with second propanal and Prins-type oxonium ion–alkene cyclization. Asymmetric synthesis of **73** is also demonstrated [26].

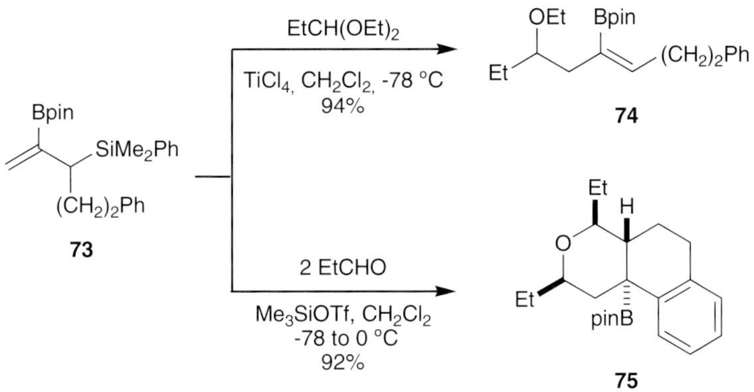

Scheme 5.18 Lewis acid-promoted reactions of β-borylallylsilane.

Rhodium-catalyzed conjugate addition of γ-borylallylsilane **76** to benzalacetone (**77**) proceeds upon heating, providing ζ-ketoallylsilane **78**, which undergoes intramolecular allylation with the aid of Bu_4NF to give vinylcyclobutanol **79** as a single isomer (Scheme 5.19) [27].

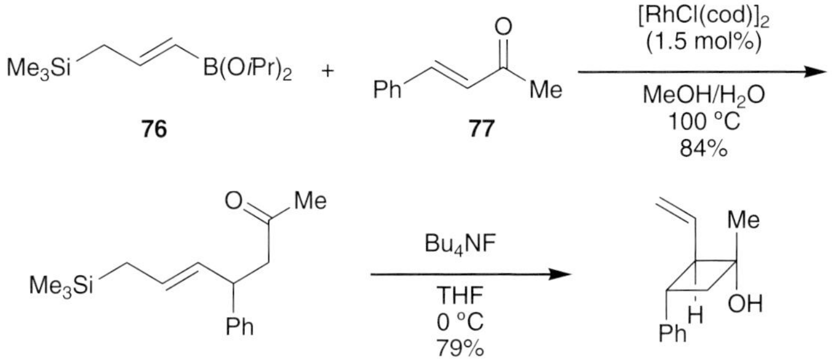

Scheme 5.19 Cyclobutanol synthesis with γ-borylallylsilane.

Hydrogenolysis of allyl acetates **80** with triethylammonium formate gives allylsilanes substituted at the α-position (**81**) (Scheme 5.20) [28]. Disilyl reagent **81a** reacts with octanal in the presence of $BF_3 \cdot OEt_2$ at −40 °C to yield *tert*-butyldimethylsilyl-substutited product **82a** solely, while allylation with **81b** proceeds at −78 °C to give **82b**. Thus, the carbon–tin bond in **81b** is cleaved much faster than the carbon–silicon bond.

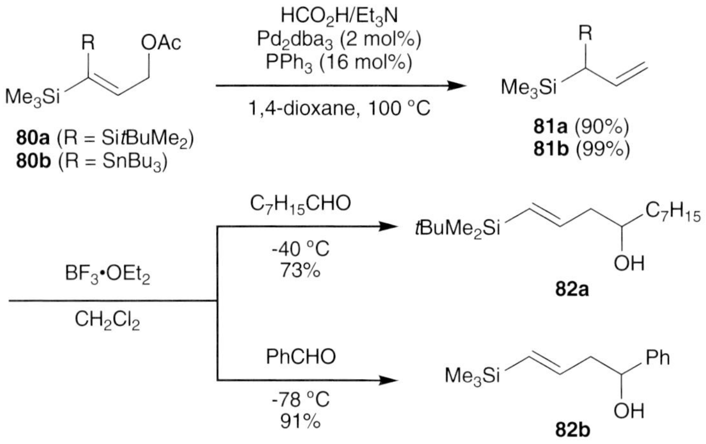

Scheme 5.20 Preparation and Lewis acid-promoted aldehyde addition of α-silyl- and -stannylallylsilanes.

Geminally silylated allylsilanes, conveniently prepared by isomerization of *gem*-disilylalkenes with 10% Pd/C in diethyl ether [29], are used for stereoselective synthesis of *trans*-2,3-disubstituted oxepanes. For example, **84** reacts with benzaldehyde to give **85** that bears *trans*-2-silylethenyl and phenyl groups, via an acetalization–cyclization sequence (Scheme 5.21).

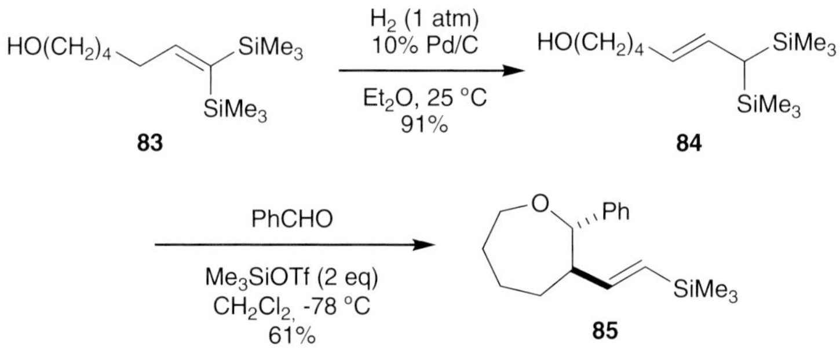

Scheme 5.21 Synthesis and cyclization of α-silylallylsilane.

One-pot stereoselective synthesis of all-*cis*-2,3,5-trisubstituted tetrahydrofurans is accomplished starting with α-silylmethyl allylic silane **86** (Scheme 5.22), which is treated with double amounts of benzyloxyacetaldehyde in the presence of $BF_3 \cdot OEt_2$ to give tetrahydrofuran **88** with high 2,2-*cis*, 2,5-*cis*-selectivity possibly by sequential intramolecular allylations [30]. Starting material **86** is readily available from disilanyl homoallyl ether **89** through intramolecular bissilylation, ring-opening with PhLi, and dehydration.

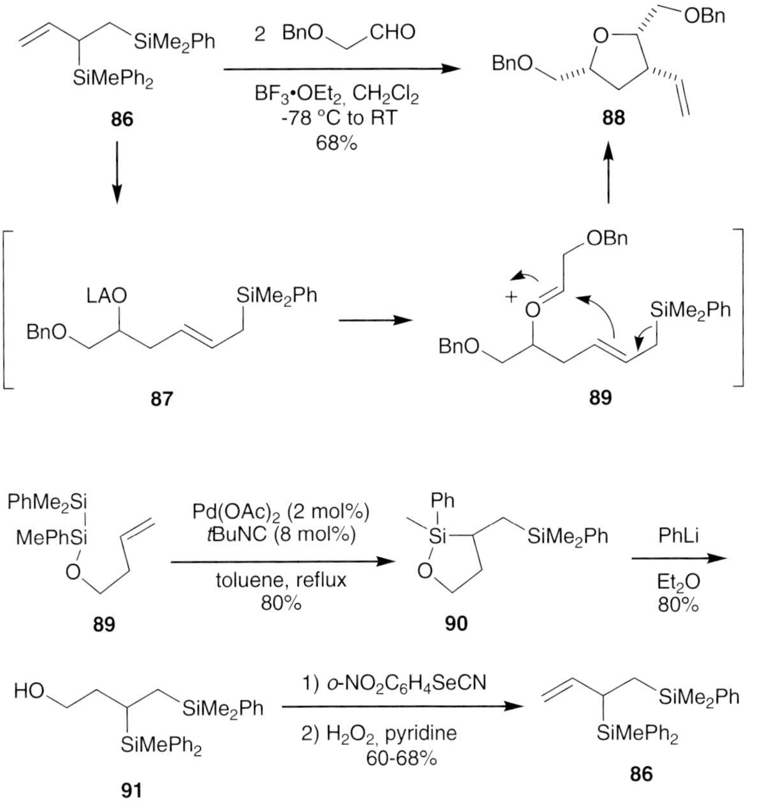

Scheme 5.22 One-pot synthesis of 2,3,5-trisubstituted tetra-hydrofurans with α-silylmethyl allylic silane.

Double deprotonation of 2-methylpropene followed by bissilylation gives (β-silyl-methyl)allylsilane **92**, which behaves as a trimethylenemethane dianion, as **92** reacts with bis(acetal) in the presence of TiCl$_4$ to provide 3,5-dimethoxy(methyle-ne)cyclohexane **93** (Scheme 5.23) [31].

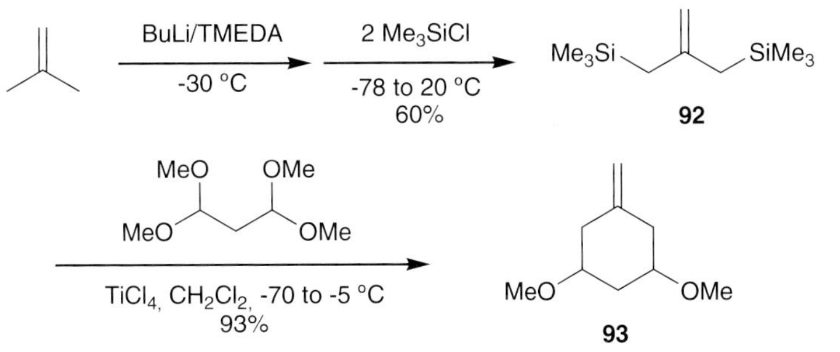

Scheme 5.23 Cyclization of (β-silylmethyl)allylsilane with bis(acetal).

Intramolecular electrophilic reaction of (β-silylmethyl)allylsilane with an imino group is an efficient approach to an 1-azabicyclo[3.2.1]octane framework [32]. For example, treatment of **94** with formaldehyde in CH_3CN at room temperature and then with trifluoroacetic acid produces **97** as a trifluoroacetate salt, presumably through domino cyclization through **95** and then of **96** (Scheme 5.24).

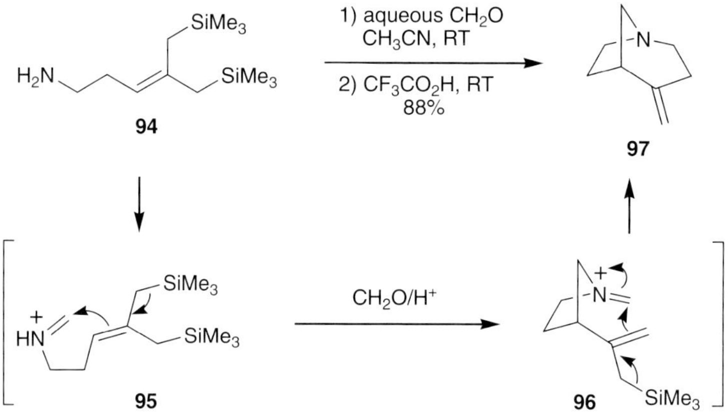

Scheme 5.24 Sequential double cyclization of amino containing (β-silylmethyl)allylsilane.

An alternative trimethylenemethane dianion equivalent is (β-stannylmethyl)allylsilane **98** [33]. Since allyltin is more reactive than allylsilane, **98** first reacts as an allyltin and then as an allylsilane. For example, asymmetric aldehyde addition of **98** utilizing a BINOL-titanium catalyst followed by the acetalization–cyclization protocol gives rise to optically active *cis*-2,6-disubstituted pyrans **100** with high enantiomeric excess (Scheme 5.25).

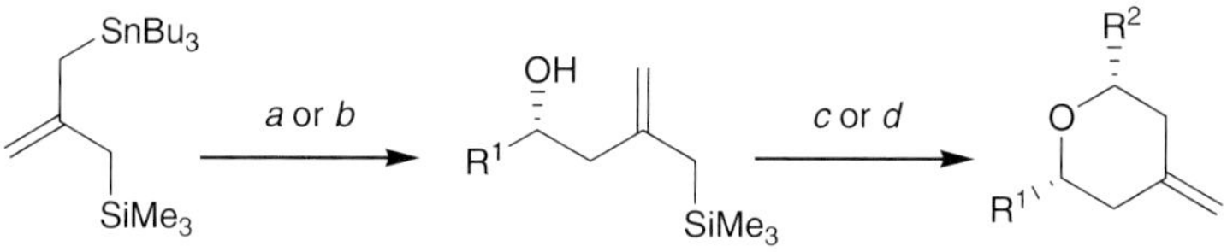

condition *a* : R^1CHO, [(*R*)-BINOL]Ti[OCH(CF$_3$)$_2$] (5 mol%), PhCF$_3$, -20 °C,
     54~94% yield, 90~97% ee
condition *b* : R^1CHO, [(*R*)-BINOL]Ti[OCH(CH$_3$)$_2$] (10 mol%), CH$_2$Cl$_2$, -20 °C,
     74~96% yield, 90~96% ee
condition *c* : R^2CHCl(OMe), *i*PrNEt$_2$, CH$_2$Cl$_2$, 0 to 23 °C, then Me$_3$SiNTf$_2$
     (10 mol%), -78 °C, 84~91% yield, 33 : 1~55 : 1 d.r.
condition *d* : R^2CHO, Me$_3$SiOTf, Et$_2$O, -78 °C, 95~98% yield, >99 : <1 d.r.

Scheme 5.25 Lewis acid-catalyzed asymmetric aldehyde addition and cyclization of (β-stannylmethyl)allylsilanes.

When the same strategy is applied to (β-carbamoyloxymethyl)allylsilane **101**, tetra-substituted pyran **104** is produced stereoselectively (Scheme 5.26) [34]. Hereby, bismuth(III) triflate monohydrate is found effective for the last cyclization.

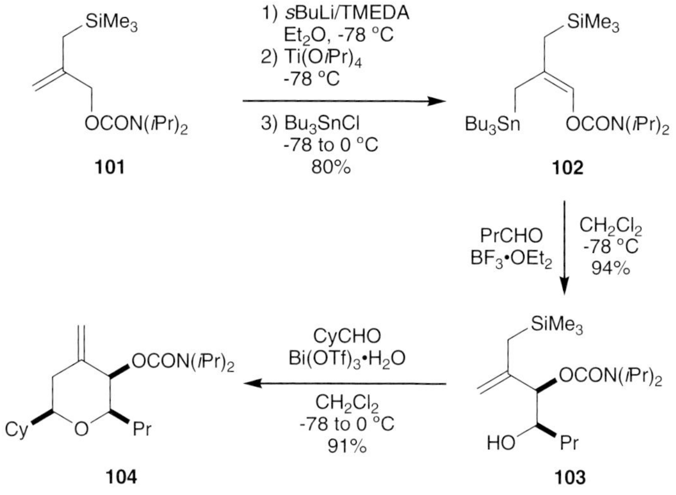

Scheme 5.26 Stereoselective synthesis of a tetrasubstituted pyran starting with (β-carbamoyloxymethyl)allylsilane.

5.3
Alkenylsilanes

5.3.1
Intermolecular Reactions of Polyfunctional Alkenylsilanes

Transition metal-catalyzed silicon-based cross-coupling reaction has emerged as a versatile carbon–carbon bond-forming process with high stereocontrol and excellent functional group tolerance [35]. For example, (α-benzoyloxy)alkenylsilanes **105**, prepared as a pure *E*-isomer by *O*-acylation of a lithium enolate derived from the corresponding acylsilane, reacts with carboxylic acid anhydrides in the presence of [RhCl(CO)₂]₂, giving rise to α-acyloxy ketones **106**, which are then converted into 1,2-diketones by acidic workup (Scheme 5.27) [36].

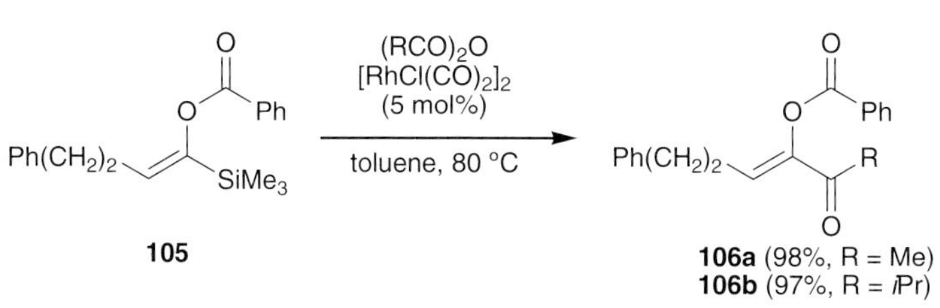

Scheme 5.27 Rh-catalyzed acylation of (α-benzoyloxy)alkenylsilane.

5.3.2
Intramolecular Reactions of Polyfunctional Alkenylsilanes

Treatment of [(2-methoxy)ethoxy]methyl (MEM) ethers **107**, derived from the corresponding bishomoallylic alcohols, with $SnCl_4$ in CH_2Cl_2 at −20 °C induces cyclization to give 3-alkylidenetetrahydropyrans **108** with retention of configuration of the silicon-substituted C=C bond (Scheme 5.28) [37]. The cyclization is applicable to the synthesis of 3-alkylidenetetrahydrofurans and -oxepanes starting from homoallylic and trishomoallylic MEM ethers, respectively.

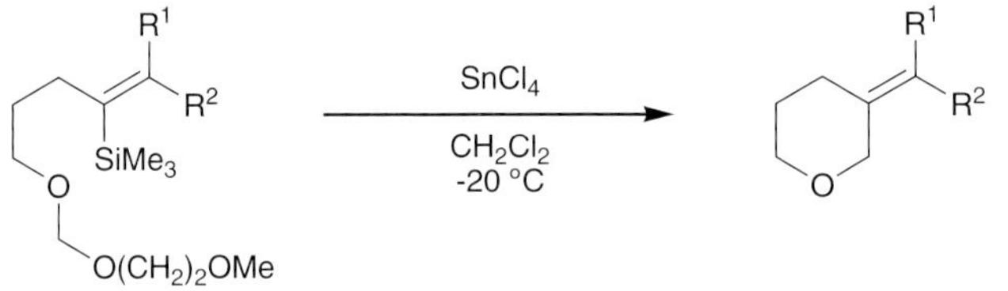

107a (R^1 = H, R^2 = Bu)
107b (R^1 = Bu, R^2 = H)

108a (89%)
108b (92%)

Scheme 5.28 Lewis acid-promoted cyclization of alkenylsilane bearing an acetal moiety.

Intramolecular alkenylsilylation of alkynes also proceeds in the presence of a Lewis acid catalyst [38]. Alkynyl-tethered alkenylsilanes **109** undergo cyclization in the presence of $EtAlCl_2$ or $AlCl_3$ and give (*E*)-cyclic dienylsilanes **110** with the trimethylsilyl group remaining in the product (Scheme 5.29). The C≡C bond has apparently inserted between the C–Si bond in **109** in a *trans* fashion.

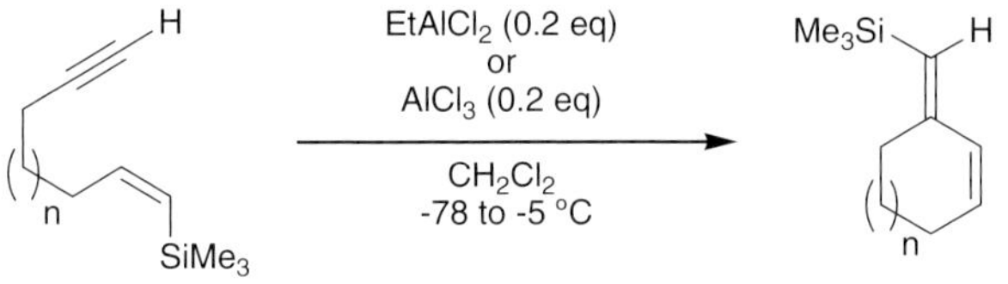

109a (n = 1)
109b (n = 2)

110a (n = 1: 92%)
110b (n = 2: 89%)

Scheme 5.29 Intramolecular alkenylsilylation of alkynes.

Intramolecular cross-coupling reaction of alkenylsilanes provides an efficient approach toward medium-sized rings having an internal 1,3-diene moiety [39]. Coupling precursors **112**, in which alkenyl iodide and cyclic silyl ether functionalities are installed at the terminal positions, are prepared by Mo-catalyzed ring-closing olefin metathesis of **111** (Scheme 5.30). The coupling reaction proceeds smoothly in THF at room temperature in the presence of [(π-allyl)PdCl]₂ (7.5 mol%) and Bu_4NF (10 eq), giving rise to 9-, 10-, 11-, and 12-membered cycloalkadienes **113–116**. The versatility of this protocol is demonstrated by a total synthesis of antifeedant (+)-brasilenyne [40].

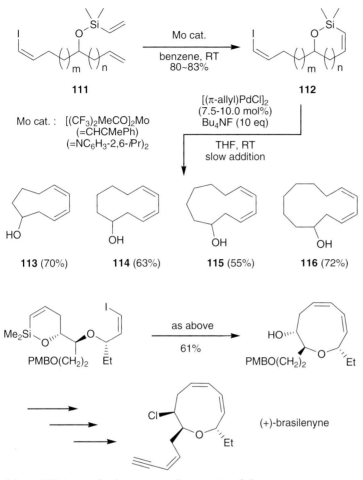

Scheme 5.30 Intramolecular cross-coupling reaction of alkenylsilanes leading to medium-sized rings.

5.3.3
Synthetic Reactions of Metal-containing Alkenylsilanes

Metallation of allyldimethylphenylsilane (**117**) with a superbase consisting of BuLi/*t*BuOK followed by quenching with (+)-*B*-methoxydiisopinocampheylborane [(+)-Ipc$_2$BOMe] gives chiral γ-silylallylborane **118**, which reacts with aldehydes to provide *anti-β*-hydroxyallylsilanes **119** with excellent enantioselectivity (Scheme 5.31) [41]. Me$_3$SiOTf-catalyzed acetalization–cyclization of **119** with an aldehyde affords optically active dihydropyran **120** with high *cis*-selectivity [42]. A boat-like transition state is proposed for the cyclization of the oxonium ion intermediate to reasonably explain the stereochemical outcome.

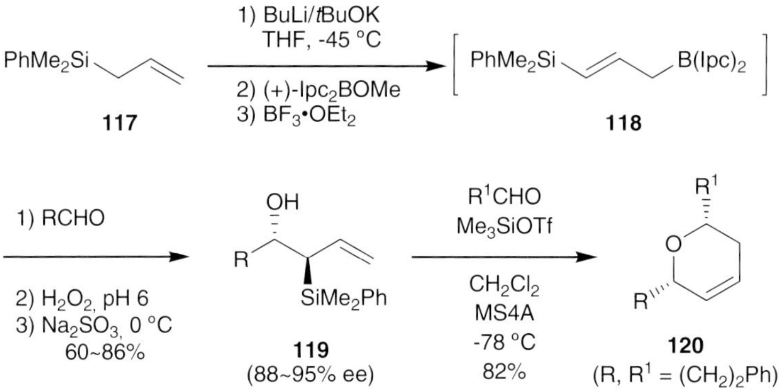

Scheme 5.31 Stereoselective synthesis of 2,6-*cis*-dihydropyrans from γ-silylallylborane.

Alkenylsilanes **123** bearing a boryl group at the α-position are readily available by the reaction of alkylidene-type lithium carbenoids **121** with silylboranes (Scheme 5.32). The reaction is considered to proceed through borate intermediate **122** with high stereospecificity [43]. Thus, when unsymmetrical carbenoids **121** are stereoselectively generated, stereodefined *gem*-silylborylethenes **123** are readily prepared. Synthetic utility of **123** is demonstrated by the subsequent Suzuki–Miyaura coupling followed by fluoride-mediated aldehyde addition as shown in the transformation from **123a** to **125**.

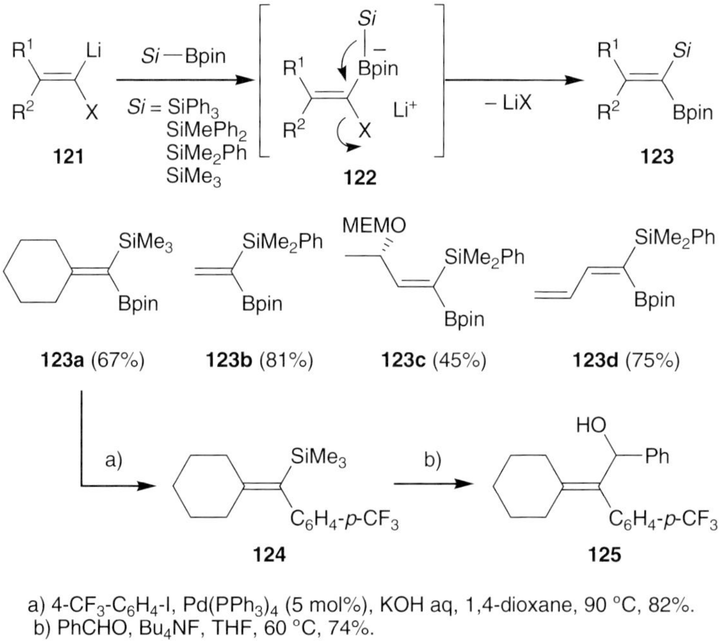

a) 4-CF$_3$-C$_6$H$_4$-I, Pd(PPh$_3$)$_4$ (5 mol%), KOH aq, 1,4-dioxane, 90 °C, 82%.
b) PhCHO, Bu$_4$NF, THF, 60 °C, 74%.

Scheme 5.32 Synthesis and reactions of *gem*-silylborylated alkenes.

Transition metal-catalyzed cleavage of silicon–silicon and silicon–heteroatom bonds followed by addition of each component to triple bonds is an efficient method for the preparation of polyfunctional alkenylsilanes [44]. For example, (Z)-1-silyl-2-stannylethene **127**, prepared by Pd-catalyzed silastannylation of ethyne with stannylsilane **126**, is allowed to couple with two different aryl iodides step by step in one pot in the presence of BnPdCl(PPh₃)₂ and CuI as catalysts, giving rise to unsymmetrical (Z)-stilbene **129** (Scheme 5.33) [45].

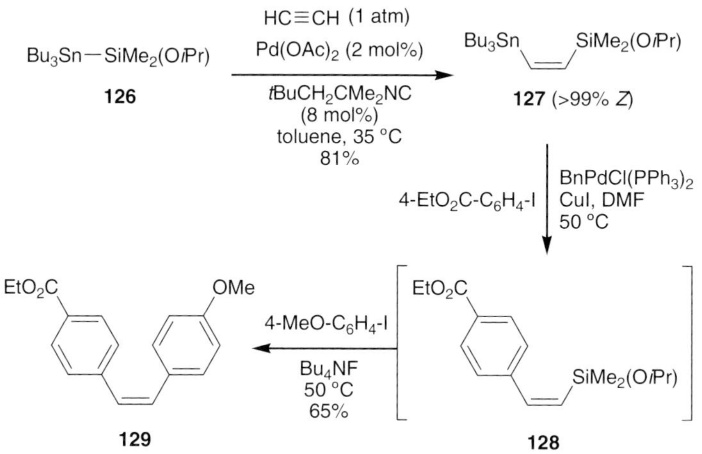

Scheme 5.33 Transition metal-catalyzed preparation and transformations of 2-stannylalkenylsilane.

5.4
Alkylsilanes

5.4.1
Synthetic Reactions of Polyhalomethylsilanes

Organometallic compounds bearing a leaving group such as halogen at the metallated carbon are called carbenoids that are thermally labile due to the ease of α-elimination taking place by intramolecular coordination of the leaving group to the metal [46]. Such intramolecular coordination is suppressed totally by the use of a nonmetallic counter-cation. The resulting anionic species may be called a naked anion. To generate naked anions, nucleophilic activation of organosilanes with an ammonium or sulfonium fluoride is extremely useful [47]. For example, (trichloromethyl)trimethylsilane **130** reacts with 3-methyl-2-butenal in the presence of tris(diethylamino)sulfonium difluorotrimethylsilicate (TASF) to give trichloromethylated alcohol **131** in high yield (Scheme 5.34) [48]. It is noteworthy that the 1,2-addition proceeds at room temperature, in sharp contrast to the reaction of polyhalomethyllithiums that needs extremely low reaction temperatures. Under the same conditions, dichlorobis(trimethylsilyl)methane **132** adds aldehyde double to afford 1,3-diol **133**.

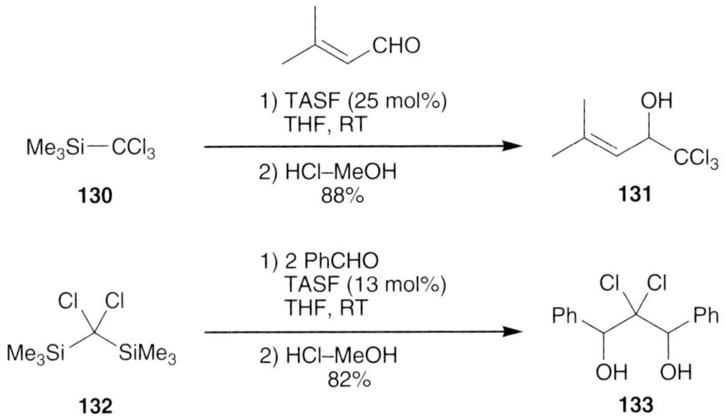

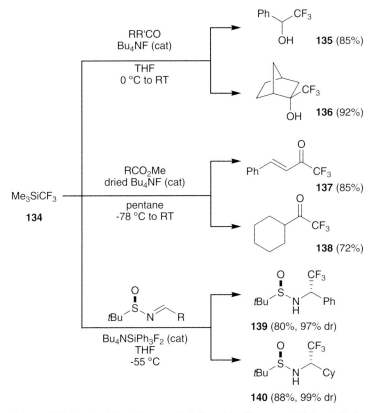

Scheme 5.34 Fluoride-catalyzed aldehyde addition of polyhalomethylsilanes.

Scheme 5.35 Nucleophilic trifluoromethylation with trifluoromethyl(trimethyl)silane.

The fluoride-induced activation strategy [49] is applicable to the generation of a trifluoromethyl anion equivalent from trifluoromethyl(trimethyl)silane (134), which, in the presence of a catalytic amount of Bu$_4$NF in THF, gives CF$_3$-aldehyde (or ketone) adducts (Scheme 5.35) [50]. Under similar conditions, methyl carboxylates are converted into the corresponding trifluoromethyl ketones [51]. Use of a commercial THF solution of Bu$_4$NF dried with activated MS4A prior to use is essential for success of transformation. Imine addition of a CF$_3$ group using 134 is effected using Bu$_4$NSiPh$_3$F$_2$ [52]. In particular, highly stereoselective trifluoromethylation of optically active sulfinylimines 139 and 140 is achieved as shown in the bottom of Scheme 5.35.

5.4.2
Synthetic Reactions of Cyclopropyl, Oxiranyl, and Aziridinylsilanes

Cyclopropyl anionic reagents are also accessible by the fluoride-based nucleophilic activation of the corresponding silylated precursors [53]. Methyl 1-trimethylcyclopropanecarboxylate (141, R = CO$_2$Me) reacts with acetaldehyde in the presence of Bu$_4$NF to give adduct 142, while the aldehyde addition of 1-cyano-1-trimethylsilylcyclopropane (141, R = CN) is smoothly mediated by benzyltrimethylammonium fluoride (Scheme 5.36).

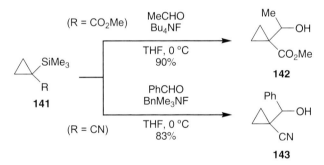

Scheme 5.36 Aldehyde addition of cyclopropylsilanes.

Upon treatment with a fluoride ion of trimethylsilyl-substituted oxiranes (144 and 146), and -aziridine (148) generate, the corresponding naked anionic species, which react with aldehydes to give the corresponding adducts (145, 147, and 149) with retention of configuration (Scheme 5.37) [54].

In the presence of TiCl$_4$, trimethylsilylmethylated cyclopropyl ketone 150 undergoes desilylative ring-opening reaction to generate (Z)-enolate 151, which then reacts with cinnamaldehyde, giving rise to syn-adduct 152 stereoselectively (Scheme 5.38) [55].

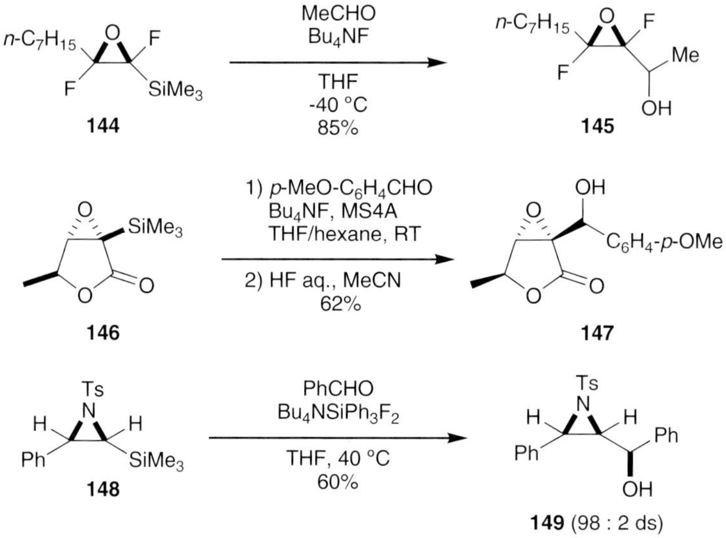

Scheme 5.37 Fluoride-mediated aldehyde addition of oxiranyl- and aziridinylsilanes.

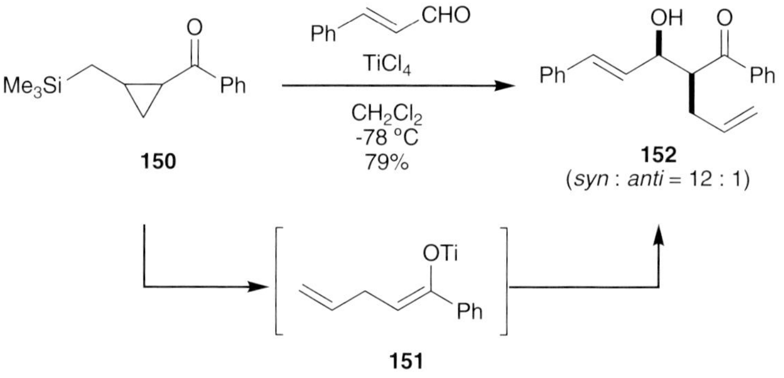

Scheme 5.38 Lewis acid-promoted reaction of cyclopropylmethylsilane with aldehyde.

5.4.3
Synthetic Reactions of Polysilylmethanes

Bis(trimethysilyl)methane derivatives react with aldehydes and ketones in the presence of a fluoride ion to afford di- and trisubstituted alkenes in one pot [56]. The reaction involves the fluoride-catalyzed carbonyl addition followed by Peterson elimination. For example, α,α-bis(trimethylsilyl)acetonitrile **153** produces β-phenylacrylonitrile **154** with high *E*-selectivity, whereas tris(trimethylsilyl)-methane **155** reacts with anisaldehyde at room temperature to give alkenylsilane **156** (Scheme 5.39).

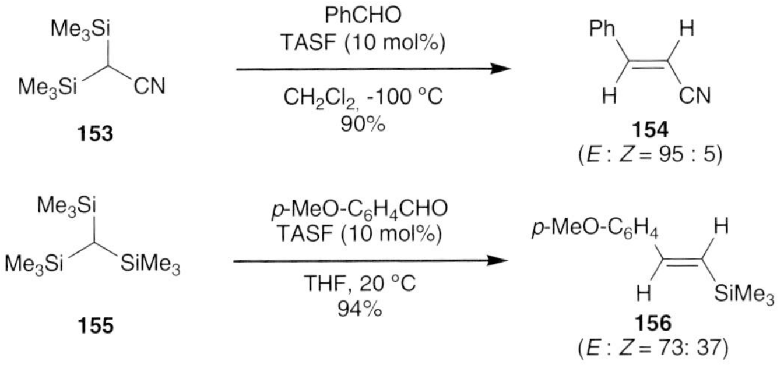

Scheme 5.39 Fluoride-catalyzed reaction of polysilylmethanes.

The strategy described above is applicable to fluorotris(trimethylsilyl)methane (**157**); it reacts with two molecules of benzaldehyde to give 2-fluoroallyl alcohol **158** (Scheme 5.40) [57]. The reaction involves five events in a single operation: generation of a naked methyl anion, aldehyde addition, Peterson elimination, generation of a naked sp^2 anion, and the second aldehyde addition.

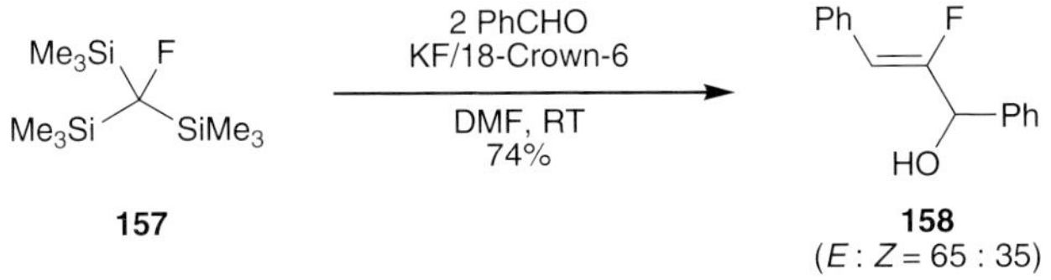

Scheme 5.40 Reaction of fluorotris(trimethylsilyl)methane with aldehyde.

5.5
Miscellaneous Preparations and Reactions of Polyfunctional Organosilicon Reagents

A silicon–silicon bond of hexamethyldisilane (**159**) is cleaved by Bu_4NF in hexamethylphosphoric triamide (HMPA) to produce a naked silyl anion and fluorotrimethylsilane [58]. The resulting silyl anion undergoes aldehyde addition to afford, after acidic workup, 1-trimethylsilyl-1-alcohols like **160** (Scheme 5.41). The silyl anion can react with 1,3-dienes to give 1,4-bis(trimethylsilyl)-2-butene **161** with high (*E*)-selectivity.

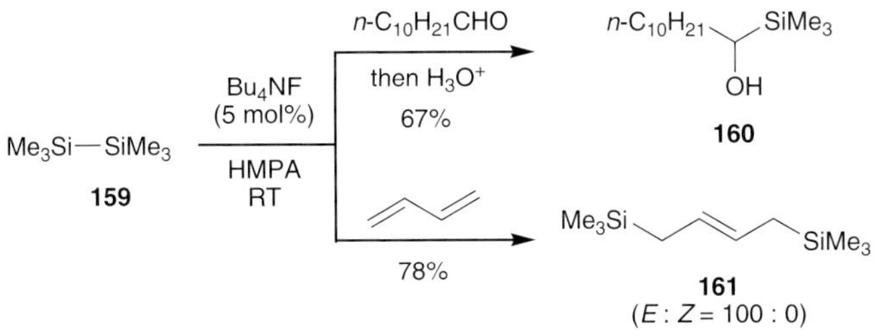

Scheme 5.41 Fluoride-catalyzed reaction of hexamethyldisilane.

The 1,4-disilylation of 1,3-dienes with phenyl-containing disilanes **162** is cata-lyzed by such a transition metal complex as Ni(PPh₃)₄ or Pt(CO)₂(PPh₃)₂ (Scheme 5.42) [59]. The presence of a phenyl group is essential for the successful addition.

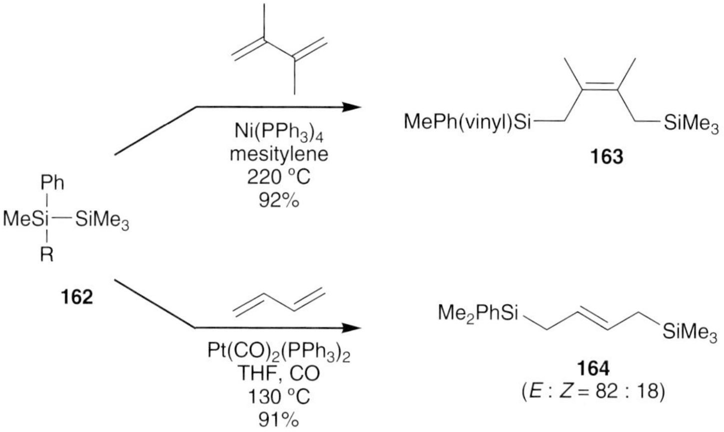

Scheme 5.42 Transition metal-catalyzed addition of disilanes to 1,3-dienes.

Direct silylation of aromatic compounds is carried out with 1,2-di-*tert*-butyl-1,1,2,2-tetrafluorodisilane (**165**) that serves as a silylating reagent in the presence of an iridium catalyst [60]. For example, the Ir-catalyzed C–H activation reaction of *o*-xylene selectively proceeds at the aromatic C–H bond rather than the benzylic one to give **166** in a high yield (Scheme 5.43). The synthetic utility of the products

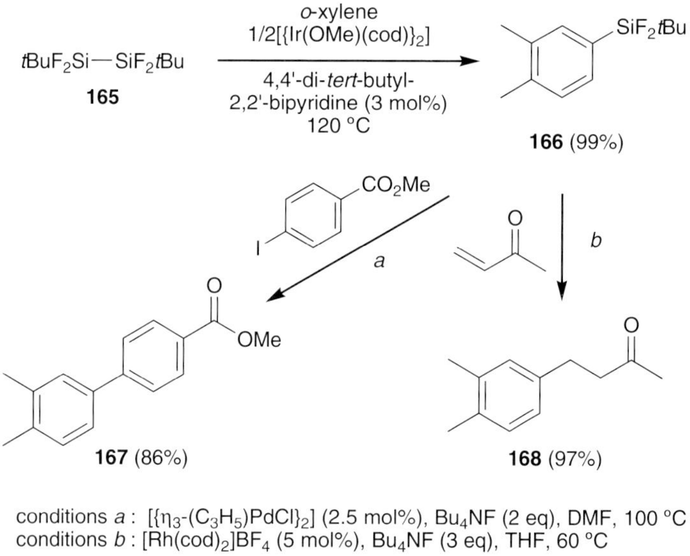

conditions *a* : [{η₃-(C₃H₅)PdCl}₂] (2.5 mol%), Bu₄NF (2 eq), DMF, 100 °C
conditions *b* : [Rh(cod)₂]BF₄ (5 mol%), Bu₄NF (3 eq), THF, 60 °C

Scheme 5.43 Ir-catalyzed synthesis and reactions of *tert*-butyldifluorosilylated arenes.

(e.g. **166**) is demonstrated by the Pd-catalyzed cross-coupling with 4-iodobenzoic acid methyl ester as well as the Rh-catalyzed 1,4-addition to methyl vinyl ketone.

Triallyl(aryl)silanes are stable and readily accessible organosilicon reagents that undergo the Pd-catalyzed cross-coupling reaction as an arylmetal in the presence of a fluoride ion [61]. The particular silanes can be used as a bifunctional reagent containing both allylsilane and arylsilane fucntionalities [62]. Thus, **169** delivers an allyl group to *p*-bromobenzaldehyde with the aid of Bu$_4$NF, and then a phenyl group by a newly added PdCl$_2$/PCy$_3$ catalyst system and additional Bu$_4$NF (Scheme 5.44).

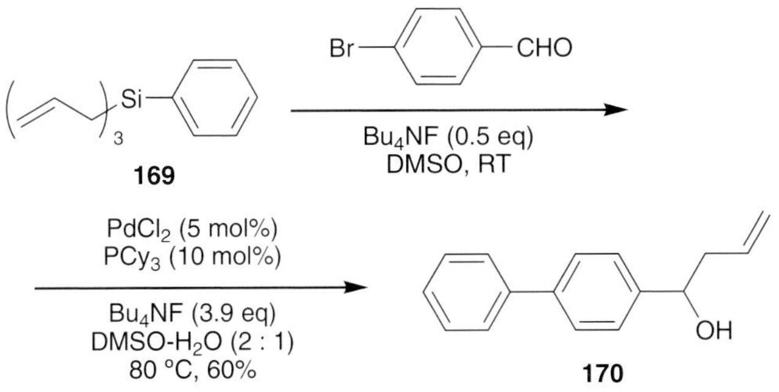

Scheme 5.44 Sequential reaction of triallyl(aryl)silane.

Acylsilanes having an alkynyl moiety undergo intramolecular cyclization in the presence of [RhCl(CO)$_2$]$_2$ to give α-alkylidenecyclopentanones and -hexanones, as shown by the example of **171** in Scheme 5.45 [63]. Use of acetic acid or trifluoroacetic acid increases the yield of product **172**.

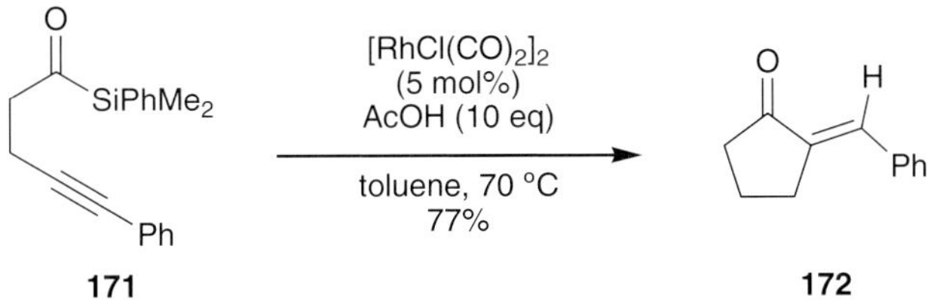

Scheme 5.45 Rh-catalyzed cyclization of acylsilane.

References

1 J. Emsley, *The Elements*, 3rd edn, Oxford University Press, Oxford, **1998**, p. 190.

2 Reviews on synthetic organosilicon chemistry: (a) T. H. Chan, I. Fleming, *Synthesis* **1979**, 761. (b) E. W. Colvin, *Silicon in Organic Synthesis*, Butterworths, London, **1981**. (c) P. D. Magnus, T. Sarkar, S. Djuric, in *Comprehensive Organometallic Chemistry, Vol. 7* (Eds. G. Wilkinson, F. G. A. Stone, E. W. Abel), Pergamon Press, Oxford, **1982**, p. 515. (d) H. Sakurai, *Pure & Appl. Chem.* **1982**, *54*, 1. (e) W. P. Weber, *Silicon Reagents for Organic Synthesis*, Springer, Berlin, **1983**. (f) E. W. Colvin, *Silicon Reagents in Organic Synthesis*, Academic Press, London, **1988**. (g) A. Hosomi, *Acc. Chem. Res.* **1988**, *21*, 200. (h) D. Schinzer, *Synthesis* **1988**, 263. (i) I. Fleming, J. Dunogues, R. H. Smithers, *Org. React.* **1989**, *37*, 57. (j) S. Patai, Z. Rappoport, *The Chemistry of Organic Silicon Compounds*, John Wiley & Sons, Chichester, **1989**. (k) G. Majetich, in *Organic Synthesis: Theory and Applications, Vol. 1* (Ed. T. Hudlicky), JAI Press Inc., Greenwich, **1989**, pp. 173. (l) I. Fleming, in *Comprehensive Organic Synthesis, Vol. 2* (Eds. B. M. Trost, I. Fleming), Pergamon Press, Oxford, **1991**, p. 563. (m) Y. Yamamoto, N. Asao, *Chem. Rev.* **1993**, *93*, 2207. (n) E. W. Colvin, in *Comprehensive Organometallic Chemistry II, Vol. 11* (Eds. E. W. Abel, F. G. A. Stone, G. Wilkinson), Pergamon, Oxford, **1995**, p. 313. (o) C. E. Masse, J. S. Panek, *Chem. Rev.* **1995**, *95*, 1293. (p) M. A. Brook, *Silicon in Organic, Organometallic, and Polymer Chemistry*, John Wiley & Sons, Inc., New York, **2000**. (q) J. A. Marshall, *Chem. Rev.* **2000**, *100*, 3163. (r) I. Fleming, *Science of Synthesis, Vol. 4*, Georg Thieme Verlag, Stuttgart, **2002**. (s) L. Chabaud, P. James, Y. Landais, *Eur. J. Org. Chem.* **2004**, 3173.

3 Review on uses of organosilicon compounds in synthesis of natural products: E. Langkopf, D. Schinzer, *Chem. Rev.* **1995**, *95*, 1375.

4 (a) J. S. Panek, M. Yang, *J. Org. Chem.* **1991**, *56*, 5755. (b) J. S. Panek, M. Yang, *J. Am. Chem. Soc.* **1991**, *113*, 6594. See also, (c) J. S. Panek, T. D. Clark, *J. Org. Chem.* **1992**, *57*, 4323.

5 Reviews on remote stereocontrol: (a) K. Mikami, M. Shimizu, H.-C. Zhang, B. E. Maryanoff, *Tetrahedron* **2001**, *57*, 2917. (b) H. Sailes, A. Whiting, *J. Chem. Soc., Perkin Trans.* **2000**, *1*, 1785. (c) H. J. Mitchell, A. Nelson, S. Warren, *J. Chem. Soc., Perkin Trans.* **1999**, *1*, 1899.

6 N. F. Jain, N. Takenaka, J. S. Panek, *J. Am. Chem. Soc.* **1996**, *118*, 12475.

7 (a) T. Hu, N. Takenaka, J. S. Panek, *J. Am. Chem. Soc.* **1999**, *121*, 9229. (b) T. Hu, N. Takenaka, J. S. Panek, *J. Am. Chem. Soc.* **2002**, *124*, 12806.

8 (a) J. S. Panek, N. F. Jain, *J. Org. Chem.* **1994**, *59*, 2674. (b) J. V. Schaus, N. Jain, J. S. Panek, *Tetrahedron* **2000**, *56*, 10263.

9 (a) C. T. Brain, E. J. Thomas, *Tetrahedron Lett.* **1997**, *38*, 2387. (b) L. C. Dias, R. Giacomini, *Tetrahedron Lett.* **1998**, *39*, 5343.

10 K. Miura, H. Saito, T. Nakagawa, T. Hondo, J.-i. Tateiwa, M. Sonoda, A. Hosomi, *J. Org. Chem.* **1998**, *63*, 5740.

11 K.-i. Imamura, E. Yoshikawa, V. Gevorgyan, Y. Yamamoto, *J. Am. Chem. Soc.* **1998**, *120*, 5339.

12 Reviews on transition metal-catalyzed carbocyclization: (a) I. Ojima, M. Tzamarioudaki, Zhaoyang Li, R. J. Donovan, *Chem. Rev.* **1996**, *96*, 635. (b) E.-i. Negishi, C. Copéret, S. Ma, S.-Y. Liou, F. Liu, *Chem. Rev.* **1996**, *96*, 365.

13 C. Fernandez-Rivas, M. Mendez, A. M. Echavarren, *J. Am. Chem. Soc.* **2000**, *122*, 1221.

14 A. M. Castano, J.-E. Bäckvall, *J. Am. Chem. Soc.* **1995**, *117*, 560.

15 I. E. Markó, A. Mekhalfia, D. J. Bayston, H. Adams, *J. Org. Chem.* **1992**, *57*, 2211.

16 (a) I. E. Markó, D. J. Bayston, *Tetrahedron Lett.* **1993**, *34*, 6595. See also, (b) T. Sano, T. Oriyama, *Synlett* **1997**,

716. (c) I. E. Markó, J.-M. Plancher, *Tetrahedron Lett.* **1999**, *40*, 5259.

17 H. Huang, J. S. Panek, *J. Am. Chem. Soc.* **2000**, *122*, 9836.

18 A. Hosomi, H. Hayashida, Y. Tominaga, *J. Org. Chem.* **1989**, *54*, 3254.

19 D. J. Kopecky, S. D. Rychnovsky, *J. Am. Chem. Soc.* **2001**, *123*, 8420.

20 (a) X. Wang, Q. Meng, A. J. Nation, J. L. Leighton, *J. Am. Chem. Soc.* **2002**, *124*, 10672. See also, (b) L. M. Frost, J. D. Smith, D. J. Berrisford, *Tetrahedron Lett.* **1999**, *40*, 2183.

21 (a) M. J. Zacuto, J. L. Leighton, *J. Am. Chem. Soc.* **2000**, *122*, 8587. (b) S. D. Dreher, J. L. Leighton, *J. Am. Chem. Soc.* **2001**, *123*, 341. (c) M. J. Zacuto, S. J. O'Malley, J. L. Leighton, *J. Am. Chem. Soc.* **2002**, *124*, 7890.

22 S. J. O'Malley, J. L. Leighton, *Angew. Chem. Int. Ed.* **2001**, *40*, 2915.

23 T. Akiyama, K. Asayama, S. Fujiyoshi, *J. Chem. Soc., Perkin Trans. 1* **1998**, 3655.

24 (a) M. Shimizu, H. Kitagawa, T. Kurahashi, T. Hiyama, *Angew. Chem. Int. Ed.* **2001**, *40*, 4283. See also, (b) Y. Yamamoto, H. Yatagai, K. Maruyama, *J. Am. Chem. Soc.* **1981**, *103*, 3229. (c) D. S. Matteson, D. Majumdar, *Organometallics* **1983**, *2*, 230.

25 (a) M. Suginome, Y. Ohmori, Y. Ito, *J. Am. Chem. Soc.* **2001**, *123*, 4601. (b) M. Suginome, Y. Ohmori, Y. Ito, *Chem. Commun.* **2001**, 1090.

26 M. Suginome, T. Ohmura, Y. Miyake, S. i. Mitani, Y. Ito, M. Murakami, *J. Am. Chem. Soc.* **2003**, *125*, 11174.

27 Y. Yamamoto, M. Fujita, N. Miyaura, *Synlett* **2002**, 767.

28 (a) M. Lautens, P. H. M. Delanghe, *Angew. Chem. Int. Ed. Engl.* **1994**, *33*, 2448. (b) M. Lautens, R. N. Ben, P. H. M. Delanghe, *Tetrahedron* **1996**, *52*, 7221.

29 D. M. Hodgson, S. F. Barker, L. H. Mace, J. R. Moran, *Chem. Commun.* **2001**, 153.

30 T. K. Sarkar, S. A. Haque, A. Basak, *Angew. Chem. Int. Ed.* **2004**, *43*, 1417.

31 B. Guyot, J. Pornet, L. Miginiac, *Tetrahedron* **1991**, *47*, 3981.

32 T. Kercher, T. Livinghouse, *J. Am. Chem. Soc.* **1996**, *118*, 4200.

33 (a) C.-M. Yu, J.-Y. Lee, B. So, J. Hong, *Angew. Chem. Int. Ed.* **2002**, *41*, 161. (b) G. E. Keck, J. A. Covel, T. Schiff, T. Yu, *Org. Lett.* **2002**, *4*, 1189.

34 I. E. Markó, B. Leroy, *Tetrahedron Lett.* **2001**, *42*, 8685.

35 Reviews on cross-coupling reaction of organosilicon compounds: (a) T. Hiyama, in *Metal-catalyzed Cross-coupling Reactions* (Eds. F. Diederich, P. J. Stang), Wiley-VCH, Weinheim, **1998**, p. 421. (b) T. Hiyama, E. Shirakawa, *Top. Curr. Chem.* **2002**, *219*, 61. (c) S. E. Denmark, R. F. Sweis, *Acc. Chem. Res.* **2002**, *35*, 835. (d) S. E. Denmark, R. F. Sweis, in *Metal-Catalyzed Cross-Coupling Reactions, Vol. 2* (Eds. A. de Meijere, F. Diederich), Wiley-VCH, Weinheim, **2004**, p. 163.

36 M. Yamane, K. Uera, K. Narasaka, *Chem. Lett.* **2004**, *33*, 424.

37 (a) L. E. Overman, A. Castaneda, T. A. Blumenkopf, *J. Am. Chem. Soc.* **1986**, *108*, 1303. See also, (b) I. E. Markó, D. J. Bayston, *Tetrahedron* **1994**, *50*, 7141.

38 (a) N. Asao, T. Shimada, Y. Yamamoto, *J. Am. Chem. Soc.* **1999**, *121*, 3797. (b) N. Asao, T. Shimada, T. Shimada, Y. Yamamoto, *J. Am. Chem. Soc.* **2001**, *123*, 10899.

39 S. E. Denmark, S.-M. Yang, *J. Am. Chem. Soc.* **2002**, *124*, 2102.

40 S. E. Denmark, S.-M. Yang, *J. Am. Chem. Soc.* **2002**, *124*, 15196.

41 (a) W. B. Roush, A. N. Pinchuk, G. C. Micalizio, *Tetrahedron Lett.* **2000**, *41*, 9413. (b) W. B. Roush, P. T. Grover, *Tetrahedron* **1992**, *48*, 1981.

42 W. B. Roush, G. J. Dilley, *Synlett* **2001**, 955.

43 (a) T. Hata, H. Kitagawa, H. Masai, T. Kurahashi, M. Shimizu, T. Hiyama, *Angew. Chem. Int. Ed.* **2001**, *40*, 790. (b) T. Kurahashi, T. Hata, H. Masai, H. Kitagawa, M. Shimizu, T. Hiyama, *Tetrahedron* **2002**, *58*, 6381. See also, (c) M. Shimizu, T. Kurahashi, H. Kitagawa, T. Hiyama, *Org. Lett.* **2002**, *5*, 225. (d) M. Shimizu, T. Kurahashi, H. Kitagawa, K. Shimono, T. Hiyama, *J. Organomet. Chem.* **2003**, *686*, 286.

44 Reviews on *vic*-dimetalation of unsaturated bonds: (a) M. Suginome, Y. Ito,

Chem. Rev. **2000**, *100*, 3221.
(b) I. Beletskaya, C. Moberg, *Chem. Rev.* **1999**, *99*, 3435.

45 M. Murakami, T. Matsuda, K. Itami, S. Ashida, M. Terayama, *Synthesis* **2004**, 1522.

46 Reviews on carbenoid chemistry:
(a) G. Boche, J. C. W. Lohrenz, *Chem. Rev.* **2001**, *101*, 697. (b) M. Braun, *Angew. Chem. Int. Ed.* **1998**, *37*, 431.
(c) K. G. Taylor, *Tetrahedron* **1982**, *38*, 2751. (d) H. Siegel, in *Topics in Current Chemistry, Vol. 106*, Springer-Verlag, Berlin, **1982**, p. 55. (e) A. Krief, *Tetrahedron* **1980**, *36*, 2531. (f) G. Köbrich, *Angew. Chem. Int. Ed. Engl.* **1972**, *11*, 473.

47 Review on synthetic reactions of organosilicon compounds with nucleophilic activation: G. G. Furin, O. A. Vyazankina, B. A. Gostevsky, N. S. Vyazankin, *Tetrahedron* **1988**, *44*, 2675.

48 M. Fujita, M. Obayashi, T. Hiyama, *Tetrahedron* **1988**, *44*, 4135.

49 Review on perfluoroalkylation with fluorinated organosilicon reagents: G. K. S. Prakash, A. K. Yudin, *Chem. Rev.* **1997**, *97*, 757.

50 (a) G. K. S. Prakash, R. Krishnamurti, G. A. Olah, *J. Am. Chem. Soc.* **1989**, *111*, 393. (b) R. Krishnamurti, D. R. Bellew, G. K. S. Prakash, *J. Org. Chem.* **1991**, *56*, 984.

51 J. Wiedemann, T. Heiner, G. Mloston, G. K. S. Prakash, G. A. Olah, *Angew. Chem. Int. Ed.* **1998**, *37*, 820.

52 G. K. S. Prakash, M. Mandal, G. A. Olah, *Angew. Chem. Int. Ed.* **2001**, *40*, 589.

53 C. Blankenship, G. J. Wells, L. A. Paquette, *Tetrahedron* **1988**, *44*, 4023.

54 (a) T. Dubuffet, R. Sauvetre, J. F. Normant, *Tetrahedron Lett.* **1988**, *29*, 5923. (b) K. Kuramochi, H. Itaya, S. Nagata, K.-i. Takao, S. Kobayashi, *Tetrahedron Lett.* **1999**, *40*, 7367. (c) V. K. Aggarwal, E. Alonso, M. Ferrara, S. E. Spey, *J. Org. Chem.* **2002**, *67*, 2335.

55 V. K. Yadav, R. Balamurugan, *Org. Lett.* **2003**, *5*, 4281.

56 C. Palomo, J. M. Aizpurua, J. M. Garcia, I. Ganboa, F. P. Cossio, B. Lecea, C. Lopez, *J. Org. Chem.* **1990**, *55*, 2498.

57 (a) M. Shimizu, T. Hata, T. Hiyama, *Tetrahedron Lett.* **1999**, *40*, 7375. (b) M. Shimizu, T. Hata, T. Hiyama, *Bull. Chem. Soc. Jpn.* **2000**, *73*, 1685.

58 T. Hiyama, M. Obayashi, I. Mori, H. Nozaki, *J. Org. Chem.* **1983**, *48*, 912.

59 (a) M. Ishikawa, Y. Nishimura, H. Sakamoto, T. Ono, J. Ohshita, *Organometallics* **1992**, *11*, 483. (b) Y. Tsuji, R. M. Lago, S. Tomohiro, H. Tsuneishi, *Organometallics* **1992**, *11*, 2353.

60 T. Ishiyama, K. Sato, Y. Nishio, N. Miyaura, *Angew. Chem. Int. Ed.* **2003**, *42*, 5346.

61 (a) Y. Nakao, T. Oda, A. K. Sahoo, T. Hiyama, *J. Organomet. Chem.* **2003**, *687*, 570. (b) A. K. Sahoo, Y. Nakao, T. Hiyama, *Chem. Lett.* **2004**, *33*, 632.

62 Y. Nakao, T. Oda, T. Hiyama, unpublished result.

63 (a) M. Yamane, T. Amemiya, K. Narasaka, *Chem. Lett.* **2001**, 1210. Reviews on acylsilane chemistry: (b) P. F. Cirillo, J. S. Panek, *Org. Prep. Proc. Int.* **1992**, *24*, 553. (c) P. C. B. Page, S. S. Klair, S. Rosenthal, *Chem. Soc. Rev.* **1990**, *19*, 147. (d) A. Ricci, A. Degl'Innocenti, *Synthesis* **1989**, 647.

6
Polyfunctional Tin Organometallics for Organic Synthesis

Eric Fouquet and Agnès Herve

6.1
Introduction

The use of functionalized organotins in organic syntheses has considerably increased in the last two decades. Some reactions, such as the Stille coupling or the nucleophilic allylation, can even be considered as cornerstones for numerous synthetic applications. The increasing popularity of organotins is mainly due to the ease of their preparation, even in the optically active form, associated to a good balance between stability and reactivity. The only drawbacks involved in the use of organotins could be related to environmental considerations, but in contrast with other heavy metals this aspect has not led organotin chemistry to decline. Nowadays organotin chemistry is still flourishing both from the methodological aspect and for the synthetic use. As this chapter cannot treat the subject in an exhaustive way, it will focus on the formation of carbon–carbon bonds by the use of organotin reagents

6.2
Metal-Catalyzed Coupling Reactions

6.2.1
The Stille Cross-Coupling Reaction

The palladium-catalyzed cross-coupling of organotin reagents with organic electrophiles is one of the most important reactions leading to the formation of a new C–C bond. The first examples were reported during the period 1976–77, by Eaborn and coworkers [1] and Kosugi et al. [2]. The following mechanistic studies and synthetic applications performed by Stille [3] made this reaction a standard method in organic synthesis. The reason for this is two fold. First, the reaction is a particularly mild process with neutral conditions and tolerates a wide variety of functional groups such as carboxylic acid, ester, amide, nitro, ether, amine, hydroxyl, ketone and even aldehyde, as well as a high degree of stereochemical com-

Organometallics. Paul Knochel
Copyright © 2005 WILEY-VCH Verlag GmbH & Co. KGaA, Weinheim
ISBN: 3-527-31131-9

plexity in either the organotin or the electrophilic coupling partner. Secondly, organostannanes are accessible by numerous methods, and easy to handle since they are relatively insensitive to moisture and oxygen.

6.2.1.1 Mechanism

In terms of mechanism, the stille coupling is characterized by a sequence of: (i) active Pd^0 species formation, (ii) oxidative addition, (iii) transmetallation and (iv) reductive elimination.

Detailed mechanisms will not be discussed here, but the reader can refer to recent reviews [4]. If both the oxidative addition and the reductive elimination are reasonably understood, the transmetallation step is still a matter of intense debate. In any case, it is noteworthy that the nature of the ligands on the palladium may exert a dramatic influence on the kinetics of the transmetallation step. Moreover, the transmetallation rate was also found to depend on the nature of the group transferred from the organotin with the generally accepted order: alkynyl>alkenyl>aryl>allyl>benzyl>>>alkyl. There is no ideal catalytic system for a given reaction and several factors (ligands, additives, cocatalyst, solvents) have to be considered.

6.2.1.1.1 Ligands and Catalysts

Pd(PPh$_3$)$_4$ is a commonly used catalyst for the Stille cross-coupling reaction. However, several studies demonstrated that the Pd/phosphine *ratio* exerts a dramatic influence on the kinetics, in the fact that an excess of ligand would inhibit the coupling by slowing the formation of the coordinatively unsaturated PdL$_2$ active species. Pd(PPh$_3$)$_4$ can be replaced by the correct combination of Pd$_2$dba$_3$/phos-

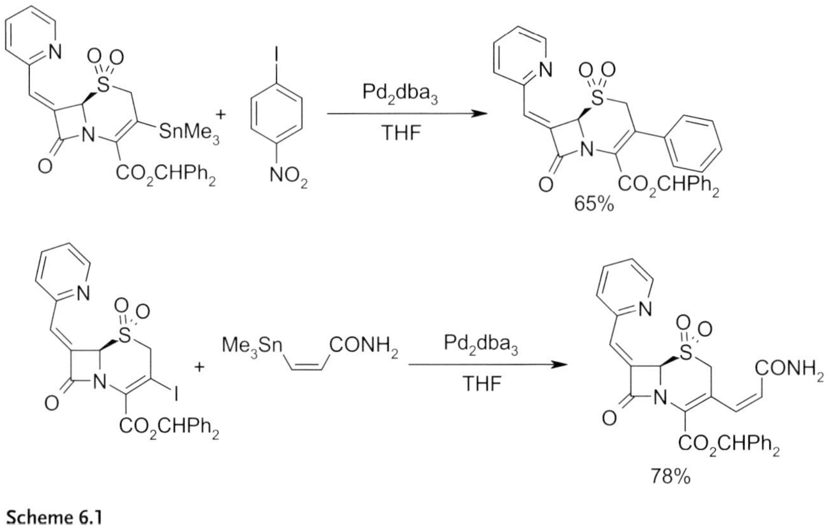

Scheme 6.1

phine ligand or Pd^{II}(PPh$_3$)$_2$Cl$_2$, reduced *in situ* to provide the active catalytic species. A large acceleration rate was observed in some cases with ligands of reduced donicity such as tri(2-furyl)phosphine (TFP) [5], phosphites and triphenylarsine [6]. Sterically demanding ligands such as tris(*o*-tollyl)phosphine-, tri(ter-butyl)phosphine and di(ter-butyl)methylphosphine were employed to favor the ligand dissociation step [7]. The use of air-stable trialkylphosphonium salts was also reported [8]. Finally, it is noteworthy that several Stille couplings were performed under "ligandless" conditions without altering the stability of the catalyst [9] as exemplified by the preparation of new cephalosporins [10] (Scheme 6.1).

6.2.1.1.2 Cocatalysts

Discovered by Liebeskind and Fengl [11], the beneficial effect of Cu(I) on the rate of sluggish Stille cross-couplings, referred as the "copper effect", is well documented [12]. The function performed by copper in the catalytic cycle is solvent dependent. In ethereal solvents, Cu^I is a scavenger for free neutral ligands that otherwise causes retardation of the rate-limiting transmetallation step. In highly polar solvents, the rate-accelerating effect of Cu^I is due to a preliminary transmetallation reaction from the organostannanes to generate more reactive organocopper species that further enter in the catalytic cycle. Other salts such as zinc [13] and cadmium [14] chlorides were also used as additives. In the case of unreactive triflates under classical conditions [15], the use of LiCl is recommended in ethereal solvents [16]. Nevertheless, the addition of such a stabilizing chloride source was observed to inhibit the reaction in polar solvents [17]. A remarkable effect of LiCl was reported in the Stille cross-coupling of allenyltins with a wide range of organic iodides to give the corresponding aryl-, alkenyl and disubstituted allenes [18]. The use of nucleophiles (amines, fluoride sources) to increase the reactivity of the organotin species, via hypercoordinate intermediates, is a well-established strategy (see Section 6.2.1.2.5).

6.2.1.1.3 Solvents and Media

Solvents used for the Stille coupling include hydrocarbons, organic chlorides, ethereal solvents, highly polar solvents and even water, but most of the reactions are conducted in THF, DMF or NMP. The reaction was also adapted to the fluorous biphasic catalysis [19]. Stille reactions were also performed in ionic liquids [20], supercritical CO$_2$ [21] and in aqueous medium [22] by using water soluble ligands.

6.2.1.2 Organotins for the Stille Reaction

6.2.1.2.1 Alkynyltins

Alkynyltins are considered to be the most reactive organotins and their coupling generally proceeds smoothly [23]. They were used for different types of reactions such as the preparation of substituted alkkynylpyrones [24], a one-pot sequence of the Stille/iodolactonization reaction furnishing 2(2H)-pyranone derivatives [25] and a tandem Stille/carbopalladation sequence affording highly substituted enynes [26]. A recent application has concerned the introduction of alkynyl substituents to the C2-position of benzodiazepin dilactams (Scheme 6.2) [27].

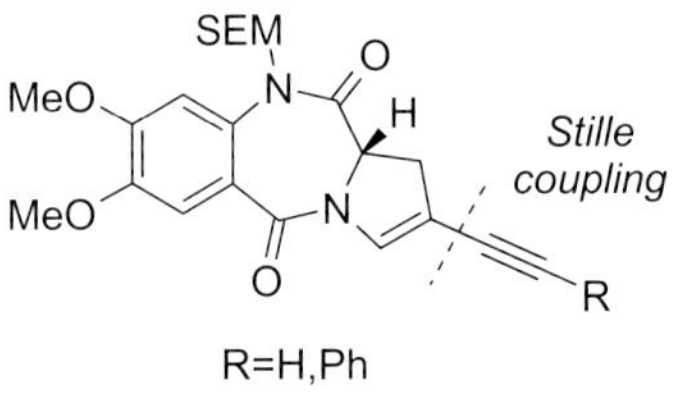

R=H,Ph

Scheme 6.2

6.2.1.2.2 Alkenyltins

The Stille cross-coupling of alkenyltins has been widely used in organic chemistry, the major synthetic application being the preparation of polyconjugated systems via vinyl–vinyl cross-couplings (see Section 6.2.1.4.1 or 6.2.1.5.1). The transfer of alkene proceeds with retention of the double-bond stereochemistry and the reaction is not affected by the steric hindrance of the electrophile. However, when α-vinylstannanes are used as coupling partners, the rate of the Stille coupling is decreased and an alternative reaction, the *cine* substitution, sets in [28]. In this case, adding copper salts in combination with LiCl may solve the problem [29]. Several fluoroalkenylstannanes [30] were used to prepare functionalized fluoroalkenes, potential intermediates in the syntheses of fluorine-containing bioactive compounds [31]. α-Stannyl enamides [32] were coupled with a range of halides such acylhalides to give the corresponding α,β-unsaturated ketones that may serve as highly functionalized Michael acceptors (Scheme 6.3).

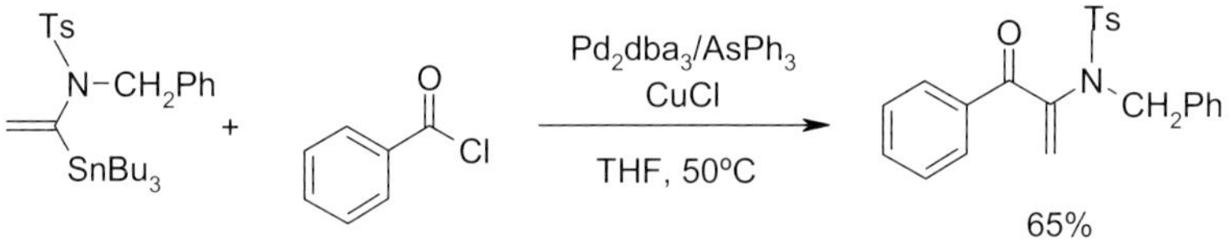

Scheme 6.3

The coupling of allenyltins is also efficient and affords allenyl [33] or propargyl [34] products depending on the nature of the substrates. An elegant example is the palladium-catalyzed regio and stereoselective annulation of allenylstannanes by β-iodo vinylic acids to give the corresponding α-pyrones via a Stille reaction/ cyclization sequence [35] (Scheme 6.4).

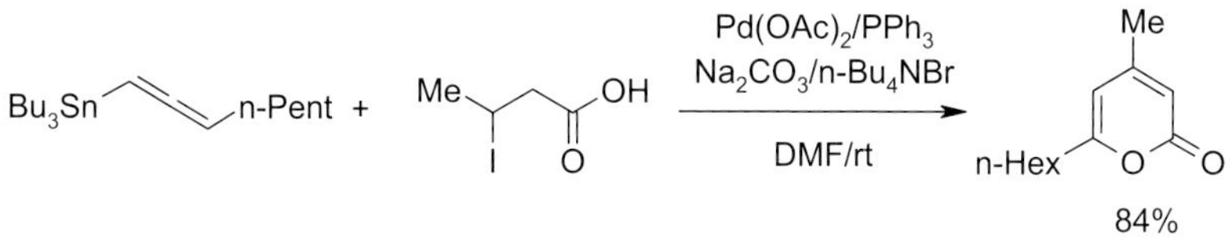

Scheme 6.4

6.2.1.2.3 Aryl- and Heteroaryltins

Both electron-deficient and electron-rich aryltins are suitable partners for the Stille reaction. The only limitation of the reaction was an ortho substitution on the organotin and the substrate until the recent use of electron-rich and sterically demanding ligands such as P(t-Bu)$_3$, which permitted the synthesis of tetra-ortho-substituted biaryls. A plethora of heterocyclic stannanes [36] were used in Stille couplings. Their major applications concern the synthesis of pharmacologically interesting compounds, as illustrated by the synthesis of an endothelin antagonist [37] (Scheme 6.5), as well the preparation of supramolecular compounds.

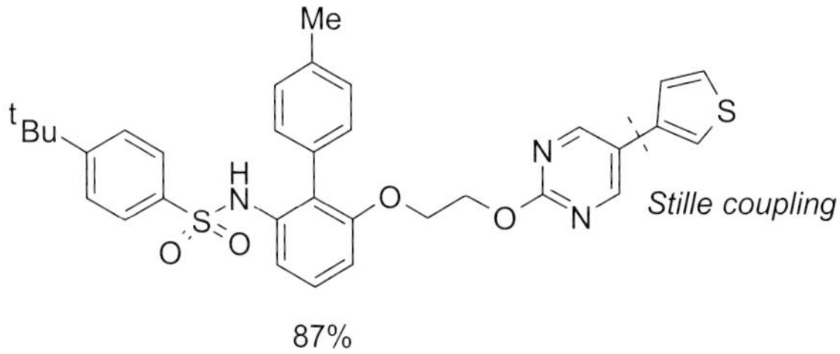

Scheme 6.5

6.2.1.2.4 Allyl and Benzyltins

Benzyl- and allyltins present a much lower reactivity in Stille coupling than organotins containing a Sn–Csp^2 bond. Allyltins have not been frequently used, due in part to the tendency of products to reconjugate after the cross-coupling when olefinic or aromatic substrates are used as electrophilic partners [38], and the frequent loss of regioselectivity observed whith γ-substituted allyltins [39]. A nice application is represented by the reaction of a β,γ-disubstituted allyltin with a naphthoquinone-bromide used in a biomimetic-type synthesis of benzo [a]naphthacene quinines related to Pradimicinone [40] (Scheme 6).

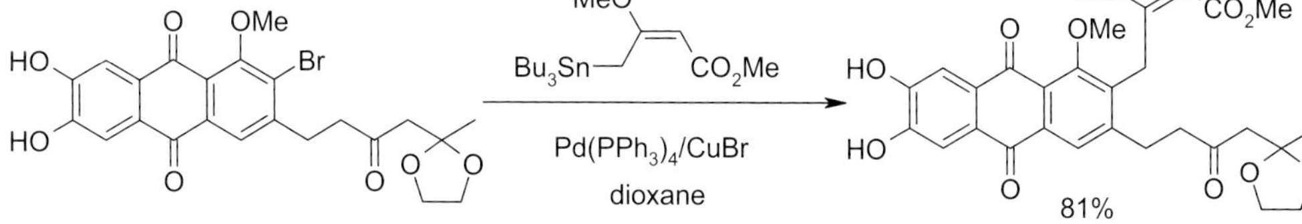

Scheme 6.6

6.2.1.2.5 **Alkyltins**

There are relatively few examples of the successful introduction of alkyl groups with the Stille cross-couplings. This is probably due to the competitive β-hydride elimination when using alkylstannanes bearing β-hydrogens. In most cases, the transfer of alkyl groups requires harsh conditions [41] and methyl, ethyl and butyl groups were transferred from the corresponding tetraalkyltins at elevated temperatures [42]. Nevertheless, it is noteworthy that alkyltins bearing a heteroatom in the α position to the tin atom cross-couple with success, the heteroatom enhancing the nucleophilicity of the carbon to be transferred [43]. In 1992, Vedejs et al. [44] and Brown et al. [45] independently reported that intramolecular coordination of amine to tin greatly accelerates the transfer of the alkyl group from tin. This methodology was applied to the transfer of radiolabelled $^{11}CH_3$ (Scheme 6.7) [46] and to the synthesis of bioactive sulfonamide [47].

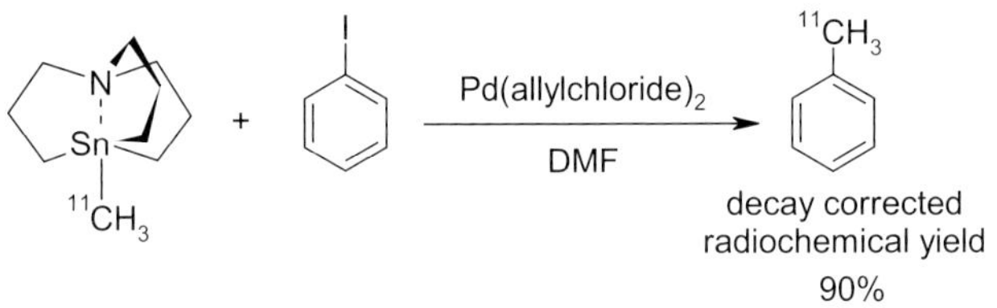

Scheme 6.7

The activation, involving the expansion of the coordination sphere at tin [48], was recently used with externally coordinating ligands such as the fluoride ion [49] and permitted the transfer of alkyl groups from alkylfluorostannates [50]. Similar activation with diethylamine was also reported [51] as well as the transfer of the synthetically useful silylmethyl group via a 2-pyridyldimethyllsilyl(triorgano)tin reagent [52].

6.2.1.3 **Substrates**

6.2.1.3.1 **Organic Halides**

Aryl iodides and bromides couple with a large range of organostannanes. Contrary to this, and for a long time, only activated aryl chlorides were thought to be suitable substrates, the oxidative addition becoming the rate-limiting step of the reaction [53]. In 1999, however, Fu and coworkers reported that, when associated with a fluorine source, a combination of $Pd/P(t\text{-}Bu)_3$ could serve as a versatile catalyst for the Stille reaction of nonactivated and very congested aryl chlorides [54]. Three other catalytic systems were also described to permit the coupling of aryl chlorides [55].

Heteroaryl halides were used in numerous reactions with aromatic-, heteroaromatic- and vinylstannanes such as in the synthesis of Deoxyvariolin B [56] (Scheme 6.8).

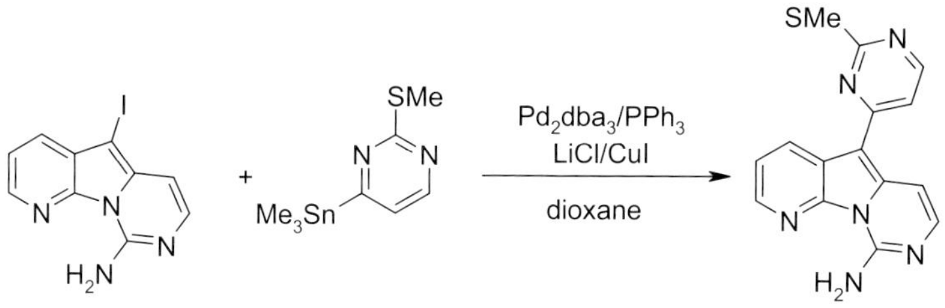

Scheme 6.8

Of particular synthetic interest is the coupling of 5-halouracyls, 5-halouridines, 2-halopurines, 6-halopurines, 5-iodocytosine and 5-iodo-2-deoxythiouridine and their applications in the synthesis of modified nucleosides [57] as exemplified by the coupling of 8-bromo-1-ribofuranosidylpurin [58] (Scheme 6.9).

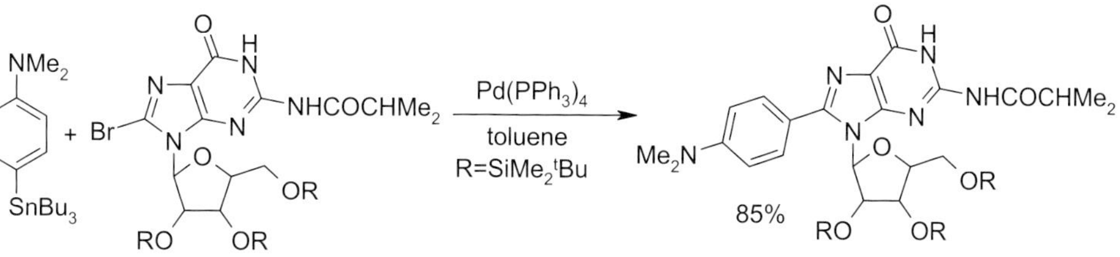

Scheme 6.9

Few examples involving Csp^3-X electrophiles are reported. If, P(tBu)$_2$Me was successfully used for the coupling of alkyl bromides with vinylstannanes, this ligand was shown to be ineffective for couplings with arylstannanes [59]. The ligand of choice for coupling aryl stannanes with alkyl bromides and iodides appeared to be the electron-rich cHex(Pyr)$_2$P [60].

Acyl chlorides, chloroformates and carbamoyl chlorides couple efficiently with organotins giving access to various aldehydes, ketones and α,β-unsaturated ketones, α,β-unsaturated esters and amides, respectively. This reaction, which represents an alternative to the carbonylative three-component coupling reaction, was applied in total synthesis of cytotoxic mycalozol [61] , 10,11-dihydroleukotriene B metabolites [62] and 9-methoxystrobilurin [63]. Recently, a convenient and general one-pot synthesis of α-substituted amides and N-protected amines by a palladium-catalyzed three-component-coupling of imines, acyl chlorides or chloroformates, and organotin reagents was also described [64].

6.2.1.3.2 Sulfonates

Aryl and heteroaryl sulfonates are extensively used as electrophilic coupling partners. The reaction of vinyl sulfonates is generally limited to triflates, even if some couplings of mesylates and tosylates are reported. It was observed that the addition of halide salts such as LiCl, is necessary when the coupling is performed in nonpolar solvents. The catalytic activity can be optimized when replacing strongly donating ligands by triphenylarsine ligands. The mechanism of the coupling with organic triflates is not well understood and remains a matter of debate. It is noteworthy that other sulfonate substrates including mesylates, [65] tosylates, [66] fluorosulfonates [67] and nonaflates [68] were employed.

6.2.1.3.3 Miscellaneous

Various salts such as hypervalent iodines [69] aryldiazonium salts [70] and sulfonium salts [71] undergo facile Stille coupling. Acetates [72], carbonates [73] and phosphates [74] are also quite reactive. Finally, heteroaromatic tioethers [75] and sulfonyl chlorides [76] were recently reported to couple under palladium-catalyzed copper-mediated catalysis.

6.2.1.4 Intermolecular Stille Cross-coupling

6.2.1.4.1 Vinyl–Vinyl Coupling

Vinyl–vinyl coupling was extensively used for the preparation of polyconjugated systems and widely applied to the preparation of complex natural molecules such as Cochlemycin A, [77] (+)-Crocacin D [78] (Scheme 6.10), Gambierol [79], Bafilomycin V_1 [80], (–)-Sanglifehrin A [81] and Apoptolidin [82]. Stille coupling of a vinylstannane with cyclic vinyl halides was used for the preparation of Himbacine derivatives [83].

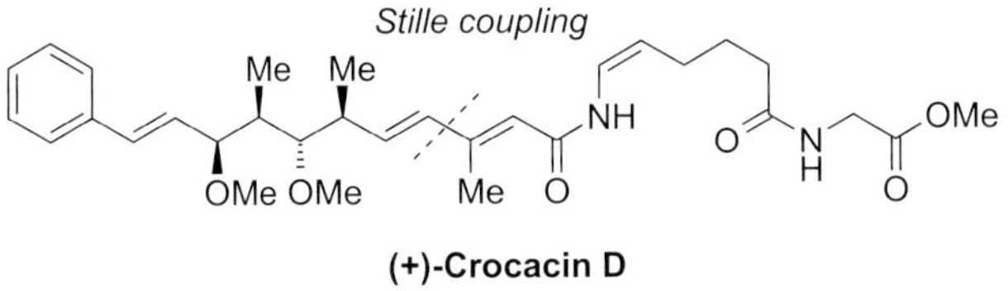

(+)-Crocacin D

Scheme 6.10

The use of (E,Z)- or (E,E)-dienyltins and polyenyltins allowed the synthesis of polyconjugated natural products such as Cytostatin [84], (+)-Fostriecin, [85] Dermostatin A (Scheme 6.11) [86], Polycephalin C [87] and (+)-Calyculin A [88].

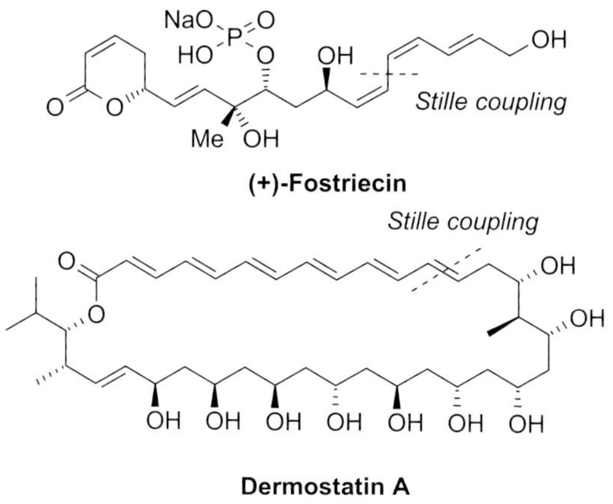

(+)-Fostriecin

Dermostatin A

Scheme 6.11

Some reactions of particular interest are some examples of tandem reactions such as (i) the Stille coupling/elimination sequence with a 1,1-dibromo-1-alkene in the preparation of Callipeltoside [89] or (ii) the Stille coupling/electrocyclization cascade in the synthesis of immunosuppressants (Scheme 6.12) [90] or (iii) distannylated reagents such as 1,2-vinylditin and 1,4-dienyltin used in double Stille reaction sequences [91] and peculiar cascade reactions [92].

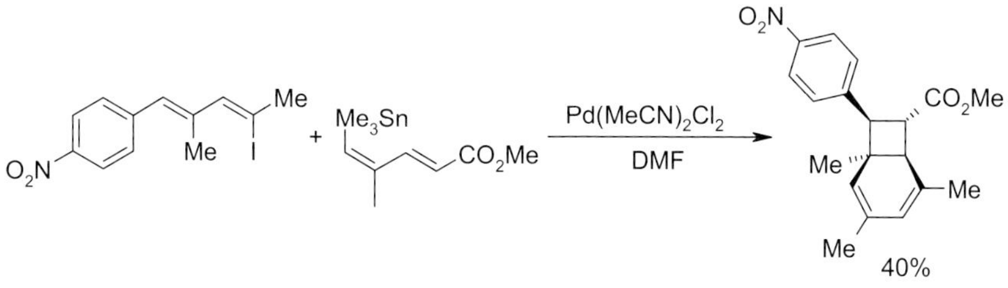

Scheme 6.12

6.2.1.4.2 Aryl–Aryl Coupling

Aryl–aryl Stille coupling is a major tool in organic synthesis and was applied to the preparation of various ligands [93], catalysts [94] and to the design of materials with electronic and optical properties such as oligopyridines [95] and oligothiophenes [96]. All types of aromatic substrates were used and the reaction was shown to be tolerant to a wide range of functional groups such as chlorine or fluorine, trifluoromethyl, acetylenes, nitriles, ethers and thioethers, esters and amides, ketones and aldehydes, ketals, nitro, unprotected amines, hydroxyles or carboxylic acids. As previously mentioned (see Section 6.2.1.2.3), an ortho-substitution may sometimes alter the reaction, whenever some recent examples described the coupling of very hindered aryl substrates [54]. Interesting aspects of this reaction are the possible polysubstitutions on aromatic rings [97] and sequential couplings depending on the intrinsic reactivity of each functionality of the substrate under a given set of conditions [98].

6.2.1.4.3 Carbonylative Coupling

When conducted under a CO atmosphere, the organopalladium(II) species resulting from the oxidative addition are able, under a moderate pressure, to give an acyl palladium(II) intermediate that can further react with the organotin by the usual route of transmetallation and reductive elimination to give the carbonylated cross-coupled compound. If aryl, heteroaryl, alkenyl and benzyl halides were used as electrophilic coupling partners, their sulfonate counterparts are the most popular substrates and Stille carbonylative couplings often take place under mild conditions. Other substrates, such as mono- and difluoroiodododecanes [99], triarylantimony(V) diacetates [100], aryl diazonium salts [101] and hypervalent iodine compounds [102], were used as well. Aryl-, alkynyl-, akenyl-, allyl- and simple alkyltins couple even if double bond-migration is the major drawback with allylstannanes.

The so-called "carbonylative Stille coupling" offers a good synthetic method for the synthesis of unsymmetrical ketones [103], α,β-unsaturated ketones [104] and esters [105]. Addition of LiCl and CuI permitted the palladium-catalyzed carbonylative cross-coupling of sterically hindered vinylstannanes to give the corresponding cross-conjugated 1,4-dien-3-ones [106] (Scheme 6.13). A practical synthesis of photoactivable 4-aroyl-L-phenylalanines from 4-iodo-L-phenylalanines was performed by using a carbonylative Stille cross-coupling as the key step [107]. A carbonylative coupling of organotin compounds with diaryliodonium salts was employed to prepare ^{11}C-labeled ketones [108].

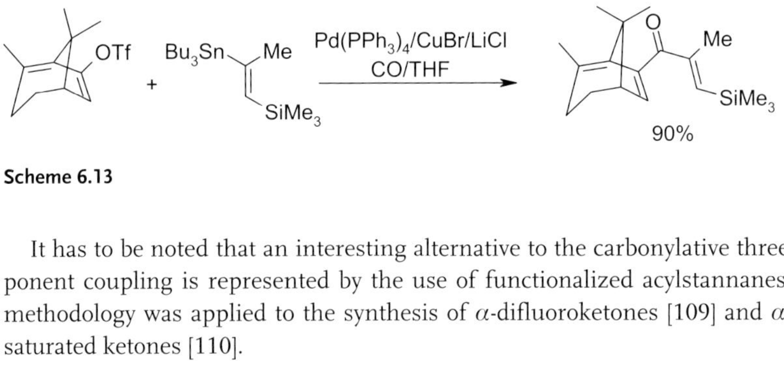

Scheme 6.13

It has to be noted that an interesting alternative to the carbonylative three-component coupling is represented by the use of functionalized acylstannanes. This methodology was applied to the synthesis of α-difluoroketones [109] and α,γ-unsaturated ketones [110].

6.2.1.5 Intramolecular Stille Cross-coupling

6.2.1.5.1 Vinyl–Vinyl Coupling

Since the first intramolecular version described by Pierce in 1985 [111], the Stille cross-coupling has proven to be an ideal reaction to get rings ranging from four to

thirty-two members [112]. The main advantage of this methodology over the classical macrocyclizations lies in the fact that high dilution techniques are not required, so that it has been widely exploited to prepare macrocycles of biological
interest. Most intramolecular Stille couplings involve alkenyl–alkenyl cyclizations,
and it is noteworthy that the stereochemistry of both the alkene electrophile and
the alkenyltin are conserved in the final cyclic structure [113]. An elegant example
is the reported synthesis of (–)Macrolactin A, an extremely cytotoxic compound,
via palladium-catalyzed Stille couplings for stereospecific constructions of the
three isolated dienes, including the ultime macrocyclization [114] (Scheme 6.14).

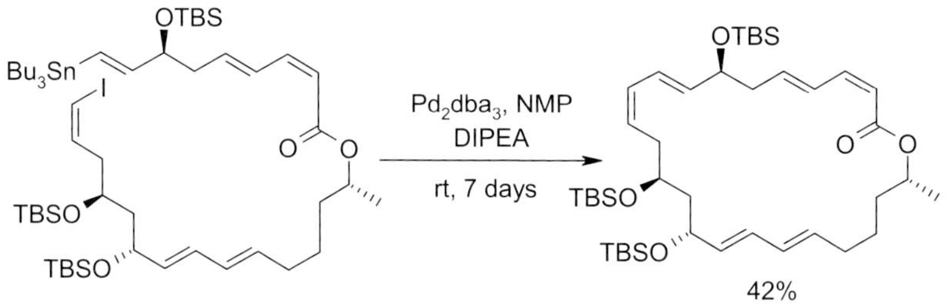

Scheme 6.14

Intramolecular alkenyl–alkenyl Stille cyclization was used in the synthesis of
macrocycles such as Bafilomycin A [115], (±)-isocembrene [116], Lankacidins
[117], Sanglifehrin [118], (+)-Concanamycin F [119], (–)-Pateamin [120], Apoptolidin [121], Concanamycin F [122], (–)-Lasonolide A [123], Rhizoxin D [124], Amphi-
dinolide A [125], and thiazole derivatives related to Leinamycin [126] (Scheme
6.15).

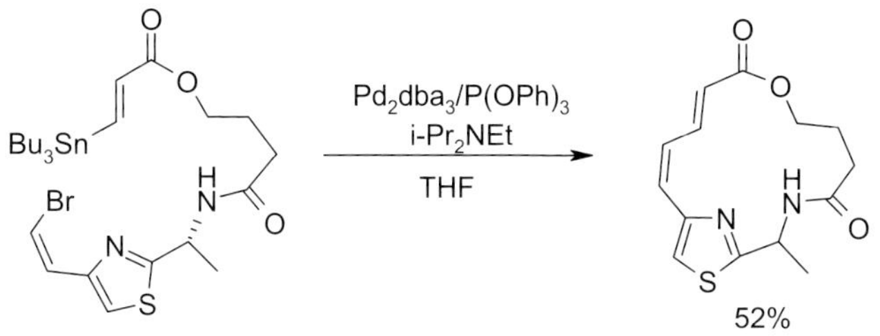

Scheme 6.15

6.2.1.5.2 Aryl–Aryl and Aryl–Vinyl Couplings

Aryl–alkenyl and aryl–aryl cyclizations via Stille coupling are far less developed.
In the case of alkenyl–aryl couplings, the olefinic moiety can be extracyclic or
intracyclic with Z or E stereochemistry. An example is the cyclization of polymer-
bound alkenylstannane with aryl iodide moiety in the total synthesis of (S)-zeara-
lenone [127] (Scheme 6.16).

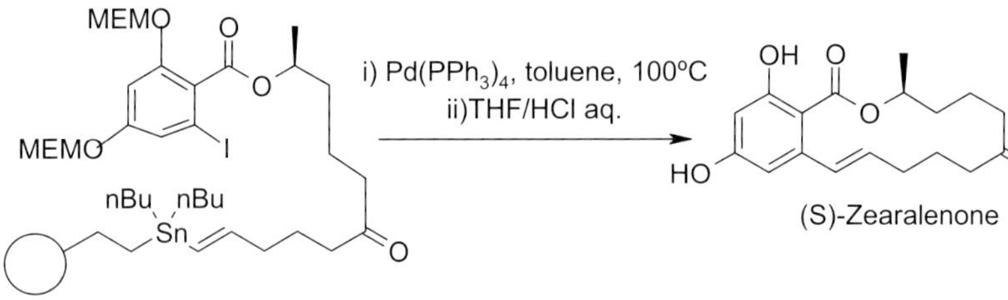

Scheme 6.16

Aryl–aryl cyclizations were used for the preparation of polyaromatic compounds, cryptands and heterocycles. A particularly interesting version is the Still–Kelly cyclization that was used for the synthesis of phenanthro [9,10-d]pyrazoles involving the intermolecular formation of the aromatic stannane followed by an intramolecular coupling with the aryl halide moiety [128]. Benzo [4,5]furopyridines [129] and dibenz [c,a]azepines [130] were prepared by related intramolecular coupling of diiodides in the presence of hexamethylditin.

The carbonylative macrocyclization using acyl chlorides and chloroformates as substrates represents an interesting tool for the synthesis of cyclic ketones, lactones, α,β-unsaturated esters. A complementary approach is based on the carbonyl insertion reaction with carbon monoxide and subsequent cross-coupling as exemplified by the approach to the core structure of Phomactins C and D, with an alkyne-enoltriflate carbonylative coupling as the key macrocyclization step [131] (Scheme 6.17).

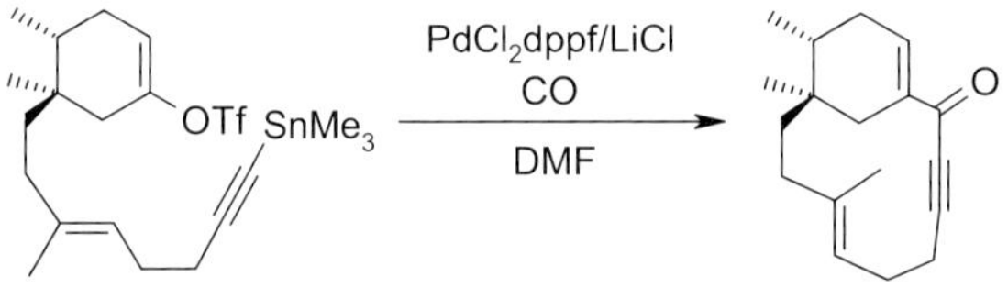

Scheme 6.17

6.2.1.6 Solid-Phase-Supported Stille Coupling

The solid phase organic synthesis has offered, in the past decade, new investigation fields in the Stille coupling reaction. The first approach, in which the organotin is grafted to the polymer via the tin atom, follows the chemistry initiated by Pereyre and coworkers [132] and Kuhn and Neumann [133] and is illustrated by the synthesis of zearalenone (Scheme 6.16). The second approach in which the organotin is grafted to the polymer via the organic moiety, was exploited for the synthesis of biaryls [134] and benzopyrans [135]. The third approach, in which the substrate is bound to the polymer, has been widely used for aryl–aryl [136], and aryl–alkenyl [137] couplings. Recent illustrative examples are the solid-phase syntheses of 2,6,8-trisubstituted purines [138], substituted pyridylpiperazines [139] or uridines [140]. This methodology was also employed in the carbonylative three-component Stille coupling as exemplified by the solid-phase synthesis of dissymmetrical diaryl ketones [141]. Finally, it is possible to perform the Stille coupling with polymer-supported palladium catalysts in order to facilitate the recovery and the reuse of palladium species [142].

6.2.1.7 Stille Coupling Catalytic in Tin

In order to overcome the reluctance of using organotins, despite their great synthetic potential, Maleczka et al. proposed a vinyl-vinyl Stille coupling catalytic in tin by using a terminal alkyne and an alkenyl halide as substrates [143].

They described a one-pot tandem-catalyzed hydrostannylation [144]/Stille reaction protocol for the stereoselective generation of vinyltins and their subsequent coupling with electrophiles, employing only catalytic amounts of Me_3SnCl. During this process, the organotin halide byproduct is recycled back to the organotin hydride by using either polymethylhydrosiloxane (PMHS) with Na_2CO_3 (the "Sn–O approach") [145] or PMHS coordinated by KF (the "Sn–F approach") [146] as the mild hydride donor. Two kinds of palladium catalysts are needed to catalyze both the hydrostannation ($PdCl_2(PPh_3)_2$) and the cross-coupling (Pd_2dba_3/TFP). The protocol was recently extended to polymer-supported dibutyltin chloride and dimethyltin chloride [147].

6.2.2
Other Metal-Catalyzed Coupling Reactions

6.2.2.1 Palladium-Catalyzed Reactions

Some palladium-catalyzed reactions of organotins, such as carbostannylations, are not related to the Stille cross-coupling. The history of the transition-metal-catalyzed carbostannylation [148] began with alkynylstannylation of alkynes catalyzed by a palladium-iminophosphine complex [149]. Thus, alkynylstannanes added to a carbon–carbon triple bond of various acetylenes, conjugated ynoates and propargyl amines and ethers in the presence of a catalytic amount of a palladium–iminophosphine complex [150]. The reaction also proceeded with arynes to afford ortho-substituted arylstannanes, which could further be converted into 1,2-substituted arenes via carbon–carbon bond-forming reactions [151].

$Me_3SnSnMe_3$ reacted with allenes and aryl iodides in a three-component palladium-catalyzed carbostannylation to afford the corresponding allylstannanes [152]. Finally, Pd_2dba_3-catalyzed allylstannylations of various alkynes by an α-methylallylstannane were reported [153].

6.2.2.2 Copper-Catalyzed Reactions

It is now well known that the use of cocatalytic Cu^I salts dramatically enhances the reaction rate of sluggish Stille couplings by transmetallating the organostannane (see Section 6.2.1.1.2). Several studies established that the resulting organocopper species was able to couple with organic elcetrophiles without any palladium. Various cross-couplings of organostannanes with organic halides [154] and triflates [155] under stoichiometric amounts of Cu^I were reported [156].

It was also demonstrated that α-heteroatom-substituted alkyltributyltins cross-coupled in the presence of catalytic amounts of Cu^I [157]. This catalytic use of copper salts was extended to various aryl–aryl, aryl–heteroaryl and aryl–vinyl cross-couplings [158].

In addition to the use of Cu^I [159], Cu^{II} salts such as CuCl$_2$ [160] and Cu(NO$_3$)$_2$ [161] efficiently mediated and catalyzed the homocoupling of aryl, alkenyl and alkynylstannanes.

Finally, a catalytic enantioselective approach for the formation of allyl α-amino acid derivatives by reaction of N-tosyl-α-imino esters with alkylstannanes catalyzed by chiral Cu^I complex was developed [162].

6.2.2.3 Nickel-Catalyzed Reactions

The three-component reaction between tetramethyltin, organic halides and carbon monoxide was the first reported example of nickel-catalyzed coupling involving organostannanes [163]. This methodology was further extended to the Ni^0 catalyzed coupling of alkynystannanes, allylchlorides and 1-alkynes for the regio- and stereoselective preparation of 3,6-dien-1-ynes [164], or 1,4-enynes [165]. Aryl mesylates were also used in cross-coupling with aryltins [166].

Nevertheless, recent synthetic applications dealt with the carbostannylation reactions of alkynes [167], 1,2-dienes [168] and 1,3-dienes [169] with alkynylstannanes, allylstannanes and acylstannanes (Scheme 6.18). Various mono- and disubstituted allenes participated in the reaction and acyl stannanes interestingly added mainly at an internal double bond.

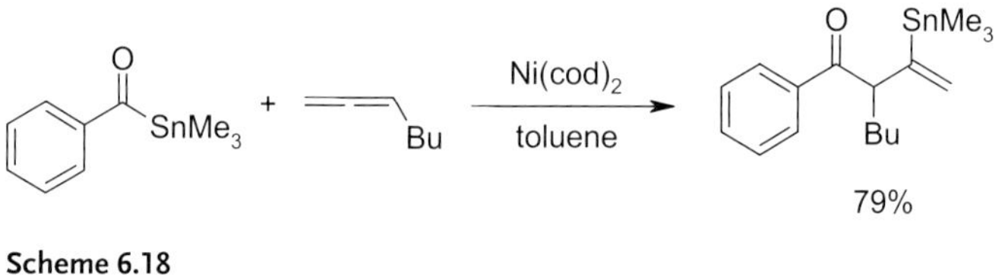

Scheme 6.18

6.2.2.4 Rhodium-Catalyzed Reactions

The addition of organometallic reagents to carbonyl compounds is the general method to synthesize secondary alcohols and organotins are promising reagents for such chemoselective reactions. The allylation of aldehydes with allylstannanes can be catalyzed by transition-metal complexes such as rhodium [170]. This methodology was further extended to the addition of arylstannanes to carbon–heteroatom bonds in the presence of catalytic amounts of a cationic rhodium complex [171]. The reaction of aldehydes [172], α-dicarbonyl compounds and imines [173] (see also Section 6.3.2.3) with arylstannanes gave the corresponding alcohols, α-hydroxycarbonyl compounds and amines, respectively. The rhodium-catalyzed conjugated addition of aryl- and vinylstannanes to α,β-unsaturated ketones and esters has been targeted also to prepare arylated ketones and esters [174,175]. This reaction was used for the synthesis of natural and unnatural amino acid derivatives in air and water [176] (Scheme 6.19).

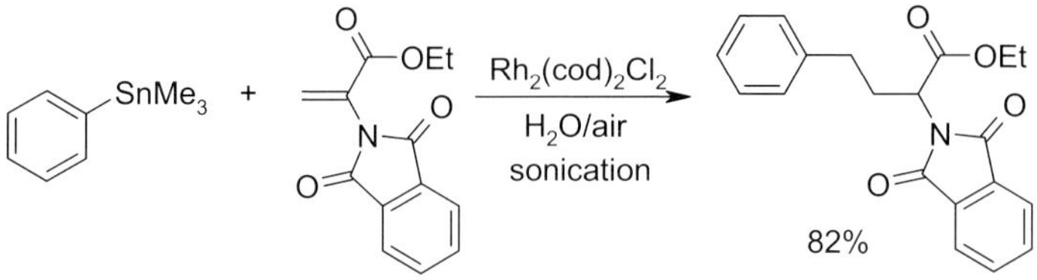

Scheme 6.19

6.3
Nucleophilic Additions

6.3.1
Nucleophilic Addition onto Carbonyl Compounds

6.3.1.1 Introduction

The addition of organotins to carbonyl compounds is a reaction of major importance in organic synthesis, especially with functionalized allyltins. There are several advantages such as (i) the easy preparation and stability of the organometallic reagents even under an enantiopure fashion, (ii) their great reactivity once the carbonyl partner is activated, (iii) the possibility to reach an excellent regio- and stereocontrol of the addition. Importantly, the organotin reagents authorize several mechanisms involving different transition states, so that the selectivity of the reaction appears to be closely dependent on the experimental conditions.

6.3.1.2 Functionalized Allyltins

6.3.1.2.1 Activation of the Carbonyl Compounds
The first activation of the carbonyl substrate by a Lewis acid was reported in 1979 [177] and initiated the increasing use of organotin reagents. The reaction proceeds via an open transition state, as the tin does not compete with the Lewis acid in the coordination of the carbonyl. The diastereoselectivity of the reaction can be induced either by the organometallic or the carbonyl partners.

Diastereoselectivity induced by γ-substituted allyltins: the reaction leads to an excellent *syn* diastereoselectivity. First reported by Maruyama and coworkers [178], this was explained by proposing an antiperiplanar transition state for the *syn* selectivity whatever the nature (E or Z) of the crotylstannane. Contrastingly, it was also reported that γ,γ-disubstituted allyltins reacted stereospecifically, (E)-reagent giving a *syn* adduct and (Z)-reagent giving an *anti* adduct [179]. The initial proposal was then completed by the synclinal acyclic transition state [180] to account for the diastereoselectivity change. This reaction was applied to various γ-substituted allyltins in the total synthesis of complex frameworks [181] with excellent diastereoselectivities (Scheme 6.20). It is worth noting that α-substitution on the allyltin reagent may affect dramatically the *syn* selectivity as well [182]. Thus, the syntheses of optically active α-oxygenated allyltins and their subsequent rearrangement to γ-alkoxyallyltins [183] allowed an easy access to stereocontrolled 1,2 *syn* diols further applied to synthetic purpose [184].

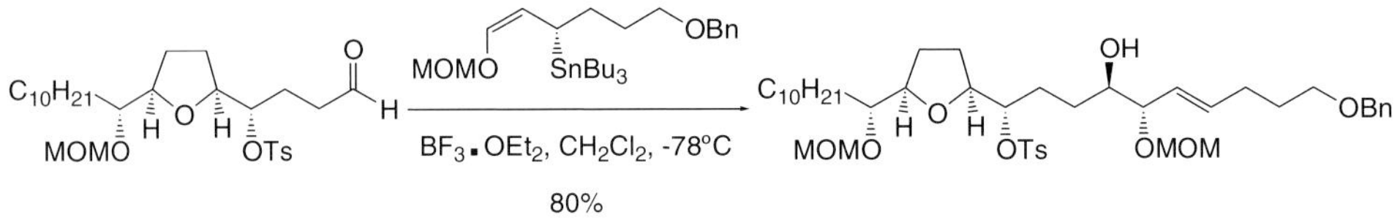

Scheme 6.20

In addition to the amount of work done with aldehydes as substrates, there is some evidence of diastereoselective addition of γ-substituted allyltins to ketones, leading to tertiary homoallylic alcohols with TiCl$_4$ or SnCl$_2$ as Lewis acid [185].

Diastereoselectivity induced by chiral aldehydes: the substrate plays an important role in the facial diastereoselection, particularly when there is an asymmetric center adjacent to the carbonyl group. In the general case, the approach of the allyltin is assumed to follow the Felkin–Anh model giving the *syn* adduct preferentially. This induction was used for the synthesis of natural products [186] even as complex as Ciguatoxin or Laulimalide [187].

Chelation control: however, a reversal of the diastereofacial selectivity may arise when the substrate has, in the α or β position of the side chain, a group prone to complexation with the Lewis acid. Then, the use of bidentate Lewis acids (Mg^{II}, Zn^{II}, Sn^{IV} or Ti^{IV}) allows the reaction to proceeding under a "chelation control", model preferentially provides the *syn* adduct for a 1,4-chelation. Various α-alkoxy aldehydes [188] were used in carbohydrate chemistry. Similarly, α-amino aldehydes were used as precursors for β-amino alcohols (Scheme 6.21) [189]. On the contrary, a 1,5-chelation control gives predominantly the *anti* adduct, so that β-alk-

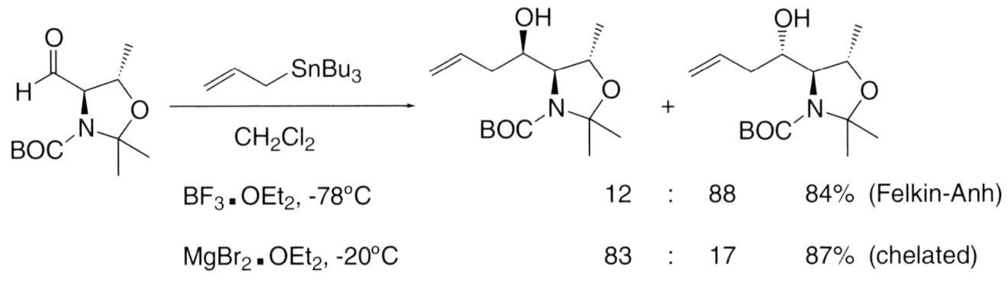

Scheme 6.21

oxy aldehydes are frequently used in total synthesis of complex targets such as Discodermolide [190] or Roxaticin macrolide [191].

Match/mismatch effect: the addition of γ-substituted allyltins to α-substituted aldehydes leads to three contiguous asymmetric centers mostly with a good control of each diastereoselectivity and is commonly applied in total synthesis such as for Erythromycin [192] (Scheme 6.22). It has to be noted that when the reaction is done with a chiral aldehyde and a chiral δ-substituted allyltin, a "matching effect" may happen when both partners imposed a convergent selectivity, or a "mismatching effect" when the facial selectivity is divergent [193]. Such a stereoconvergent effect was used in the synthesis of the antitumor agent Azinomycin [194].

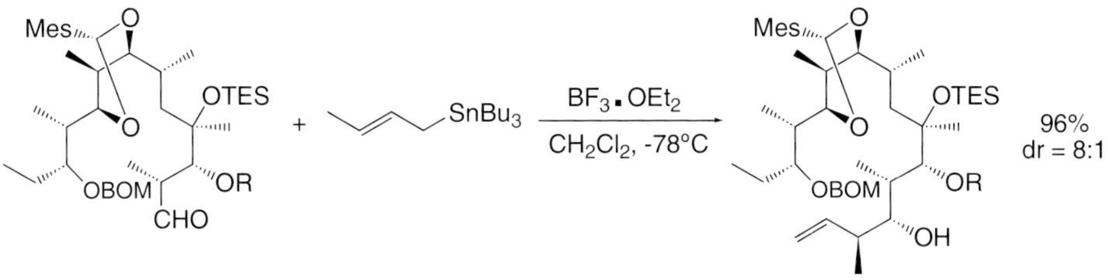

Scheme 6.22

As an alternative to the use of Lewis acids, Brönsted acids, such as trifluoromethane sulfonic acid, can be used for the allylation of aldehydes in EtOH/H$_2$O [195]. Using the same idea, a chelation control can be carried out without any Lewis acid simply by adding lithium salts such as LiClO$_4$ to the reaction mixture [196].

6.3.1.2.2 Activation of the Allyltin Reagent

In situ transmetallation: the activation of the allyltin can be ensured by a transmetallation with the Lewis acid prior to the addition on the carbonyl (also called "reversed" addition). The first example was reported with SnCl$_4$ [197] but other Lewis acids such as TiCl$_4$, AlCl$_3$, InCl$_3$ can be used. The difference lies in the nature of the transient allyl metal, which becomes a stronger Lewis acid, implying a cyclic 6-membered transition state already involved in thermal or high-pressure activated allylations. This changes the diastereoselectivity pattern of the reaction, (E)-crotyltin giving an *anti* selectivity, when (Z)-crotyltin leads to *syn* selectivity. It has to be noted that the transmetallation proceeds via the kinetic allylmetal, which turns to the thermodynamic crotyltin. Under thermodynamic control, the homoallylic alcohol is usually obtained with a high level of *anti* selectivity. The transmetallation can be used with more sophisticated Lewis acids such as the C-2 symmetry Corey's bromoborane, in order to induce a stereoselectivity dominated by the chiral auxiliary [198] (Scheme 6.23).

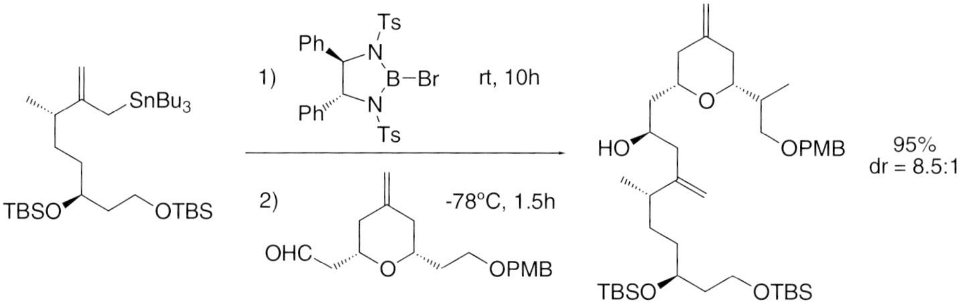

Scheme 6.23

Stereochemical outcomes: the easy and efficient access to enantiopure α-substituted allyltins associated to the transmetallation process is at the origin of an impressive progress in the field of enantioselective synthesis [199]. This is due to the "chirality transfer" that occurs in the two steps of the reaction. Thus, an efficient enantioselective synthesis of 1,2 diols can be achieved starting from enantio-enriched α-alkoxytins. However, in some case the use of a strong Lewis acid caused a premature decomposition of the allyltin. Marshall et al. circumvented this by using $InCl_3$ [200] and applied it to the enantioselective synthesis of sugar related compounds [201] (Scheme 6.24).

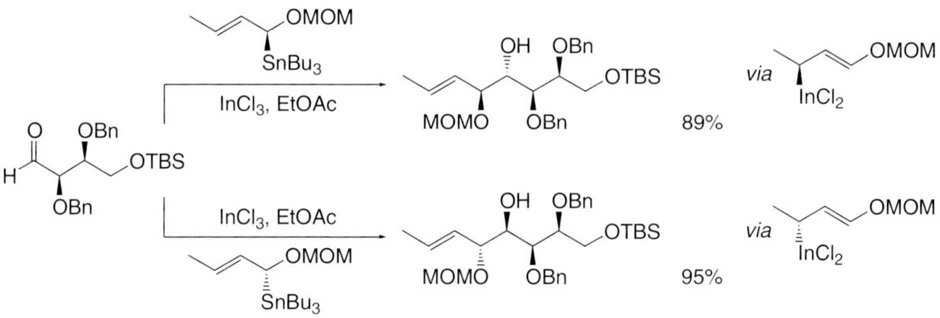

Scheme 6.24

The enantiocontroled preparation of γ-substituted allyltins was also used for the enantioselective synthesis of functionalized homoallylic alcohols. A major contribution is given by a comprehensive study of Thomas on various alkoxylated allyltins, which upon transmetallation with $SnCl_4$ gave intramolecularly coordinated trihalogenotins [202]. This coordination appears to be essential either for its stabilization via a rigid cyclic structure, and for the transfer of the chirality by directing the addition on the less hindered face of the trihalogenotin (Scheme 6.25). Interestingly, when using α-chiral aldehydes, the stereochemical induction of the tin reagent prevails over that of the substrate [203] effect. Efficient 1,4-, 1,5-, 1,6-, and 1,7-asymmetric inductions were achieved in that way [204], and have found application in the total synthesis of Epipatulolide and Epothylones macrolides [205], or complex tetrahydrofurans [206].

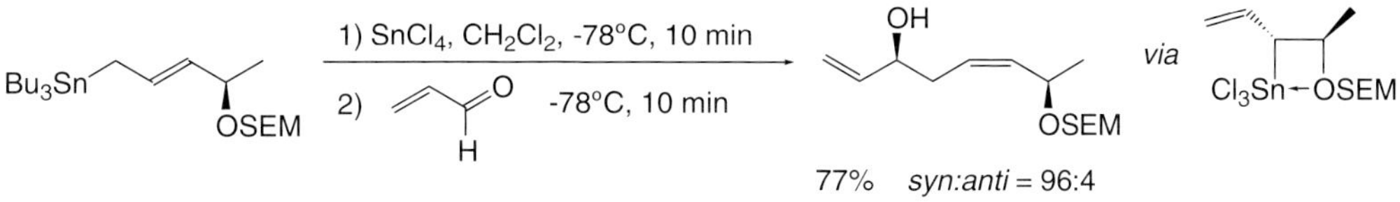

Scheme 6.25

Finally, the stoichiometry of the added Lewis acid has to be taken into account. Indeed, this factor is often underestimated but may cause a complete reversal of the stereoselectivity of the addition, by creating a chelation control that can compete with the cyclic 6-membered transition state [207].

6.3.1.3 Catalytic Use of Lewis Acid

There is a great effort underway to find a catalytic version of the Lewis-acid-activated reaction. Bulky aluminum reagents may be used (5 to 10 mol%), for which the development of unfavorable interactions with the resulting tributyltin alkoxide moiety is accounted for the decomplexation reaction [208]. Lanthanide triflates (2 mol%) were also used, in the presence of stoichiometric amounts of benzoic acid in order to regenerate the Yb(OTf)$_3$ catalyst [209]. InCl$_3$ can perform allylation and alkynylation addition reactions in a similar way, when associated to trimethylsilylchloride as regenerating reagent [210]. The transmetallation of allyltributyltin with catalytic quantities of dialkyldichlorotin leads to a much more reactive allyldialkylchlorotin. The turnover of the catalytic system is maintained by HCl [211] or Me$_3$SiCl [212]. The catalytic effects of B(C$_6$F$_5$)$_3$ or PhB(C$_6$F$_5$)$_2$ were also evidenced with a good chemoselectivity [213] and appeared to be influenced by the acidity of the Lewis acid [214]. Lastly, an association of triarylmethyl chloride as Lewis acid and chlorosilane as regenerating agent, was described as giving promising results [215].

6.3.1.4 Enantioselectivity

In parallel with the search for catalytic systems, has emerged an impressive amount of results in the field of enantioselective allylation. The pioneering work of Marshall using a chiral (acyloxy)borane (CAB) system [216] was readily followed by titanium/BINOL catalysts [217], leading to homoallylic alcohols with enantiomeric excess up to 98%. An extension of this work in fluorous phase was also developed with 6,6′-perfluoroalkylated BINOLs [218]. Replacing the titanium by zirconium (IV) salts, led to more reactive catalyst for the allylation of aromatic and aliphatic aldehydes [219]. One of the more active catalyst is the zirconium-BINOL system associated with 4-tert-butylcalix [4]arene, which remains active with only 2% of the chiral inductor [220]. The use of activators, such as iPrSSiMe$_3$, iPrSBEt$_2$, iPrSAlEt$_2$ or B(OMe)$_3$ was reported for both systems. They are believed to accelerate the reaction by regenerating Ti [221] or Zr [222] catalysts.

It was demonstrated that BINOL catalysts authorize the β- and γ-functionalizations of the allyltin reagents without lowering the enantioselectivity level [223], and such a strategy was used in the total syntheses of macrolides [224] or substituted tetrahydropyran units [225]. It was noteworthy that the BINOL-Ti catalysis was extended to the enantioselective allylation of alkyl and aromatic ketones in good yields with up to 96% ee [226]. Silver/BINAP was used as well, with a marked *anti* selectivity, when using crotyltins whatever is the nature, (E) or (Z) of the double bond [227]. This reaction was extended to other organometallics such as 2,4-pentadienylstannanes [228] or buta-2,3-dienylstannanes [229] (Scheme 6.26).

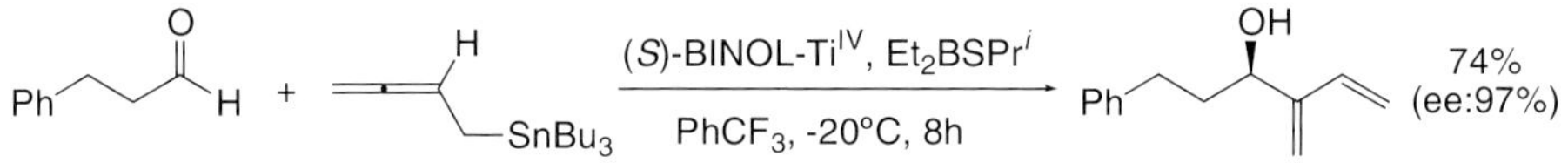

Scheme 6.26

Alternatively, catalysts with nitrogen ligands such as bisoxazolines [230] were introduced as Lewis acids, as well as air-stable and water-resistant (Phebox)rhodium(III) complexes that gave up to 80% ee [231].

6.3.1.5 **Others Organotin Reagents**

6.3.1.5.1 **Activated Allyltins**
Monoallyltrihalogenotins prepared without transmetallation [232] by oxidative addition organic halides to stannous halides or stannylene reagents, are interesting because all the tin side products are inorganic, allowing an easy purification and giving a solution to the toxicity problem of tributyltin residues. They have been used for the preparation of homoallylic alcohols [233], and the synthesis of α-methylene-γ-butyrolactones and α-methylene-γ-butyrolactames when starting from β-functionalized allyltins [234]. The *anti/syn* selectivity was about 90:10 and is consistent with a cyclic transition state. Adjacent groups to the carbonyl can affect the stereochemistry by hexacoordinating the tin atom and directing the reaction under a chelation control [235].

This ability of allylhalogenotins to extend their coordination sphere allowed the preparation of chiral hypervalent complexes with diamine ligands, which were efficient in the asymmetric synthesis of homoallylic alcohols with up to 82% ee [236]. Similarly, a chiral hypervalent allyltin was prepared from low valent tin[II] catecholate, chiral dialkyltartrate and allylic halide [237]. The allylation of aldehydes and activated ketones proceeded with high enantiomeric excess.

Other developments involving allyltins supported on solid phase are under study and are shown to proceed with the similar diastereoselectivity to that observed in liquid phase [238].

6.3.1.5.2 **Allenyl- and Propargyltins**

Allenyl- and propargyltins are peculiar reagents due to a possible interconversion, catalyzed by Lewis acids, between the two forms, thus leading to a mixture of homopropargylic and homoallenic alcohols. Complementary studies showed that it is possible to get selectively homoallenyl alcohols and homopropargyl alcohols [239] when preparing *in situ* the organotin reagent. The asymmetric approaches include the preparation of the chiral allenyltin configurationally stable, starting from enantio-enriched propargylic precursors. When submitted to transmetallation with Sn, Bi, or In Lewis acids, prior addition to the aldehyde, the homopropargylic alcohol was obtained in a 95:5 *anti/syn* fashion and a 90% ee [240]. On the other hand, the use of chiral allenyltin reagent without former transmetallation gave the *syn* adduct selectively (95:5) [241] (Scheme 6.27). Similarly to what is observed with allyltins, the use of chiral allenyltins and chiral aldehydes may lead to the same "match/mismatch" effect [242]. Both approaches were applied to the synthesis of Discodermolide polyketide or Aplyronine macrolide [243].

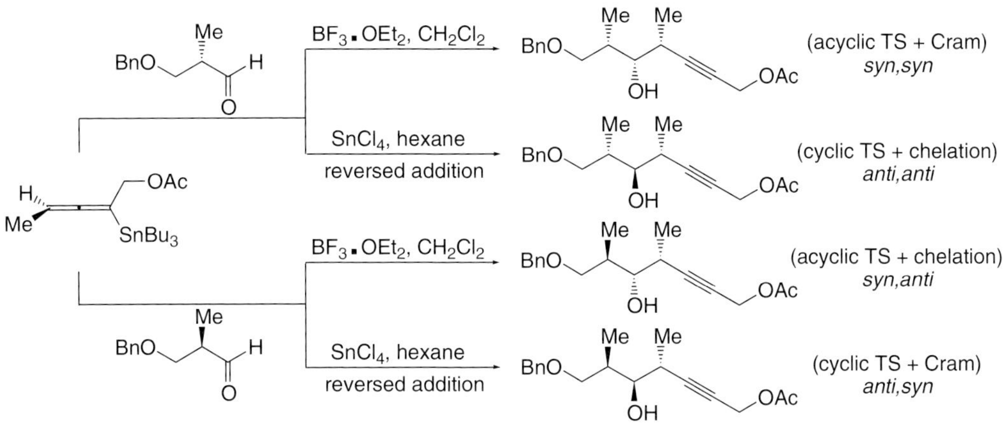

Scheme 6.27

Catalytic asymmetric allenylation was explored with the BINOL/Ti system giving selectively the homoallenyl alcohol with up to 95% ee [244]. Nevertheless, the lower reactivity of allenyltins compared to allyltins necessitated a nearly stoichiometric amount of the catalyst. Recently, the system BINOL/Ti/iPrSBEt$_2$, overcame this limitation, making the system truly catalytic, with ee in the range of 81 to 97% [245]. Interestingly, this reaction gave exclusively the allenylation adduct irrespective to the propargyl or allenyl structure of the organotin reagent [246].

6.3.1.5.3 **Alkynyltins**

The nucleophilic addition onto aldehydes was also extended to alkynyltributyltins, which were found to be reactive upon transmetallation with catalytic amount of InCl$_3$ [247]. The direct addition of alkynyltrimethyltins onto β-alkoxy aldehydes was also reported with a 1,3-*anti*//*syn* selectivity reaching 94:6 when using MeAlCl$_2$ as Lewis acid [248].

6.3.2
Nucleophilic Addition onto Imines and Related Compounds

6.3.2.1   Reactions with Imines

6.3.2.1.1   Introduction
Similarly to what was observed with carbonyl compounds, allyltins need the help of Lewis acid to react with imines. This was reported with TiCl$_4$ and BF$_3$.Et$_2$O [249] and extended to various Lewis acids such as MgBr$_2$, Et$_2$AlCl, NbCl$_5$ [250]. It was also demonstrated that TiCl$_4$ participated exclusively by activating the imine, so that contrary to what was observed with aldehydes, the "reverse addition" failed to give any homoallylic amine with the noticeable exception of SnCl$_4$. Other activating agents such as Selectfluor™ mediated smooth allylstannation of aldehydes and imines [251]. Similarly to what was evidenced with carbonyl compounds, the activation of the iminyl group can be achieved catalytically by using La(OTf)$_3$ as Lewis acid [252]. Such catalytic activation associated to benzoic acid was applied to the three-component synthesis of homoallylic amines, by the *in situ* formation of various imines [253]. Such a multicomponent version of the reaction was realized in water with SnCl$_2$.2H$_2$O as Lewis acid [254].

6.3.2.1.2   Mechanisms and Diastereoselectivity
Allyltins are the only allylmetals, with allylboron, to add regioselectively onto imines providing exclusively γ-adducts. The *syn/anti* selectivity of the addition can be explained with two sets of models involving cyclic or open transition states [255]. Crotyltins react regioselectively with α-alkylimines to give exclusively branched products with an excellent *syn/anti* selectivity up to 30:1, when the imine activation is done at −78 °C prior the addition of the crotyltin. This selectivity, consistent with an acyclic transition state, rapidly fades when operating at higher temperature, as a result of an equilibrium between the two imine/Lewis acid complexes.

All the models already discussed for the diastereofacial selectivity in the case of carbonyl compounds are still valid for the imines. However, due to the substitution on the nitrogen atom, imines can possess an additional chiral auxilliary capable of influencing the diastereoselectivity. Carbohydrates, for instance, were used as chiral templates, for the synthesis of N-glycosyl-N-homoallylamines and β-aminoacids [256]. As a consequence, the introduction of chiral centers both on the carbonyl and the amine moieties of the substrate may cause matching or mismatching effects [257].

The first exemple of 1,2-asymmetric induction, reported by Yamamoto et al., involved N-propylaldimines derived from α-phenylpropionaldehyde. The reaction gave mainly the *anti* product [258], consistent with a Felkin–Ahn approach. The 1,3-asymmetric induction was studied with the imine, prepared from 1-phenylethylamine and isovaleraldehyde, leading to a somewhat lower 7:1 diastereoselectivity. Thomas and coworkers were able to extend its 1,5-asymmetric induction concept (see Section 6.3.1.2.2) to the allylation of imines, by using SnCl₄ as transmetallating agent [259]. The stereochemistry of the reaction was usually imposed by the organotin partner (Scheme 6.28), whenever, in some cases, a match/mismatch effect could occur between the facial selectivity of the imine and the 1,5-stereoselectivity of the stannane.

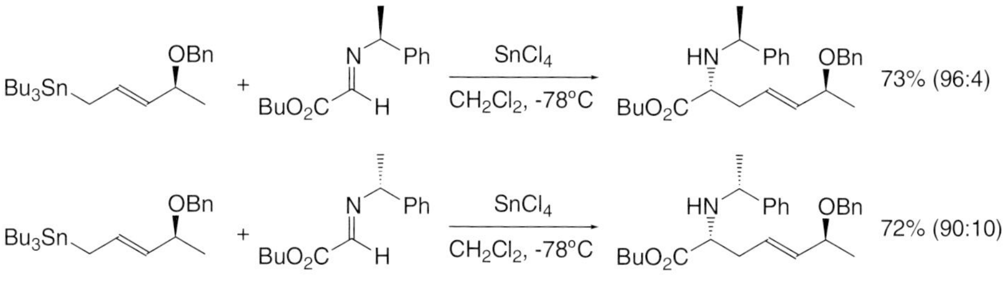

Scheme 6.28

There are very few examples with α,β-unsaturated aldimines, but it has to be noted that under TiCl₄ activation, they are able to undergo a 1,4-nucleophilic addition of ketene silyl acetal followed by a 1,2-addition of the allyltributyltin on the resulting imine giving the homoallylic amine in good yield [260].

6.3.2.2 Other Imino Substrates

6.3.2.2.1 Reactions with Iminium Salts

Iminium salts are widely used substrates to overcome the lack of reactivity of imines. Most of the iminium salts are prepared *in situ*, such as acyliminiums, generated from the corresponding α-ethoxycarbamates, which were shown to react with γ-alkoxyallyltins to give α-amino alcohols in good yields. The *syn/anti* selectivity is dependent on the nature of the iminium substituents [261]. This reaction was extended to various cyclic α-alkoxycarbamates with high diastereoselectivities [262]. Imines can also be activated by Me₃SiCl to give the corresponding iminium salt reactive enough to undergo the allylation reaction with allyltributyltin [263]. An interesting activation by organoaluminums in the presence of benzoyl peroxide was used to achieve tandem N-alkylation-C-allylation or N-silylation-C-allylation reactions [264] (Scheme 6.29).

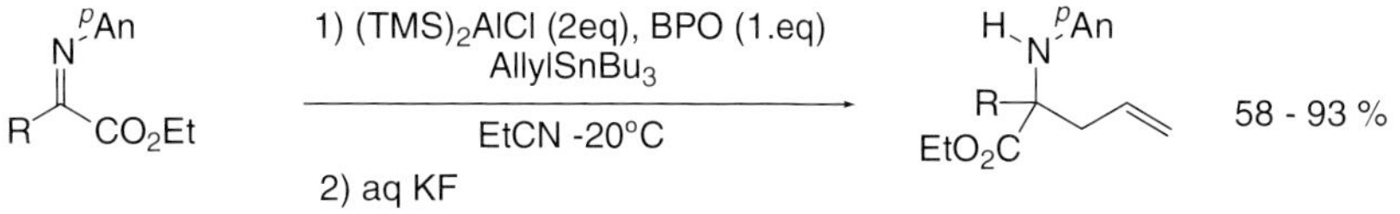

Scheme 6.29

In situ formed iminiums were also able to react with alkyltins in an intramolecular fashion, leading to the formation of cyclopropane [265] or cyclopentane rings (Scheme 6.30). Similarly, the intramolecular reaction of γ-alkoxystannane with hydrazones, activated by a Lewis acid, was used to prepare 5- or 6-membered β-amino cyclic ethers, for which the trans preference for the cyclization was consistent with an acyclic transition state [266].

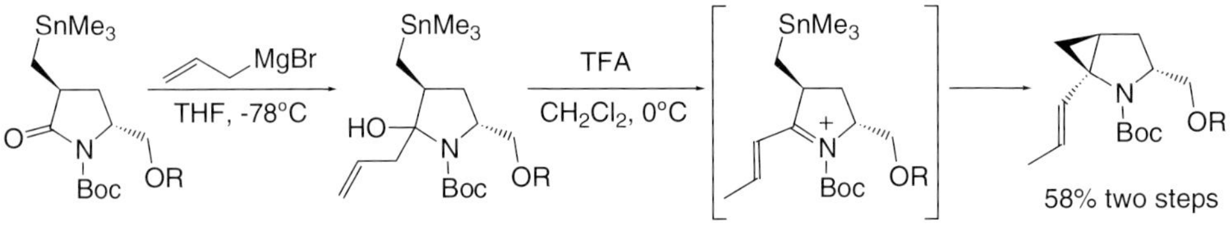

Scheme 6.30

Chiral acyliminiums were used for the preparation of enantiopure piperidines [267]. Recently, the use of enantio-enriched γ-alkoxyallyltins onto chiral acyl iminiums [268] provided a new entry into the synthesis of potential precursors of α-amino-β-hydroxy acids or aminosugars, with a total control of the stereochemistry.

6.3.2.2.2 Reactions with N-heterosubstituted Imines

These reagents are used as "protected" imines, which upon allylation and subsequent deprotection give an access to primary homoallylic amines. For instance, the use of benzoyl- and acylhydrazones as stable surrogates of imines were exploited in allylation reactions with tetraallyltin [269]. Finally, nitrones can be used as substrates for allylation reactions giving access to homoallylic hydroxylamines [270].

6.3.2.2.3 Reactions with Pyridines and Pyridiniums

Allyltins also react with similar substrates such as pyridines or pyridiniums selectively to the α position [271]. An enantioselective approach was done with a chiral acyl chloride as activator and enantioselectivity inductor [272]. This approach was applied to heterocycles such as β-carboline, leading to both enantiomers depending on the nature of the allyltin engaged in the reaction (Scheme 6.31) [273]. Similarly, oxazolidinones were used as chiral auxiliaries, to promote the synthesis of chiral 1,4-dihydropyridines [274].

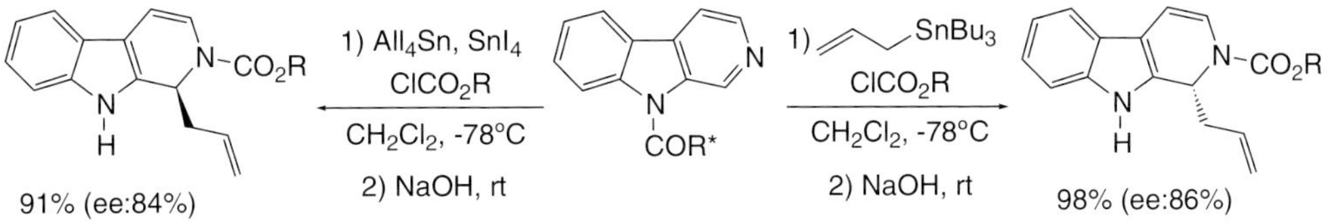

91% (ee:84%)   98% (ee:86%)

Scheme 6.31

6.3.2.3 Catalytic Enantioselective Addition

Organotins are involved in the increasing work related to the catalytic enantiose-lective addition to imines [275]. The first example of catalytic, enantioselective ally-lation of imines was reported by Yamamoto and coworkers by using 5% of bis π-allyl palladium complex [276]. Contrary to the BINAP ligand, which was found to be totally ineffective under these conditions, β-pinene ligands used as nontrans-ferable allyl ligands gave up to 81% ee. Nevertheless, it was shown that Tol-BINAP-Cu^I catalysts were also efficient for the allylation of N-tosyl imines [277] giving access to α-amino acids with up to 98% ee. Finally, a polymer-supported π-allyl palladium catalyst was developed, showing promising results in terms of stability and reusability, although leading to a moderate ee (13–47%) [278].

6.4
Radical Reactions of Organotins

6.4.1
Introduction

The radical chemistry of organotins is overwhelmed by the tin hydride chemistry. In the past decades, the knowledge of kinetic parameters authorized the expedi-tious construction of complex molecules by using cascade radical reaction based on Bu$_3$Sn° methodology. Moreover, these strategies also offered an excellent dia-stereocontrol, especially for the construction of polycyclic skeletons. These syn-thetic applications of Bu$_3$SnH, which will not be covered in this chapter, were reviewed in recent years [279]. In addition to the tin hydride chemistry, there are several applications of organotins in radical syntheses involving mainly allylstan-nanes.

6.4.2
Allyltins

6.4.2.1 Mechanistic Overview

Whereas the demonstration of the ability of allyltin reagents to undergo homolytic cleavage of the carbon–tin bond goes back to the early 1970s [280], it was only ten years later that Keck and Yates evidenced the synthetic potential of this reaction

[281]. This remains nowadays a particularly useful way for introducing various functionalized allyl groups under mild and neutral conditions. The radical chain mechanism involving allyltins is a S_H2' mechanism that can be schematized as following (Scheme 6.32): the initiation step (i), producing the tributyltin radical, is done usually by thermolysis of radical initiators or photochemical irradiation of Bu_6Sn_2. The reaction (ii), forming the initial carbon radical, admits various substrates such as iodides, bromides, selenides, dithio- and thiocarbonates. Less reactive substrates such as chlorides, benzoates or phenylthioethers can be used as well, because the competitive addition of tributyltin radicals to the allyltin reagent is degenerated due to the reverse β-fragmentation of the resulting radical. By the way, less reactive substrates such as chlorides, benzoates or phenylthioethers can be used efficiently as well. The additions of radicals to the allyltin (iv) are approximately a hundred times slower than the hydride transfer [282], avoiding a premature quenching of the carbon radicals, so that the evolution of the primarily formed radical, via several intra- or intermolecular elementary steps (iii), is technically possible, without using any slow addition or high dilution techniques [283]. This authorizes multicomponent intermolecular coupling reactions, involving activated olefins [284], or carbon monoxide [285] for instance. Finally, a rapid β-scission (v) with k_f being likely over $10^6\,s^{-1}$ occurs to regenerate the tributyltin radical.

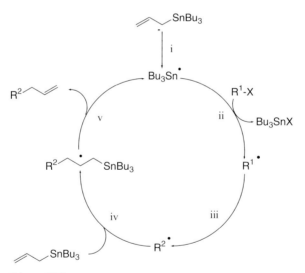

Scheme 6.32

The addition of carbon radicals to allylstannane is not severely affected by their nature, whenever the electrophilic radicals attack the allyltin π bond more rapidly than nucleophilic alkyl radicals [286], thus authorizing a wide range of substrates to react. As a consequence, the radical allylic transfer was applied to the synthesis of complex structures such as β-lactams [287], steroids [288], alkaloids [289] or

C-glycosides [290]. Moreover, polycyclic substrates permitted the reaction to proceed with an excellent diastereoselectivity (Scheme 6.33) [291].

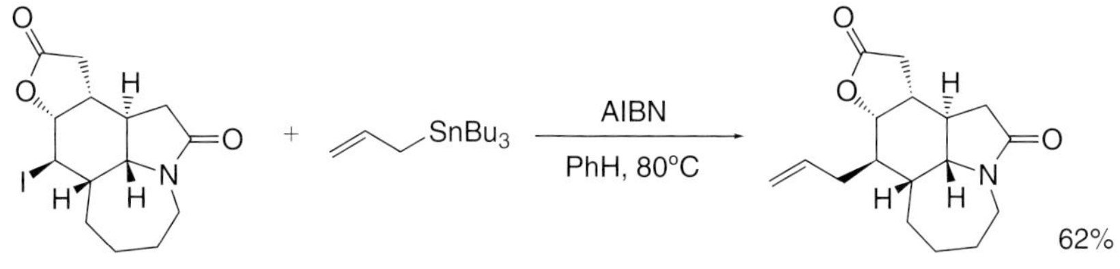

Scheme 6.33

6.4.2.2 Functionalized Allyltins

The synthesis and use of functionalized allyltins has been widely explored and gave contrasting results depending on the substitution position. Baldwin evidenced that the substitution in α-position has to be avoided, as the competitive tin radical addition to the double bond is no longer degenerated and results in an isomerization to the nonreactive γ-substituted allyltin.

By contrast, the use of allyltins β-substituted by an electron-withdrawing group represents a particularly attractive option, due to their enhanced reactivity towards nucleophilic carbon radicals. For this purpose several allyltins were prepared with amide, ester, chloride, nitrile, trimethylsilyl and sulfones functionalities [292]. They have been used for the synthesis of 1,4-dienes [293] or 10–15 membered α-methylene lactones [294], in aminoacids [295] or carbohydrate chemistry [296]. The $\tilde{\beta}$-functionalization by nonactivating alkyl groups is tolerated as well and was used in the synthesis of prostaglandins [297] or β-lactams [298]. This was also applied to radical cascade reactions with up to four elementary steps with an excellent diastereoselectivity control (Scheme 6.34) [299].

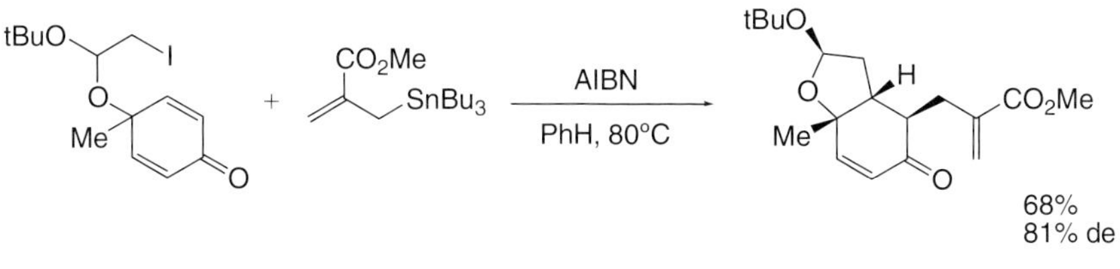

Scheme 6.34

Substitution in the γ-position, led to poorly reactive allylstannanes, due to the decreasing rate of the radical additions to the double bond. It has been established that, generally, the competitive allylic hydrogen abstraction became predominant, leading to diene side-products [300]. There are very few successful examples using γ-substituted allyltins in an intermolecular fashion [301], however, this limitation can be overcome for the intramolecular cyclization processes [302]. Recent advances showed that α-carbonyl radicals added efficiently to crotylstannanes at low temperature with an Et₃B/O₂ initiation process giving interesting diastereo-selectivities (Scheme 6.35) [303].

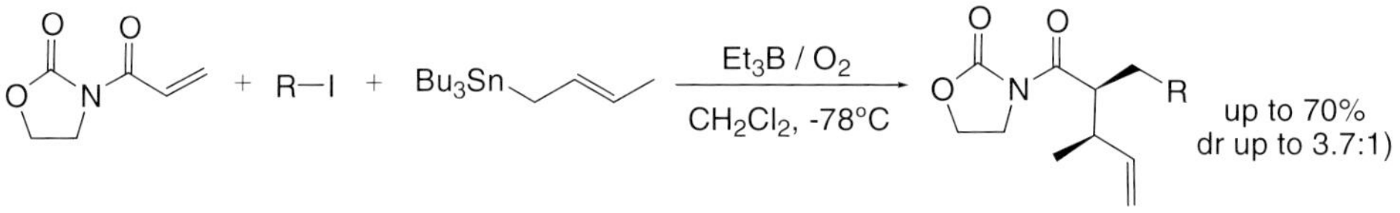

Scheme 6.35

6.4.3
Other Organotin Reagents

6.4.3.1 Tetraorganotins

Some related reagent, the 2,4-pentadienyltin, was shown to be reactive as well [304]. Propargyltin was equally found to be efficient for transferring an allene group [305]. However, a larger excess of propargyltin is needed, due to the radical isomerization of the propargyltin to the less reactive allenyltin. This was used in the synthesis of modified nucleosides [306].

Vinyltins were used for synthetic purpose in radical addition/elimination sequences. The main limitation comes from the necessity to functionalize the olefin by suitable groups, such as phenyl [307] or esters [308], in order to stabilize the transient carbon–centerd radical. This was applied to the stereoselective preparation of 1′-C branched nucleosides (Scheme 6.36) [309]. An intramolecular version was also developed, giving access to methylene cyclopentane units [310].

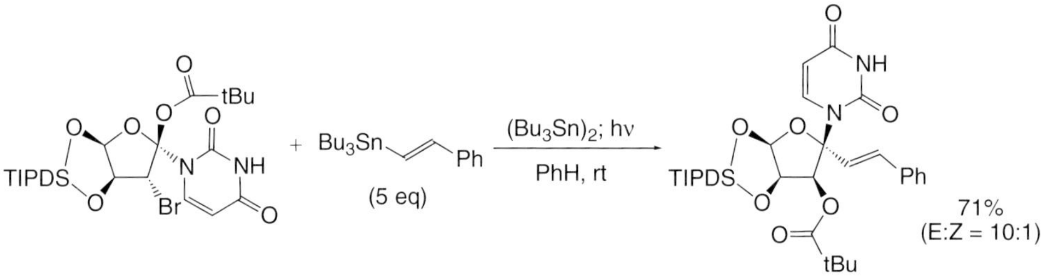

Scheme 6.36

Tin enolates are in metalotropic equilibrium between the O- and C-stannylated forms, so that the enolate form can be considered as the oxygenated analog of an allyl tin. Thus, the S_H2' reaction can be extended to tin enolates, used as electron-rich scavengers for carbon-centerd radicals [311]. A synthetically useful extension of this reaction proposed the carbostannylation of alkenes with tin enolates [312], which can be associated to cascade radical process (Scheme 6.37).

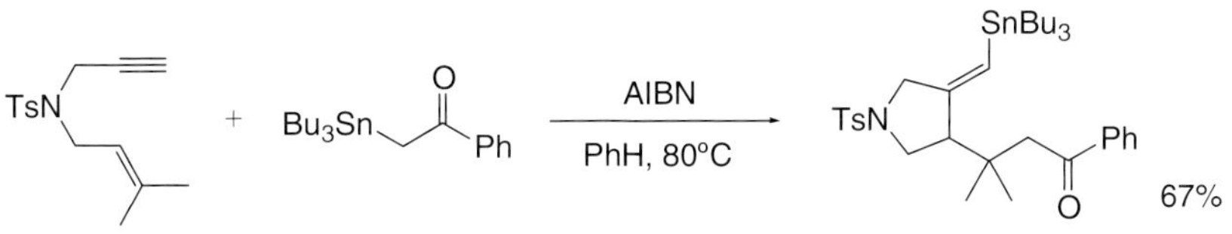

Scheme 6.37

Finally, alkyltins can participate in radical chemistry, especially when the β-elimination is thermodynamicaly favored, leading, for instance, to carbocyclic ring expansions [313]. In a similar way, radical reactions involving a 1,3-stannyl shift could afford 5-*exo* cyclizations [314].

6.4.3.2 Modified Organotins

In contrast with the important amount of work done to solve the purification problems caused by organotin side products in tin-hydride chemistry, very little attention has been paid to the allyl transfer process. Nevertheless, as most of the radical reactions are conducted with an excess of the allyltin, new allyltin reagents were proposed to optimize the purification process.

An alternative was developed using monoallyltins, giving after reaction hydrolysable, inorganic tin side products [315]. They were able to transfer efficiently the functionalized allyl group via a radical chain mechanism using $XSn [N(TMS)_2]_2$° as the chain-carrier agent [316]. Allyltin reagent supported on polymer underwent free radical allylic transfer with a marked preference for electron-poor carbon radicals [317]. Finally, the fluorous method developed by Curran was recently successfully extended to a four components radical reaction, using fluorinated allyltin reagents [318].

6.4.4
The Stereoselective Approach

Radical chemistry was long considered to be unable to achieve stereoselective reactions in acyclic reactions. There is actually a continuous interest in finding systems allowing radical chemistry to proceed with high levels of stereoselectivity. With that aim, the Et_3B/O_2 initiation system is commonly used at low temperature. Most of the work done with allyltributyltin is relevant to the 1,2-asymmetric induction. The first approach consists in making rigid the acyclic transient radical to induce a facial diastereoselection. This can be done either by favoring hydrogen bonding [319], or by adding bidentates Lewis acids, which permit work to be done under chelation control. The use of $MgBr_2.OEt_2$ for the allylation of α-iodo-β-alkoxyesters at −78 °C gave *de* over 100:1 [320]. Lanthanide triflates are able to give up to 10:1 *de* at refluxing dichloromethane [321]. The "chelation control" approach was also used in 1,3-asymmetric induction of α-bromoketones (Scheme 6.38) [322]. Importantly, this methodology can be extended to achiral substrates by using chiral Lewis acids, prepared from $Zn (OTf)_2$ and Pfaltz ligands, reaching up to 90% ee [323]. Recent developments with MgI_2 and bisoxazoline ligands permitted the successive creation of two chiral centers with the control of relative and absolute stereochemistry [324].

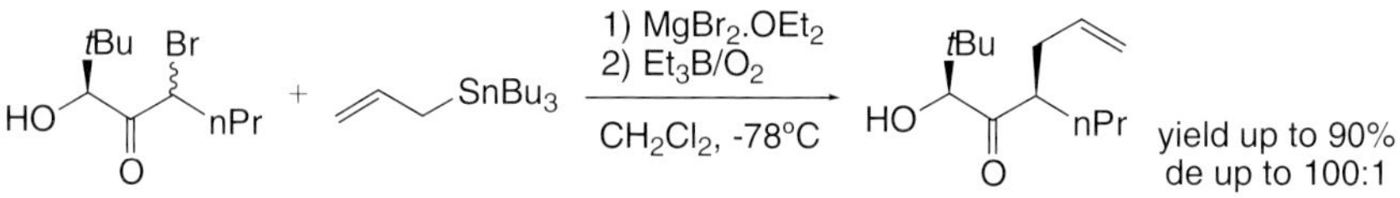

Scheme 6.38

The stereocontrolled introduction of the allyl group was studied with various chirality inductors such as chiral sulfoxides [325]. The use of Lewis acid or arylurea additives, to complex the sulfoxide, enhanced the stereoselection up to 50:1 [326]. Chiral auxiliaries such as oxazolidinones were reported to act efficiently when the allylation is done on the oxazolidinone ring [327]. The lower selectivity usually obtained when the reaction is done on the tethering chain of the oxazolidinyl nitrogen [328] was overcome by using a Lewis acid in order to proceed under a chelation control, raising the selectivity up to 100:1 [329]. A recent example of carbon-centerd radical generated in the α position to the nitrogen showed a good diastereoselectivity even without any Lewis acid. The *de*, up to 98:2, remained, however, strongly dependent on the radical nature [330].

6.5
Transmetallations

6.5.1
Introduction

Discovered by Seyferth and Weiner in 1959 [331], the transmetallation reaction consists of the replacement of the tin atom by another metal and has partly been dealt in this chapter for the activation of allylstannanes by Lewis acids in the nucleophilic addition reactions (Section 6.2.1.2.2) and for coupling reactions mediated by copper (Section 6.1.1.1.2). That apart, tin-to-lithium exchange is by far the most important process from the synthetic point of view and it was widely applied to the synthesis of compounds of biological interest. This popularity can be explained by the reactivity of the tin–carbon bond, which makes organotin reagents better candidates for this reaction than their silicon counterparts as well as by the compatibility of the process with a wide range of functional groups. Other elements including boron and copper are also regularly used in transmetallation reactions.

6.5.2
Tin-to-lithium Exchange

6.5.2.1 α-Heterosubstituted Alkyltins

It was established very early that both oxygen and nitrogen atoms considerably increased the stability of a α-carbanion [332] so that α-heterosubstituted organolithiums have gained considerable importance as normal synthetic intermediates. A great deal of knowledge was accumulated about their mode of generation, their stability and their reactivity with various electrophiles [333]. This explains the amount of literature on the use of α-heterosubstituted organotins in tin-to-lithium transmetallation [334]. Another striking point of the process is the complete retention of configuration at the carbanion.

6.5.2.1.1 Oxygen-substituted Organotins

Since Still introduced the use of α-alkoxyorganostannanes as precursors of α-alkoxyorganolithium by tin–lithium exchange [335], this chemistry was exploited in a variety of applications including the stereocontrolled synthesis of complex natural molecules, such as Zoanthamine [336] and Aspidospermin [337] alkaloids and (+)-Taxusin [338].

The transient alkyloxymethyllithium can react intermolecularly with electrophilic subtrates as illustrated by the diastereoselective addition to aldehydes of an α-alkoxyorganolithium prepared from a chiral derivative of tributylstannylmethanol by lithiodestannylation [339].

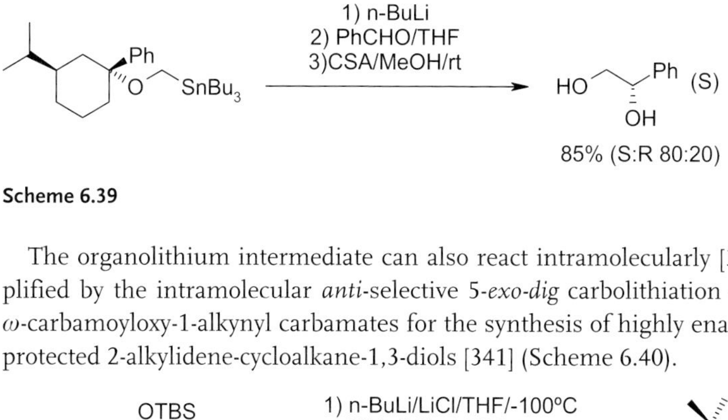

Scheme 6.39

The organolithium intermediate can also react intramolecularly [340] as exemplified by the intramolecular *anti*-selective 5-*exo-dig* carbolithiation of α-lithiated ω-carbamoyloxy-1-alkynyl carbamates for the synthesis of highly enantio-enriched protected 2-alkylidene-cycloalkane-1,3-diols [341] (Scheme 6.40).

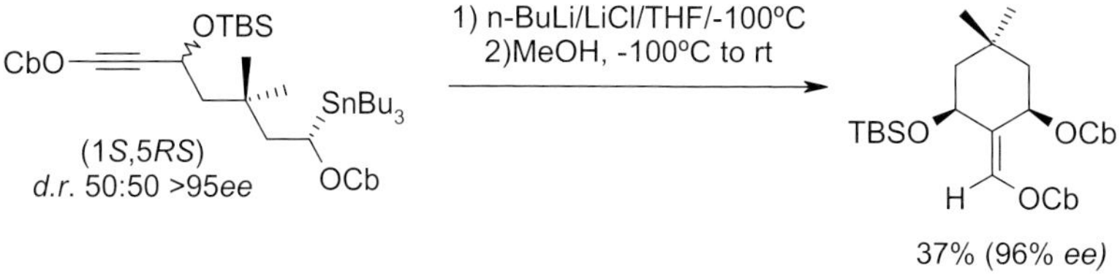

Scheme 6.40

6.5.2.1.2 **Nitrogen-substituted Organotins**

It has been known since 1971 that α-amino organostannanes can serve as useful precursors of α-amino organolithiums [342]. The configurational stability of α-amino organolithiums was applied to the transfer of chiral aminomethyl units via an (aminomethyl)lithium intermediate, thus making the organostannane precursors interesting tools for the enantiosynthesis of β-amino alcohols, α-amino ketones, and unusual α-aminoacids. For synthethic applications, much attention has been given to cyclic systems, in particular to piperidines [343], pyrrolidines [344], pyrrolidinones [345], oxazolidines [346] and oxazolidinones [347]. Gawley and Zhangreported the synthesis of 2-lithiopyrrolidines and piperidines from the corresponding 2-stannyl derivatives and their reaction with different electrophiles [348]. The stereochemistry of the reaction depended on the nature of the electrophile, as ketones, aldehydes and esters reacted with retention, alkyl halides reacted with inversion, while electrophiles such as benzyl bromide, ethyl bromoacetate and benzophenone gave complete racemization. An interesting application is the preparation of N-(α-lithioalkyl) oxazolidinones from the corresponding organotin derivatives and their carboxylation to provide enantiopure α-amino acids with the possibility of labelling the carboxyl group by ^{11}C (Scheme 6.41) [349].

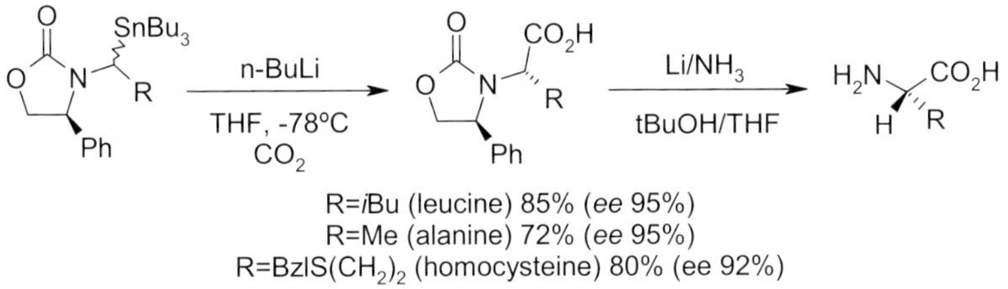

R=*i*Bu (leucine) 85% (*ee* 95%)
R=Me (alanine) 72% (*ee* 95%)
R=BzlS(CH$_2$)$_2$ (homocysteine) 80% (ee 92%)

Scheme 6.41

The progress has been much slower with acyclic systems. However, the transmetallation of N-benzyl-N-Cbz- [350], N-Boc-N-methyl- [351], and N-Boc-N-tertbutylthiomethyl- [352] protected aminotins provided the corresponding aminolithium derivatives that may be trapped with electrophiles with complete retention of configuration. This chemistry was applied to the preparation of β-aminoalcohols, N-methyl-β-aminoalcohols and N-methyl-β-amino acids. It can be noted that the preparation of enantio-enriched α- or β-amino carbanions can be achieved starting from linear racemic precursors when the tin-to-lithium exchange is done in the presence of (−)-sparteine [353].

Alternatively, tin-to-lithium exchange was found to be a method of choice for the preparation of 2-azaallyllithium species, which can undergo efficient [3+2], [4+2] and [6+4] cycloaddition reactions with alkenes [354], dienes [355] and trienes [356], respectively. Varous functionalized organotins can be used, such as stannylimines, stannylimidates, stannylthioamidates or stannylamidins. This azaallyl anion route to pyrolidines was used for the synthesis of different alkaloids, such as (±)-Lapidilectine B [357], (+)-Cocaine [358], (±)-Crinine (Scheme 6.42), (±)-6-Epicrinine, (−)-Amabiline and (−)-Augustamine [359].

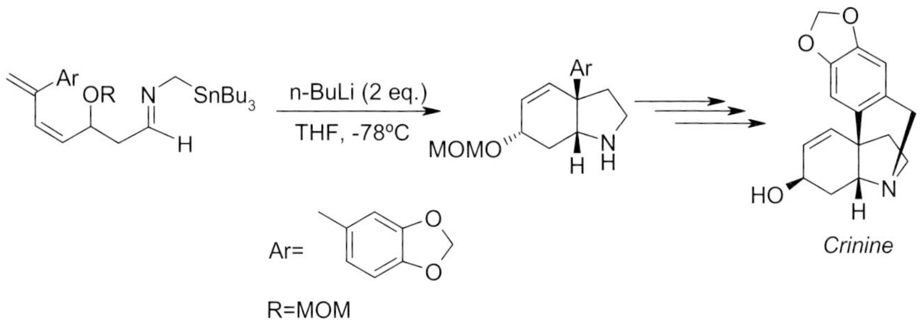

Scheme 6.42

The tin-to-lithium exchange can be applied to aziridin chemistry, as illustrated by the reported synthesis of the Aziridinomitrosene skeleton (Scheme 6.43), by an intramolecular Michael addition after lithiodestannylation followed by a spontaneous aromatization [360].

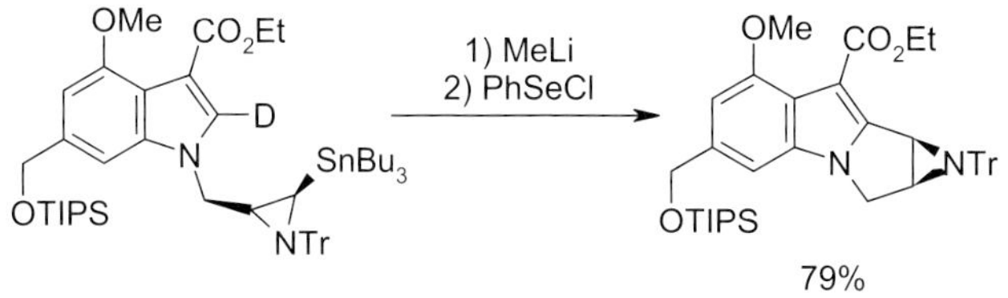

Scheme 6.43

Finally, an aza-Wittig rearrangement of acyclic enantio-enriched N,N-diallyllic α-amino alkyllithium prepared via a tin-to-lithium exchange was reported, the process proceeding predominantly with inversion of configuration at the lithium-bearing carbon terminus [361].

6.5.2.2 Alkenyltins

The tin-to-lithium exchange in alkenyltins is characterized by the preservation of the alkene stereochemistry and is compatible with several functionalities on the alkene moiety such as amino, hydroxyl and ester groups. The transmetallation of functionalized vinyltins has found an application in the synthesis of complex molecules such as unsaturated fatty acids [362], prostaglandin side-chain [363], various antibiotics [364] and a biotoxin [365].

1,1- and 1,2-distannnylalkenes are also lithiated. An interesting example involved the bis stannylated enyne, prepared from an unsymmetrical butadiene. A subsequent regioselective lithiation of the internal tin residue followed by a 1,4-retro-Brook rearrangement afforded the functionalized vinylsilane in a stereoselective fashion [366] (Scheme 6.44), which can also be considered as a masked triyne.

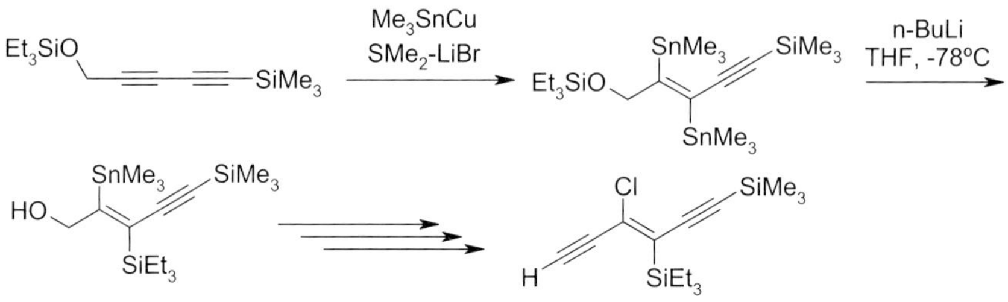

Scheme 6.44

6.5.3
Tin to Other Metal Exchanges

Until recently tin-to-copper transmetallation involved a transient organolithium species. The first example of direct tin-to-copper exchange was reported in 1988 with the formation of a mixed cuprate by treatment of an alkenylstannane with $R_2Cu(CN)Li_2$ [367]. The reaction was applied to allylic cuprates [368] and extended to alkynyltin and aryltin reagents, which were used in coupling reactions. Indeed, as mentioned in Section 6.2.1.1.2, it was observed that in polar solvents, the addition of Cu^I salts resulted, presumably, in the transient formation of an organocopper species, thus allowing sluggish Stille-coupling reactions to proceed.

The transmetallation was applied to the preparation of vinyl, alkynyl and arylboranes via a tin-to-boron exchange. Noticeable examples include the preparation of organoborane Lewis acids [369], the synthesis of alkynyldihaloboranes and their Diels–Alder reaction with 1,3-dienes [370], the formation of cycloheptatrienyl(dipropyl)borane [371] and the preparation of 1-benzoborepines [372]. In addition to this, the transmetallation tin-to-boron can be applied to allylboron reagents as mentioned in Section 6.3.1.2.2.

Finally, several other transmetallations such as tin-to-magnesium [373], tin-to-indium [374], tin-to-stibin [372] and tin-to-arsenic [375] are reported as well.

6.6
Conclusion

To be complete one should add the extremely rich tin hydride chemistry, or the halodestannylation reaction as well as the creation of various carbon-heteroelement bonds. We should also add the use of tin oxides or hydroxides as protecting groups in polyol chemistry, the use of organotins as catalysts for many reactions such as transesterifications or amide formation. Finally, we could also mention the use of optically active organotin Lewis acids as efficient chirality inductors. All of these reactions, which are not covered in this chapter tend to prove that the use of organotins for organic synthesis is now undoubtedly established and is not limited to the carbon–carbon bond formation with the aforementioned reactions.

References

1 D. Azarian, S. S. Dua, C. Eaborn, D. R. M. Walton, *J. Organomet. Chem.*, **1976**, *117*, C55–C75.

2 M. Kosugi, K. Sasazawa, Y. Shimizu, T. Migita, *Chem. Lett.*, **1977**, 301–302. For an historical account of the Stille reaction see: M. Kosugi, K. Fugami, *J. Organomet.*, **2002**, *653*, 50–53.

3 D. Milstein, J. K. Stille, *J. Am. Chem. Soc.*, **1978**, *100*, 3636–3638. J. K. Stille, *Pure Appl. Chem.*, **1985**, *57*, 1771.

4 V. Farina, G. P. Roth, *Adv. Met.-Org. Chem.*, **1996**, *5*, 1–53. P. Espinet, A. M. Echavarren, *Angew. Chem. Int. Ed.*, **2004**, *43*, 4704–4734. A. Ricci, F. Angelucci, M. Bassetti, C. Lo Sterzo, *J. Am. Chem. Soc.*, **2002**, *124*, 1060–1071.

5 For a reviw on 2-furyl phosphines see: N. G. Andersen, B. A. Keay, *Chem. Rev.*, **2001**, *101*, 997–1030

6 V. Farina, B. Krishnan, D. R. Marshall, G. P. Roth, *J. Org. Chem.*, **1993**, *58*, 5434–5444; V. Farina, *Pure Appl. Chem.*, **1996**, *68*, 73–78.

7 S. Ito, T. Okujima, N. Morita, *J. Chem. Soc., Perkin Trans. 1*, **2002**, 1896–1905

8 M. R. Netherton, G. C. Fu, *Org. Lett.*, **2001**, *3*, 4295–4298

9 V. Farina, G. P. Roth, *Tetrahedron Lett.*, **1991**, *32*, 4243–4246; A. M. Echavarren, O. de Frutos, N. Tamayo, P. Nokeda, P. Calle, *J. Org. Chem.*, **1997**, *62*, 4524–4527

10 J. D. Buynak, L. Vogeti, H. Chen, *Org. Letters*, **2001**, *3*, 2953–2956

11 L. S. Liebeskind, R. W. Fengl, *J. Org. Chem.*, **1990**, *55*, 5359–5364

12 V. Farina, S. Kapadia, B. Krishnan, C. Wang, L. S. Liebeskind, *J. Org. Chem.*, **1994**, *59*, 5905–5911; X. Han, B. M. Stolz, E. J. Corey, *J. Am. Chem. Soc.*, **1999**, *121*, 7600–7605; A. L. Casado, P. Espinet, *Organometallics*, **2003**, *22*, 1305–1309; P. H. Mee, J. E. Baldwin, *Angew. Chem. Int. Ed.*, **2004**, *43*, 1132–1136

13 W. J. Scott, J. K. Stille, *J. Am. Chem. Soc.*, **1986**, *108*, 3033–3040.

14 D. A. Evans, W. C. Black, *J. Am. Chem. Soc.*, **1993**, *115*, 4497–4513.

15 J. Scott, J. K. Stille, *J. Am. Chem. Soc.*, **1986**, *108*, 3033–3040

16 E. Piers, R. W. Friesen, B. A. Keay, *Tetrahedron*, **1990**, *47*, 4555–4570; C. M. Hettrick, J. K. kling, W. J. Scott, *J. Org. Chem.*, **1991**, *56*, 1489–1492

17 V. Farina, B. Krishnan, D. R. Marshall, G. P. Roth, *J. Org. Chem.*, **1993**, *58*, 5434–5444

18 C.-W. Huang, M. Shanmugasundaram, H.-M. Chang, C.-H. Cheng, *Tetrahedron*, **2003**, *59*, 3635–3641.

19 For reviews see: (a) G. Pozzi, I. Shepperson, *Coordination Chemistry Reviews*, **2003**, *242*, 115–124; E. Wolf, G. van Koten, B.-J. Deelman, *Chem. Soc. Rev.*, **1999**, *28*, 37–41; E. G. Hope, A. M. Stuart, *J. Fluorine Chem.*, **1999**, *100*, 75–83

20 S. T. Handy, X. Zhang, *Org. Lett.*, **2001**, *3*, 233–236; C. Chaippe, G. Imperato, E. Napolitano, D. Pieraccini, Green. Chem., **2004**, *6*, 33–36

21 T. Osswald, S. Scneider, S. Wang, W. Bannwarth, *Tetrahedron Lett.*, **2001**, *42*, 2965–2967; N. Shezad, R. S. Oakes, A. A. Clifford, C. M. Rayner, *Tetrahedron Lett.*, **1999**, *40*, 221–2224; D. K. Morita, D. R. Pesiri, S. A. David, W. H. Glaze, W. Tumas, *Chem. Commun.*, **1998**, 1397–1398

22 S. Venkatraman, T. Huang, C.-J. Li, *Adv. Synth. Catal.*, **2002**, *344*, 399–405

23 J. K. Stille, J. H. Simpson, *J. Am. Chem. Soc.*, **1987**, *109*, 2138–2152

24 L. R. Marrison, J. M. Dickinson, R. Ahmed, I. J. S. Fairlamb, *Tetrahedron Lett.*, **2002**, *43*, 8853–8857

25 M. Biagetti, F. Bellina, A. Carpita, P. Stabile, R. Rossi, *Tetrahedron*, **2002**, *58*, 5023–5038; M. Biagetti, F. Bellina, A. Carpita, S. Viel, L. Mannina, R. Rossi, *Eur. J. Org. Chem.*, **2002**, 1063–1076.

26 L. R. Pottier, J.-F. Peyrat, M. Alami, J. D. Brion, *Tetrahedron Lett.*, **2004**, *45*, 4035–4038.

27 A. C. Tiberghien, D. Hagan, P. W. Howard, D. E. Thurston, *Bioorg. Med. Chem. Lett.*, **2004**, *14*, 5041–5044

28 E. Fillion, N. J. Taylor, *J. Am. Chem. Soc.*, **2003**, *125*, 12700–12701; A. Flohr,

Tetrahedron Lett., **1998**, *39*, 5177–5180;
P. Quayle, J. Wang, J. Xu, *Tetrahedron
Lett.*, **1998**, *39*, 482–492; C. A. Busacca,
J. Swestock, R. E. Johnson, T. R. Bailay,
L. Musza, C. A. Rodger, *J. Org. Chem.*,
1994, *59*, 7553–7556; G. Stork,
R. C. A. Issacs, *J. Am. Chem. Soc.*, **1990**,
112, 7399–7400

29 R. D. Mazzola, Jr., S. Giese, C. L. Benson,
F. G. West, *J. Org. Chem.*, **2004**, *69*,
220–223; G. D. Allred, L. S. Liebeskind,
J. Am. Chem. Soc., **1996**, *118*, 2748–2749

30 C. Chen, K. Wilcoxen, Y.-F. Zhu,
K.-I. Kim, J. R. McCarthy, *J. Org. Chem.*,
1999, *64*, 3476–3482; L. Lu,
S. D. Pedersen, Q. Liu, R. M. Narrske,
D. J. Burton, *J. Org. Chem.*, **1997**, *62*,
1064–1071; I. H. Jeong, Y. S. Park,
B. T. Kim, *Tetrahedron Lett.*, **2000**, *41*,
8917–8921; Y. Shen, G. Wang, *J. Fluorine Chem.*, **2004**, *125*, 91–94.

31 For a review on the use fluorinated organometallics in organic synthesis see:
D. J. Burton, Z. Y. Yang, P. A. Morton,
Tetrahedron, **1994**, *50*, 2993–3063

32 S. Minière, J.-C. Cintrat, *J. Org. Chem.*,
2001, *66*, 7385–7388; L. Timbart,
J.-C. Cintrat, *Chem. Eur. J.*, **2002**, *8*,
1637–1640

33 I. S. Aidhen, R. Braslau, *Synth. Commun.*, **1994**, *24*, 789–790; C.-W. Huang,
M. Shanmugasundaram, H.-M. Chang,
C.-H. Cheng, *Tetrahedron*, **2003**, *53*,
3635–3641

34 E. Keinan, M. Peretz, *J. Org. Chem.*,
1983, *48*, 5302–5309

35 S. Rousset, M. Abarbbri, J. Thibonnet,
A. Duchêne J.-L. Parrain, *Chem. Commun.*, **2000**, 1987–1988

36 For a recent review on metallated heterocycles and their application in Stille
coupling see: R. Chinchillla, C. Nájera,
M. Yus, *Chem. Rev.*, **2004**, *104*,
2667–2722

37 H. Morimoto, H. Shimadzu,
E. Kuushiyama, H. Kawanishi,
T. Hosaka, Y. Kawase, K. Yasuda,
K. Kushiyama, R. Yamauchi-Kohno,
K. Yamada, *J. Med. Chem.*, **2001**, *44*,
3355–3368

38 J. W. Labadie, D. Tueting, J. K. Stille,
J. Org. Chem., **1983**, *48*, 4634–4642;
A. M. Echawarren, J. K. Stille, *J. Am.
Chem. Commun.*, **1987**, *109*, 5478–5486

39 S. Labadie, *J. Org. Chem.*, **1989**, *54*,
2496–2498

40 K. Krohn, S. Bernhard, U. Flörke,
N. Hayat, *J. Org. Chem.*, **2000**, *65*,
3218–3222

41 W. G. Peet, W. Tan, *J. Chem. Soc., Chem.
Commun.*, **1983**, 853–854

42 H. Xie, Y. Shao, J. M. Becker, F. Naider,
R. A. Gibbs, *J. Org. Chem.*, **2000**, *65*,
8552–8563

43 R. J. Lindermann, J. M. Siedlecki, *J. Org.
Chem.*, **1996**, *61*, 6492–6493

44 E. Veddejs, A. . Haight, W. O. Moss,
J. Am. Chem. Soc., **1992**, *114*, 6556–6558

45 J. M. Brown, M. Pearson,
J. T. B. H. Jastrrrzebski, G. van Koten,
Chem. Commun., **1992**, 1440–1441

46 T. Forngren, L. Samuelsson,
B. Långström, *J. Label. Comp. Radiopharm.*, **2004**, *47*, 71–78

47 G. A. Grasa, S. P. Nolan, *Org. Lett.*,
2001, *3*, 119–122.

48 E. Napolitano, V. Farina, M. Persico,
Organometallics, **2003**, *22*, 4030–4037

49 E. Abele, K. Rubina, M. Fleicher,
J. Popelis, P. Arsennyan, L. Edmuns,
Appl. Organomet. Chem., **2002**, *16*,
141–147; G. A. Grasa, S. P. Nolan, *Org.
Lett.*, **2001**, *3*, 119–122

50 E. Fouquet, M. Pereyre, A. L. Rodriguez,
J. Org. Chem., **1997**, *62*, 5242–5243;
E. Fouquet, A. L. Rodriguez, *Synlett*
1998, 1323–1324. A. Herve,
A. L. Rodriguez, E. Fouquet, *J. Org.
Chem.*, **2005**, *70*, 1953–1956.

51 M. T. Barros, C. D. Maycock,
M. I. Madureira, M. R. Ventura, *Chem.
Commun.*, **2001**, 1662–1663

52 K. Itami, T. Kamei, J.-I Yoshida" *J. Am.
Chem. Soc.*, **2001**, *123*, 8773–8779;
K. Itami, M. Mineno, T. Kamei,
J.-I. Yoshida, *Org. Lett.*, **2002**, *4*,
3635–3638

53 For a general review on palladium-catalyzed coupling reactions of aryl chlorides see: A. F. Littke, G. C. Fu, *Angew.
Chem. Int. Ed.*, **2002**, *41*, 4176–4211

54 A. F. Littke, G. C. Fu, *Angew. Chem. Int.
Ed.*, **1999**, *38*, 2411–2413; A. F. Littke,
L. Schwarz, G. C. Fu, *J. Am. Chem. Soc.*,
2002, *124*, 6343–6348

55 G. A. Grasa, S. P. Nolan, *Org. Lett.*,
2001, *3*, 119–122; R. B. Bedford,
C. S. J. Cazin, S. L. Hazelwood, *Chem.*

Commun., **2002**, 2608–2609; W. Su, S. Urggaonkar, J. G. Verkade, *Org. Lett.*, **2004**, *6*, 1421–1424

56 M. Alvarez, D. Fernandez, J. A. Joule, *Tetrahedron Lett.*, **2001**, *42*, 315–317

57 L. A. Agrofolio, I. Gillaizeau, Y. Saito, *Chem. Rev.*, **2003**, *103*, 1875–1916

58 J. L. Sessler, M. Sathiosatham, C. T. Brown, T. A. Rhodes, G. Wiederrecht, *J. Am. Chem. Soc.*, **2001**, *123*, 36555–3660

59 K. Menzel, G. C. Fu, *J. Am. Chem. Soc.*, **2003**, *125*, 3718–3719

60 H. Tang, K. Menzel, G. C. Fu, *Angew. Chem. Int. Ed.*, **2003**, *42*, 5079–5082

61 T. V. Hansen, L. Skattebøl, *Tetrahedron Lett.*, **2004**, *45*, 2809–2811

62 Y. Nakayama, G. B. Kumar, Y. Kobayashi, *J. Org. Chem.*, **2000**, *65*, 707–715

63 H. Uchiro, K. Nagasawa, Y. Aiba, S. Kobayashi, *Tetrahedron Lett.*, **2000**, *41*, 4165–4168

64 J. L. Davis, R. Dhawan, B. A. Arntsen, *Angew. Chem. Int. Ed.*, **2004**, *43*, 590–594

65 C. M. Hettrick, J. K. Kling, W. J. Scott, *J. Org. Chem.*, **1991**, *56*, 1489–1492

66 J. P. Marino, J. K. Long, *J. Am. Chem. Soc.*, **1988**, 110, 7916–7917; L. Schio, F. Chatreaux, M. Klich, *Tetrahedron Lett.*, **2000**, *41*, 1543–1547

67 D. Badone, C. Cecchi, U. Guzzi, *J. Org. Chem.*, **1992**, *57*, 6321–6323

68 T. Okauchi, T. Yano, T. Fukamachi, J. Ichikawa, T. Minami, *Tetrahedron Lett.*, **1999**, *40*, 5337–5340; B. M. Stoltz, T. Kano, E. J. Corey, *J. Am. Chem. Soc.*, **2000**, *122*, 9044–9045

69 S.-K. Kang, H.-W. Lee, J.-S. Kim, S.-C. Choi, *Tetrahedron Lett.*, **1996**, *37*, 3723–3726; K. W. Stagliano, H. C. Malinakova, *Tetrahedron Lett.*, **1997**, *38*, 6617–6620; K. W. Stagliano, H. C. Malinakova, *J. Org. Chem.*, **1999**, *64*, 8034–8040; S.-K. Kang, Y.-T. Lee, S.-H. Lee, *Tetrahedron Lett.*, **1999**, 40, 3573–3576

70 K. Kikukawa, K. Kono, F. Wada, T. Matsuda, *J. Org. Chem.*, **1983**, *48*, 1333–1336; D. J. Koza, Y. A. Nsiah, *Bioorg. Med. Chem. Lett.*, **2002**, *112*, 2163–2165

71 S. Zhang, D. Marshall, L. S. Liebeskind, *J. Org. Chem.*, **1999**, *64*, 2796–2804

72 Z. Ni, A. Padwa, *Synlett* **1992**, 869–872; K. Itami, T. Koike, J.-I. Yoshida, *J. Am. Chem. Soc.*, **2001**, *123*, 6957–6958; W. D. Shipe, E. J. Sorensen, *Org. Lett.*, **2002**, *4*, 2063–2066; H. W. Lam, G. Pattenden, *Angew. Chem. Int. Ed.*, **2002**, *41*, 508–511

73 J. Justicia, J. E. Oltra, J. M. Cuerva, *J. Org. Chem.*, **2004**, *69*, 5803–5806.

74 K. C. Nicolaou, J. L. Gunzer, G.q. Shi, A. Agrios, P. Gärtner, Z. Yang, *Chem. Eur. J.*, **1999**, *5*, 646–658; C. Buon, L. Chacun-Lefèvre, R. Rabot, P. Bouyssou, G. Coudert, *Tetrahedron*, **2000**, *56*, 605–614

75 M. Egi, L. Libeskind, *Org. Lett.*, **2003**, *5*, 801–802; F.-A. Alphonse, F. Suzenet, A. Keromnes, B. Lebret, G. Guillaumet, *Org. Lett.*, **2003**, *5*, 803–805.

76 S. R. Dubbaka, P. Vogel, *J. Am. Chem. Soc.*, **2003**, *125*, 15292–15293

77 T. A. Dineen, W. R. Roush, *Org. Lett.*, **2004**, *6*, 2043–2046

78 J. T. Feutrill, M. J. Lilly, M. A. Rizzacasa, *Org. Lett.*, **2002**, *4*, 525–527

79 H. Fuwa, M. Sasaki, M. Satake, K. Tachibana, *Org. Lett.*, **2002**, *4*, 2981–2984

80 J. A. Marshall, N. D. Adams, *J. Org. Chem.*, **2002**, *67*, 733–740

81 L. A. Paquette, M. Duan, C. Kempmann, *J. Am. Chem. Soc.*, **2002**, *124*, 4257–4270

82 K. C. Nicolaou, Y. Li, K. C. Fylakakidou, H. J. Mitchell, H.-X. Wei, B. Weyershausen, *Angew. Chem. Int. Ed.*, **2001**, *40*, 3849–3853 and 3854–3857

83 G. Van Cauwenberge, L.-J. Gao, D. Van Haver, M. Milanesio, D. Viterbo, P. J. De Clercq, *Org. Lett.*, **2002**, *4*, 1579–1582; L. S.-M. Wong, L. A. Sharp, N. M. C. Xavier, P. Turner, M. S. Sherburn, *Org. Lett.*, **2002**, *4*, 1955–1957

84 L. Bialy, H. Waldmann, *Angew. Chem. Int. Ed.*, **2002**, *41*, 1748–1751

85 K. Miyashita, M. Ikejiri, H. Kawasaki, S. Maemura, T. Imanishi, *Chem. Commun.*, **2002**, 742–743; T. Esumi, N. Okamoto, S. Hatakkeyama, *Chem. Commun.*, **2002**, 3042–3043

86 C. J. Sinz, S. D. Rychnovsky, *Angew. Chem. Int. Ed.*, **2001**, *40*, 3224–3227

87 D. A. Longbottom, A. J. Morrison, D. J. Dixon, S. V. Ley, *Angew. Chem. Int. Ed.*, **2002**, *41*, 2786–2789

88 O. P. Anderson, A. G. M. Barrett, J. J. Edmunds, S.-I. Hachiya, J. A. Hendrix, K. Horita, J. W. Malecha, C. J. Parkinson, A. VanSicke, *Can. J. Chem.*, **2001**, *79*, 1562–1592

89 H. F. Olivo, F. Velázquez, H. C. Trevisan, *Org. Lett.*, **2000**, *2*, 4055–4058; B. M. Trost, O. Dirat, J. L. Gunzer, *Angew. Chem. Int. Ed.*, **2002**, *41*, 841–843

90 C. M. Beaudry, D. Trauner, *Org. Lett.*, **2002**, *4*, 2221–2224

91 D. Nozawa, H. Takikawa, K. Mori, *J. Chem. Soc., Perkin Trans. 1*, **2000**, 2043–2043

92 J. Suffert, B. Salem, P. Klotz, *J. Am. Chem. Soc.*, **2001**, *123*, 12107–12108

93 M. Janka, G. K. Anderson, N. P. Rath, *Inorg. Chim. Acta.*, **2004**, *357*, 2339–2344; P. D. Jones, T. E. Glass, *Tetrahedron Lett.*, **2001**, *42*, 2265–2267

94 U. S. Schubert, C. Eschbaumer, M. Heller, *Org. Lett.*, **2000**, *2*, 3373–3376

95 For nonexhaustive recent examples of oligopyridines syntheses via Stille coupling see: M. M. M. Raposo, A. M. C. Fonseca, G. Kirsch, *Tetrahedron*, **2004**, *60*, 4071–4078; A. Puglisi, M. Benaglia, G. Roncan, *Eur. J. Org. Chem.*, **2003**, 1552–1558; P. F. H. Schwab, F. Fleischer, J. Milch, *J. Org. Chem.*, **2002**, *67*, 443–449; M. Heller, U. S. Shubert, *J. Org. Chem.*, **2002**, *67*, 8269–8272; C. R. Woods, M. Benaglia, S. Toyota, K., Hardcastle, J. S. Siegel, *Angew. Chem. Int. Ed.*, **2001**, *40*, 749–751; G. Ulrich, S. Bedel, C. Picard, P. Tisnes, *Tetrahedron Lett.*, **2001**, *42*, 6113–6115; A. Elghayoury, R. Ziessel, *J. Org. Chem.*, **2000**, *65*, 7753–7763

96 For nonexhaustive recent examples of oligothiophenes synthesis via Stille coupling see: H. Yu, B. Xu, T. M. Swager, *J. Am. Soc.*, **2003**, *125*, 1142–1143; B. Jousselme, P. Blanchard, E. Levillain, J. Delaunay, M. Allain, P. Richomme, D. Rondeau, N. Gallego-Planas, J. Roncali, *J. Am. Chem. Soc.*, **2003**, *125*, 1363–1370; G. Sotgiu, M. Zambianchi, G. Barbarella, F. Aruffo, F. Cipriani, A. Ventola, *J. Org. Chem.*, **2003**, *68*, 1512–1520; P. A. van Hal, E. H. A. Beckers, S. C. J. Meskers, R. A. J. Janssen, B. Jousselme, P. Blanchard, J. Roncali, *Chem. Eur. J.*, **2002**, *8*, 5415–5429; C. Xia, X. Fan, J. Locklin, R. C. Advincula, *Org. Lett.*, **2002**, *4*, 2067–2070; J. Frey, A. D. Bond, .A. B. Holmes, *Chem. Commun.*, **2002**, *20*, 2424– 2425.

97 C. R. Woods, M. Benaglia, S. Toyota, K. Hardcastel, J. S. Siegel, *Angew. Chem. Int. Ed.*, **2001**, *40*, 749–751

98 T. Bach, S. Heuser, *J. Org. Chem.*, **2002**, *67*, 5789–5795; T. C. Govaerts, I. Vogels, F. Campernolle, G. Hoornaert, *Tetrahedron Lett.*, **2002**, *43*, 799–802; W.-S. Kim, H.-J. Kim, C.-G. Cho, *Tetrahedron Lett.*, **2002**, *43*, 9015–9017; C. Wu, H. Nakumura, A. Murai, O. Shimomura, *Tetrahedron Lett.*, **2001**, *42*, 2997–3000; W. M. De Borggraeve, F. J. R. Rombouts, E. V. Van der Eycken, S. M. Toppet, G. J. Hoornaert, *Tetrahedron Lett.*, **2001**, *42*, 5693–5695

99 R. Shimizu, T. Fuchikami, *Tetrahedron Lett.*, **2001**, *42*, 6891–6894.

100 S.-K. Kang, H.-C. Ryu, S.-W. Lee, *J. Organomet. Chem.*, **2000**, *610*, 38–41.

101 K. Kikukawa, T. Idemoto, A. Katayama, K. Kono, F. Wada, T. Matsuda, *J. Chem. Soc. Perkin Trans. 1*, **1987**, 1511–1514

102 S.-K. Kang, T. Yamaaguchi, T.-H. Kim, P.-S. Ho, *J. Org. Chem.*, **1996**, *61*, 9082–9083

103 F. Garrido, S. Raeppel, A. Mann, M. Lautens, *Tetrahedron Lett.*, **2001**, *42*, 265–266 R. Shimizu, T. Fuchikami, *Tetrahedron Lett.*, **2001**, *42*, 6891–6894

104 S. Ceccarelli, U. Piarulli, C. Gennari, *J. Org. Chem.*, **2000**, *65*, 6254–6256

105 R. J. Franks, K. M. Nicholas, *Orgnometallics*, **2000**, *19*, 1458–1460.

106 R. D. Mazzola, Jr., S. Giese, C. L. Benson, F. G. West, *J. Org. Chem.*, **2004**, *69*, 220–223

107 E. Morera, G. Ortar, *Bioorg. Med. Chem. Lett.*, **2000**, *10*, 1815–1818

108 M. H. Al-Qaahtani, V. W. Pike, *J. Chem. Soc., Perkin Trans. 1*, **2000**, 1033–1036; M. W. Nader, F. Oberdorfer, *Appl. Rad. Isotopes*, **2002**, *57*, 681–685

109 M. R. Garayt, J. M. Percy, *Tetrahedron Lett.*, **2001**, *42*, 6377–6380

110 Y. Obora, M. Nakanishi, M. Tokunaga, Y. Tsuji, *J. Org. Chem.*, **2002**, *67*, 5835–5837

111 E. Piers, R. W. Friesen, B. A. Keay, *Chem. Commun.*, **1985**, 809–810

112 M. A. J. Duncton, G. Pattenden, *J. Chem. Soc., Perkin Trans. 1*, **1999**, 1235–1246; M. A. J. Duncton, G. Pattenden, *J. Chem. Soc., Perkin Trans. 1*, **1999**, 1235–1246

113 E. Marsaults, P. Deslonngchamps, *Org. Lett.*, **2000**, *2*, 3317–3320

114 A. B. Smith, III, G. R. Ott, *J. Am. Chem. Soc.*, **1998**, *120*, 3925–3948; A. B. Smith, III, G. R. Ott, *J. Am. Chem. Soc.*, **1996**, *118*, 13095–13096; Y. Kim, R. A. Singer, E. M. Carreira, *Angew. Chem. Int. Ed.*, **1998**, *37*, 1261–1263

115 E. Quéron, R. Lett, *Tetrahedron Lett.*, **2004**, *45*, 4533–4537 and 4539–4543

116 L. Peng, F. Zhang, T. Mei, T. Zao, Y. Li, *Tetrahedron Lett.*, **2003**, *44*, 5921–5923

117 A. Chen, A. Nelson, N. Tanikkul, E. J. Thomas, *Tetrahedron Lett.*, **2001**, *42*, 1251–1254

118 K. C. Nicolau, F. Murphy, S. Barluenga, T. Ohshima, H. Wei, J. Xu, D. L. F. Gray, O. Baudoin, *J. Am. Chem. Soc.*, **2000**, *122*, 3830–3838

119 I. Patterson, V. A. Doughty, M. D. McLeod, T. Trieselmann, *Angew. Chem. Int. Ed.*, **2000**, *39*, 1308–1312

120 M. J. Remuiñán, G. Pattenden, *Tetrahedron Lett.*, **2000**, *41*, 7367–7371

121 K. Toshima, T. Arita, K. Kato, D. Tanaka, S. Matsumura, *Tetrahedron Lett.*, **2001**, *42*, 8873–8876

122 K. Toshima, T. Jyojima, N. Miyamoto, M. Katohno, M. Nakata, S. Matsumura, *J. Org. Chem.*, **2001**, *66*, 1708–1715

123 E. Lee, H. Young Song, J. Won Kang, D.-S. Kim, C.-K. Jung, J. Min Jao, *J. Am. Chem. Soc.*, **2002**, *124*, 384–385

124 I. S. Mitchell, G. Pattenden, J. P. Stonehouse, *Tetrahedron Lett.*, **2002**, *43*, 493–497

125 H. W. Lam, G. Pattenden, *Angew. Chem. Int. Ed.*, **2002**, *41*, 508–511

126 L. Breydo, H. Zang, K. S. Gates, *Tetrahedron Lett.*, **2004**, *45*, 5711–5716

127 K. C. Nicolaou, N. Winssinger, J. Pastor, F. Murphy, *Angew. Chem. Int. Ed.*, **1998**, *37*, 2534–2537.

128 R. Olivera, R. SanMartin, E. Dominguez, *J. Org. Chem.*, **2000**, *65*, 7010–7019

129 S. Yue, J. J. Li, *Org. Lett.*, **2002**, *4*, 2201–2203

130 L. A. Saudan, G. Bernardinelli, E. P. Kündig, *Synlett* **2000**, 483–486

131 T. J. Houghton, S. Choi, V. Rawal, *Org. Lett.*, **2001**, *3*, 3615–3617

132 G. Ruel, N. Ke The, G. Dumartin, B. Delmond, M. Pereyre, *J. Organomet. Chem.*, **1993**, *444*, C18–C19

133 H. Kuhn, W. P. Neumann, *Synlett* **1994**, 123–124

134 M. S. Brody, M. G. Finn, *Tetrahedron Lett.*, **1999**, *40*, 415–418

135 K. C. Nicolaou, J. A. Pfefferkorn, A. J. Roecker, G.-Q. Cao, S. Barluenga, H. J. Mitchell, *J. Am. Chem. Soc.*, **2000**, *122*, 9939–9953

136 X. Wang, J. J. Parlow, J. A. Porco, Jr., *Org. Lett.*, **2000**, *2*, 3509–3512; F. Stieber, U. Grether, H. Waldmann, *Angew. Chem. Int. Ed.*, **1999**, *38*, 1073–1077; Y. Hu, S. Baudart, J. A. Porco, Jr., *J. Org. Chem.*, **1999**, *64*, 1049–1051

137 U. Grether, H. Waldmann, *Angew. Chem. Int. Ed.*, **2000**, *39*, 1629–1632; S. Tan, M. A. Foley, B. R. Stockwell, M. D. Shair, S. L. Schreiber, *J. Am. Chem. Soc.*, **1999**, *121*, 9073–9087

138 W. K.-D. Brill, C. Riva-Toniolo, *Tetrahedron Lett.*, **2001**, *42*, 6515–6518

139 F. Louërat, P. Gros, Y. Fort, *Tetrahedron Lett.*, **2003**, *44*, 5613–5616

140 V. Aucagne, S. Berteina-Raboin, P. Guenot, L. A. Agrofoglio, *J. Comb. Chem.*, **2004**, *6*, 717–723.

141 W. Yun, S. Li, B. Wang, L. Chen, *Tetrahedron Lett.*, **2001**, *42*, 175–177

142 J. K. Stille, H. Su, D. H. Hill, P. Schneider, M. Tanaka, D. L. Morrison, L. S. Hegedus, *Organometallics*, **1991**, *10*, 1993–2000; S.-K. Kang, T.-G. Baik, S.-Y. Song, *Synlett* **1999**, 327–329

143 R. E. Maleczka, Jr., W. P. Gallagher, I. Tierstiege, *J. Am. Chem. Soc.*, **2000**, *122*, 384–385

144 For a recent review on metal-catalyzed hydrostannations see: N. D. Smith, J. Mancuso, M. Lautens, *Chem. Rev.*, **2000**, *100*, 3257–3282.

145 W. P. Gallagher, I. Terstiege, R. E. Maleczka, Jr., *J. Am. Chem. Soc.*, **2001**, *123*, 3194–3294

146 R. E. Maleczka, Jr., W. P. Gallagher, *Org. Lett.*, **2001**, *3*, 4173–4176; R. E. Malleczka, Jr., J. M. Lavis, D. H. Clarck, W. P. Gallagher, *Org. Lett.*, **2000**, *2*, 3655–3658.

147 A. G. Hernán, V. Guillot, A. Kuvshinov, J. D. Kilburn, *Tetrahedron Lett.*, **2003**, *44*, 8601–8603

148 E. Shirakawa, H. Yoshida, T. Kurahashi, Y. Nakao, T. Hiyama, *J. Am. Chem. Soc.*, **1998**, *120*, 2975–2976; E. Shirakawa, T. Hiyama, *J. Organomet. Chem.*, **1999**, *576*, 169–178

149 H. Yoshida, E. Shirakawa, T. Kurahashi, Y. Nakao, T. Hiyama, *Organometallics*, **2000**, *19*, 5671–5678

150 E. Shirakawa, T. Hiyama, *J. Organomet. Chem.*, **2002**, *653*, 114–121

151 T. Matsubara, *Organometallics*, **2003**, *22*, 4297–4304; H. Yoshida, Y. Honda, E. Shirakawa, T. Hiyama, *Chem. Commun.*, **2001**, 1880–1881

152 F.-Y. Yang, M.-Y. Wu, C.-H. Cheng, *Tetrahedron Lett.*, **1999**, *40*, 6055–6058

153 E. Shirakawa, H. Yoshida, Y. Nakao, T. Hiyama, *Org. Lett.*, **2000**, *2*, 2209–2211

154 H. Tanaka, S.-I. Sumida, S. Torii, *Tetrahedron Lett.*, **1996**, *37*, 5967–5970; T. Takeda, K.-I. Matsunaga, T. Uruga, M. Takakura, T. Fujiwara, *Tetrahedron Lett.*, **1997**, *38*, 2879–2882

155 S.-K. Kang, J.-S. Kim, S.-C. Choi, *J. Org. Chem.*, **1997**, *62*, 4208–4209

156 E. Piers, T. Wong, *J. Org. Chem.*, **1993**, *58*, 3609–3610; T. Takeda, K. Matsunaga, Y. Kabasawa, T. Fujiwara, *Chem. Lett.*, **1995**, 771–772; G. D. Allred, L. S. Liebeskind, *J. Am. Chem. Soc.*, **1996**, *118*, 2748–2749

157 J. R. Falck, R. K. Bhatt, J. Ye, *J. Am. Chem. Soc.*, **1995**, *117*, 5973–5982

158 N. S. Nudelman, C. Carro, *Synlett* **1999**, 1942–1944

159 E. Piers, P. L. Gladstone, J. G. K. Yee, E. J. McEachern, *Tetrahedron*, **1998**, *54*, 10609–10626

160 S.-K. Kang, T.-G. Baik, X. H. Jiao, Y.-T. Lee, *Tetrahedron Lett.*, **1999**, *40*, 2383–2384

161 K. Mitsukura, S. Korekiiyo, T. Itoh, *Tetrahedron Lett.*, **1999**, *40*, 5739–5742

162 X. Fang, M. Johannsen, S. Yao, N. Gathergood, R. G. Hazell, K. A. Jorgensen, *J. Org. Chem.*, **1999**, *64*, 4844–4849

163 M. Tanaka, *Chem. Commun.*, **1981**, 47–48

164 S.-I. Ikeda, D.-M. Cui, Y. Sato, *J. Org. Chem.*, **1994**, *59*, 6877–6878

165 D.-M. Cui, N. Hashimoto, S.-I. Ikeda, Y. Sato, *J. Org. Chem.*, **1995**, *60*, 5752–5756.

166 V. Percec, J.-Y. Bae, D. H. Hill, *J. Org. Chem.*, **1995**, *60*, 6895–6903

167 E. Shirakawa, Y. Yamamoto, Y. Nakao, S. Oda, T. Tsuchimoto, T. Hiyama, *Angew. Chem. Int. Ed.*, **2004**, *43*, 3448–3451; E. Shirakawa, K. Yamasaki, H. Yoshida, T. Hiyama, *J. Am. Chem. Soc.*, **1999**, *121*, 10221–10222

168 Y. Nakao, E. Shirakawa, T. Tsuchimoto, T. Hiyama, *J. Organomet. Chem.*, **2004**, in press; E. Shirakawa, Y. Nakao, T. Hiyama, *Chem. Commun.*, **2001**, 263–264; E. Shirakawa, Y. Nakao, T. Tsuchimoto, T. Hiyama, *Chem. Commun.*, **2002**, 1962–1963

169 E. Shirakawa, Y. Nakao, H. Yoshida, T. Hiyama, *J. Am. Chem. Soc.*, **2000**, *122*, 9030–9031

170 J. M. Nuss, R. A. Rennels, *Chem. Lett.*, **1993**, 197–198

171 S. Oi, M. Moro, H. Fukuhara, T. Kawanishi, Y. Inoue, *Tetrahedron*, **2003**, *59*, 4351–4361

172 C.-J. Li, Y. Meng, *J. Am. Chem. Soc.*, **2000**, *122*, 9538–9539

173 S. Oi, M. Moro, H. Fukuhara, T. Kawanishi, Y. Inoue, *Tetrahedron Lett.*, **1999**, *40*, 9259–9262; T. Hayashi, M. Ishigedani, *J. Am. Chem. Soc.*, **2000**, *122*, 976–977; M. Ueda, N. Miyaura, *J. Organomet. Chem.*, **2000**, *595*, 31–35

174 S. Oi, M. Moro, S. Ono, Y. Inoue, *Chem. Lett.*, **1998**, 83–84; S. Oi, M. Moro, H. Ito, Y. Honma, S. Miyano, Y. Inoue, *Tetrahedron*, **2002**, *58*, 91–97

175 T. Huang, Y. Meng, S. Venkatraman, D. Wang, C.-J. Li, *J. Am. Chem. Soc.*, **2001**, *123*, 7451–7452; S. Venkatraman, Y. Meng, C.-J. Li, *Tetrahedron Lett.*, **2001**, 4459–4462

176 T. S. Huang, C.-J. Li, *Org. Lett.*, **2001**, *3*, 2037–2039

177 Y. Naruta, K. Maruyama, *Chemistry Lett.*, 881 (1979)

178 Y. Yamamoto, H. Yatagai, Y. Naruta, K. Maruyama, *J. Am. Chem. Soc.,* **1980**, *102*, 7107–7109

179 Y. Nishigaichi, A. Takuwa, *Tetrahedron Lett.,* **1999**, *40*, 109–112

180 G. E. Keck, D. E. Abbott, E. P. Borden, E. J. Enholm, *Tetrahedron Lett.,* **1984**, *25*, 3927–3931; G. E. Keck, K. A. Savin, E. N. K. Cressman, D. E. Abbott, *J. Org. Chem.,* **1994**, *59*, 7889–7896; T. Hayashi, K. Kabeta, I. Hamachi, M. Kumada, *Tetrahedron Lett.,* **1983**, *23*, 2865–2869

181 J. A. Marshall, H. Jiang, *J. Org. Chem.,* **1999**, *64*, 971–975; G. C. Micalizio, W. R. Roush, *Tetrahedron Lett.,* **1999**, *40*, 3351–3354; B. Leroy, E. Marko, *Tetrahedron Lett.,* **2001**, *42*, 8685–8688

182 S. Watrelot-Bourdeau, J. L. Parrain, J. P. Quintard, *J. Org. Chem.,* **1997**, *62*, 8261–8263

183 J. A. Marshall, G. S. Welmaker, B. W. Gung, *J. Am. Chem. Soc.,* **1991**, *113*, 647–656

184 J. A. Marshall, G. S. Welmaker, *J. Org. Chem.,* **1994**, *59*, 4122–4125; J. A. Marshall, J. A. Jablonowski, H. Jiang, *J. Org. Chem.,* **1999**, *64*, 2152–2154

185 M. Yasuda, K. Hirata, M. Nishino, A. Yamamoto, A. Baba, *J. Am. Chem. Soc.,* **2002**, *124*, 13442–13447; D. Basavaiah, B. Sreenivasulu, *Tetrahedron Lett.,* **2002**, *43*, 2987–2990

186 K. Y. Lee, C. Y. Oh, J. H. Kim, J. E. Joo, W. H. Ham, *Tetrahedron Lett.,* **2002**, *43*, 9361–9363

187 A. Tatami, M. Inoue, H. Uehara, M. Hirama, *Tetrahedron Lett.,* **2003**, *44*, 5229–5233; M. T. Crimmins, M. G. Stanton, S. P. Allwein, *J. Am. Chem. Soc.,* **2002**, *124*, 5958–5959

188 S. J. Jarosz, B. Fraser-Reid, *J. Org. Chem.,* **1989**, *54*, 4011–4013; J. A. Marshall, D. V.Yashunsky, *J. Org. Chem.,* **1991**, *56*, 5493–5495; B. W. Gung, J. P. Melnick, M. A. Wolf, J. A. Marshall, S. Beaudoin, *J. Org. Chem.,* **1994**, *59*, 5609–5613; M. Adamcyeski, E. Quinoa, P. Crews, *J. Org. Chem.,* **1990**, *55*, 240–242; J. A. Marshall, G. P. Luke, *J. Org. Chem.,* **1993**, *58*, 6229–6234

189 J. A. Marshall, B. M. Seletzky, P. S. Coan, *J. Org. Chem.,* **1994**, *59*, 5139–5140

190 S. S. Harried, C. P. Lee, G. Yang, T. I. H. Lee, D. C. Myles, *J. Org. Chem.,* **2003**, *68*, 6646–6660

191 D. A. Evans, B. T. Connell, *J. Am. Chem. Soc.,* **2003**, *125*, 10899–10905

192 P. J. Hergenrother, A. Hodgson, A. S. Judd, W. C. Lee, S. F. Martin, *Angew. Chem. Int. Ed.* **2003**, *42*, 3278–3281

193 J. A. Marshall, J. A. Jablonowski, G. P. Luke, *J. Org. Chem.,* **59**, 7825 (1994)

194 R. S. Coleman, J. S. Kong, T. E. Richardson, *J. Am. Chem. Soc.,* **1999**, *121*, 9088–9095

195 T. P. Loh, J. Xu, *Tetrahedron Lett.,* **1999**, *40*, 2431–2434; T. P. Loh, J. Xu, Q. Y. Hu, J. J. Vittal, *Tetrahedron Asym.,* **2000**, *11*, 1565–1569

196 D. J. Dixon, A. C. Foster, S. V. Ley, *Org. Lett.,* **2000**, *2*, 123–125.

197 M. Fishwick, M. G. H. Wallbridge, *J. Organomet. Chem.,* **1970**, *25*, 69

198 D. R. Williams, S. V. Plummer, S. Patnaik, *Angew. Chem. Int. Ed.* **2003**, *42*, 3934–3938

199 J. A. Marshall, *Chem. Rev.,* **1996**, *96*, 31–47

200 J. A. Marshall, K. W. Hinkle, *J. Org. Chem.,* **1995**, *60*, 1920–1921

201 J. A. Marshall, K. W. Hinkle, *J. Org. Chem.,* **1996**, *61*, 105–108

202 E. J. Thomas, *Chemtracts,* **1994**, *7*, 207; E. J. Thomas, *Chem. Commun.,* **1997**, 411–412

203 J.S. Carey, E.J. Thomas, *Synlett* **1992**, 585–586

204 Y. Nishigaichi, M. Yoshikawa, Y. Takigawa, A. Takuwa, *Chem. Lett.,* **1996**, 961–962; A. H. McNeil, E. J. Thomas, *Tetrahedron Lett.,* **1990**, *31*, 6239–6242; J. S. Carey, E. J. Thomas, *Tetrahedron Lett.,* **1993**, *34*, 3935–3938; J. S. Carey, E. J. Thomas, *J. Chem. Soc., Chem. Commun.,* **1994**, 283–284

205 E. K. Dorling, E. J. Thomas, *Tetrahedron Lett.,* **1999**, *40*, 471–474; E. K. Dorling, A. P. Thomas, E. J. Thomas, *Tetrahedron Lett.,* **1999**, *40*, 475–476; N. Martin, E. J. Thomas, *Tetrahedron Lett.,* **2001**, *42*, 8373–8377

206 L. Arista, M. Gruttadauria, E. J. Thomas, *Synlett* **1997**, 627–628

207 B. Leroy, I. Marko, *Org. Lett.*, **2002**, *4*, 47–50

208 A. Marx, H. Yamamoto, *Synlett* **1999**, 584–586

209 H. C. Aspinall, A. F. Browning, N. Greeves, P. Ravenscroft, *Tetrahedron Lett.*, **1994**, *35*, 4639–4642; H. C. Aspinall, N. Greeves, E. G. McIver, *Tetrahedron Lett.*, **1998**, *39*, 9283–9286; H. C. Aspinall, J. S. Bissett, N. Greeves, D. Levin, *Tetrahedron Lett.*, **2002**, *43*, 319–321

210 M. Yasuda, Y. Sugawa, A. Yamamoto, I. Shibata, A. Baba, *Tetrahedron Lett.*, **1996**, *37*, 5951–5954

211 J. K. Whitesell, R. Apodaca, *Tetrahedron Lett.*, **1996**, *37*, 3955–3958

212 A. Yanagisawa, M. Morodome, H. Nakashima, H. Yamamoto, *Synlett* **1997**,1309–1310

213 T. Ooi, D. Uraguchi, N. Kagushima, K. Maruoka, *J. Am. Chem. Soc.*, **1998**, *120*, 5327; K. Maruoka, T. Ooi, *Chem. Eur. J.*, **1999**, *5*, 829–833

214 J.M. Blackwell, W.E. Piers, M. Parvez, *Org. Lett.*, **2000**, *2*, 695–698; D. J. Morisson, W. E. Piers *Org. Lett.*, **2003**, *5*, 2857–2860

215 C. T. Chen, S. D. Chao, *J. Org. Chem.*, **1999**, *64*, 1090–1091

216 J. A. Marshall, Y. Tang, *Synlett* **1992**, 653–654

217 A. L. Costa, M. G. Piazza, E. Tagliavini, C. Trombini, A. Umani-Ronchi, *J. Am. Chem. Soc.*, **1993**, *115*, 7001–7002; G. E. Keck, K. H. Tarbet, L. S. Geraci, *J. Am. Chem. Soc.*, **1993**, *115*, 8467–8468; G. E. Keck, L. S. Geraci, *Tetrahedron Lett.*, **1993**, *34*, 7827–7830

218 Y. Yin, G. Zhao, Z. Qian, W. Yin, *J. Fluor. Chem.*, **2003**, *120*, 117–120

219 P. Bedeschi, S. Casolari, A. L. Costa, E. Tagliavini, A. Umani-Ronchi, *Tetrahedron Lett.*, **1995**, *36*, 7897–7901; M. Kurosu, M. Lorca, *Tetrahedron Lett.*, **2002**, *43*, 1765–1769

220 S. Casolari, P. G. Cozzi, P. Orioli, E. Tagliavini, A. Umani-Ronchi, *J. Chem. Soc. Chem. Commun.*, **1997**, 2123–2124

221 C. M. Yu, H. S. Choi, W. H. Jung, S. S. Lee, *Tetrahedron Lett.*, **1996**, *37*, 7095–7098; C. M. Yu, H. S. Choi, W. H. Jung, H. J. Kim, J. Shin, *J. Chem. Soc. Chem. Commun.*, **1997**, 761–762; C. M. Yu, H. S. Choi, S. K. Yoon, W. H. Jung, *Synlett* **1997**, 889–891

222 C. M. Yu, H. S. Choi, W. H. Jung, H. J. Kim, J. K. Lee, *Bull. Korean Chem. Soc.*, **1997**, *18*, 471

223 G. E. Keck, D. Krishnamurthy, M. C. Grier, *J. Org. Chem.*, **1993**, *58*, 6543–6544; G. E. Keck, T. Yu, *Org. Lett.*, **1999**, *1*, 289–291; S. Weigand, R. Brückner, *Chem. Eur. J.*, **1996**, *2*, 1077

224 A. Fürstner, K. Langemann, *J. Am. Chem. Soc.*, **1997**, *119*, 9130–9136

225 G. E. Keck, J. A. Covel, T. Schiff, T. Yu, *Org. Lett.*, **2002**, *4*, 1189–1192; C. M. Yu, J. Y. Lee, B. So, J. Hong, *Angew. Chem. Int. Ed.*, **2002**, *41*, 161–163

226 S. Casolari, D. Addario, E. Tagliavini, *Org. Lett.*, **1999**, *1*, 1061–1064; K. M. Waltz, J. Gavenonis, P. J. Walsh, *Angew. Chem. Int. Ed.*, **2002**, *41*, 3697–3699; S. Kii, K. Maruoka, *Chirality*, **2003**, *15*, 68–70

227 A. Yanagisawa, H. Nakashima, A. Ishiba, H. Yamamoto, *J. Am. Chem. Soc.*, **1996**, *118*, 4723–4724; A. Yanagisawa, A. Ishiba, H. Nakashima, H. Yamamoto, *Synlett* **1997**, 88–90

228 A. Yanagisawa, Y. Nakatsuka, H. Yamamoto, *Synlett* **1997**, 933–935

229 M. Luo, Y. Iwabuchi, S. Hatakeyama, *Synlett* **1999**, 1109–1111; M. Luo, Y. Iwabuchi, S. Hatakeyama, *Chem. Commun.*, **1999**, 267–268; C. M. Yu, S. J. Lee, M. Jeon, *J. Chem. Soc., Perkin Trans.1*, **1999**, 3557–3558

230 P. G. Cozzi, P. Orioli, E. Tagliavini, A. Umani-Ronchi, *Tetrehedron Lett.*, **1997**, *38*, 145–148

231 Y. Motoyama, H. Narusawa, H. Nishiyama, *Chem. Commun.*, **1999**, 131–132

232 T. Mukaiyama, T. Harada, S. Shoda, *Chem. Lett.*, **1980**, 1507; S. A. S. David, *Tetrahedron Lett.*, **1983**, *24*, 4009–4012; C. Petrier, J. Einhorn, J. L. Luche, *Tetrahedron Lett.*, **1985**, *26*, 1449–1452

233 Y. Masuyama, R. Hayashi, K. Otake, Y. Kurusu, *J. Chem. Soc., Chem. Commun.*, **1988**, 44–45; Y. Masuyama, J. P. Takaraa, Y. Kurusu, *J. Am. Chem. Soc.*, **1988**, *110*, 4473–4474; T. Imai, S. Nishida, *Synthesis*, **1993**, 395–399; Y. Masuyama, M. Kishida, Y. Kurusu,

J. Chem. Soc., Chem. Commun., **1995**, 1405–1406

234 E. Fouquet, A. Gabriel, B. Maillard, M. Pereyre,*Tetrahedron Lett.*, **1993**, *34*, 7749–7752; E. Fouquet, M. Pereyre, A. L. Rodriguez, T. Roulet, *Bull. Soc. Chim. Fr.*, **1997**, *134*, 959–968

235 E. Fouquet, A. Gabriel, B. Maillard, M. Pereyre, *Bull. Soc. Chim. Fr.*, **1995**, *132*, 590–598

236 S. Kobayashi, K. Nishio, *Tetrahedron Lett.*, **1995**, *36*, 6729–6732

237 K. Yamada, T. Tozawa, M. Nishida, T. Mukaiyama, *Bull. Chem. Soc. Jpn.*, **1997**, *70*, 2301–2308

238 J. Cossy, C. Rasamison, D. Gomez Pardo, *J. Org. Chem.*, **2001**, *66*, 7195–7198

239 A. Kundu, S. Prabhakar, M. Vairamani, S. Roy, *Organometallics*, **1999**, *18*, 2782–2785

240 J. A. Marshall, C. M. Grant, *J. Org. Chem.*, **1999**, *64*, 8214–8219

241 J. A. Marshall, X. J. Wang, *J. Org. Chem.*, **1990**, *55*, 6246–6248; J. A. Marshall, J. F. Perkins, M. A. Wolf, *J. Org. Chem.*, **1995**, *60*, 5556–5559

242 J. A. Marshall, X. J. Wang, *J. Org. Chem.*, **1991**, *56*, 3211–3213; J. A. Marshall, X. J. Wang, *J. Org. Chem.*, **1992**, *57*, 1242–1252

243 J. A. Marshall, B. A. Johns, *J. Org. Chem.*, **1998**, *63*, 817–823; J. A. Marshall, B. A. Johns, *J. Org. Chem.*, **1998**, *63*, 7885–7892; J. A. Marshall, B. A. Johns, *J. Org. Chem.*, **2000**, *65*, 1501–1510

244 G. E. Keck, D. Krishnamurthy, X. Chen, *Tetrahedron Lett.*, **1994**, *35*, 8323–8326

245 C. M. Yu, S. K. Yoon, S. J. Lee, J. Y. Lee, S. S. Kim, *Chem. Commun.*, **1998**, 2749–2750

246 C. M. Yu, S. K. Yoon, K. Baek, J. Y. Lee, *Angew. Chem. Int. Ed.* **1998**, *37*, 2392–2395

247 M. Yasuda, T. Miyai, I. Shibata, A. Baba, *Tetrahedron Lett.*, **1995**, *36*, 9497–9500

248 D. A. Evans, D. P. Halstead, B. D. Allison, *Tetrahedron Lett.*, **1999**, *40*, 4461–4461

249 B. M. Trost, P. J. Bonk, *J. Am. Chem. Soc.*, **1985**, *107*, 1778–1781; G. E. Keck, E. J. Enholm, *J. Org. Chem.*, **1985**, *50*, 146–147

250 C. K. Z. Andrade, G. R. Oliveira, *Tetrahedron Lett.*, **2002**, *43*, 1935–1937

251 J. Liu, C.-H. Wong, *Tetrahedron Lett.*, **2002**, *43*, 3915–3917

252 C. Bellucci, P. G. Cozzi, A. Umani-Ronchi, *Tetrahedron Lett.*, **1995**, *36*, 7289–7292

253 H. C. Aspinall, J. S. Bissett, N. Greeves, D. Levin, *Tetrahedron Lett.*, **1995**, *43*, 323–325

254 T. Akiyama, Y. Onuma, *J. Chem. Soc. Perkin Trans. I*, **2002**, 1157–1158

255 Y. Yamamoto, T. Komatsu, K. Maruyama, *J. Org. Chem.*, **1985**, *50*, 3115

256 S. Laschat, H. Kunz, *J. Org. Chem.*, **1991**, *56*, 5883–5889

257 S. Masamune, W. Choy, J. S. Petersen, L. R. Sita, *Angew. Chem., Int. Ed. Engl.*, **1985**, *24*, 1–30

258 Y. Yamamoto, S. Nishii, K. Maruyama, T. Komatsu, W. Ito, *J. Am. Chem. Soc.*, **1986**, *108*, 7778–7786

259 D. J. Hallett, E. J. Thomas, *J. Chem. Soc., Chem. Commun.*, **1995**, 657–658; G. W. Bradley, D. J. Hallett, E. J. Thomas, *Tetrahedron Asymmetry*, **1995**, *6*, 2579–2582; D. J. Hallett, E. J. Thomas, *Tetrahedron Asymmetry*, **1995**, *6*, 2575–2578

260 M. Shimizu, A. Morita, T. Kaga, *Tetrahedron Lett.*, **1999**, *40*, 8401–8405

261 Y. Yamamoto, M. Schmid, *J. Chem. Soc., Chem. Commun.*, **1989**, 1310–1312

262 Y. Yamamoto, H. Sato, J. Yamada, *Synlett* **1991**, 339–341

263 D. K. Wang, L. X. Dai, X. L. Hou, *Tetrahedron Lett.*, **1995**, *36*, 8649–8652

264 Y. Niwa, M. Shimizu *J. Am. Chem. Soc.*, **2003**, *125*, 3720–3721

265 S. Hanessian, R. Buckle, M. Bayrakdarian, *J. Org. Chem.*, **2002**, *67*, 3387–3397; S. Hanessian, U. Reinholdt, G. Gentile, *Angew. Chem. Int. Ed.*, **1997**, *36*, 1881–1884

266 J. Y. Park, I. Kadota, Y. Yamamoto, *J. Org. Chem.*, **1999**, *64*, 4901–4908

267 K. T. Wanner, E. Wadenstorfer, A. Kärtner, *Synlett* **1991**, 797–798

268 J. A. Marshall, K. Gill, B. M. Seletsky, *Angew. Chem. Int. Ed.*, **2000**, *39*, 953–956

269 S Kobayashi, K. Sugita, H. Oyamada, *Synlett* **1999**, 138–140; K. Manabe,

H. Oyamada, K. Sugita, S. Kobayashi, *J. Org. Chem.*, **1999**, *64*, 8054–8057

270 M. Lombardo, S. Spada, C. Trombini, Eur. *J. Org. Chem.*, **1998**, 2361–2364 M. Gianotti, M. Lombardo, C. Trombini, *Tetrahedron Lett.*, **1998**, *39*, 1643–1646

271 R. Yamaguchi, M. Moriyasu, M. Yoshioka, M. Kawanisi, *J. Org. Chem.*, **1988**, *53*, 3507–3512

272 T. Itoh, Y. Matsuya, Y. Enomoto, K. Nagata, M. Miyazaki, A. Ohsawa, *Synlett* **1999**, 1799–1801

273 T. Itoh, Y. Matsuya, Y. Enomoto, K. Nagata, M. Miyazaki, A. Ohsawa, *Tetrahedron*, **2001**, *57*, 7277–7289

274 S. Yamada, M. Ichikawa, *Tetrahedron Lett.*, **1999**, *40*, 4231–4234

275 S. Kobayashi, H. Ishitani, *Chem. Rev.*, **1999**, *99*, 1069–1094

276 H. Nakamura, N. Asao, Y. Yamamoto, *Chem. Commun.*, **1995**, 1273–1274; H. Nakamura, N. Asao, Y. Yamamoto, *J. Am. Chem. Soc.*, **1996**, *118*, 6641

277 X. Fang, M. Johannsen, S. Yao, N. Gathergood, R. G. Hazell, K. A. Jorgensen, *J. Org. Chem.*, **1999**, *64*, 4844–4849

278 M. Bao, H. Nakamura, Y. Yamamoto, *Tetrahedron Lett.*, **2000**, *41*, 131–134

279 M. W. Carland, K. H. Schiesser, in *The Chemistry of Organic Ge, Sn and Pb compounds*,Pataï, vol 2, 1401–1484, Wiley Chichester, **2001**

280 H. Kosugi, K. Kurino, K. Takayama, T. Migita, *J. Organomet. Chem.*, **1973**, *56*, C11–C13; J. Grignon, M. Pereyre, *J. Organomet. Chem.*, **1973**, *61*, C33–C35

281 G. E. Keck, J. B. Yates, *J. Org. Chem.*, **1982**, *47*, 3590–3591; G. E. Keck, J. B. Yates, *J. Am. Chem. Soc.*, **1982**, *104*, 5829–5831

282 D. P. Curran, P. A. van Elburg, B. Giese, S. Gilges, *Tetrahedron Lett.*, **1990**, *31*, 2861–2864

283 C. V. Ramana, S. M. Baquer, R. G. Gonnade, M. K.Gurjar, *Chem. Commun.*, **2002**, 614–615

284 K. Mizuno, M. Ikeda, S. Toda, Y. Otsuji, *J. Am. Chem. Soc.*, **1988**, *110*, 1288–1290; G. E. Keck, C.P. Kordik, *Tetrahedron Lett.*, **1993**, *34*, 6875–6876

285 I. Ryu, H. Yamazaki, K. Kusano, A. Ogawa, N. Sonoda, *J. Am. Chem. Soc.*, **1991**, *113*, 8558–8560

286 T. Migita, K. Nagai, M. Kosugi, *Bull. Chem. Soc. Jpn.*, **1983**, *56*, 2480

287 T. Toru, Y. Yamada, T. Ueno, E. Maekawa, Y. Ueno, *J. Am. Chem. Soc.*, **1988**, *110*, 4815–4817; L. C. Blaszczak, H. K. Armour, N. G. Halligan, *Tetrahedron Lett.*, **1990**, *31*, 5693–5696

288 D. J. Hart, K. Krishnamurthy, *Synlett* **1991**, 412–413

289 J. A. Campbell, D. J. Hart, *Tetrahedron Lett.*, **1992**, *33*, 6247–6250

290 G. E. Keck, E. J. Enholm, D. F. Kachensky, *Tetrahedron Lett.*, **1984**, *25*, 1867–1870; J. Dupuis, B. Giese, D. Rüegge, H. Fischer, H. G. Korth, R. Sustmann, *Angew. Chem. Int. Ed. Engl.*, **1984**, *23*, 896–898

291 J. D. Ginn, A. Padwa, *Org. Lett.*, **2002**, *4*, 1515–1517

292 J. E. Baldwin, R. M. Adlington, D. J. Birch, J. A. Crawford, J. B. Sweeney, *J. Chem. Soc., Chem. Commun.*, **1986**, 1339–1340; J. E. Baldwin, R. M. Adlington, C. Lowe, I. A. O'Neil, G. L. Sanders, C. J. Schofield, J. B. Sweeney, *J. Chem. Soc., Chem. Commun.*, **1988**, 1030–1032; E. Lee, S. G. Yu, C. U. Hur, S. M. Yang, *Tetrahedron Lett.*, **1988**, *29*, 6969–6970; A. Padwa, S. Shaun-Murphree, P. E. Yeske *Tetrahedron Lett.*, **1990**, *31*, 2983–2986

293 K. Miura, H. Saito, D. Itoh, A. Hosomi, *Tetrahedron Lett.*, **1999**, *40*, 8841–8844

294 J. E. Baldwin, R. M. Adlington, M. B. Mitchell, J. Robertson, *J. Chem. Soc., Chem. Commun.*, **1990**, 1574–1575

295 A. Sutherland, J. F. Caplan, J. C. Vederas, *Chem. Commun.*, **1999**, 555–556

296 B. Giese, T. Linker, *Synthesis*, **1992**, 46–47

297 Y. Yoshida, F. Ono, F. Sato, *J. Org. Chem.*, **1994**, *59*, 6153–6155; S. Hanessian, M. Alpegiani, *Tetrahedron*, **1989**, *45*, 941–950

298 J. E. Baldwin, R. Fieldhouse, A. T. Russell, *Tetrahedron Lett.*, **1993**, *34*, 5491–5494

299 M. P. Sibi, J. Chen, T. R. Rheault, *Org. Lett.* **2001**, *3*, 3679–3681; F. Villar, T. Kolly-Kovac, O. Equey, P. Renaud, *Chem. Eur. J.*, **2003**, *9*, 1566–1570

300 G. E. Keck, J. B. Yates, *J. Organomet. Chem.*, **1983**, *248*, C21–C28

301 D. P. G. Hamon, R. A. Massy-Westropp, P. Razzino, *Tetrahedron* **1995**, *51*, 4183–4194; A. Takuwa, J. Shiigi, Y. Nishigaichi, *Tetrahedron Lett.*, **1993**, *34*, 3457–3460

302 G. E. Keck, E. N. K. Cressman, E. J. Enholm, *J. Org. Chem.*, **1989**, *54*, 4345–4349; S. J. Danishefsky, J. S. Panek, *J. Am. Chem. Soc.*, **1987**, *109*, 917–918

303 M. P. Sibi, H. Miyabe, *Org. Lett.* **2002**, *4*, 3435–3498; M. P. Sibi, M. Aasmul, H. Hasegawa, T. Subramanian, *Org. Lett.* **2003**, *5*, 2883–2888

304 G. A. Kraus, B. Andersh, Q. Su, J. Shi, *Tetrahedron Lett.*, **1993**, *34*, 1741–1744

305 J. E. Baldwin, R. M. Adlington, A. Basak, *J. Chem. Soc., Chem. Commun.*, **1984**, 1284–1285

306 M. Ethève-Quelquejeu, J. M. Valéry, *Tetrahedron Lett.*, **1999**, *40*, 4807–4810

307 J. E. Baldwin, D. R. Kelly, *J. Chem. Soc., Chem. Commun.*, **1985**, 682–683

308 J. E. Baldwin, D. R. Kelly, C. B. Ziegler, *J. Chem. Soc., Chem. Commun.*, **1984**, 133–134

309 K. Haraguchi, Y. Itoh, K. Matsumoto, H. Tanaka, *J. Org. Chem.*, **2003**, *68*, 2006–2009

310 D. P. Curran, P. A. van Elburg, B. Giese, S. Gilges, *Tetrahedron Lett.*, **1989**, *30*, 2501–2504

311 G. A. Russel, L. L. Herold, *J. Org. Chem.*, **1985**, *50*, 1037–1040; Y. Watanabe, T. Yoneda, Y. Ueno, T. Toru, *Tetrahedron Lett.*, **1990**, *31*, 6669–6672

312 K. Miura, H. Saito, N. Fujisawa, D. Wang, H. Nishikori, A. Hosomi, *Org. Lett.*, **2001**, *3*, 4055–4057

313 J. E. Baldwin, R. M. Adlington, J. Robertson, *J. Chem. Soc., Chem. Commun.*, **1988**, 1404–1405

314 S. Y. Chang, Y. F. Shao, S. F. Chu, G. T. Fan, Y. M. Tsai, *Org. Lett.*, **1999**, *1*, 945–948

315 E. Fouquet, M. Pereyre, T. Roulet, *J. Chem. Soc., Chem. Commun.* **1995**, 2387–2388

316 E. Fouquet, M. Pereyre, J-C. Rayez, M-T. Rayez, T. Roulet, *C. R. Acad. Sci.*, **2001**, *4*, 2C, 641–648

317 E. J. Enholm, M. E. Gallagher, K. M. Moran, J. S. Lombardi, J. P. Schulte, *Org. Lett.*, **1999**, *1*, 689–691

318 I. Ryu, T. Niguma, S. Minakata, M. Komatsu, *Tetrahedron Lett.*, **1999**, *40*, 2367–2370

319 S. Hanessian, H. Yang, R. Schaum, *J. Am. Chem. Soc.*, **1996**, *118*, 2507–2508

320 Y. Guindon, B. Guérin, C. Chabot, N. Makintosh, W. W. Ogilvie, *Synlett* **1995**, 449–451

321 H. Nagano, Y. Kuno, Y. Omori, M. Iguchi, *J. Chem. Soc., Perkin Trans.*, **1996**, 389–391

322 E. J. Enholm, S. Lavieri, T. Cordova, I. Ghiviriga, *Tetrahedron Lett.*, **2003**, *44*, 531–534

323 J. Hongliu Wu, R. Rakinov, N. A. Porter, *J. Am. Chem. Soc.*, **1995**, *117*, 11029–11030; N. A. Porter, J. Hongliu Wu, G. Zhang, A. D. Reed, *J. Org. Chem.*, **1997**, *62*, 6702–6703

324 M. P. Sibi, J. Chen, *J. Am. Chem. Soc.*, **2001**, *123*, 9472–9473

325 P. Renaud, T. Bourquard, *Tetrahedron Lett.*, **1994**, *35*, 1707–1710

326 P. Renaud, M. Ribezzo, *J. Am. Chem. Soc.*, **1991**, *113*, 7803–7805; P. Renaud, N. Moufid, L. Huang Kuo, D. P. Curran, *J. Org. Chem.*, **1994**, *59*, 3547–3552; D. P. Curran, L. Huang Kuo, *J. Org. Chem.*, **1994**, *59*, 3259–3261

327 S. Kano, Y. Yuasa, S. Shibuya, *Heterocycles*, **1990**, *31*, 1597–1601; S. Kano, T. Yokomatsu, S. Shibuya, *Heterocycles*, **1990**, *31*, 13–16

328 R. Radinov, C. L. Mero, A. T. McPhail, N. A. Porter, *Tetrahedron Lett.*, **1995**, *36*, 8183–8186

329 M. P. Sibi, J. Ji, *Angew. Chem. Int. Ed. Engl.*, **1996**, *35*, 191–192

330 L. Giraud, P. Renaud, *J. Org. Chem.*, **1998**, *63*, 9162–9163

331 D. Seyferth, M. A. Weiner, *J. Org. Chem.*, **1959**, *24*, 1395–1396; D. Seyferth, M. A. Weiner, *J. Am. Chem. Soc.*, **1961**, *83*, 3583–3586

332 J. S. Sawyer, A. Kucerovy, T. L. Macdonald, G. J. McGarvey, *J. Am. Chem. Soc.*, **1988**, *110*, 842–853

333 P. Beak, A. Basu, D. J. Gallagher, Y. S. Park, S. Thayumanavan, *Acc. Chem. Res.*, **1996**, *29*, 552–560; S. V. Kessar, P. Singh, *Chem. Rev.*, **1997**, *97*, 721–737; A. Basu, S. Thayumanavan, *Angew. Chem. Int. Ed.*, **2002**, *41*, 716–738; P. Beak, W. J. Zajdel,

D. B. Reitz, *Chem. Rev.*, **1984**, *84*, 471–523

334 P. Graña, M. R. Paleo, F. J. Sardina, *J. Am. Chem. Soc.*, **2002**, *124*, 12511–12514

335 W. J. Still, *J. Am. Chem. Soc.*, **1978**, *100*, 1481–1487; W. C. Still, C. Sreekumar, *J. Am. Chem. Soc.*, **1980**, *102*, 1201–1202

336 G. Hirai, Y. Koiizumi, S. M. Moharram, H. Oguri, M. Hirama, *Org. Lett.*, **2002**, *4*, 1627–1630

337 M. E. Kuehne, F. Xu, *J. Org. Chem.*, **1998**, *63*, 9427–9433

338 R. Hara, T. Furukawa, H. Kashima, Y. Horigushi, I. Kuwajima, *J. Am. Chem. Soc.*, **1999**, *121*, 3072–3082

339 R. P. Smmyj, J. M. Chong, *Org. Lett.*, **2001**, *3*, 2903–2906

340 G. Gralla, B. Wibbeling, D. Hoppe, *Org. Lett.*, **2002**, *4*, 2193–2195; G. Christoph, D. Hoppe, *Org. Lett.*, **2002**, *4*, 2189–2192

341 G. Gralla, B. Wibbeling, D. Hoppe, *Tetrahedron Lett.*, **2003**, *44*, 8979–8982

342 D. J. Petersen, *J. Am. Chem. Soc.*, **1971**, *93*, 4027–4031

343 G. Chambournier, R. E. Gawley, *Org. Lett.*, **2000**, *2*, 1561–1564; R. E. Gawley, E. Low, G. Chambournier, *Org. Lett.*, **1999**, *1*, 653–655; R. E. Gawley, Q. Zhang, *J. Am. Chem. Soc.*, **1993**, *115*, 7515–7516

344 M. E. Kopach, A. I. Meyers, *J. Org. Chem.*, **1996**, *61*, 6764–6765.

345 D. M. Iula, R. E. Gawley, *J. Org. Chem.*, **2000**, *65*, 6196–6201

346 J.-C. Cintrat, V. Léat, J.-L. Parrain, E. Le Grognec, I. Beaudet, L. Toupet, J.-P. Quintard, *Organometallics*, **2004**, *23*, 943–945

347 W. H. Pearson, A. C. Lindbeck, J. W. Kampf, *J. Am. Chem. Soc.*, **1993**, *115*, 2622–2636

348 R. E. Gawley, Q. Zhang, *J. Org. Chem.*, **1995**, *60*, 5763–5769

349 F. Jeanjean, G. Fournet, D. Le Bars, J. Goré, *Eur. J. Org. Chem.*, **2000**, 1297–1305; F. Jeanjean, N. Perol, J. Goré, G. Fournet, *Tetrahedron Lett.*, **1997**, *38*, 7547–7550

350 W. H. Pearson, A. C. Lindbeck, *J. Org. Chem.*, **1989**, *54*, 56651–5654

351 J. M. Chong, S. B. Park, *J. Org. Chem.*, **1992**, *57*, 2220–2222

352 A. Ncube, S. B. Park, J. M. Chong, *J. Org. Chem.*, **2002**, *67*, 3625–3636

353 G. Christoph, D. Hoppe, *Org. Lett.*, **2002**, *4*, 2189–2192; J. M. Laumer, D. D. Kim, P. Beak, *J. Org. Chem.*, **2002**, *67*, 6797–6804

354 W. H. Pearson, Y. Mi, *Tetrahedron Lett.*, **1997**, *38*, 5441–5444; W. H. Pearson, E. P. Stevens, *J. Org. Chem.*, **1998**, *63*, 9812–9827

355 W. H. Pearson, D. M. Mans, J. W. Kampf, *J. Org. Chem.*, **2004**, *69*, 1235–1247

356 W. H. Pearson, D. M. Mans, J. W. Kampf, *Org. Lett.*, **2002**, *4*, 3099–3102

357 W. H. Pearson, Y. Mi, I. Y. Lee, P. Stoy, *J. Am. Chem. Soc.*, **1998**, *63*, 3607–3617

358 M. Mans, W. H. Pearson, *Org. Lett.*, **2004**, *6*, 3305–3308

359 W. H. Pearson, F. E. Lovering, *J. Org. Chem.*, **1998**, *63*, 3607–3617.

360 M. Kim, E. Vedejs, *J. Org. Chem.*, **2004**, *69*, 7262–7265; E. Vedejs, J. D. Little, *J. Org. Chem.*, **2004**, *69*, 1794–1799; E. Vedejs, J. D. Little, L. M. Seaney, *J. Org. Chem.*, **2004**, *69*, 1788–1793

361 T. Tomoyasu, K. Tomooka, T. Nakai, *Tetrahedron Lett.*, **2003**, *44*, 1239–1242

362 E. J. Corey, T. M. Eckrich, *Tetrahedron Lett.*, **1984**, *25*, 2419–2422; E. J. Corey, J. R. Cashman, T. M. Eckrich, D. R. Corey, *J. Am. Chem. Soc.*, **1985**, *107*, 713–715

363 E. J. Corey, R. H. Wollenberg, *J. Org. Chem.*, **1975**, *40*, 2265–2266

364 E. J. Corey, B. C. Pan, D. H. Pan, D. R. Deardorff, *J. Am. Chem. Soc.*, **1982**, *104*, 6816–6818; E. J. Corey, D. H. Hua, B. C. Pan, S. P. Seitz, *J. Am. Chem. Soc.*, **1982**, *104*, 6818–6820

365 J. D. White, G. Wang, L. Quaranta, *Org. Lett.*, **2003**, *5*, 4983–4986

366 S. M. Simpkins, B. M. Kariuki, C. S. Aaricó, L. R. Cox, *Org. Lett.*, **2003**, *5*, 3971–3974

367 J. R. Behling, K. A. Babiak, J. S. Ng, A. L. Campbell, R. Moretti, M. Koerner, B. H. Lipshutz, *J. Am. Chem. Soc.*, **1988**, *110*, 2641–2643

368 B. H. Lipschutz, R. Crow, S. H. Dimock, *J. Am. Chem. Soc.*, **1990**, *112*, 4063–4064

369 J. J. Eisch, B. W. Kotowicz, *Eur. J. Inorg. Chem.*, **1998**, 761–769

370 S.-W. Leung, D. A. Singleton, *J. Org. Chem.*, **1997**, *62*, 1955–1960.

371 I. D. Gridnev, O. L. Tok, N. A. Gridneva, Y. N. Bubnov, P. R. Schreiner, *J. Am. Chem. Soc.*, **1998**, *120*, 1034–1043

372 H. Sashida, A. Kuroda, *J. Chem. Soc., Perkin Trans. 1*, **2000**, 1965–1969

373 R. I. Yousef, T. Rüffer, H. Schmidt, D. Steinborn, *J. Organomet. Chem.*, **2002**, *655*, 111–114

374 J. D. Hoefelmeyer, M. Schulte, F. P. Gabaï, *Inorg. Chem.*, **2001**, *40*, 3833–3834

375 A. J. Ashe, III, X. Fang, J. W. Kampf, *Organometallics*, **2001**, *20*, 2109–2113

7
Polyfunctional Zinc Organometallics for Organic Synthesis

Paul Knochel, Helena Leuser, Liu-Zhu Gong, Sylvie Perrone, and Florian F. Kneisel

7.1
Introduction

Organozincs have been known since the preparation of diethylzinc by Frankland in 1849 in Marburg (Germany) [1]. These organometallic reagents were fairly often used to form new carbon–carbon bonds until Grignard [2] discovered in 1900 a convenient preparation of organomagnesium compounds. These reagents were found to be more reactive species toward a broad range of electrophiles and afforded generally higher yields compared to organozincs. However, some reactions were still performed with zinc organometallics such as the Reformatsky reaction [3] or the Simmons–Smith cyclopropanation [4]. The intermediate organometallics (zinc enolate and zinc carbenoid) were more easy to handle and more selective than the corresponding magnesium organometallics. Remarkably, Hunsdiecker reported in a German patent of 1943 that organozinc reagents bearing long carbon chains terminated by an ester function can be prepared [5]. This functional-group tolerance remained largely ignored by synthetic chemists and it became clear only recently that organozinc compounds are prone to undergo a large range of transmetallations due to the presence of empty low-lying p-orbitals that readily interact with the *d*-orbitals of many transition metal salts leading to highly reactive intermediates [6]. One can wonder why unreactive zinc reagents can produce highly reactive organometallic intermediates reacting with many electrophiles that are unreactive toward organozincs. This can be explained by the presence of d-orbitals at the transition-metal center that makes a number of new reaction pathways available that were not accessible to the zinc precursors since the empty d-orbitals of zinc are too high in energy to participate to most organic reactions. It is the combination of the high tolerance of functionalities of organozinc derivatives with a facile transmetallation to many transition metal complexes, which makes organozincs such valuable reagents. Especially important are the transmetallation of RZnX reagents to organocopper compounds [6] and to palladium intermediates [7] which allow the performance of cross-coupling reactions with high efficiency (Negishi reaction [7]; Scheme 7.1).

Organometallics. Paul Knochel
Copyright © 2005 WILEY-VCH Verlag GmbH & Co. KGaA, Weinheim
ISBN: 3-527-31131-9

$$RCu \cdot ZnX_2 \xleftarrow{\ CuX\ } RZnX \xrightarrow{\ ArPdX\ } \underset{Ar}{R-Pd}$$

Scheme 7.1 Transmetallation of organozinc reagents.

Furthermore, the highly covalent character of the carbon–zinc bond [8] affords organozincs configurationally stable at temperatures where the corresponding organo-magnesiums and -lithiums undergo a racemization. This property makes them good candidates for the preparation of chiral organometallics [9]. In this chapter, we will describe the preparation methods of polyfunctional organozinc compounds followed by a detailed presentation of their reactivity in the absence and in the presence of transition-metal catalysts.

7.2
Methods of Preparation of Polyfunctional Organozinc Reagents

7.2.1
Classification

There are three important classes of organozinc reagents: (i) organozinc halides of the general formula RZnX (ii) diorganozincs of the general formula R^1ZnR^2 in which R^1 and R^2 are two organic groups and (iii) zincates of the general formula $R^1(R^2)(R^3)ZnMet$ in which the metal (Met) is usually Li or MgX. The reactivity of these zinc reagents increases with the excess of negative charge of the zinc center (Scheme 7.2). The reactivity of organozinc halides strongly depends on the electronegativity of the carbon attached to zinc and on the aggregation of the zinc reagent. A stabilization of the negative carbanionic charge by inductive or mesomeric effects leads to a more ionic carbon–zinc bond and to a higher reactivity:

alkynyl < alkyl < alkenyl ≤ aryl < benzyl < allyl

$RZnX < R_2Zn < R_3ZnMgX < R_3ZnLi$

Scheme 7.2 Reactivity order of zinc organometallics.

7.2.2
Preparation of Polyfunctional Organozinc Halides

7.2.2.1 Preparation by the Oxidative Addition to Zinc Metal

The oxidative addition of zinc dust to functionalized organic halides allows the preparation of a broad range of polyfunctional organozinc iodides such as **1–5** [10–14]. Several functional groups such as nitro or azide groups inhibit the radical-transfer reaction leading to the zinc reagent. On another hand, hydroxyl groups form zinc alkoxides that coat the zinc surface and therefore hamper the reaction (a similar behavior was observed for other acidic hydrogen atoms (carboxylic acids, imidazoles)). As a general rule, the nature of the zinc dust is less important than its activation. Finely cut zinc foil or zinc dust (commercially available source (ca. 325 mesh)) can be used. Zinc slowly oxidizes in air and is covered by an oxide layer. Its activation is of great importance; this is done by removing the oxide layer via chemical methods. A very efficient procedure consists of treating zinc with 1,2-dibromoethane (5 mol%) in THF (reflux for 0.5 min) followed by the addition of Me_3SiCl (1–2 mol%; reflux for 0.5 min) [10, 15–17].

$$FG-R-I \xrightarrow[\text{THF}]{\text{Zn dust}} FG-R-ZnI$$

FG = CO_2R, enoate, CN, halide, $(RCO)_2N$, $(TMS)_2N$, RNH, NH_2, RCONH, $(RO)_3Si$, $(RO)_2PO$, RS, RSO, RSO_2, PhCOS

R = alkyl, aryl, benzyl, allyl

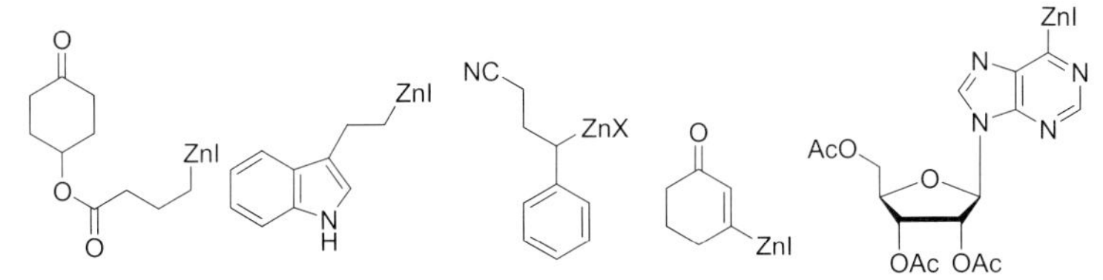

Scheme 7.3 Functionalized organozinc compounds prepared by oxidative addition.

Under these conditions, a broad range of polyfunctional alkyl iodides are converted to the corresponding organozinc halides in high yields [6]. In the case of primary alkyl iodides, the insertion occurs at 40–50 °C, whereas secondary alkyl iodides already react at 25–30 °C. Secondary alkyl bromides also react under these conditions [18], but primary alkyl bromides are usually inert with this type of activation and much better results are obtained by using Rieke-zinc [19–21]. Thus, the reduction of zinc chloride with finely cut lithium and naphthalene produces within 1.5 h highly reactive zinc (Rieke-zinc).

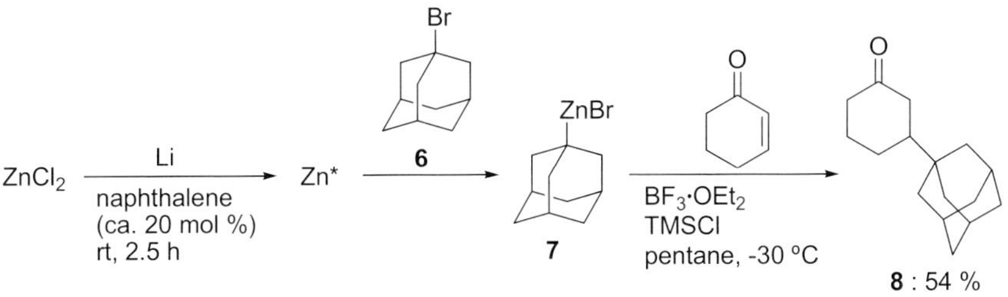

Scheme 7.4 Preparation of tertiary alkylzinc reagents using Rieke-zinc.

This activated zinc [22] readily inserts in secondary and tertiary alkyl bromides. Adamantly bromide (**6**) is converted into the corresponding organozinc reagent (**7**) and its reaction with cyclohexenone in the presence of $BF_3 \cdot OEt_2$ and TMSCl furnishes the 1,4-addition product **8** (Scheme 7.4). Rieke-zinc proves also to be very useful for preparing aryl- and heteroaryl-zinc halides. Thus, the reaction of Rieke-zinc with *p*-bromobenzonitrile (**9**) in refluxing THF provides after 3 h the corresponding zinc reagent (**10**), which is benzoylated leading to the ketone **11** in 73% yield (Scheme 7.5) [18a].

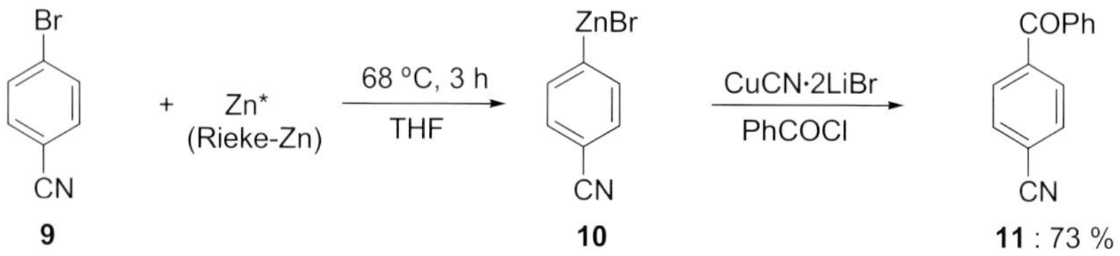

Scheme 7.5 Preparation of functionalized arylzinc bromides using Rieke-zinc.

Interestingly, many electron-deficient heterocyclic and aryl bromides or iodides are sufficiently activated to react with commercially available zinc powder [14]. In the case of benzylic halides, bromides and even chlorides can be used [12]. Thus, for the functionalized benzylic bromide **12a**, the formation of the corresponding benzylic zinc bromide (**13a**) by the direct insertion of zinc dust is complete within 2 h at 5 °C.

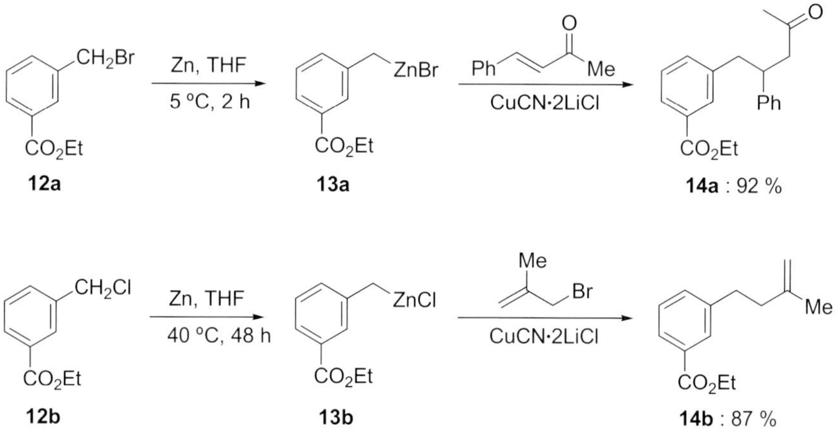

Scheme 7.6 Preparation of functionalized benzylic zinc reagents.

After a Michael addition, the expected conjugated addition product **14a** is formed in 92% yield. The corresponding benzylic chloride (**12b**) requires a reaction time of 48 h leading to the benzylic zinc chloride (**13b**). Allylation of **13b** provides the aromatic benzoate **14b** in 87% yield (Scheme 7.6). The use of DMSO/THF mixture has a favorable effect allowing the synthesis of substituted benzylic reagents such as **3** [9]. Similarly, the preparation of alkylzinc iodides is facilitated if the reaction is performed in THF and NMP mixtures. Such solutions of $MeO_2C(CH_2)_4ZnI$ add to benzaldehyde in the presence of TMSCl (2 equiv) in 70% yield [23]. The use of ultrasound also promotes the formation of organozinc compounds [24]. This procedure proved to be especially useful for the preparation of the Jackson reagent (**15**) derived from serine. The reaction of this zinc derivative with various electrophiles, either in the presence of a copper(I) or palladium(0) catalyst, leads to products of type **16–18** [25]; Scheme 7.7.

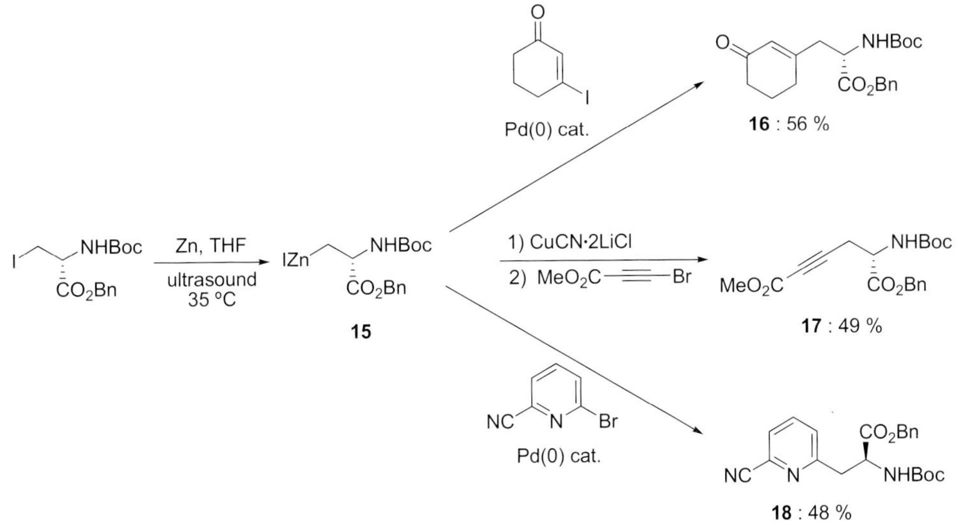

16 : 56 %

15

1) CuCN·2LiCl
2) MeO_2C——Br

17 : 49 %

Pd(0) cat.

18 : 48 %

Zn, THF
ultrasound
35 °C

Pd(0) cat.

Scheme 7.7 Ultrasound-mediated preparation of the Jackson reagent **15**.

19   **20**   **21**

19a   **19b**

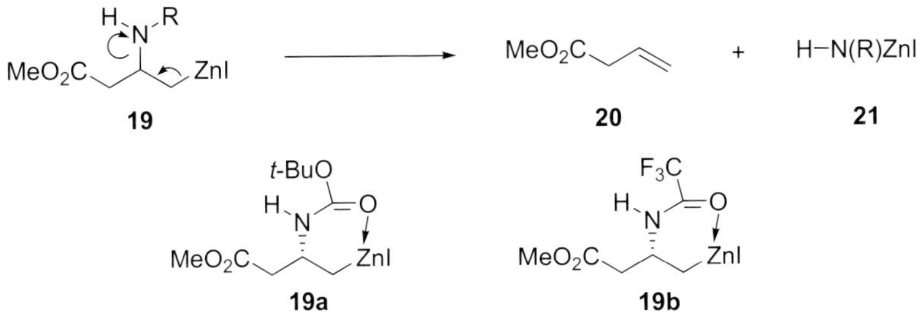

Scheme 7.8 Stability of α-amino alkylzinc reagents.

The decomposition of the zinc reagent **19** leading to methyl but-3-enoate **20** and the zinc amide **21** has been extensively studied by Jackson et al. [26]. It was found that the zinc species **19a** undergoes the elimination ca. three times faster than the zinc reagent **19b**. This might be surprising since -NHBoc is not as good a leaving group as -NHCOCF$_3$. This may be explained by the chelation of the Boc-group with the zinc metallic center, which enhances the ate-character of the metal as well as the electron-density of the C–Zn bond and that favors therefore the elimination. This factor seems to be more important than the leaving-group ability of -NHR [27]. Interestingly, a free phenolic function is tolerated in cross-coupling reactions.

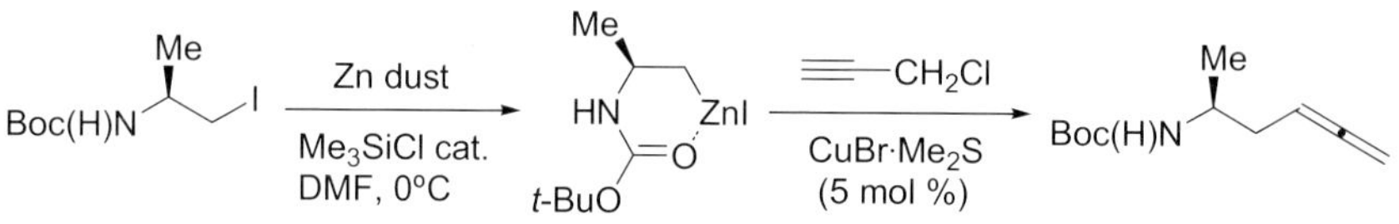

Scheme 7.9 Generation of a β-amino alkylzinc reagent in DMF.

Organozinc reagents bearing a free NH-function in β-position such as **22** can be readily prepared by the direct insertion of zinc dust previously activated with TMSCl in DMF in to the corresponding β-iodoamino derivative **23**. Interestingly, the best reactivity of this chelate-stabilized zinc species can be obtained by using catalytic amounts of CuBr·Me$_2$S (5 mol%). In the case of the reaction with propargyl chloride the corresponding allene **24** is obtained in 60% yield via an S$_N$2'-mechanism (Scheme 7.9) [28].

During the preparation of allylic zinc reagents, the formation of Wurtz-coupling products may be observed, especially if the intermediate allylic radical is well stabilized. However, the direct insertion of zinc foil to allyl bromide in THF at 5 °C is one of the best methods for preparing an allylic anion equivalent. Allylic zinc reagents are more convenient to prepare and to handle than their magnesium- and lithium counterparts [15]. Similarly, electron-rich benzylic bromides such as **25a** often lead to homo-coupling products. The use of the corresponding phosphate **25b** and catalytic amounts of LiI in dimethyltetrahydropyrimidinone (DMPU) provides the corresponding zinc reagent in quantitative yield (Scheme 7.10) [29]. The presence of LiI generates small concentrations of the benzylic iodide, which is converted to the zinc reagent. Little homo-coupling is observed under these conditions.

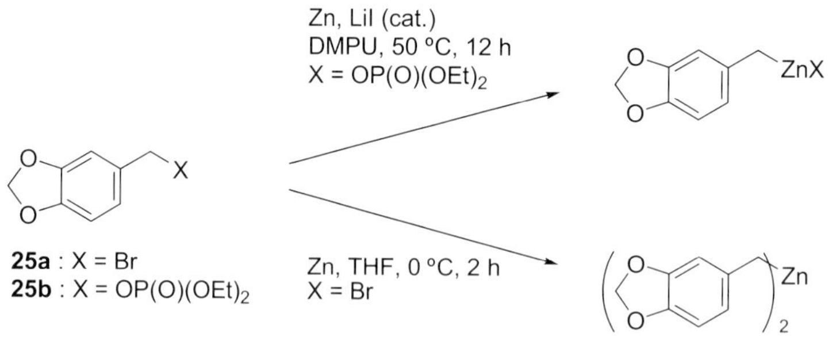

Scheme 7.10 Importance of the precursor for the preparation of benzylic zinc reagents.

The addition of lithium iodide and bromide mixtures allows also the performance of the zinc insertion with primary alkyl chlorides, tosylates or mesylates as starting material (Scheme 7.11) [29]. Thus, the alkyl tosylate **26** is converted in

N,N-dimethylacetamide (DMAC) in the presence of lithium iodide (0.2 equiv) and lithium bromide (1.0 equiv) after heating at 50 °C for 12 h to the zinc organometallic **27**. After transmetallation with CuCN · 2LiCl, the zinc reagent **27** undergoes an addition-elimination to 3-iodo-2-cyclohexenone leading to the enone **28** in 85% yield. The addition of both lithium iodide and lithium bromide is necessary in order to observe fast reactions. The direct exchange of a sulfonate to the corresponding iodide with LiI is slow and a stepwise reaction first with LiBr leading to the corresponding alkyl bromide, then a reaction with LiI leading to the corresponding iodide is a faster reaction pathway. The chloroalkyl mesylate **29** is converted to the zinc species **30**, which undergoes a substitution reaction with the unsaturated nitro derivative **31** leading to the tetra-substituted nitroolefin **32** in 85% yield; Scheme 7.11 [29, 30].

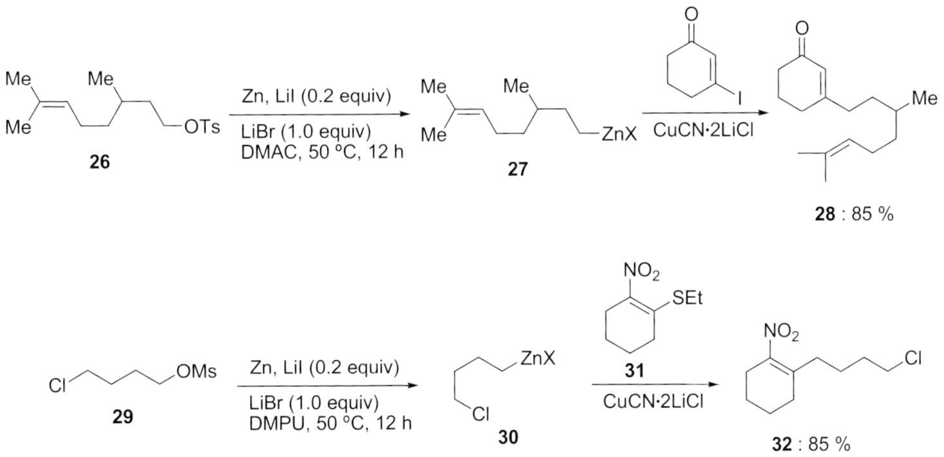

Scheme 7.11 Preparation of alkylzinc derivatives starting from alkyl sulfonates.

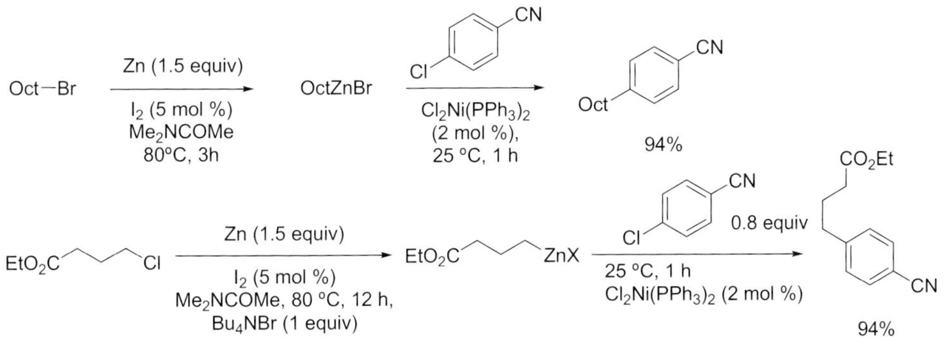

Scheme 7.12 Iodine-catalyzed formation of organozinc bromides.

Alkylzinc bromides bearing various functional groups [31] can be readily prepared by the direct insertion of zinc metal (dust, powder or shot) to alkyl bromides by performing the reaction in the presence of iodine (1–5 mol%) in a polar solvent like DMAC. It is also possible to use alkyl chlorides as starting material. In this case, the reaction is best performed in presence of Bu_4NBr (1 equiv). The resulting zinc reagent undergoes a smooth Ni-catalyzed cross-coupling [32] with various aryl chlorides (Scheme 7.12). For polyfluorinated organozinc halides, the zinc insertion is conveniently done with a zinc–copper couple (Scheme 7.13) [33,34]. The preparation of trifluoromethylzinc halides (**33**) is best achieved using the method of Burton [35], which involves the reaction of CF_2Cl_2 or CBr_2F_2 with zinc in DMF. This reaction produces a mixture of CF_3ZnX (**33**) and bis-trifluoro-methylzinc (**34**); Scheme 7.13 [35].

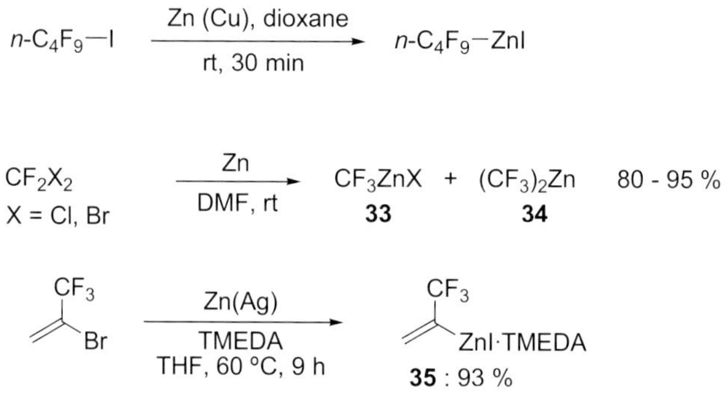

$$n\text{-}C_4F_9\text{---I} \xrightarrow[\text{rt, 30 min}]{\text{Zn (Cu), dioxane}} n\text{-}C_4F_9\text{---ZnI}$$

$$\begin{array}{c} CF_2X_2 \\ X = Cl, Br \end{array} \xrightarrow[\text{DMF, rt}]{\text{Zn}} \begin{array}{c} CF_3ZnX \\ \mathbf{33} \end{array} + \begin{array}{c} (CF_3)_2Zn \\ \mathbf{34} \end{array} \quad 80 \text{ - } 95\,\%$$

$$\underset{\text{Br}}{\overset{CF_3}{\diagup\!\!\diagdown}} \xrightarrow[\substack{\text{TMEDA} \\ \text{THF, 60 °C, 9 h}}]{\text{Zn(Ag)}} \underset{\mathbf{35}\,:\,93\,\%}{\overset{CF_3}{\diagup\!\!\diagdown}}\text{ZnI·TMEDA}$$

Scheme 7.13 Preparation of perfluorinated zinc reagents.

Interestingly, the presence of a CF_3-substituent facilitates considerably the zinc insertion. Thus, 2-bromo-trifluoropropene reacts with Zn/Ag couple in the presence of TMEDA leading to the expected zinc reagent **35** in 93% yield [35c–e].

The formation of arylzinc reagents can also be accomplished by using electrochemical methods. With a sacrificial zinc anode and in the presence of nickel 2,2-bipyridyl, polyfunctional zinc reagents of type **36** can be prepared in excellent yields (Scheme 7.14) [36]. An electrochemical conversion of aryl halides to arylzinc compounds can also be achieved by a cobalt catalysis in DMF/pyridine mixture [37]. The mechanism of this reaction has been carefully studied [38]. This method can also be applied to heterocyclic compounds such as 2- or 3-chloropyridine and 2- or 3-bromothiophenes ([39] and [36d,e]). Zinc can also be electrochemically activated and a mixture of zinc metal and small amounts of zinc formed by electroreduction of zinc halides are very reactive toward α-bromoesters and allylic or benzylic bromides [36f,g].

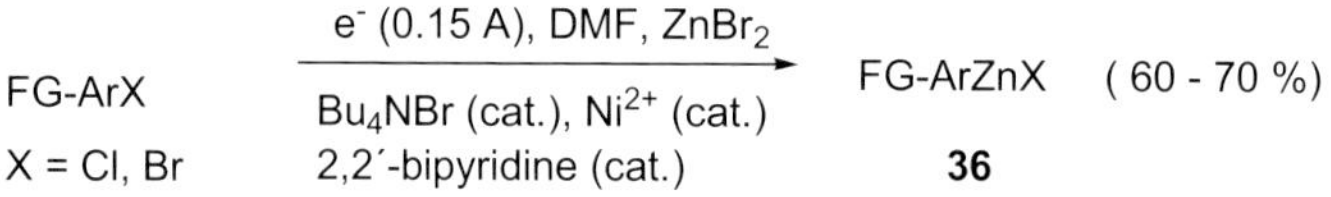

$$\begin{array}{c} \text{FG-ArX} \\ X = Cl, Br \end{array} \xrightarrow[\substack{Bu_4NBr \text{ (cat.), Ni}^{2+} \text{ (cat.)} \\ \text{2,2'-bipyridine (cat.)}}]{e^-\,(0.15\,A),\,\text{DMF, }ZnBr_2} \begin{array}{c} \text{FG-ArZnX} \\ \mathbf{36} \end{array} \quad (60\text{ - }70\,\%)$$

Scheme 7.14 Electrochemical preparation of polyfunctional arylzinc halides.

The previous results suggest that transition metals may catalyze the zinc insertion reaction. This proves to be the case and the reaction of octyl iodide with Et$_2$Zn in the presence of PdCl$_2$(dppf) (1.5 mol%) in THF at 25 °C produces OctZnI within 2 h of reaction time in 75–80% yield [40]. A detailed mechanism is given in Scheme 7.15.

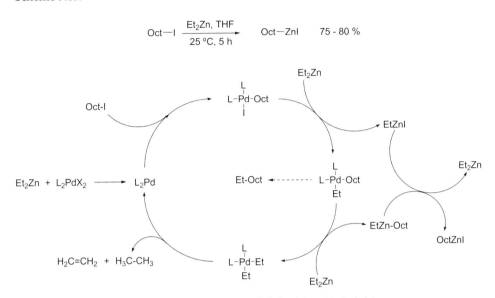

Scheme 7.15 Mechanism of the Pd-catalyzed reaction of alkyl iodides with diethylzinc.

These palladium- or nickel-catalyzed reactions are radical reactions leading to an organometallic product. By using a precursor such as **37** as a 1:1 mixture of diastereoisomers, the palladium-catalyzed cyclization provides, in a stereoconvergent way, the cyclopentylmethylzinc derivative **38** that after allylation produces the unsaturated ester **39** in 71% yield [41]. The intermediate radical cyclizes via a transition state **A** where all the substituents are in an equatorial position. Interestingly, the analogous reaction using Ni(acac)$_2$ as a catalyst allows the preparation of heterocyclic compounds such as **40**. The relative stereochemistry of up to three contiguous centers is set up in this cyclization (Scheme 7.16) [41].

Application of these methods toward the preparation of natural products, such as (−)-methylenelactocin (**41**) or *cis*-methyl jasmonate (**42**) has been accomplished [42]; Scheme 7.17.

This reaction can be applied to the preparation of benzylic zinc reagents [40a, 43]. A range of benzylic halides has been reduced with Et$_2$Zn in the presence of Pd(PPh$_3$)$_4$ as a catalyst [43]. Other metallic salts catalyze the I/Zn-exchange reaction. Thus, mixed-metal catalysis using manganese(II) bromide and copper(I) chloride allows the performance of a Br/Zn-exchange with various functionalized alkyl bromides of type **43**; Scheme 7.18 [44]. The reaction proceeds in a polar solvent such as DMPU [45] under very mild conditions.

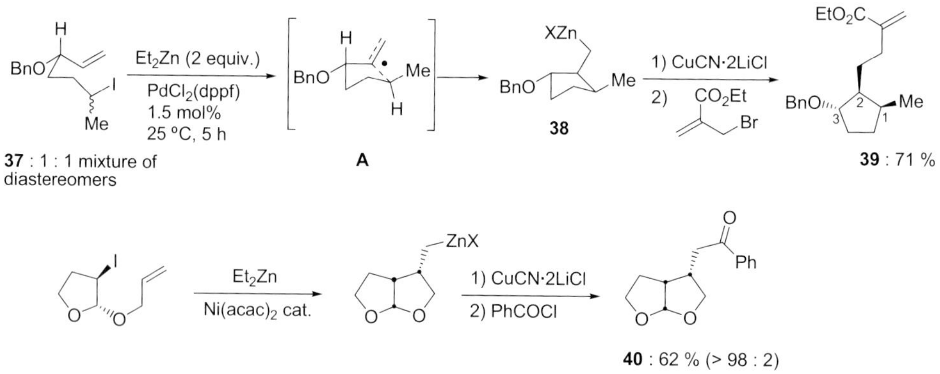

Scheme 7.16 Pd- and Ni-catalyzed radical cyclization leading to zinc organometallics.

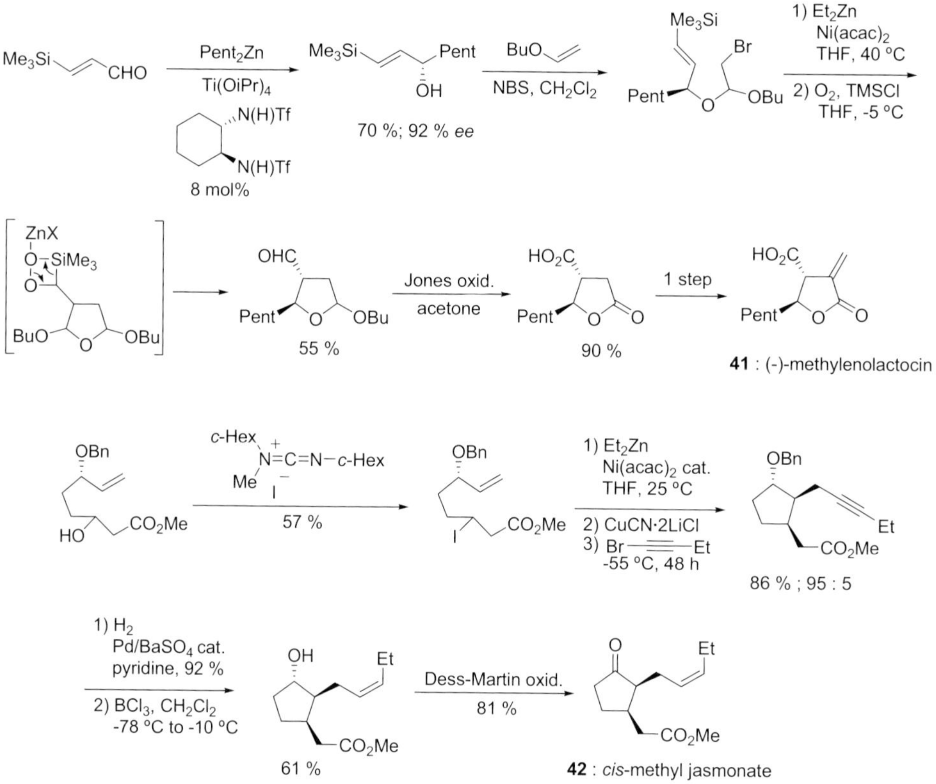

Scheme 7.17 Preparation of (−)-methylenolactocin (**41**) and *cis*-methyl jasmonate (**42**).

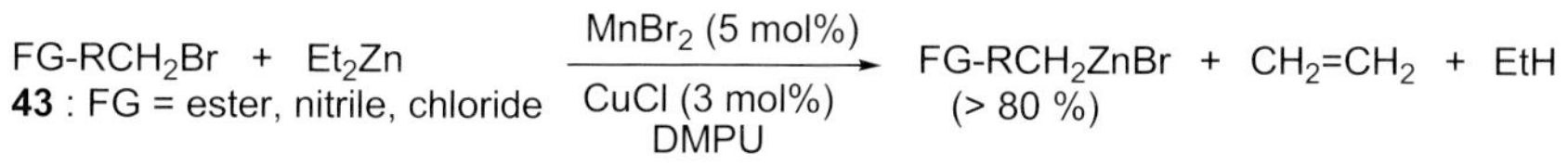
FG-RCH2Br + Et2Zn
43 : FG = ester, nitrile, chloride
MnBr2 (5 mol%)
CuCl (3 mol%)
DMPU
FG-RCH2ZnBr + CH2=CH2 + EtH
(> 80 %)

Scheme 7.18 Mixed Mn/Cu-catalyzed alkylzinc bromides synthesis.

Interestingly, low-valent cobalt species obtained by the *in situ* reduction of $CoBr_2$ with zinc catalyze the reaction of aryl bromides with zinc dust. The reaction allows the preparation of a range of functionalized arylzinc halides such as **44** (Scheme 7.19) [45].

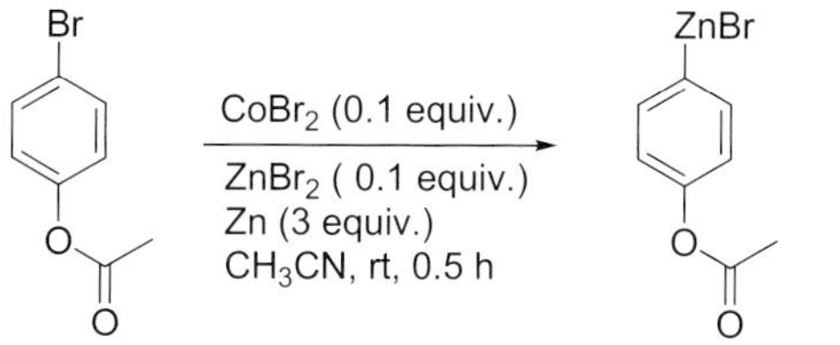
Br
CoBr2 (0.1 equiv.)
ZnBr2 (0.1 equiv.)
Zn (3 equiv.)
CH3CN, rt, 0.5 h
ZnBr
O
O
O
O

44 : > 80 %

Scheme 7.19 Cobalt-mediated insertion of zinc to aryl bromides.

In summary, the direct insertion of zinc dust to organic halides is an excellent method for preparing a broad range of polyfunctional organozinc halides bearing various functional groups like an ester [46], an ether, an acetate [47], a ketone, cyano [48], halide [49], *N,N*-bis(trimethylsilyl)amino [50], primary and secondary amino, amide phthalimide [51], sulfide, sulfoxide and sulfone [52], boronic ester [53], enone [54] or a phosphonate [55]. An alternative method is based on transmetallation reactions.

7.2.2.2 Preparation of Organozinc Halides using Transmetallation Reactions

A number of transmetallation procedures leading to zinc organometallics can be performed. Many organometallics having a polar C–Met bond are readily transmetallated by the reaction with a zinc salt to the more covalent organozinc compounds. The synthetic scope of these transmetallations depends on the availability of the starting organometallic species and on its compatibility with functional groups. Although organolithiums are highly reactive organometallic species, it is possible to prepare aryllithium species bearing cyano groups [56] or nitro groups [57] at very low temperature (–100 °C to –90 °C). By performing a halogen–lithium exchange reaction, followed by a transmetallation with $ZnBr_2$, functionalized organometallics are prepared that cannot be obtained by the insertion of zinc to the corresponding organic halide. Azides inhibit the direct zinc insertion to an organic halide. However, the reaction of the alkenyl iodide **45** bearing an azide group with *n*-BuLi at –100 °C followed by the transmetallation with $ZnBr_2$ in THF at –90 °C provides the expected zinc reagent **46** in > 85% yield (Scheme 7.20) [58].

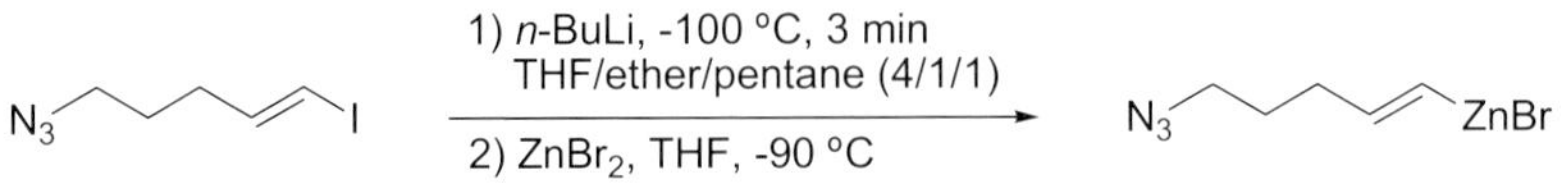

Scheme 7.20 Preparation of an alkenylzinc reagent bearing an azide function.

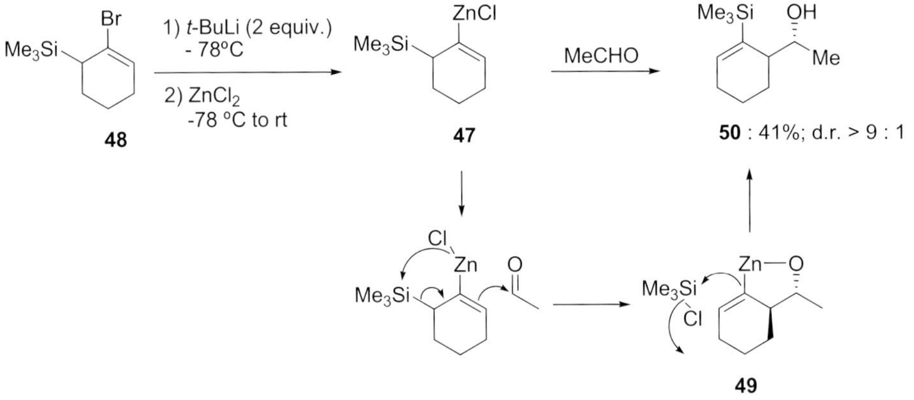

Scheme 7.21 Preparation of a zinc/silicon bimetallic of type **47**.

The mixed 1,2-bimetallic Zn/Si-reagent **47** is a versatile species that reacts with aldehydes in high diastereoselectivity [59]. It is prepared by a bromine–lithium-exchange reaction starting from **48** followed by a transmetallation with ZnCl$_2$. The reaction with acetaldehyde is leading initially to the alkenylzinc species **49**, which reacts with Me$_3$SiCl, providing the alkenylsilane **50** in 41% yield and a diastereoselectivity > 9:1.

2-Lithiated oxazoles are unstable and readily undergo a ring opening to the tautomeric isonitriles. This ring cleavage can be avoided by preparing the corresponding 2-zincated oxazole (**51**) that is much more stable toward a fragmentation reaction (Scheme 7.22) [60]. The lithiation of the *O*-vinyl carbamate (**52**) with *sec*-BuLi followed by transmetallation with zinc bromide provides the convenient acyl anion derivative **53** which undergoes smooth Pd(0)-catalyzed cross-coupling reactions; Scheme 7.22 [61].

This reaction sequence has been extended to lithium enolates. The deprotonation of the aminoester **54** with LDA followed by a transmetallation with zinc bromide in ether furnishes a zinc enolate that readily adds to the double bond providing the proline derivative **55** in high diastereoselectivity and enantioselectivity (Scheme 7.23) [62].

Similarly, zincated hydrazone derivatives of type **56** undergo an intermolecular carbozincation of strained cyclopropene rings such as **57** leading to the adduct **58** with 92% yield [63]. This type of addition can be extended to ethylene [63c]. It proceeds with an excellent stereoselectivity allowing the enantioselective synthesis of α-substituted ketones. Allylic zinc species also add to cyclopropenone acetals allowing an enantioselective allylzincation to take place [63f]. This reaction provides an entry to quaternary centers with good stereocontrol. Fluorine-substituted alkenes can be readily lithiated by the reaction with a strong base (Scheme 7.24).

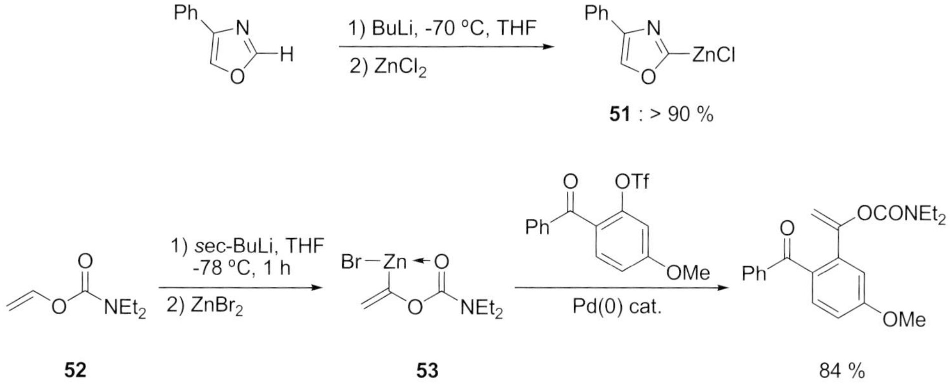

Scheme 7.22 Preparation of zinc reagents via a I/Li transmetallation sequence.

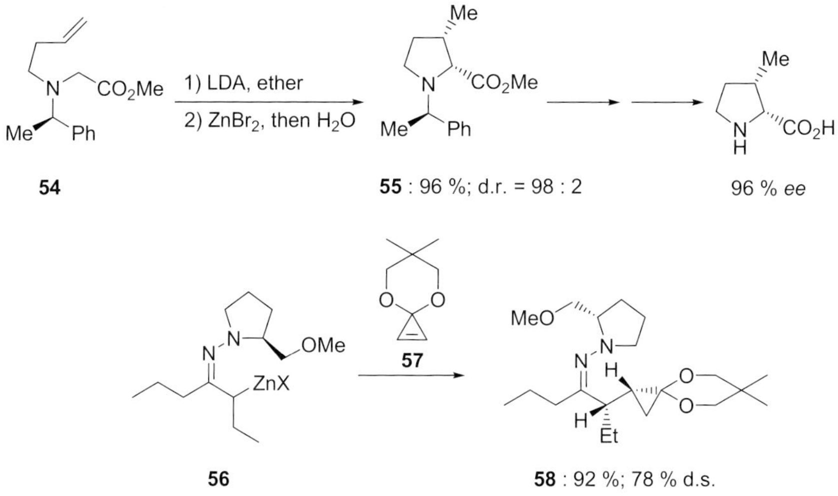

Scheme 7.23 Conversion of zinc enolates to organozinc reagents.

1-Chloro-2,2-difluorovinylzinc chloride **59** opens the access to a range of fluorine-containing molecules via cross-couplings. Normant and coworkers have prepared this zinc reagent by the deprotonation of 1-chloro-2,2-difluoroethene **60** and transmetallation [64]. These two steps can be combined in one and the lithiation of **60** with *sec*-BuLi in the presence of ZnCl$_2$ provides the corresponding dialkylzinc **61** as a colorless clear solution [65]. Percy and coworkers [66] and Anilkumar and Burton [67] reported that the deprotonation of 1-chloro-2,2,2-trifluoroethane **62** produces, after elimination and transmetallation, the zinc reagent **59**. Especially convenient is the deprotonation of liquid halothane (**63**) with *sec*-BuLi in the presence of ZnCl$_2$.

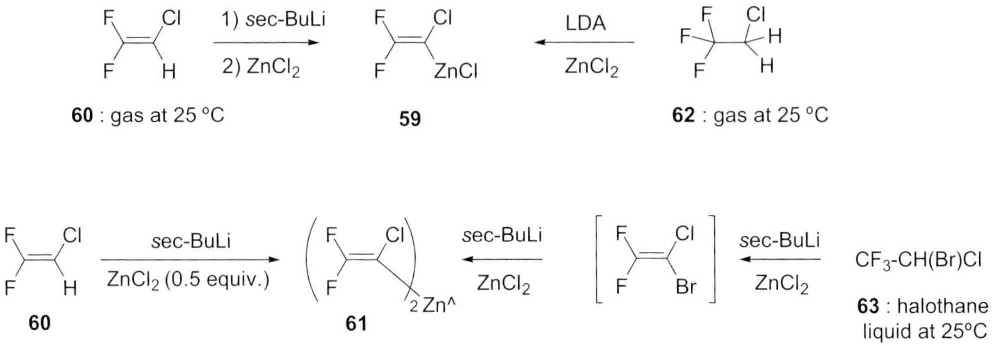

Scheme 7.24 Preparation of fluoro-substituted alkenylzincs via lithium intermediates.

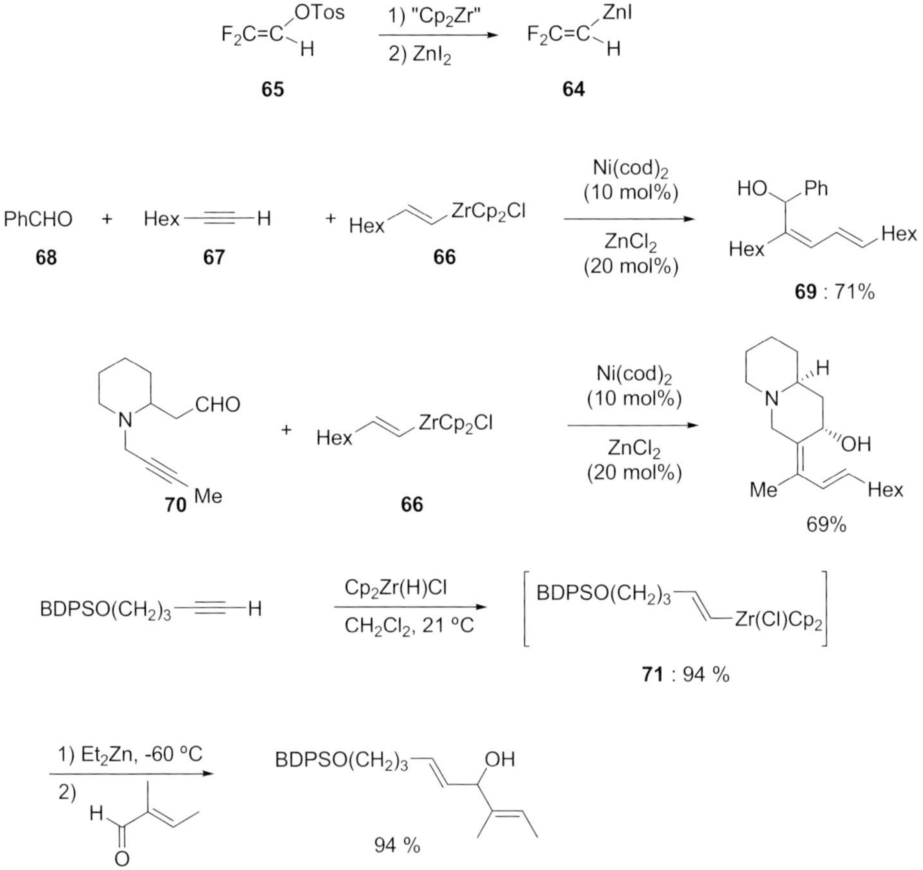

Scheme 7.25 Alkenylzinc species obtained from alkenylzirconium derivatives.

Transmetallations starting from alkenylzirconium species that are obtained by hydrozirconation using H(Cl)ZrCp$_2$ are readily accomplished [68,69]. Ichikawa and Minami have elegantly shown that diflurovinylzinc iodide **64** is obtained by the addition of "ZrCp$_2$" to the alkenyl tosylate **65** [68]. *In situ* transmetallation reactions have also been reported.

A new three-component reaction [70] has been made possible by treating an alkenylzirconium reagent of type **66** with an alkyne **67** and an aldehyde **68** in the presence of catalytical amounts of Ni(cod)$_2$ (10 mol%) and ZnCl$_2$ (20 mol%). The resulting pentadienyl alcohols like **69** are obtained in satisfactory yield. The transmetallation of the alkenylzirconium species **66** to the corresponding zinc species is essential for the success of the carbometallation reaction [71]. An intramolecular version of the reaction is possible showing the high affinity of the intermediate alkenylzinc derived from **66** for adding to the alkyne **70**. The competitive alternative addition to the aldehyde is not observed. The most general application has been reported by Wipf and Xu [69] who have demonstrated that a range of alkenylzirconium species is readily transmetallated to zinc organometallics. Thus, the reaction of the alkenylzirconium **71** with Et$_2$Zn produces a zinc reagent that adds to an unsaturated aldehyde furnishing the expected allylic alcohol in excellent yield (Scheme 7.25) [69]. Organotin compounds have also occasionally been converted to zinc and then copper compounds by generating first an organolithium derivative (Scheme 7.26) [72]. α-Aminostannanes of type **72** undergo a low temperature Sn/Li-exchange reaction with BuLi in THF and lead after a transmetallation to an organozinc species displaying a moderate reactivity. After a further transmetallation with CuCN·2LiCl, a copper–zinc species such as **73** is obtained. The reaction of **73** with electrophiles affords products of type **74** with variable enantioselectivities. Quenching of the copper–zinc reagent **73** with reactive electrophiles proceeds with retention of configuration with up to 95% *ee*.

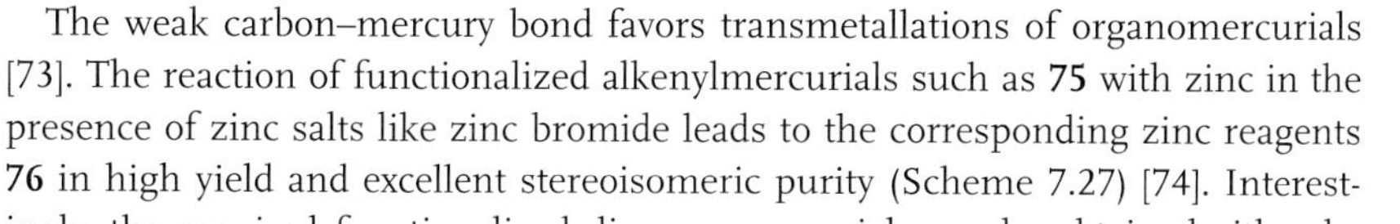

Scheme 7.26 Preparation of zinc organometallics starting from tin reagents.

The weak carbon–mercury bond favors transmetallations of organomercurials [73]. The reaction of functionalized alkenylmercurials such as **75** with zinc in the presence of zinc salts like zinc bromide leads to the corresponding zinc reagents **76** in high yield and excellent stereoisomeric purity (Scheme 7.27) [74]. Interestingly, the required functionalized diorganomercurials can be obtained either by the reaction of functionalized alkylzinc iodides with mercury(I) chloride or by methylene homologation reaction using (ICH$_2$)$_2$Hg (Scheme 7.27) [74].

A range of polyfunctional organomagnesium species are available via an iodine– or a bromine–magnesium exchange reaction [75]. Since the carbon-magnesium bond is less polar than a carbon–lithium bond, considerably more func-

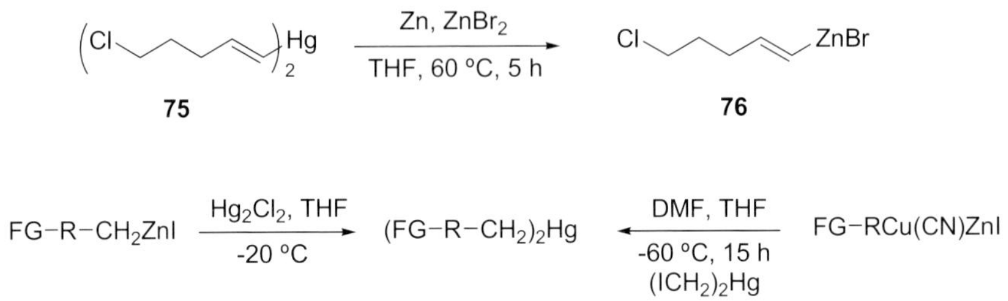

Scheme 7.27 Preparation of alkenylzinc bromides from organomercurials.

tional groups are tolerated in these organometallics and experimentally more convenient reaction conditions can be used. Thus, the reaction of the aryl iodide **77** with *i*-PrMgBr in THF at –10 °C for 3 h provides an intermediate magnesium reagent that after transmetallation with ZnBr₂ furnishes the zinc reagent **78**. Its palladium-catalyzed cross-coupling with the bromofurane **79** provides the cross-coupling product **80** in 52% yield; Scheme 7.28 [76].

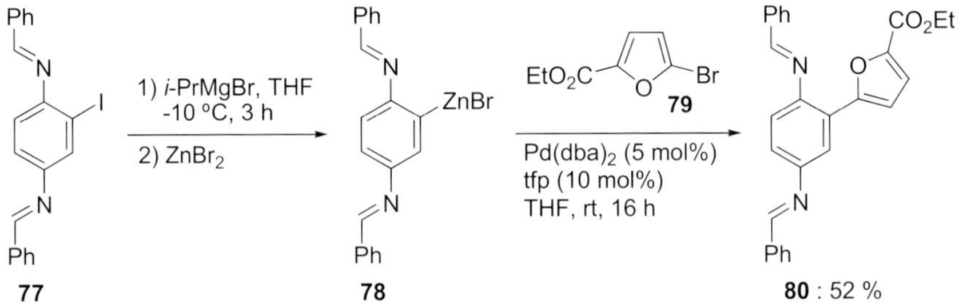

Scheme 7.28 Preparation of arylzinc halides via an I/Mg-exchange.

This exchange reaction allows the preparation of various zinc reagents bearing numerous functional groups [75]. Special methods are available for the preparation of organozinc halides such as insertion reactions using zinc carbenoids like (iodomethyl)zinc iodide [77]. The reaction of an organocopper reagent (**81**) with ICH₂ZnI

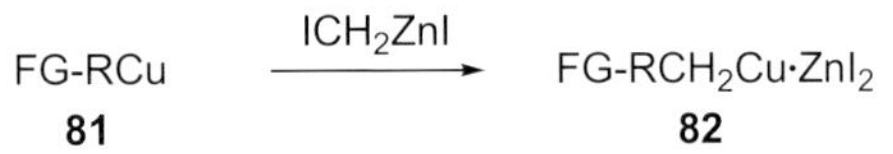

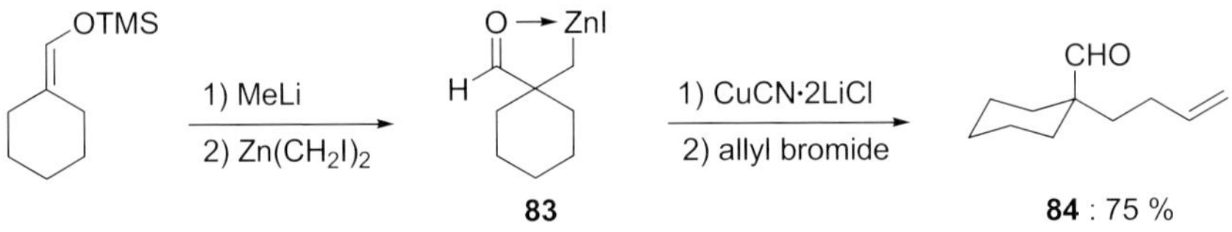

Scheme 7.29 Homologation of zinc enolate using a zinc carbenoid.

provides a copper–zinc reagent **82** that reacts with numerous electrophiles (Scheme 7.29) [78]. This reaction is quite general and allows for example to homologate lithium enolates leading to the zinc homoenolate **83** that can be readily allylated in the presence of a copper catalyst leading to the aldehyde **84** in 75% yield (Scheme 7.29) [78, 79]. A one-pot synthesis of γ-butyrolactones such as **85** can be realized via an intermediate allylic zinc species **86** generated by the homologation of an alkenylcopper reagent obtained by the carbocupration of ethyl propiolate (Scheme 7.30).

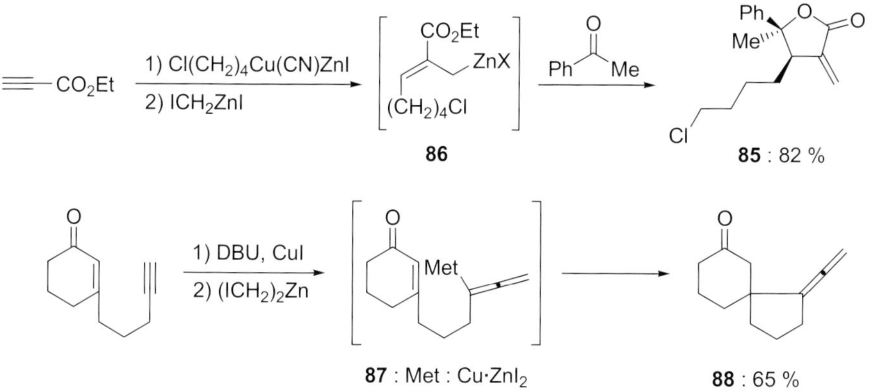

Scheme 7.30 Allylic and allenic zinc reagents via methylene homologation.

Interestingly, an intramolecular trapping of an allenylzinc–copper species **87** generated by the homologation of an alkynylcopper can be achieved leading to the spiro-product **88** in 65% yield (Scheme 7.30) [79]. This method allows the homologation of silylated lithium carbenoids [80] and can be extended to the performance of polymethylene homologations [78c]. The use of lithium tributylzincates allows the synthesis of alkylated allenylzinc species such as **89** that react with aldehydes leading to alcohols of type **90** with high diastereoselectivity (Scheme 7.31) [81, 82].

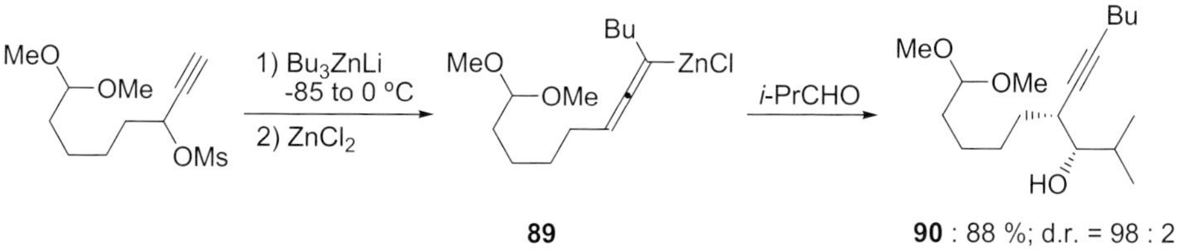

Scheme 7.31 Preparation of allenic zinc reagents using lithium trialkylzincates.

The reaction of stereomerically well-defined alkenylcopper species **91** obtained by a carbocupration with $(ICH_2)_2Zn$ leads to a selective double methylene insertion providing the chelate-stabilized alkylzinc reagent **92** that leads, after deuteration with D_2O, to the unsaturated sulfoxide **93** in 80% yield. This method has been elegantly extended by Marek (Scheme 7.32) [83].

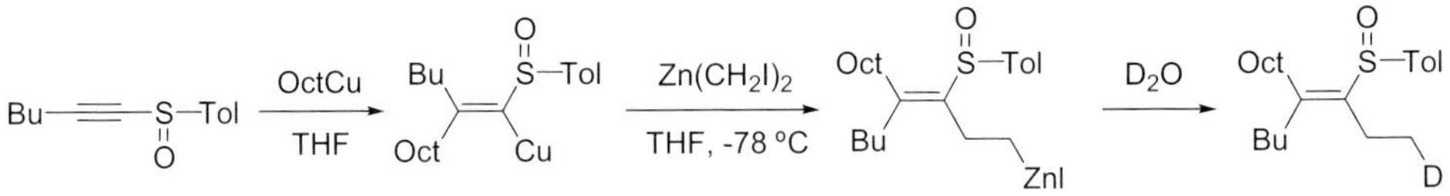

Scheme 7.32 Double homologation of alkenyl sulfoxide derivatives.

A selective reaction of 1,4-bimetallic alkanes with CuCN · 2LiCl allows the preparation of a range of new polyfunctional zinc–copper reagents [84]. Thus, the reaction of 1,4-dizincated butane (**94**) with CuCN · 2LiCl, followed by cyclohexenone in the presence of TMSCl (2 equiv) provides the new zinc–copper reagent **95** that reacts with 3-iodo-2-cyclohexenone furnishing the diketone **96** in 64% yield (Scheme 7.33) [84].

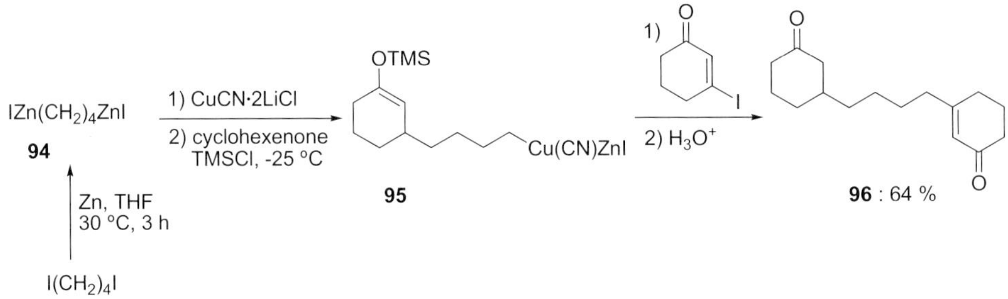

Scheme 7.33 Preparation of alkylzinc–copper reagents via the selective reaction of 1,4-dizincabutane.

The preparation of chiral alkylzinc halides by the direct insertion of zinc is complicated due to the radical nature of the zinc insertion. Nevertheless, the strained secondary alkyl iodide **97** is converted to the corresponding chiral secondary organozinc reagent (**98**) with high retention of configuration leading, after stannylation with Me$_3$SnCl, to the tin derivative **99** in 72% yield [85]. Interestingly, the *trans*-β-iodoester **100** is stereoselectively converted to the *cis*-ester **101** leading, after acylation, to the amino-ketone **102** [86]. The chelation with the ester group may be responsible for the *cis*-configuration of the zinc reagent **101** (Scheme 7.34). ^{1}H-NMR-studies [87] confirm that secondary dialkylzincs should display a high configurational stability, although it was noticed that the presence of an excess of zinc(II) salts epimerizes secondary alkylzinc reagents [87].

The importance of chelation for the configurational stability of organozinc reagents has been recently demonstrated by Normant and Marek. The reaction of the bimetallic reagent **103**, prepared by the allylzincation of the alkenyllithium **104**, leads after stereoselective sequential quenching with two electrophiles (MeOD and I$_2$), to the primary iodide **105** as a 34:66 mixture of diastereomers (Scheme 7.34) [88]. Also, alkenylzinc reagents such as **106** display a relatively high configurational stability (little racemization is observed at –65 °C (2 h)). A kinetic resolution with (*R*)-mandelic imine derivate **107** is possible. Thus, treatment of a racemic mixture of the zinc reagent **106** with half of an equivalent of the imine **107** led to a highly preferential reaction with the (S_a)-enantiomer of **106** leading to the adduct **108**. The remaining unreacted (R_a)-**106** can now be trapped with an aldehyde such as pivalaldehyde giving the chiral homopropargylic alcohol **109** in 75% yield and 87.5% *ee* [89]. This kinetic resolution has been used to prepare *anti,- anti*-vicinal amino diols in > 95% *ee* and d.r. > 40:1 [90]. In general diorganozincs are more easily prepared in optically pure form. This will be discussed in detail in the next section. On the other hand, secondary zinc reagents prepared by the direct zinc insertion to secondary alkyl iodides are obtained without stereoselectivity [91].

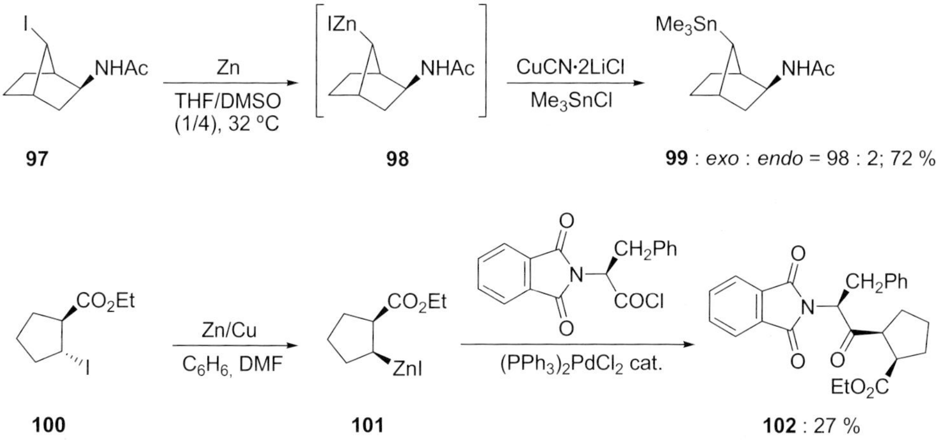

Scheme 7.34 Stereoselective preparation of secondary alkyl-zinc iodides by the direct insertion of zinc.

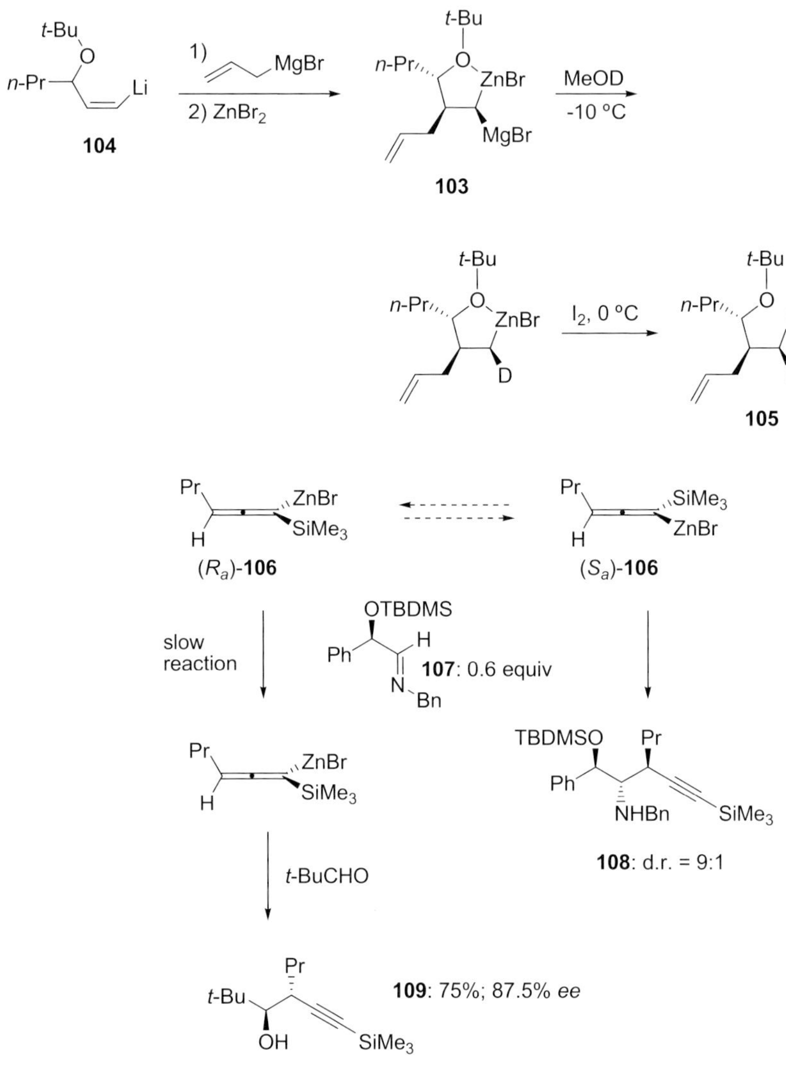

Scheme 7.35 Synthesis of chiral organozinc halides.

7.2.3
Preparation of Diorganozincs

7.2.3.1 Preparation via an I/Zn Exchange

Diorganozincs are usually more reactive toward electrophiles than organozinc halides. A wide range of methods is available for their preparation. The oldest being the direct insertion of zinc to an alkyl halide (usually an alkyl iodide) leading to an alkylzinc intermediate that, after distillation, provides the liquid diorganozinc (R_2Zn) [1]. This method is applicable only to nonfunctionalized diorganozincs bearing lower alkyl chains (up to hexyl) due to the thermic instability of higher homologs. The I/Zn exchange reaction using Et_2Zn allows the preparation of a broad range of diorganozincs. The ease of the exchange reaction depends on the stability of the newly produced diorganozincs. Thus, diiodomethane smoothly reacts with Et_2Zn in THF at –40 °C providing the corresponding mixed ethyl(iodomethyl)zinc reagent (**110**) with quantitative yield [92]. The I/Zn exchange is catalyzed by the addition of CuI (0.3 mol%). After the evaporation under vacuum of excess Et_2Zn and ethyl iodide formed during the reaction, the resulting diorganozincs **111a–d** are obtained in excellent yields; Scheme 7.36 [93].

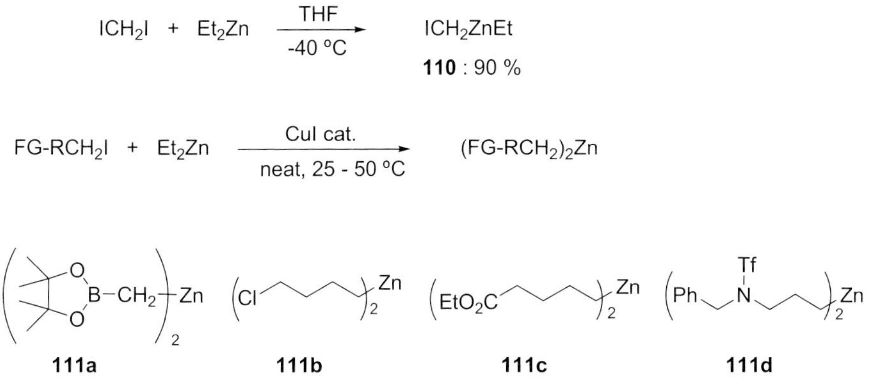
ICH2I + Et2Zn
THF
-40 °C
ICH2ZnEt
110 : 90 %
FG-RCH2I + Et2Zn
CuI cat.
neat, 25 - 50 °C
(FG-RCH2)2Zn
B–CH2
Zn
Cl
EtO2C
Ph
Tf
N
111a
111b
111c
111d

Scheme 7.36 Diorganozincs obtained by I/Zn exchange.

Interestingly, this iodine–zinc exchange can also be initiated by light [94]. Thus, the irradiation (> 280 nm) of an alkyl iodide in CH_2Cl_2 in the presence of Et_2Zn (1 equiv) provides the desired diorganozinc with excellent yields. A more reactive exchange reagent (i-Pr_2Zn) can be used instead of Et_2Zn. This organometallic has to be free of salts if optically active diorganozincs need to be prepared [95]. However, the presence of magnesium salts has a beneficial effect on the rate of exchange and a range of mixed zinc reagents of the type RZn(i-Pr) (**112**) can be prepared under mild conditions (Scheme 7.37) [96].

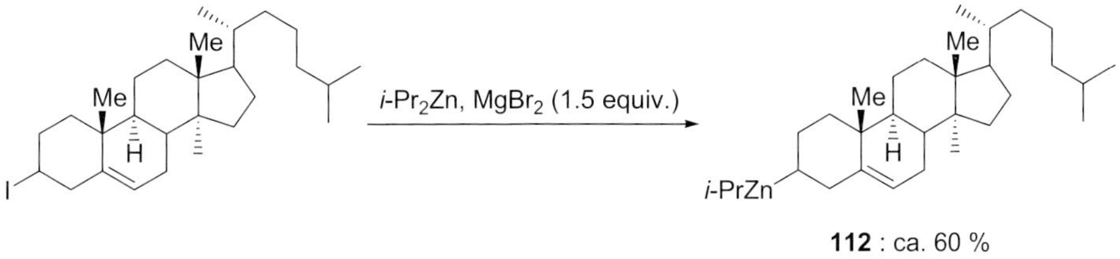
Me
Me
H
I
i-Pr2Zn, MgBr2 (1.5 equiv.)
i-PrZn
112 : ca. 60 %

Scheme 7.37 Preparation of mixed diorganozincs via an I/Zn exchange.

Since the isopropyl group is also transferred in the reaction with an electrophile at a comparable rate as the second R group, an excess of electrophile has to be added and tedious separations may be required. A more straightforward approach is possible for diarylzincs. In this case, the I/Zn-exchange can be performed under very mild conditions [97].

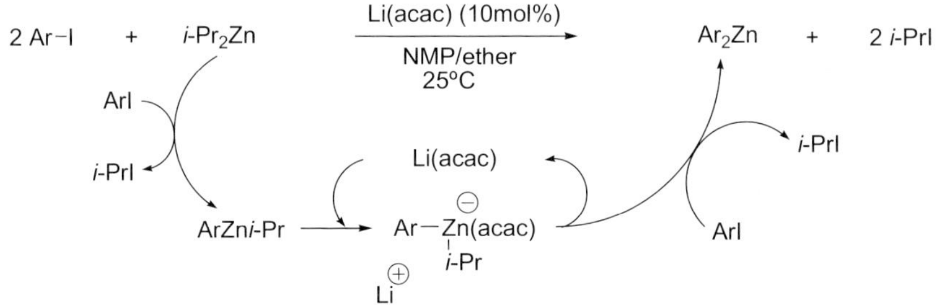

Scheme 7.38 Li(acac) catalyzed synthesis of diarylzincs.

The intermediate formation of a zincate enhances the nucleophilic reactivity of the substituents attached to the central zinc atom and makes it more prone to undergo an iodine–zinc exchange reaction. Thus, the addition of catalytic amounts of Li(acac) to an aryl iodide and i-Pr$_2$Zn allows the transfer of the two i-Pr groups with the formation of Ar$_2$Zn and i-PrI (2 equiv). This method allows the preparation of highly functionalized diarylzinc reagents bearing an aldehyde function like **113** or an isothiocyanate group like **114** (Scheme 7.39). These diarylzinc reagents undergo typical reactions of diorganozincs. Thus, the acylation of **113** with an acid chloride in the presence of Pd(0) provides the polyfunctional ketone **115** in 75% yield, whereas the reaction of the zinc reagent **114** with Me$_3$SnCl gives the aryltin derivative **116** in 66% yield.

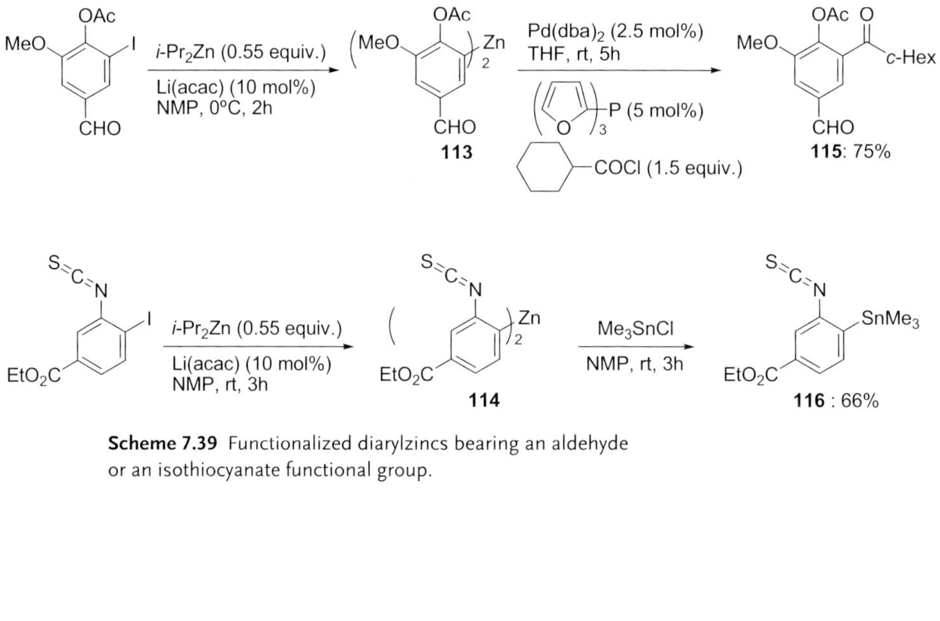

Scheme 7.39 Functionalized diarylzincs bearing an aldehyde or an isothiocyanate functional group.

Mixed diorganozincs are synthetically useful intermediates, especially if one of the group attached to zinc is preferably transferred. The trimethylsilylmethyl group is too unreactive toward most electrophiles and plays the role of a dummy ligand. The preferential transfer of the second R-group attached to zinc is therefore possible in many cases. The reaction of 4-chlorobutylzinc iodide with Me$_3$Si-CH$_2$Li in THF at $-78\,^{\circ}$C provides the mixed diorganozinc reagent **117** that readily undergoes a Michael addition with butyl acrylate in THF:NMP mixtures (Scheme 7.40) [98]. Barbier-type reactions are also well suited for the synthesis of diarylzincs although the functional-group tolerance of this method has not been investigated in detail [99,100].

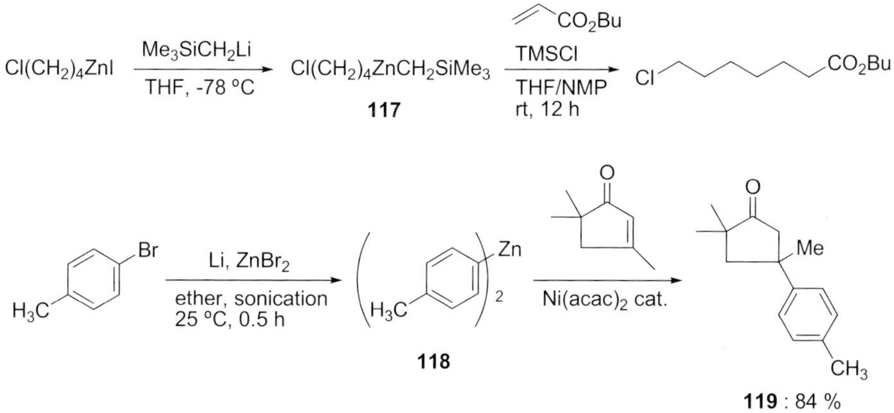

Scheme 7.40 Various syntheses and Michael additions of organozincs.

The reaction of 4-bromotoluene with lithium in the presence of zinc bromide in ether affords the corresponding zinc reagent **118** that undergoes a smooth 1,4-addition to sterically hindered enones and leads to the cyclopentanone **119** in 67% yield [99].

7.2.3.2 The Boron–Zinc Exchange

Various organoboranes react with Et$_2$Zn or i-Pr$_2$Zn providing the corresponding diorganozinc. Pioneered by Zakharin and Okhlobystin [101] and Thiele and coworkers [102], the method provides a general entry to a broad range of diorganozincs. The exchange reaction proceeds usually under mild conditions and tolerates a wide range of functional groups. It is applicable to the preparation of allylic and benzylic diorganozincs as well as secondary and primary dialkylzincs [103] and dialkenylzincs [104]. Remarkably, functionalized alkenes bearing a nitro group or a alkylidenemalonate function are readily hydroborated with Et$_2$BH [105] prepared *in situ* and converted to diorganozinc reagents such as **120** and **121**. After a copper-catalyzed allylation the expected allylated products **122** and **123** are obtained in high yields (Scheme 7.41) [103].

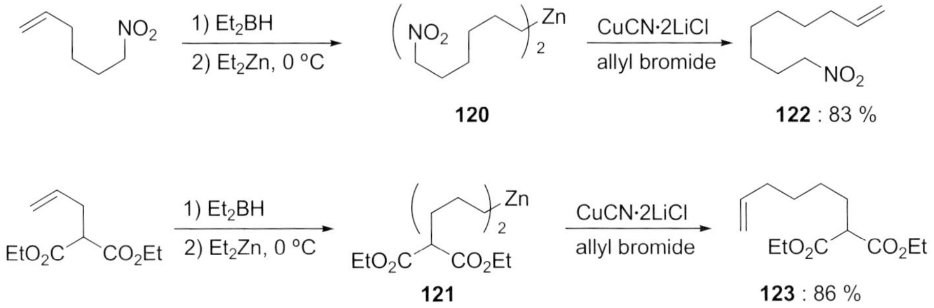

Scheme 7.41 Preparation of functionalized diorganozincs using a B/Zn exchange.

The hydroboration of dienic silyl enol ethers, such as **124** with Et$_2$BH leads to organoboranes that can be converted to new diorganozincs, such as **125**; Scheme 7.42 [106]. More importantly, this method allows the preparation of chiral secondary alkylzinc reagents. Thus, the hydroboration of 1-phenylcyclopentene with (–)-IpcBH$_2$ (99% *ee*) [107] produces, after crystallization, the chiral organoborane **126** with 94% *ee*. The reaction of **126** with Et$_2$BH replaces the isopinocampheyl group with an ethyl substituent (50 °C, 16 h) and provides after the addition of *i*-Pr$_2$Zn (25 °C, 5 h), the mixed diorganozinc **127**. Its stereoselective allylation leads to the *trans*-disubstituted cyclopentane **128** in 44% yield (94% *ee*; *trans:cis* = 98:2); Scheme 7.43 [108]. This sequence can be extended to open-chain alkenes and *Z*-styrene derivative **129** is converted to the *anti*-zinc reagent **130** that provides, after allylation, the alkene **131** in 40% yield and 74% *ee* (d.r. = 8:92).

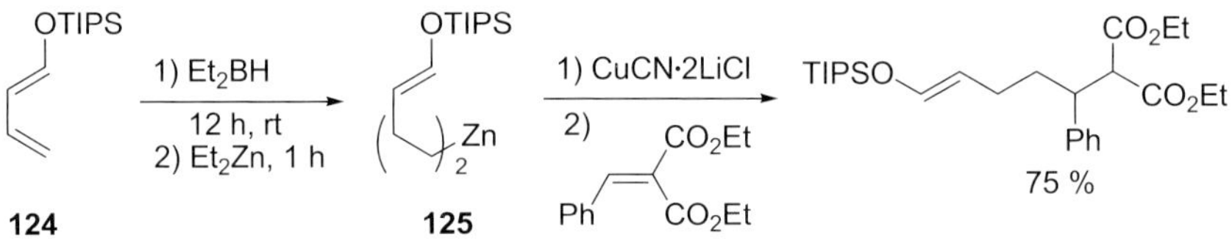

Scheme 7.42 Preparation of diorganozincs bearing a silyl ether using a boron–zinc exchange.

Similarly, the indene derivative **132** is converted by asymmetric hydroboration and B/Zn-exchange to the *trans*-indanylzinc reagent **133** that undergoes a Pd-catalyzed cross-coupling with an *E*-alkenyl iodide leading to the *trans-E*-product **134** in 35% yield (Scheme 7.43) [108b]. Several functionalized alkenes have been converted in chiral secondary alkylzinc reagents [109]. Especially interesting are unsaturated acetals, such as **135** that can be hydroborated with (–)-IpcBH$_2$ with high enantioselectivity (91% *ee*) providing, after B/Zn-exchange, the mixed zinc reagent **136**.

Its trapping with various electrophiles provides chiral products, such as **137–139**. The deprotection of **139** furnishes the *β*-alkynylaldehyde **140** in 88% *ee*. The *exo*-alkylidene acetal **141** is converted similarly to the zinc reagent **142** which can be allylated with an excellent diastereoselectivity (d.r. = 96:4) leading to the ketal **143**

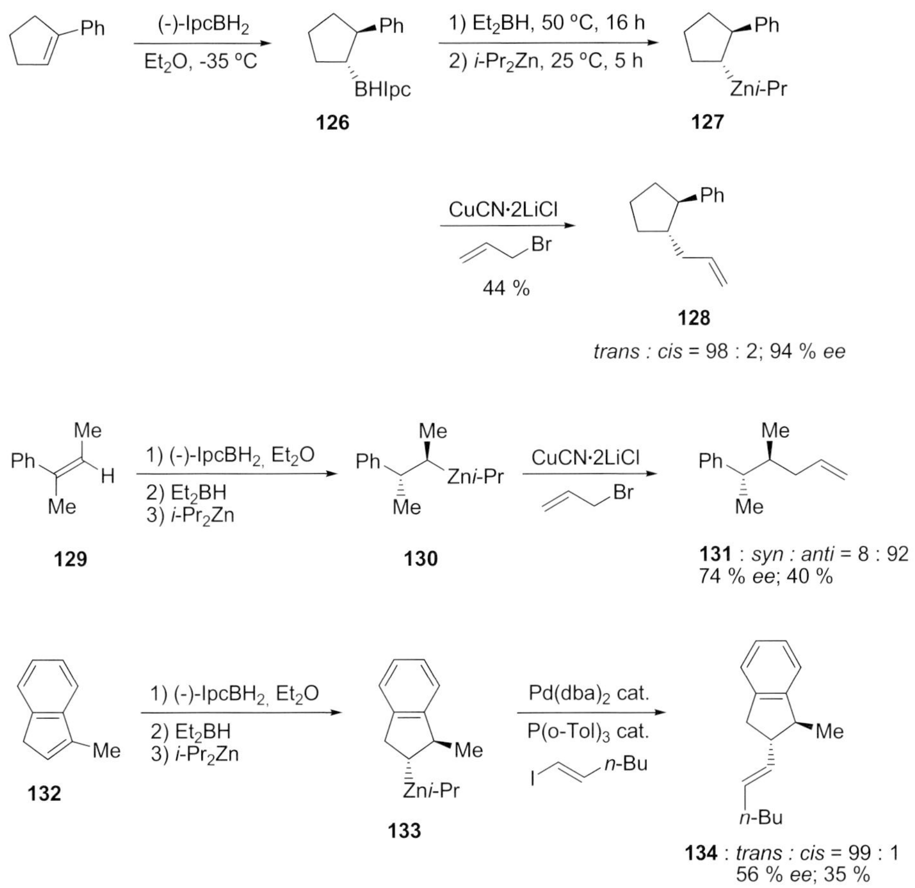

Scheme 7.43 Synthesis of chiral secondary alkylzincs via B/Zn-exchange.

(Scheme 7.44) [109]. A substrate-controlled hydroboration can also be achieved. Thus, the hydroboration of the unsaturated bicyclic olefin **144** occurs with high diastereoselectivity. B/Zn-exchange leads to the zinc reagent **145** that can be acylated with retention of configuration at C(3) leading to the bicyclic ketone with a control of the relative stereochemistry of 4 contiguous chiral centers as shown in product **146**; Scheme 7.45 [110]. The hydroboration of allylic silanes, such as **147** proceeds with high diastereoselectivity as demonstrated by Fleming and Lawrence [111]. It is difficult to use the newly formed carbon–boron bond for making new carbon–carbon bonds due to its moderate reactivity. However, the B/Zn-exchange converts the unreactive carbon–boron bond to a reactive carbon–zinc bond as in compound **148**. A further transmetallation with the THF soluble salt CuCN·2LiCl provides copper reagents which can be allylated (**149a**), alkynylated (**149b**) or acylated (**149c**); Scheme 7.45.

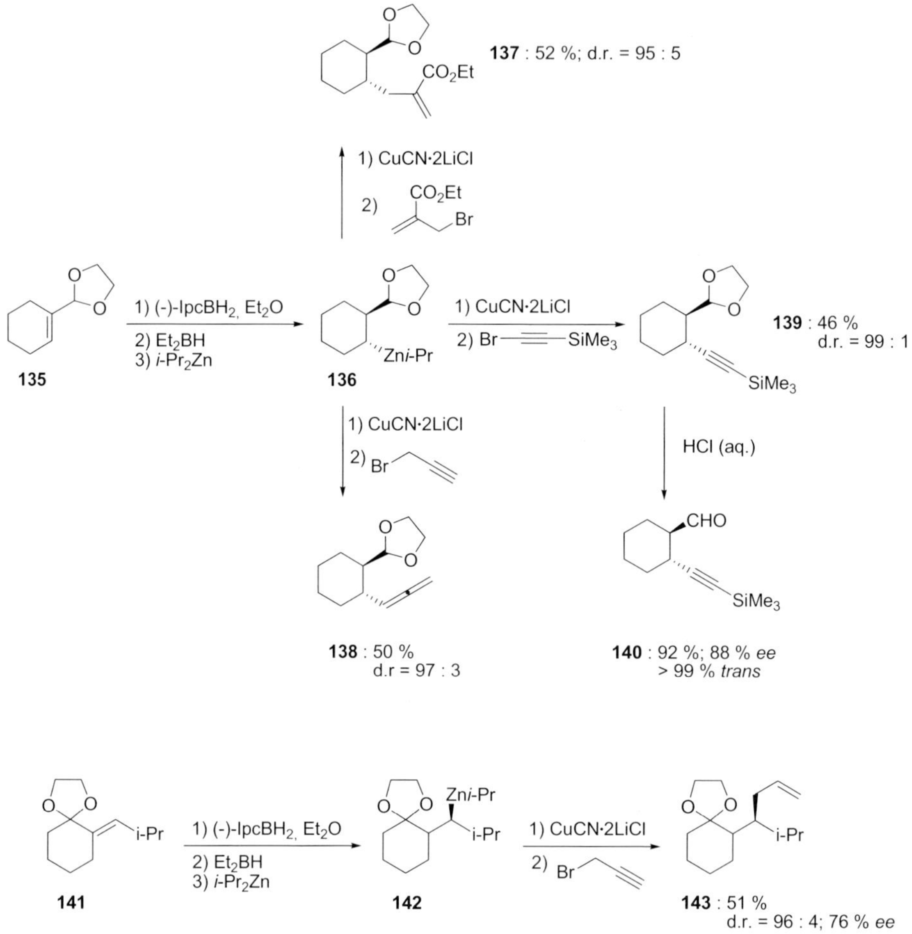

137 : 52 %; d.r. = 95 : 5

1) CuCN·2LiCl

2) CO₂Et / Br

135

1) (-)-IpcBH₂, Et₂O
2) Et₂BH
3) *i*-Pr₂Zn

136

1) CuCN·2LiCl
2) Br━━━SiMe₃

139 : 46 %
d.r. = 99 : 1

1) CuCN·2LiCl
2) Br⟋⟍

138 : 50 %
d.r = 97 : 3

HCl (aq.)

140 : 92 %; 88 % *ee*
> 99 % *trans*

141

1) (-)-IpcBH₂, Et₂O
2) Et₂BH
3) *i*-Pr₂Zn

142

1) CuCN·2LiCl
2) Br⟋⟍

143 : 51 %
d.r. = 96 : 4; 76 % *ee*

Scheme 7.44 Preparation of chiral functionalized zinc organometallics via a boron–zinc exchange.

The hydroboration of allylic amine or alcohol derivatives can be used for the preparation of alkylzinc reagents with excellent diastereoselectivity (Scheme 7.46) [112]. Thus, the diastereoselective hydroboration of the allylic sulfonamide **150** [113] affords after B/Zn-exchange the zinc reagent **151** that leads, after a copper-catalyzed acylation, to the ketone **152** with 62% yield [112]. Rhodium-catalyzed hydroborations are also compatible with the boron–zinc exchange reaction and the *exo*-methylene silylated alcohol **153** is readily hydroborated with catecholborane [114] in the presence of ClRh(PPh₃)₃ [115] affording after boron–zinc exchange the zinc reagent **154** leading, after allylation, to the *cis*-substituted products **155** and **156** [112]. The boron–zinc exchange can be extended to aromatic systems. The required aromatic boron derivatives can be readily prepared from the corresponding arylsilanes such as **157** by using BCl₃ [116].

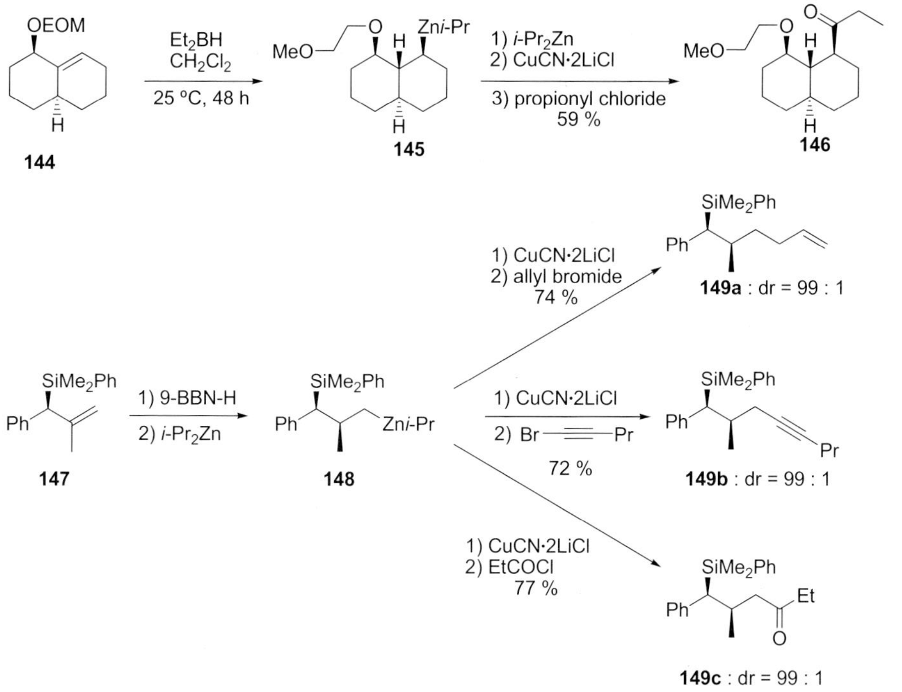

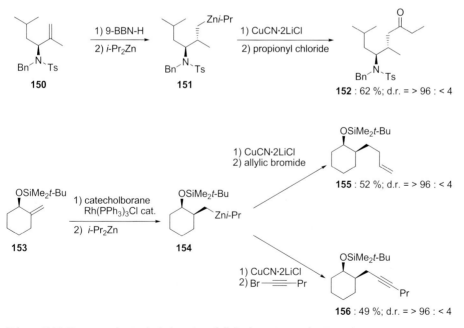

Scheme 7.45 Diastereoselective hydroboration and B/Zn-exchange.

Scheme 7.46 Diastereoselective hydroboration of allylic derivatives and B/Zn-exchange.

The resulting functionalized arylborane **158** readily undergoes a B/Zn-exchange leading to the zinc reagent **159** that can be trapped by various electrophiles, such as propargyl bromide or propionyl chloride leading, respectively, to the allene **160** (73%) and to the ketone **161** (72%); Scheme 7.47. [116].

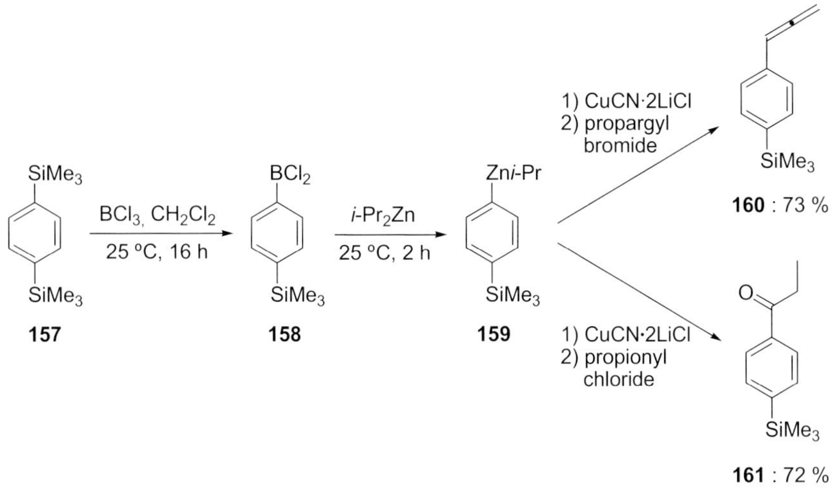

Scheme 7.47 Synthesis of arylzinc derivatives via a Si/B/Zn-exchange sequence.

7.2.3.3 Hydrozincation of Alkenes

Diorganozincs can also be prepared by a nickel-catalyzed hydrozincation. The reaction of Et$_2$Zn with Ni(acac)$_2$ may produce a nickel hydride that adds to an alkene leading after transmetallation with Et$_2$Zn, to a diakylzinc. This reaction proceeds in the absence of solvent and at temperatures between 50–60 °C. A number of functionalized olefins, like allylic alcohols or amines can be directly used. They afford the expected products in 60–75% yield; Scheme 7.48. [117]. This method is especially well suited for the preparation of functionalized diorganozincs for the asymmetric addition to aldehydes [117].

7.2.4
Diverse Methods of Preparation of Allylic Zinc Reagents

Several methods have been described for preparing allylic zinc derivatives. In contrast to alkylzincs, allylic zinc reagents are much more reactive due to the greater ionic nature of the carbon–zinc bond in these organometallics. The chemistry displayed by these reagents is not representative of the usually moderate reactivity of organozinc derivatives. Tamaru and coworkers have converted various allylic benzoates to the corresponding organozinc intermediates in the presence of palladium(0) as catalyst.

The resulting allylic zinc reagents of the tentative structure **162** reacts with aldehydes with high stereoselectivity depending on the substitution pattern

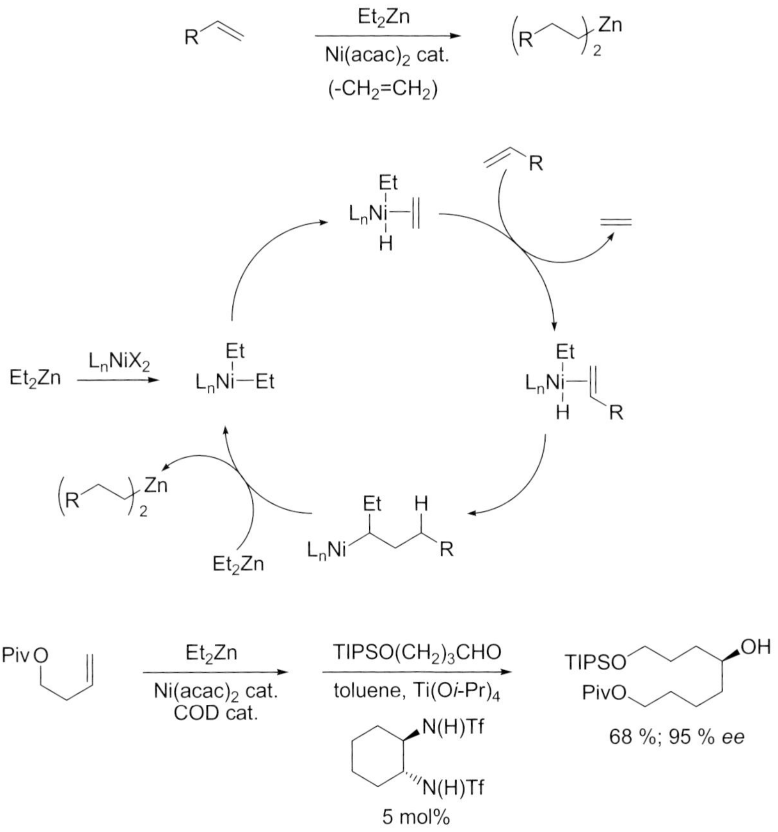

Scheme 7.48 Ni-catalyzed hydrozincation of alkenes.

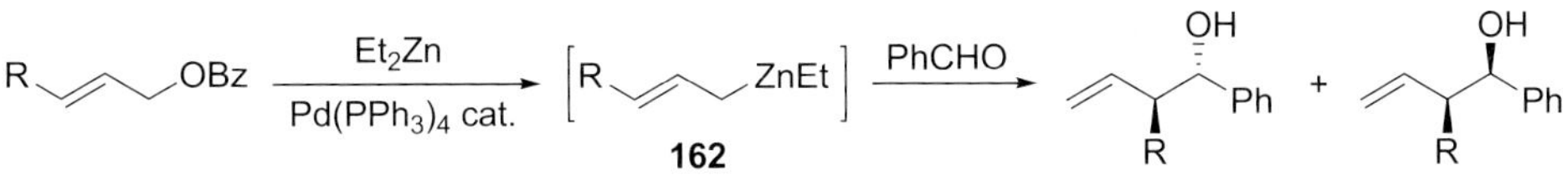

Scheme 7.49 Umpolung of the reactivity of allylic systems.

(Scheme 7.49) [118,119]. Substituted allylic zinc reagents can be prepared by the fragmentation of sterically hindered homoallylic alcoholates. This method allows the first access to functionalized allylic reagents. Thus, the treatment of the lithium homoallylic alcoholates **163a,b** with zinc chloride leads to a fragmentation and produces the new allylic reagents **164a,b** that have to be immediately trapped with benzaldehyde providing the new homoallylic alcohols **165a,b** in 56–60% yield [120].

The addition of substituted allylic zinc reagents to aldehydes is usually unselective [121]. Furthermore, the direct zinc insertion to substituted allylic halides is complicated by radical homocoupling reactions. Both of these problems are solved by the fragmentation of homoallylic alcohols. Thus, the ketone **166** reacts with

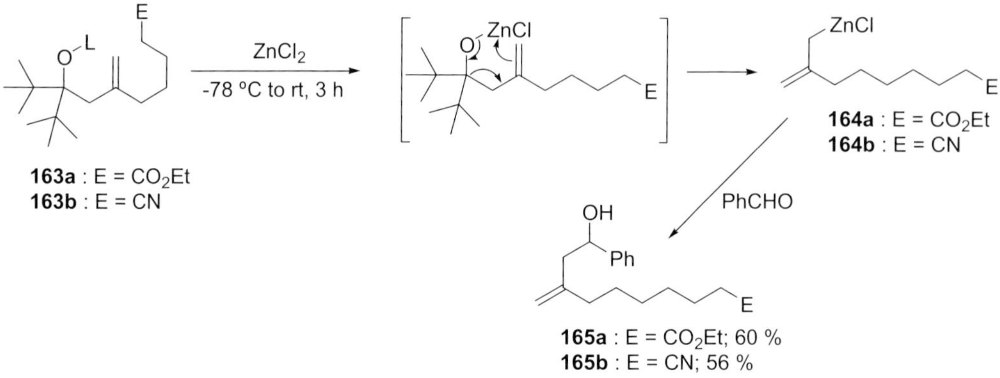

Scheme 7.50 Preparation of allylic zinc reagents by fragmentation reaction.

BuLi providing a lithium alcoholate that, after the addition of $ZnCl_2$ and an aldehyde, provides the expected addition product **167** with an excellent diastereoselectivity [120]. Oppolzer et al. have shown that the magnesium-ene reaction is a versatile method for adding allylic magnesium reagents to alkenes in an intramolecular fashion [122]. A zinc-ene [123] reaction can be initiated by the addition of BuLi to *tert*-butyl ketone **168** followed by the addition of zinc chloride. The resulting zincated spiro-derivative **169** is quenched with an acid chloride leading to the ketone **170** in 60% and > 98% *syn*-diastereoselectivity (Scheme 7.51) [120d].

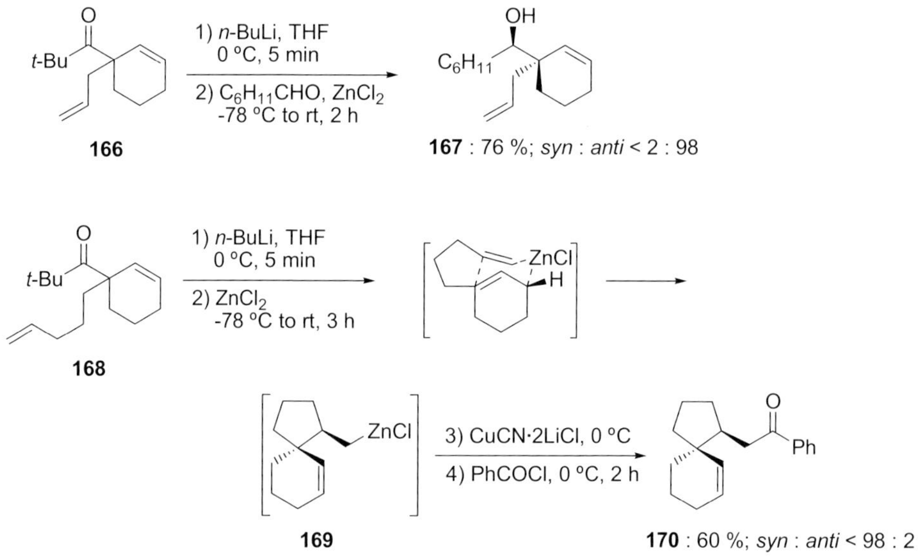

Scheme 7.51 Diastereoselective reactions of substituted allylic zinc reagents generated by fragmentation.

7.2.5
Preparation of Lithium Triorganozincates

Lithium triorganozincates are best prepared by the reaction of an alkyllithium (3 equiv) with zinc chloride or by the addition of an alkyllithium to a dialkylzinc in an etheral solvent [124]. Lithium and magnesium trialkylzincates are more reactive compared to dialkylzincs or alkylzinc halides due to the excess of negative charge at the metallic zinc center that confers a higher nucleophilicity to the organic substituents. Thus, lithium trialkylzincates readily undergo 1,4-addition reactions to enones [125]. A methyl group can serve as a dummy ligand allowing a somewhat selective transfer of the alkyl substituent (Scheme 7.52) [124]. In the presence of chiral nitrogen-chelating ligands, asymmetric 1,4-addition reactions can be performed with moderate enantioselectivity [126]. Interestingly, the addition of Bu_3ZnLi to nitrostyrene in a chiral solvent mixture of pentane and (*S,S*)-1,4-dimethylamino-2,3-dimethoxybutane leads to an optically enriched nitroalkane [127]. Triorganozincates can also be obtained by the reaction of the sulfonate **171** with Me_3ZnLi leading to the cyclopropylidene-alkylzinc reagent **172**. Its reaction with an aldehyde provides the allylic alcohol **173** in 57% yield (Scheme 7.52) [124d].

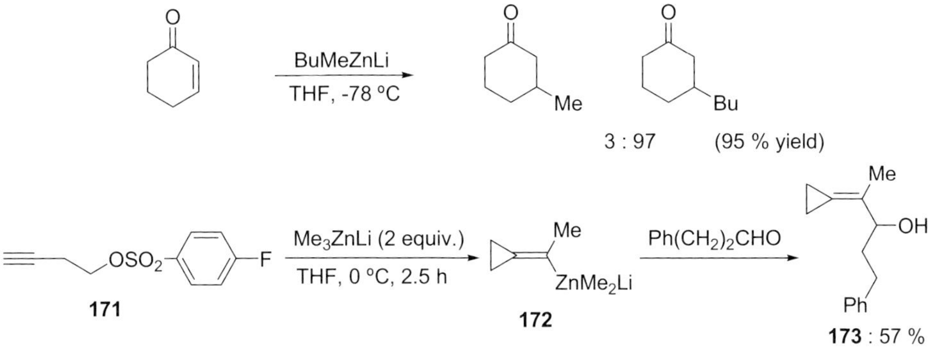

Scheme 7.52 Preparation and reactions of lithium triorganozincates.

Lithium triorganozincates can also be prepared via an I/Zn-exchange reaction. The exchange is highly chemoselective and tolerates sensitive functional groups like an epoxide or an ester. The reaction of aryl iodides **174a,b** with Me_3ZnLi provides the functionalized lithium zincates **175a,b** that undergo respectively a ring closure and an addition to PhCHO leading to the products **176a,b** in satisfactory yields (Scheme 7.53) [128].

Immobilized zincates such as **177** can be prepared by treating serine-bound 4-iodobenzoate with *t*-Bu_3ZnLi at 0 °C. They react readily with aldehydes. Transmetallation with lithium (2-thienyl)cyanocuprate provides the copper species **178** that undergoes 1,4-additions. Lithium trialkylzincates can be used for the preparation of benzylic zinc reagents using a very elegant approach of Harada. Thus, the treat-

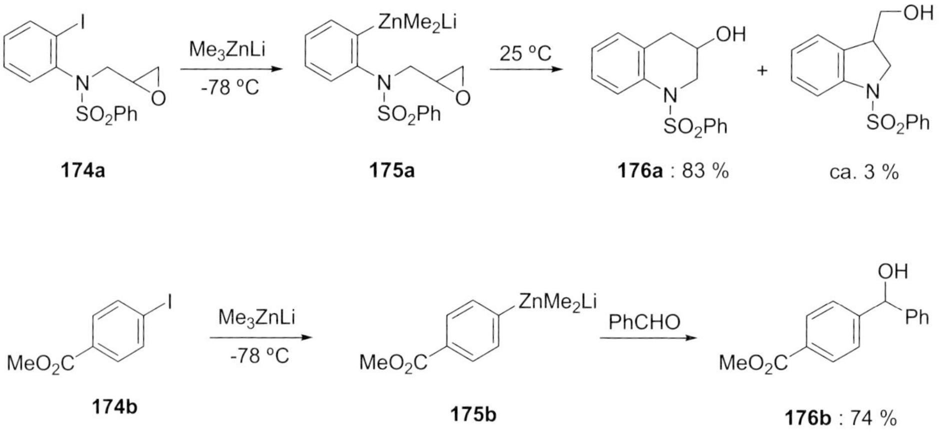

Scheme 7.53 Preparation of arylzincates via an I/Zn-exchange reaction.

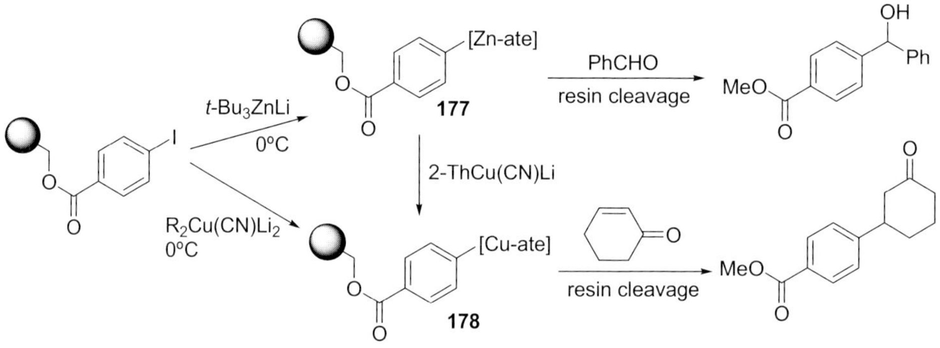

Scheme 7.54 Preparation of zincates on the solid phase.

ment of the iodomesylate **179** with Bu₃ZnLi leads to the new zincate **180** that undergoes a 1,2-migration leading to the benzylic zinc reagent **181**. It is readily quenched with an aldehyde leading to the alcohol **182** in 80% yield (Scheme 7.55) [130]. Interestingly, lithium and magnesium triarylzincates add to α,β-unsaturated sulfoxides in the presence of catalytic amounts of Ni(acac)₂ with good diastereo-selectivity (Scheme 7.55) [131].

7.3
Reactions of Organozinc Reagents

The high covalent degree of the carbon–zinc bond and the small polarity of this bond leads to a moderate reactivity of these organometallics towards many electrophiles. Only powerful electrophiles react in the absence of a catalyst. Thus, bromolysis or iodolysis reactions are high-yield reactions. In general, a direct reaction of

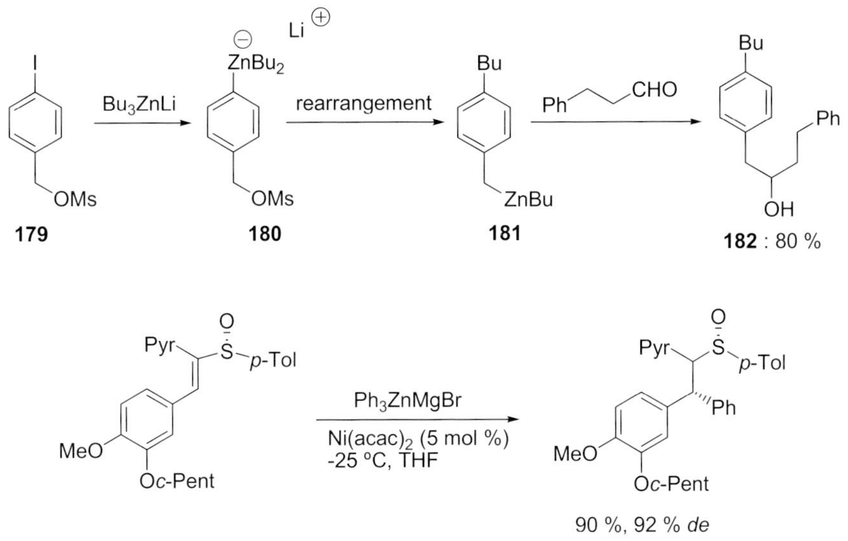

Scheme 7.55 Reactivity of lithium trialkylzincates.

organozincs with carbon electrophiles is not efficient and low yields are obtained. The addition of a catalyst is usually needed. The presence of empty *p*-orbitals at the zinc center facilitates transmetallations and a number of transition metal organometallics can be prepared in this way. These reagents are usually highly reactive towards organic electrophiles since the low-lying *d*-orbitals are able to coordinate and activate many electrophilic reagents. Many catalytic or stoichiometric transmetallations using zinc organometallics have been developed in recent years.

7.3.1
Uncatalyzed Reactions

Only reactive electrophiles react directly with organozinc derivatives. Allylic zinc reagents, and to some extent propargylic zinc reagents, are much more reactive. They add readily to carbonyl compounds or imines [132,133]. Thus, the reaction of the ester-substituted allylic zinc derivative **183** with the chiral imine **184** provides the lactam **185** with excellent diastereoselectivity (Scheme 7.56) [133]. An *in situ* generation of the allylic zinc reagent starting from the corresponding bromide **186** allows the addition to alkynes leading to skipped 1,3-dienes of type **187** [134].

Propargylic zinc derivatives react with aldehydes or ketones with variable selectivity affording a mixture of allenic and homopropargylic alcohols [135]. However, under appropriate reaction conditions, high enantioselectivities and diastereoselectivities can be achieved. Marshall and coworkers have shown that chiral propargylic mesylates such as **188** are converted to allenylzinc reagents **189** through treatment with a Pd(0)-catalyst. Their addition to an aldehyde such as **190**

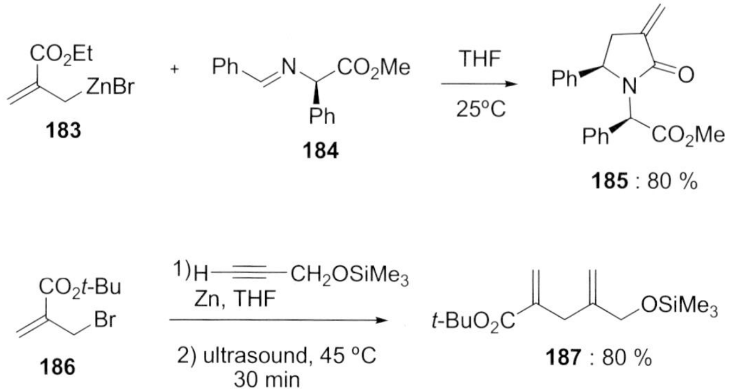

Scheme 7.56 Reactions of functionalized allylic zinc reagents.

provides the *anti*-homopropargylic alcohol **191** in 70% yield as a main diastereo-isomer (d.r. = 85 : 15) (Scheme 7.57) [136].

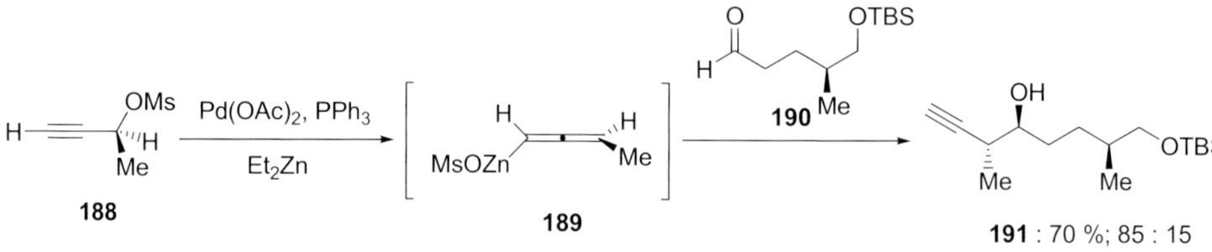

Scheme 7.57 Preparation and reactions of chiral allenyl zinc reagents.

Interestingly, 1-trimethylsilyl-propargyl zinc reagents add to aldehydes with high regio- and diastereoselectivity leading to *anti*-homopropargylic alcohols [137]. The direct oxidation of organozincs with oxygen is an excellent method for preparing hydroperoxides [138]. Recently, a new synthesis of propargylic hydroperoxides has been developed by Harada and coworkers using allenylzinc intermediates [139]. The reaction of the mesylate **192** with the lithium zincate (Bu$_3$ZnLi; 2 equiv) produces, after the addition of ZnCl$_2$ (0.5 equiv), the allenylzinc species **193** that reacts at –40 °C with O$_2$ in the presence of trimethylsilyl chloride providing the hydroperoxide **194** in 60% yield (Scheme 7.58) [139].

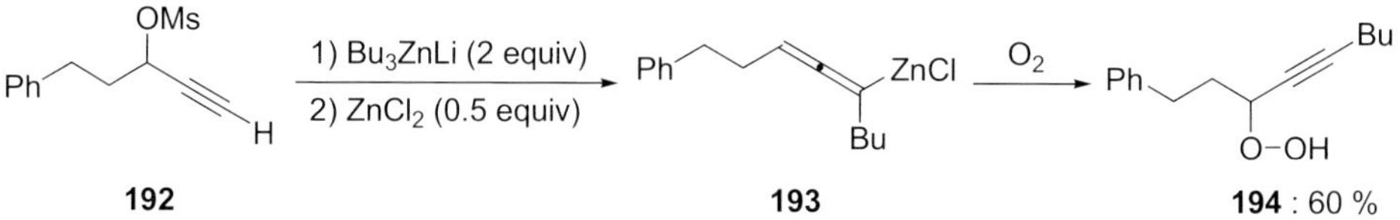

Scheme 7.58 Preparation of propargylic hydroperoxides.

The moderate reactivity of zinc organometallics is compatible with the preparation of various hydroperoxides with good selectivity. The use of perfluorinated solvents allows the oxygenation reaction to be performed at low temperature owing to the exceptionally high oxygen solubility in these media [140]. Functionalized organozincs prepared by hydrozincation, carbozincation or by boron–zinc exchange can be oxidized to the corresponding alcohols or hydroperoxides depending on the reaction conditions [141]. Tosyl cyanide reacts with a range of zinc organometallics providing the corresponding nitriles in excellent yields [142]. The functionalized alkylzinc species **195** is smoothly converted to the corresponding nitrile **196** in 67% yield. Interestingly, whereas the reaction of tosyl cyanide with benzylic bromide produces selectively 1-methylbenzonitrile in 76% yield, the cyanation of the corresponding copper reagent furnishes benzyl cyanide as sole product in 80% yield (Scheme 7.59) [142]. The reaction of 1,1-mixed bimetallics of magnesium and zinc, such as **197** [143] prepared by the addition of allylzinc bromide to (Z)-octenylmagnesium bromide with tosyl cyanide produces the nitrile **198** in 93% yield (Scheme 7.59) [142].

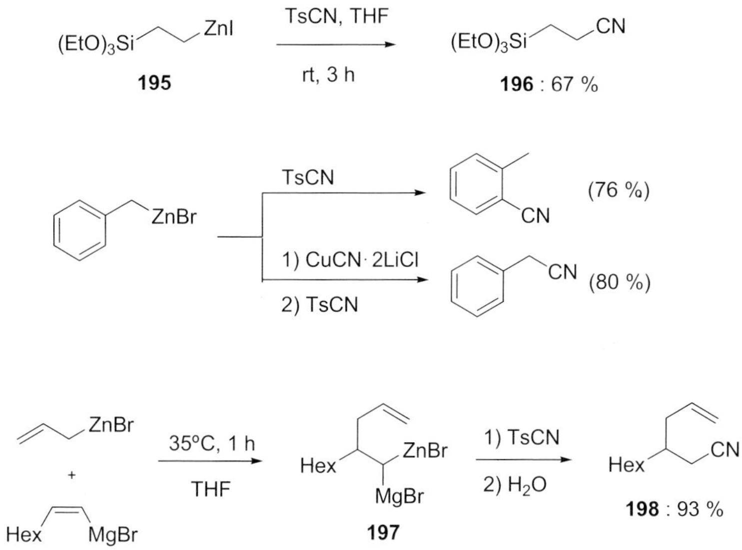

Scheme 7.59 Electrophilic cyanation of organozinc reagents.

The reaction with various chlorophosphine derivatives leads to polyfunctional phosphines with high yield [144]. The hydroboration of β-pinene with $BH_3 \cdot Me_2S$ gives *bis*-myrtanylzinc **199** in quantitative yield (Scheme 7.60) [144b]. The coupling of **199** with various chlorophosphines or PCl_3 provides new chiral phosphines, such as **200** and **201**. Polyfunctional chlorophosphines protected as their BH_3-complex, such as **202** can be prepared in two steps by the reaction of Et_2NPCl_2 with polyfunctional organozinc reagents. After protection with BH_3, the borane-complexes **203** are obtained in excellent yields. Treatment of **203** with HCl in ether leads to the borane-protected chlorophosphines of type **202** (Scheme 7.60) [145].

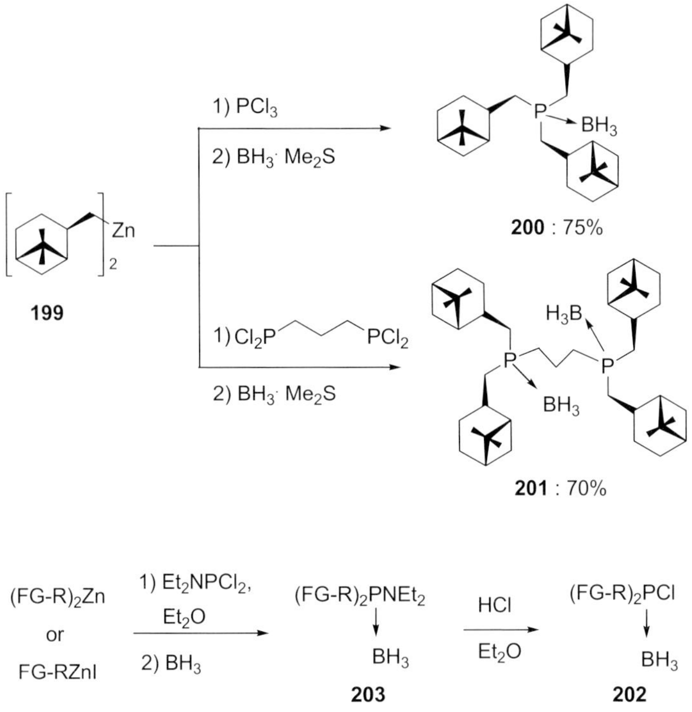

Scheme 7.60 Preparation of polyfunctional phosphines.

Similarly functionalized organometallics such as the serine-derived zinc–copper derivate **204** react under mild conditions with ClPPh$_2$. The resulting phosphine was protected as a sulfide providing enantiomerically pure **205** in 75% yield. Modification of the protecting groups furnishes the selectively protected diphenylphosphinoserine **206** in 80% yield (Scheme 7.61) [146a].

A combination of a substitution reaction with a chlorophosphine followed by a hydroboration, boron–zinc exchange allows the preparation of the mixed 1,2-diphosphine **207** in good yield (Scheme 7.62) [145].

Reactive organometallic reagents, such as Cr(CO)$_5$·THF readily add diorganozincs leading in the presence of CO (1 atm) and Me$_3$O$^+$BF$_4^-$ (rt, 2 h) to functionalized Fischer-carbene complexes [146b]. Excellent addition reactions are also obtained with iminium trifluoroacetates, such as **208**. The reaction of the aminal **209** with trifluoroacetic anhydride in CH$_2$Cl$_2$ at 0 °C gives the iminium trifluoroacetate **208** with quantitative yield. Its reaction with phenylzinc chloride furnishes the expected amine **210** in 85% yield [147]. Interestingly, this approach can be extended to functionalized organomagnesium reagents [148]. Functionalized diarylzincs such as **211** add to the activated Schiff base **212** leading to amino-acid **213** in 62% yield [149].

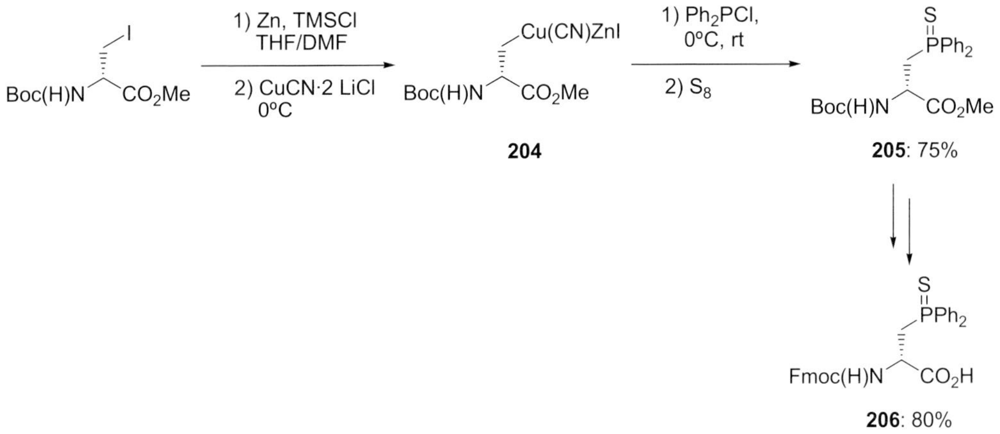

Scheme 7.61 Preparation of chiral functionalized phosphines.

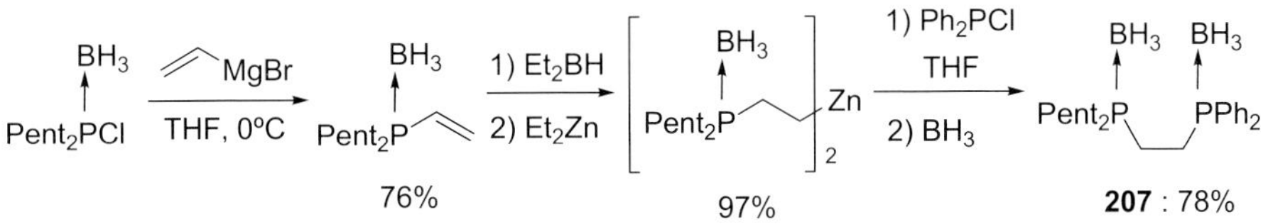

Scheme 7.62 Preparation of mixed diphosphines using functionalized zinc organometallics.

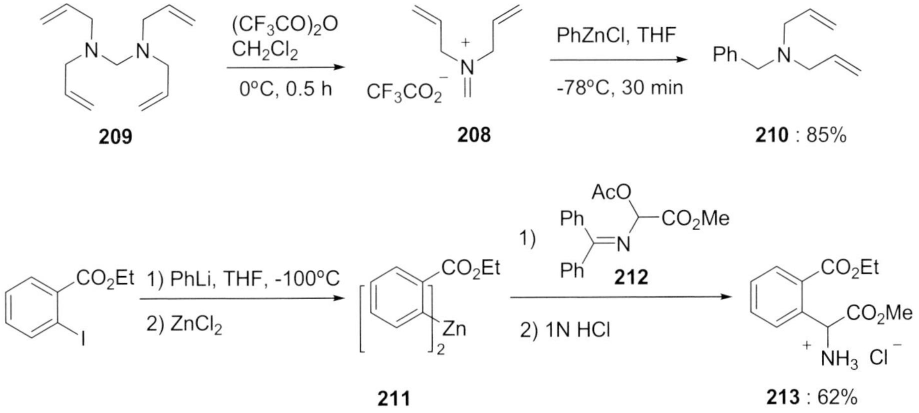

Scheme 7.63 Reaction of zinc organometallics with iminium intermediates.

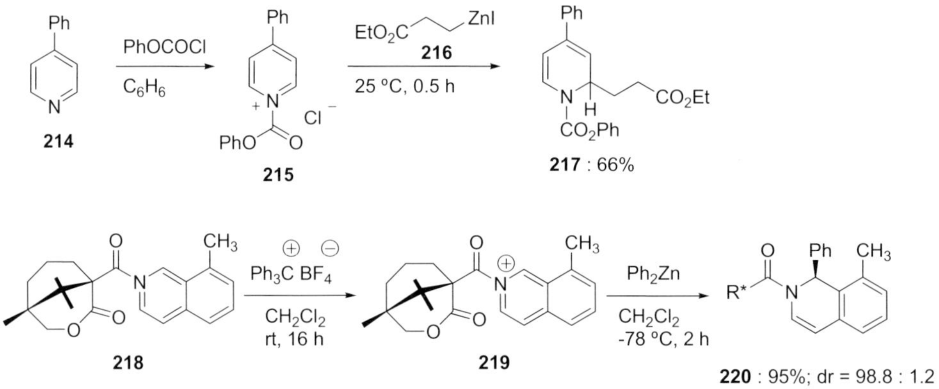

Scheme 7.64 Reaction of zinc organometallics with pyridinium derivatives.

The formation of activated iminium intermediates derived from nitrogen-heterocycles has been reported by Comins and coworkers [150]. The activation of pyridine derivative, such as **214** with phenyl chloroformate provides the pyridinium salt **215** that smoothly reacts with the zinc homoenolate **216** [151] leading to the addition product **217** in 66% yield [150]. The reaction of the unsaturated amide **218** with $Ph_3C^+BF_4^-$ produces *N*-acyliminium ions of type **219** which react with Ph_2Zn in CH_2Cl_2 producing the desired α-substituted amine **220** in 95% yield (d.r. = 98.8:1.2), Scheme 7.64 [152]. Benzotriazole is an excellent leaving group and the readily available imidazolidin-2-ones of type **221** react with aryl-, alkenyl- or alkyl-zinc derivates via an elimination-addition mechanism providing the *trans*-products of type **222** with > 99% diastereoselectivity (Scheme 7.65) [153].

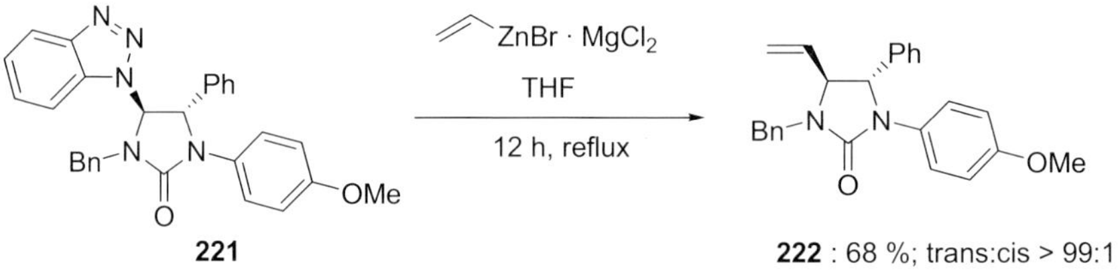

Scheme 7.65 Addition of vinylzinc bromide to iminium salts generated from benzotriazole derivatives.

The addition of zinc organometallics to *in situ* generated oxenium ions by the reaction of mixed acetals with TMSOTf [154] allows a highly stereoselective preparation of protected *anti*-1,3-diols of type **223** (Scheme 7.66). The addition of TMSOTf also triggers the allylic substitution of glycal derivatives providing the substitution products of type **224** with excellent regio- and diastereoselectivity.

The opening of epoxides with zinc reagents is a difficult reaction. However, activated epoxides such as the glycal epoxides **225** react with diorganozincs in the presence of CF_3CO_2H [155]. Presumably the reaction of R_2Zn with CF_3CO_2H produces the highly Lewis-acidic species $RZnOCOCF_3$ (226). This reaction proceeds smoothly in CH_2Cl_2 and furnishes the α-C-glycoside 227 in 58% yield.

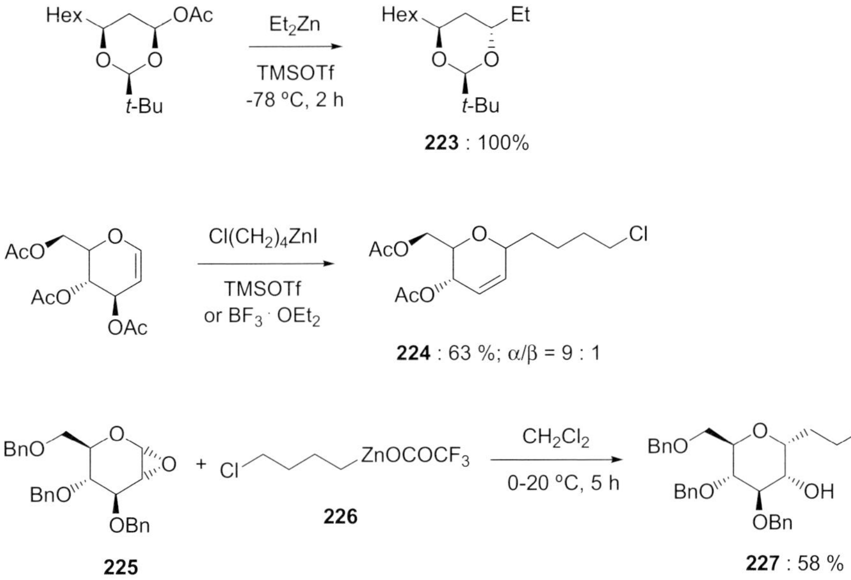

223 : 100%

224 : 63 %; $\alpha/\beta = 9 : 1$

225

226

227 : 58 %

Scheme 7.66 Reaction of zinc reagents with acetals and related compounds.

Although alkylzinc derivatives add only slowly to aldehydes, alkenylzinc derivatives display a higher reactivity. Thus, the addition of vinylzinc chloride 228 to the amino-aldehyde 229 provides the allylic alcohol 230 in 60% yield [156]. The addition rate can be increased by performing the reaction in the presence of a Lewis acid. Thus, the addition of the homoenolate 216 to the amino-aldehyde 231 provides the aminoalcohol 232 with good diastereoselectivity (Scheme 7.67) [157]. Reactive benzylic or related zinc reagents, such as 233 smoothly add to aldehydes providing the allylic alcohol 234 in almost quantitative yield (Scheme 7.67) [158]. In a noncomplexing solvent, such as dichloromethane, functionalized alkylzinc halides add to α-functionalized aldehydes leading to the addition product 235 again with a remarkable diastereoselectivity (Scheme 7.67) [158].

Benzylic acetates are unreactive toward organozinc compounds. However, various ferrocenyl acetates, such as 236 react with alkylzinc halides in the presence of $BF_3 \cdot OEt_2$ with retention of configuration leading to the chiral ferrocenyl derivatives like 237 (Scheme 7.68) [159].

The reactivity of zinc organometallics can be dramatically increased by adding polar solvents like N-methylpyrrolidinone (NMP). Under these conditions various diorganozincs add to a range of Michael acceptors like α,β-unsaturated ketones aldehydes, nitriles or nitro derivatives (Scheme 7.69) [160]. The preparation of mixed diorganozincs bearing nontransferable Me_3SiCH_2-groups allows a more efficient transfer of the functionalized group to the Michael acceptor (Scheme 7.70) [98].

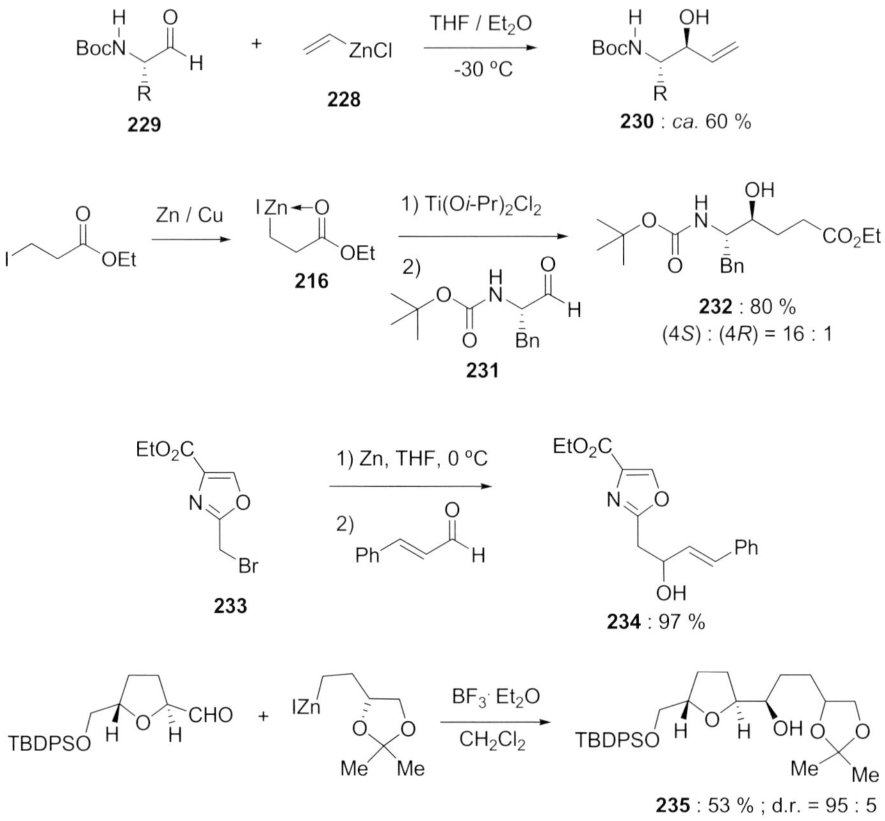

Scheme 7.67 Addition of organozinc halides to aldehydes.

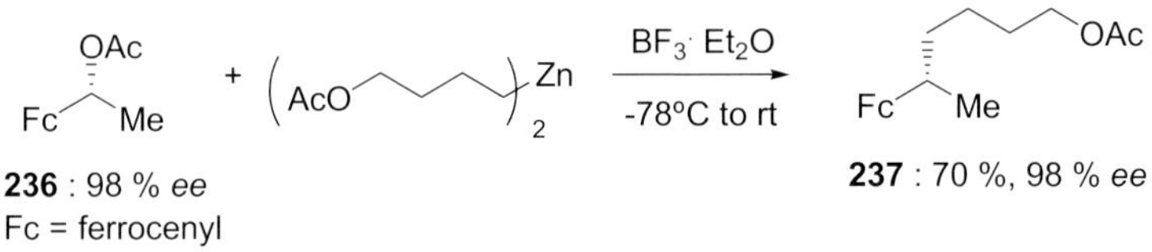

Scheme 7.68 Substitution with retention on ferrocenyl derivatives.

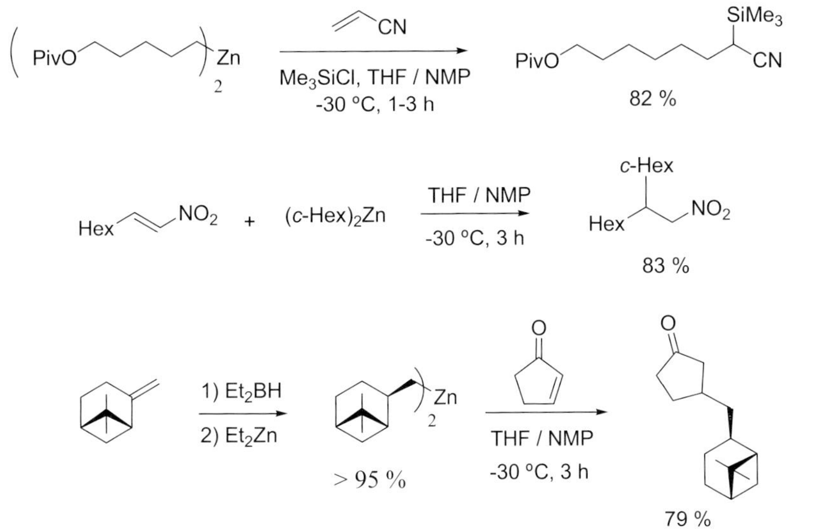

Scheme 7.69 NMP promoted Michael additions of diorganozincs.

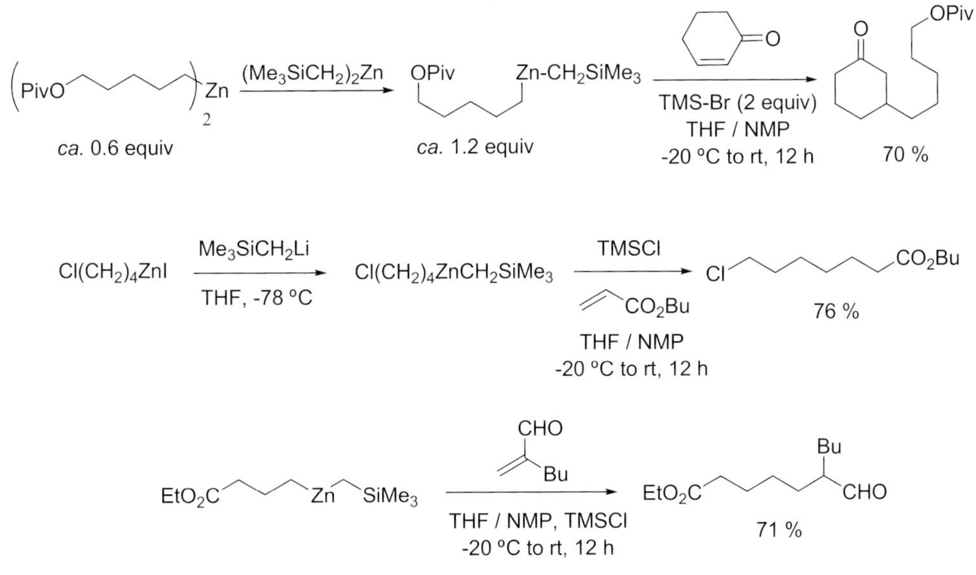

Scheme 7.70 Mixed diorganozincs for conjugate addition reactions.

In the case of an intramolecular 1,4-addition, no activation is required [161]. The iodoenone **238** is readily converted into the corresponding alkylzinc iodide that undergoes an intramolecular addition at 25 °C in THF affording the bicyclic ketone **239** in 65–67% yield [161]. The addition of alkynylzinc halides to enones is best performed in the presence of *tert*-butyldimethylsilyl triflate. Under these conditions, very high yields of the conjugated adducts, such as **240** are obtained (Scheme 7.71) [162].

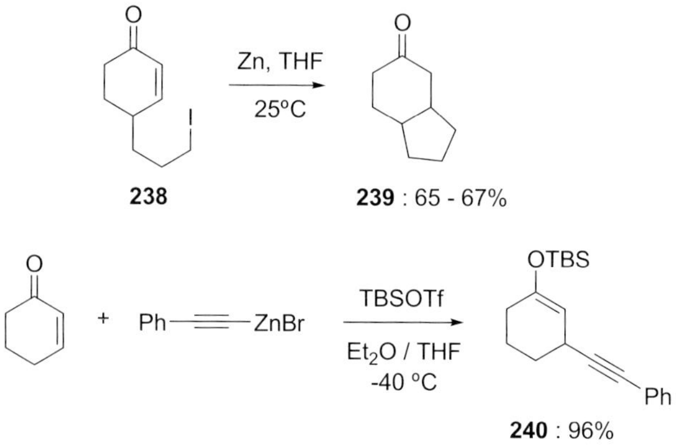

238   **239** : 65 - 67%

240 : 96%

Scheme 7.71 Michael addition of organozinc halides.

The addition of the electrophilic silyl reagent strongly activates the enone. It has also been found that Lewis acids, such as AlCl₃, accelerate the reaction of di-alkylzincs with acid chlorides in CH₂Cl₂ [163].

7.3.2
Copper(I)-catalyzed Reactions

The moderate intrinsic reactivity of zinc organometallics can be increased by transmetallation with various transition metal salts. Especially useful is the trans-metallation of diorganozincs or organozinc halides with the THF-soluble complex of copper(I) cyanide and lithium chloride (CuCN · 2LiCl) [10]. The simple mixing of organozinc compounds with CuCN · 2LiCl produces the corresponding copper

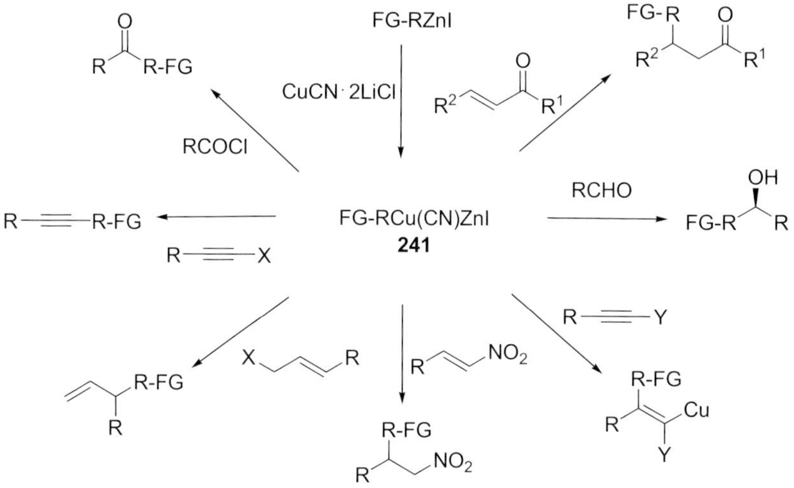

241

Scheme 7.72 General reactivity pattern of zinc–copper reagents.

species tentatively represented as RCu(CN)ZnX (**241**). Their reactivity is similar, but somewhat reduced compared to copper species prepared from organomagnesium or lithium compounds [164]. The structure of the mixed copper–zinc reagents is only known by EXAFS spectroscopy, indicating that the cyanide ligand is coordinated to the copper center [165]. In contrast to lithium- or magnesium-derived organocopper reagents, they display an increased thermal stability and alkylzinc–copper reagents can be heated in 1,2 dimethoxyethane (DME) or DMPU at 60–85 °C for several hours without appreciable decomposition [166]. A wide range of electrophiles react very efficiently with the copper–zinc compounds **241** leading to polyfunctional products (Scheme 7.72) [6].

7.3.2.1 Substitution Reactions

Substitution reactions of zinc- or zinc–copper organometallics with R₃SiCl are usually difficult, but R₃SnCl reacts much more readily. By using DMF as solvent, the heterocyclic zinc reagent **242** reacts with Ph₃SnCl, providing the tin derivative **243** in 37% yield [167]. The β-zincated phosphonate **244** provides, under mild conditions, the expected stannylated product **245** in 81% yield (Scheme 7.73) [55]. Allylation reactions with zinc organometallics proceed in the presence of copper salts with very high S_N2'-selectivity. This is in contrast to the palladium- or nickel-catalyzed reactions that proceed via a π-allyl palladium intermediate and afford generally the coupling product at the less substituted end of the allylic system.

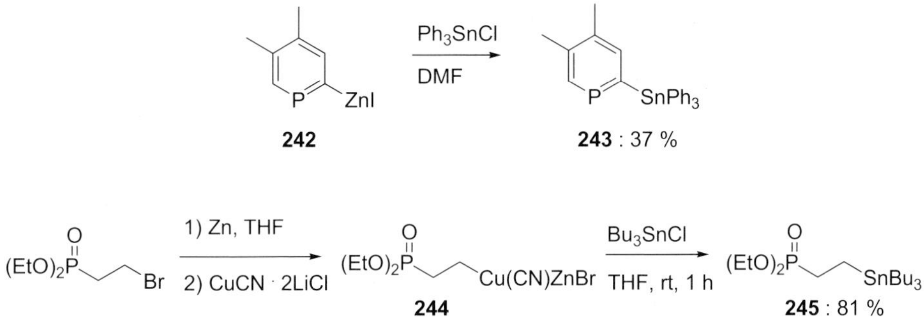

Scheme 7.73 Stannylation of zinc–copper reagents.

Thus, the zincated propionitrile **246** [168] reacts with cinnamyl bromide in the presence of CuCN·2LiCl leading to the S_N2'-substitution product **247**, whereas in the presence of catalytical amounts of Pd(PPh₃)₄ the formal S_N2-substitution product **248** is obtained (Scheme 7.74). The presence of a functional group, such as an ester like in **249** does not modify this regioselectivity and affords the S_N2'-product **250** in 80% yield (Scheme 7.74) [168b]. Interestingly, the zinc–copper carbenoid ICH₂Cu(CN)ZnI affords as an exception the S_N2-substitution product **251**, allowing an homologation of an allylic bromide by a CH₂I unit (Scheme 7.74) [47b]. A range of polyfunctional zinc–copper reagents are readily allylated [6].

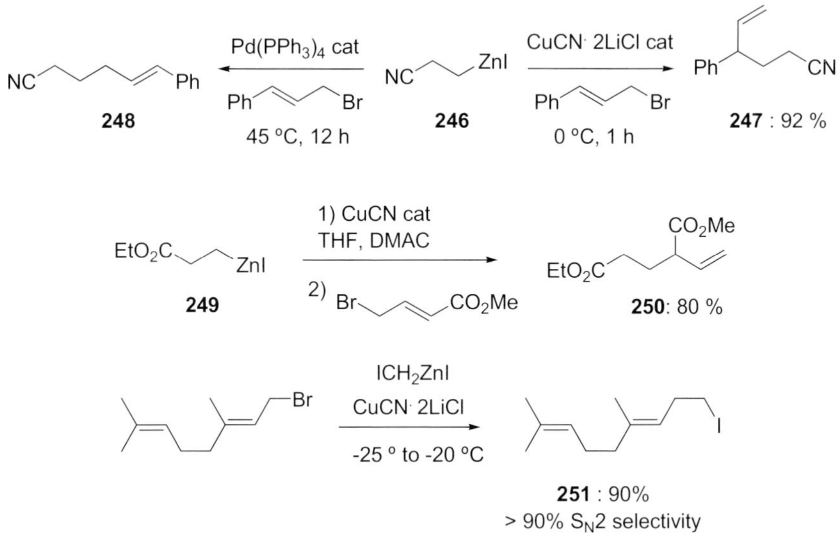

Scheme 7.74 Regioselective allylations with zinc–copper reagents.

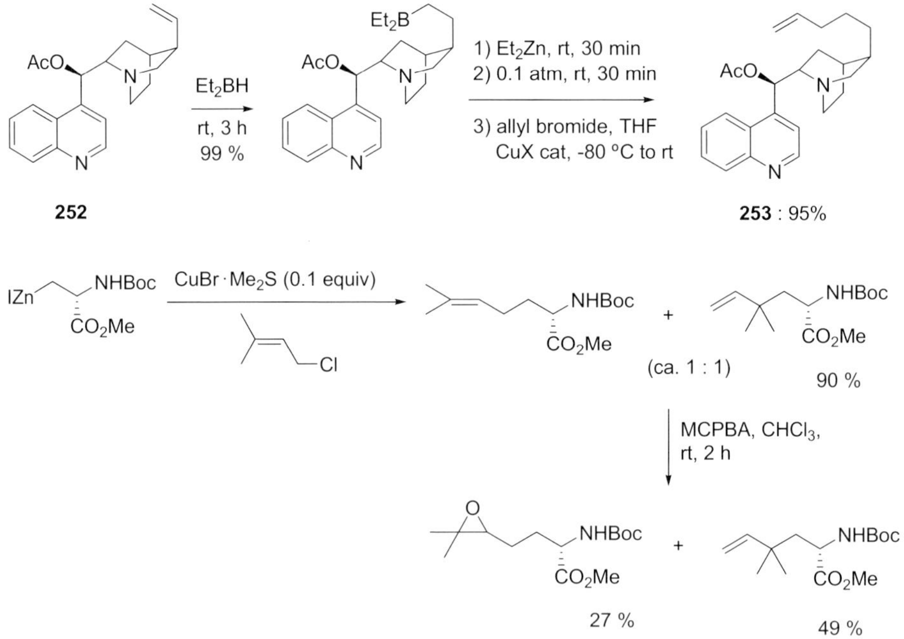

Scheme 7.75 Allylic substitutions with zinc–copper reagents.

A quinine alkaloid derivative containing a vinyl group, such as **252**, has been hydroborated, undergo a boron–zinc exchange and copper(I)-catalyzed allylation, leading to the alkaloid derivative **253** [103b]. A new route to hydrophobic aminoacids is possible by using the reaction of the zinc reagent with prenyl chloride in the presence of $CuBr \cdot Me_2S$ (0.1 equiv). Under these conditions, a mixture of S_N2 and S_N2'-products (55:45) is obtained. They can be readily separated by taking advantage of the higher reactivity of trisubstituted alkenes compared with terminal alkenes towards MCPBA (Scheme 7.75) [169].

Although zinc–copper reagents do not open epoxides, the more reactive α,β-unsaturated epoxides react readily with various functionalized zinc–copper organometallics [170] or functionalized zinc reagents in the presence of a catalytic amount of MeCu(CN)Li [171]. Thus, the opening of δ-epoxy-α,β-unsaturated esters such as **254** with $Me_2Zn \cdot CuCN$ proceeds with high *anti*-stereoselectivity leading to the allylic alcohol **255** in 96% yield. Also the opening of the unsaturated epoxide (**256**) affords the allylic alcohol **257** in 98% yield (Scheme 7.76).

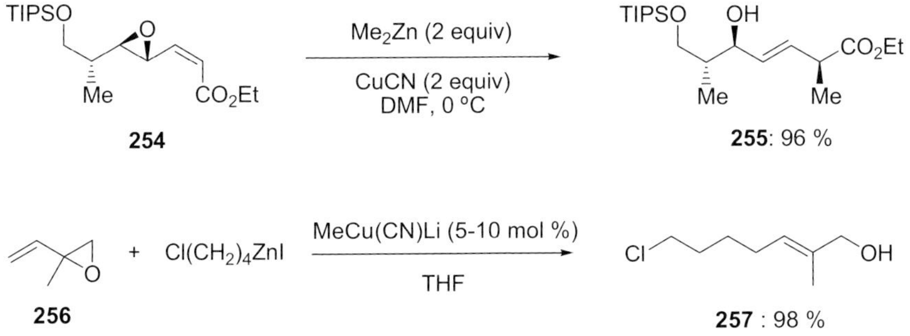

Scheme 7.76 Opening of α,β-unsaturated epoxides with copper–zinc reagents.

Interestingly, the high S_N2'-selectivity of organozinc–copper derivatives allows the performance of multiple allylic substitutions with excellent results. Thus, the reaction of the multicoupling reagent [172] **258** with an excess of copper–zinc reagent provides the double S_N2'-reaction product **259** in 89% yield (Scheme 7.77) [48b]. Propargylic halides or sulfonates react with zinc–copper reagents leading to the S_N2'-substitution product like the allenic amino-acid derivative **260**. Interestingly, the regioselectivity is reversed by performing the reaction in the presence of catalytic amounts of palladium(0). Thus, the insertion of zinc powder into the Z-alkenyl iodide **261** is complete in THF at 45 °C within 21 h. The resulting Z-zinc reagent **262** (Z:E > 99:1) reacts with the propargylic carbonate **263** in the presence of $Pd(PPh_3)_4$ (5 mol%) leading formally to the S_N2-substitution product **264** in 58% yield (Scheme 7.77) [173].

Propargylic mesylates such as fluorine-substituted derivative **265** react with PhZnCl in the presence of $Pd(PPh_3)_4$ (5 mol%) in THF at 0 °C within 2 h and provide the *anti*-S_N2'-product in excellent yield and complete transfer of the stereochemistry leading to the allene **266** (Scheme 7.78) [174]. Copper(I)-catalyzed allylic substitutions with functionalized diorganozincs proceed with high S_N2'-selectivity. Thus, the reaction of the chiral allylic phosphate **267** [175] with 3-carbethoxypropylzinc iodide in the presence of $CuCN \cdot 2LiCl$ (2 equiv) furnishes the anti-S_N2'-

substitution product **268** in 68% yield. By the addition of *n*-BuLi (1.2 equiv) and TMSCl (1.5 equiv) the bicyclic enone **269** is obtained in 75% yield and 93% *ee* (Scheme 7.79) [176].

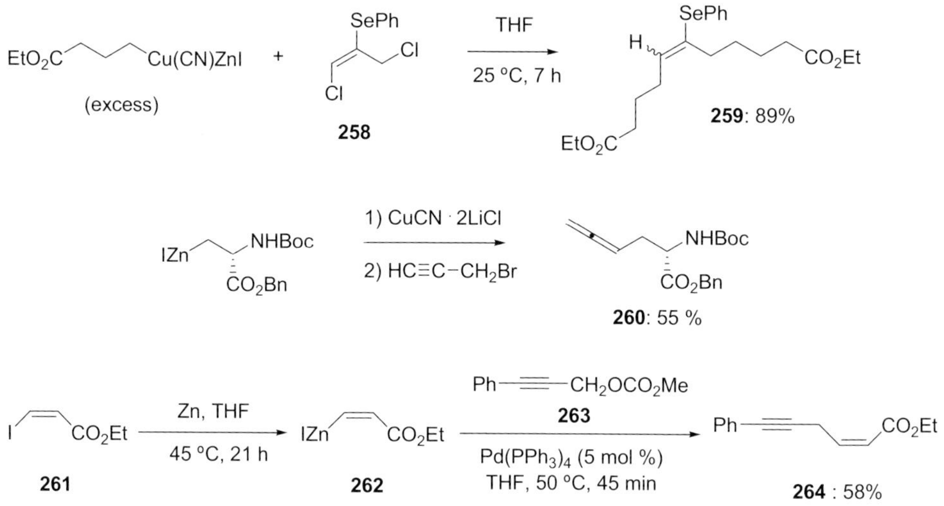

Scheme 7.77 Reaction of zinc organometallics with unsaturated halides.

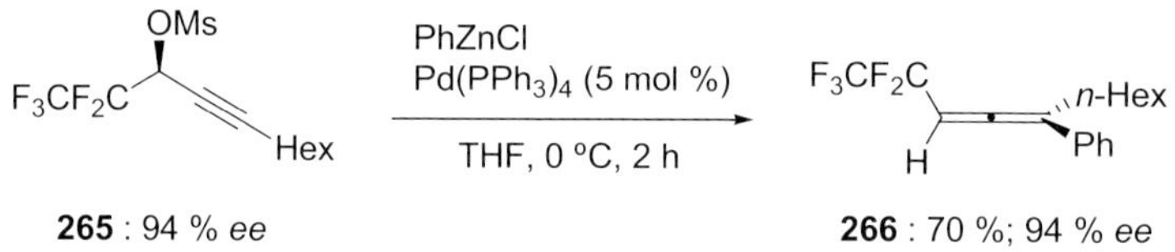

Scheme 7.78 Stereoselective propargylic substitution reaction.

This reaction can also be extended to open-chain systems. In this case, chiral allylic alcohols have been converted into pentafluorobenzoates that proved to be appropriate leaving groups. Both (*E*)- and (*Z*)- allylic pentafluorobenzoates undergo the S_N2'-substitution. In the case of (*E*)-substrates, two conformations **270** and **271** are available for an *anti*-substitution providing, besides the major *trans*-product (*trans*-**272**), also ca. 10% of the minor product *cis*-**272** (Scheme 7.79). By using the (*Z*)-allylic pentafluorobenzoates, only *trans*-substitution products are produced since the conformation **273** leading to a *cis*-product is strongly disfavored due to allylic 1,3-strain [177]. Thus, the *cis*-allylic pentafluorobenzoates (*R*,*Z*)-**274** reacts with Pent$_2$Zn furnishing only the trans-S_N2'-substitution product (*R*,*E*)-**275** in 97% yield with 93% *ee* (Scheme 7.79) [178].

Interestingly, this substitution reaction can be applied to the stereoselective assembly of chiral quaternary centers. The trisubstituted allylic pentafluorobenzoates (*E*)- and (*Z*)-**276** readily undergo a substitution reaction at −10 °C with Pent$_2$Zn furnishing the enantiomeric products (*S*)- and (*R*)-**277** with 94% *ee*.

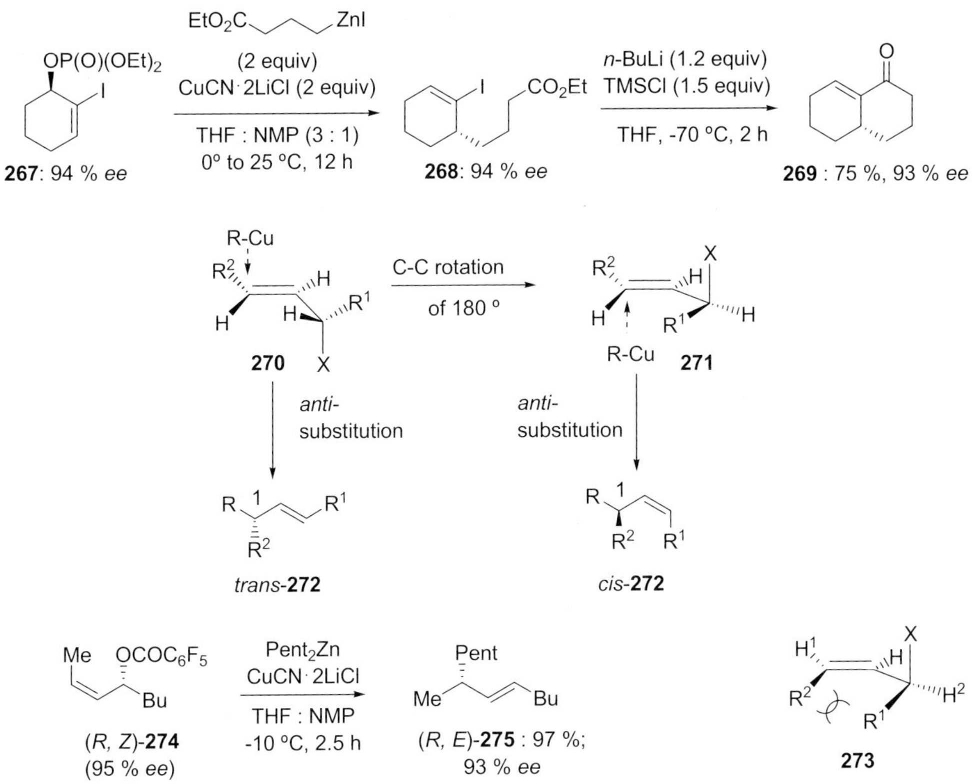

EtO$_2$C⌒⌒ZnI

OP(O)(OEt)$_2$

267: 94 % ee

(2 equiv)
CuCN·2LiCl (2 equiv)
────────────────
THF : NMP (3 : 1)
0° to 25 °C, 12 h

⌒⌒CO$_2$Et

268: 94 % ee

n-BuLi (1.2 equiv)
TMSCl (1.5 equiv)
────────────────
THF, -70 °C, 2 h

269 : 75 %, 93 % ee

R-Cu

R^2 — H
H — R^1 X

270 X

C-C rotation
of 180 °

R^2 — X
H — R^1 H
R-Cu **271**

anti-
substitution

anti-
substitution

R $\overset{1}{\underset{\overline{R^2}}{\quad}}$ R^1

trans-**272**

R $\overset{1}{\quad}$
R^2 R^1

cis-**272**

Me OCOC$_6$F$_5$
⌇
Bu

(*R*, *Z*)-**274**
(95 % ee)

Pent$_2$Zn
CuCN·2LiCl
────────────
THF : NMP
-10 °C, 2.5 h

Pent
⌇
Me⌒Bu

(*R*, *E*)-**275** : 97 %;
93 % ee

H^1 — X H
R^2 — R^1 H^2

273

Scheme 7.79 Stereoselective *anti*-S$_N$2′-substitutions.

The ozonolysis of (*R*)-**277** gives, after reductive work-up, the chiral aldehyde (*S*)-**278** in 71% yield with 94% ee. The *anti*-selectivity is observed with a wide range of diorgano-zincs like primary and secondary dialkylzincs, as well as diaryl- and dibenzyl-zinc reagents [178]. It has been applied to an enantioselective synthesis of (+)-ibuprofen **279** (Scheme 7.80). Even sterically hindered allylic substrates like the allylic phospho-nate **280** react with mixed diorganozinc reagents of the type RZnCH$_2$SiMe$_3$ in the presence of CuCN·2LiCl providing only the S$_N$2′substitution product regardless of the presence of the two methyl groups adjacent to the allylic center. The reaction with EtO$_2$C(CH$_2$)$_2$ZnCH$_2$SiMe$_3$ provides the *anti*-substitution product **281** in 81% with 97% ee. It is readily converted in (*R*)-α-ionone **282** (45%; 97% ee) The *ortho*-diphenyl-phosphanyl ligand orients the S$_N$2′ substitution in a *syn*-manner with high regio- and stereoselectivity to the *cis*-product (Scheme 7.81) [179, 180].

Allylic substitution using organozinc reagents can also be performed using a chiral catalyst [181]. The use of a modular catalyst is an especially versatile strategy and has been applied to the stereoselective preparation of quaternary centers with great success [182]. In the presence of 10 mol% of the modular ligand **283** highly enantioselective substitutions of allylic phosphates like **284**, leading to the fish deterrent sporochnol (**285**: 82% yield, 82% ee), have been performed.

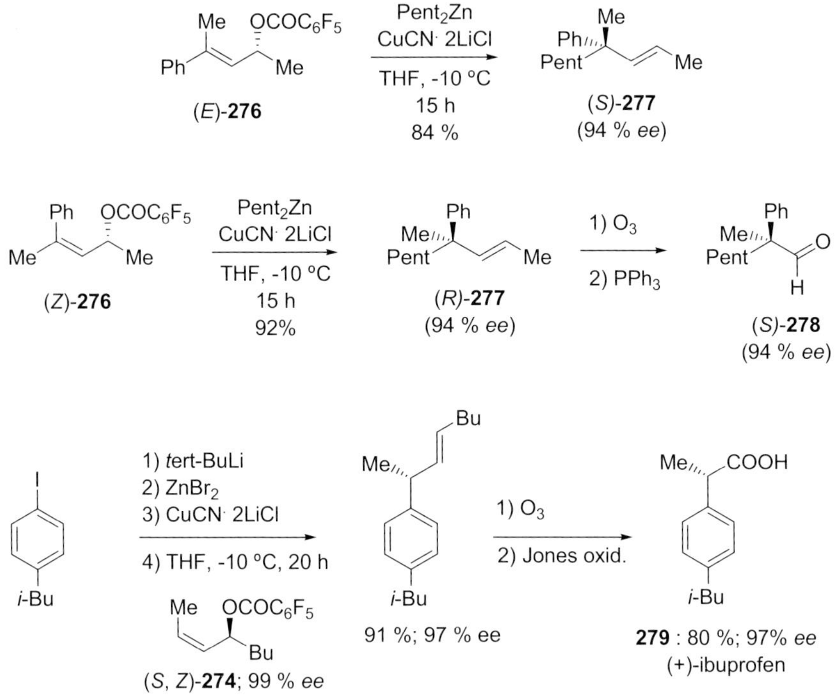

Scheme 7.80 Preparation of quarternary chiral centers and (+)-ibuprofen synthesis.

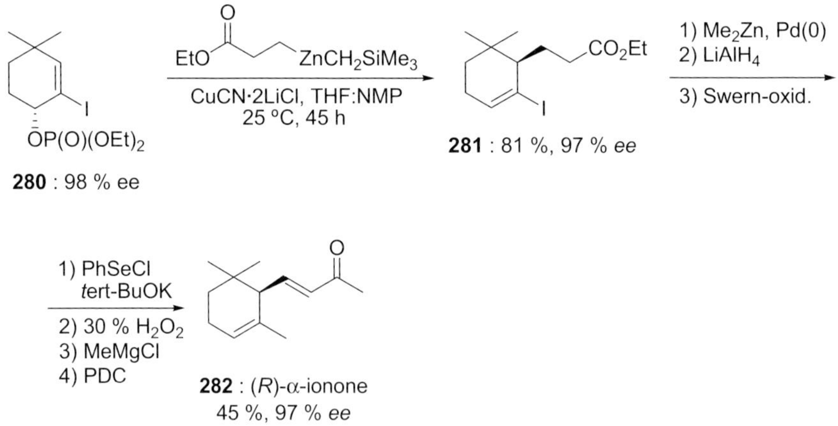

Scheme 7.81 Synthesis of (R)-α-ionone using a stereoselective *anti*-S$_N$2′-substitution.

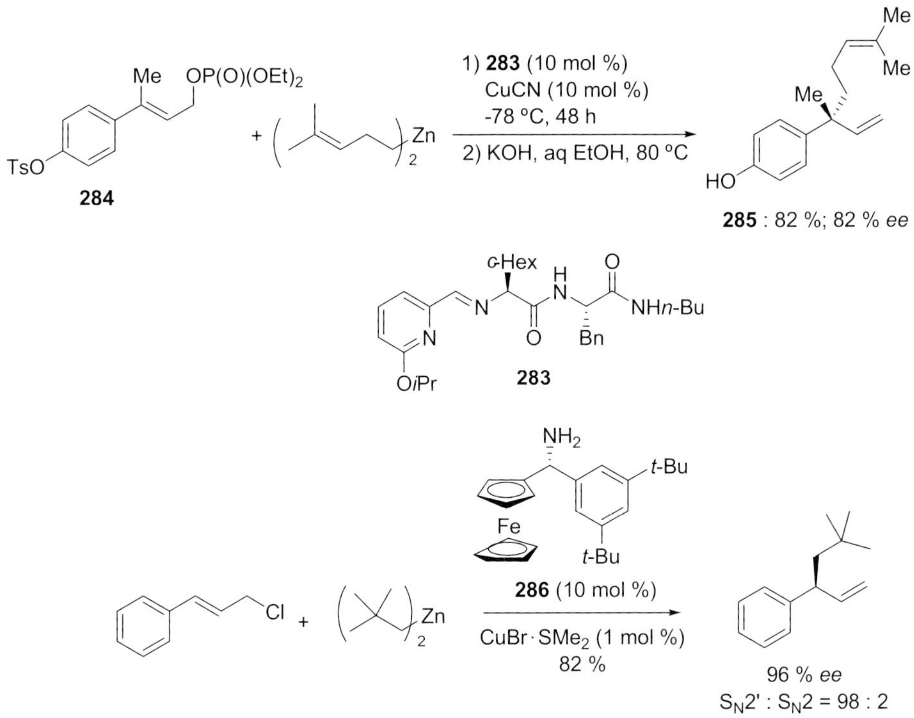

Scheme 7.82 Catalytic S_N2'-allylic substitutions.

Sterically very hindered diorganozincs like dineopentylzinc react enantioselectively with allylic chlorides in the presence of the chiral ferrocenylamine **286** with up to 98% *ee* (Scheme 7.82) [183]. The addition of zinc–copper organometallics to unsaturated cationic metal complexes derived for example from pentadienyliron and pentadienylmolybdenum cations affords the corresponding dienic complexes. Thus, the addition of a zinc–copper homoenolate to the cationic η^5-cycloheptadienyliron complex **287** leads to the polyfunctional iron-dienic complex **288** in 65% yield [184]. This chemistry has been extensively developed by Yeh and coworkers [185]. The addition of allylic and benzylic zinc reagents to (η^6-arene)-Mn(CO)$_3$ cations of type **289** provides with excellent stereoselectivity the neutral (η^5-cyclohexadienyl)Mn(CO)$_3$ complexes such as **290** in 92% yield. Less-reactive functionalized alkylzinc compounds show a more complicated reaction pathway due to an isomerization of the organozinc species (Scheme 7.83) [186]. Rigby and Kirova-Snover [187] have applied these reactions to the synthesis of several natural products. Thus, the alkylation of the cationic tropylium complex **291** with the functionalized zinc reagent **292** furnishes the chromium complex **293** that was an intermediate for the synthesis of β-cedrene (Scheme 7.83).

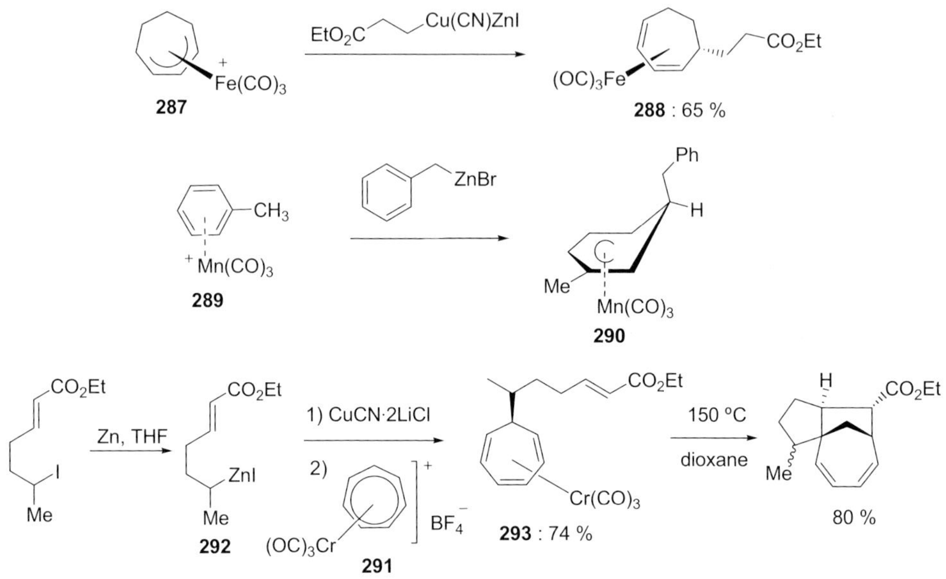

Scheme 7.83 Reaction of zinc–copper reagents with cationic transition metal complexes.

Yeh and Chuang have shown that such addition reactions provide polyfunctional cationic chromium species, such as **294** that can be converted to highly functionalized bicyclic ring systems (**295**) which are difficult to prepare otherwise (Scheme 7.84) [49].

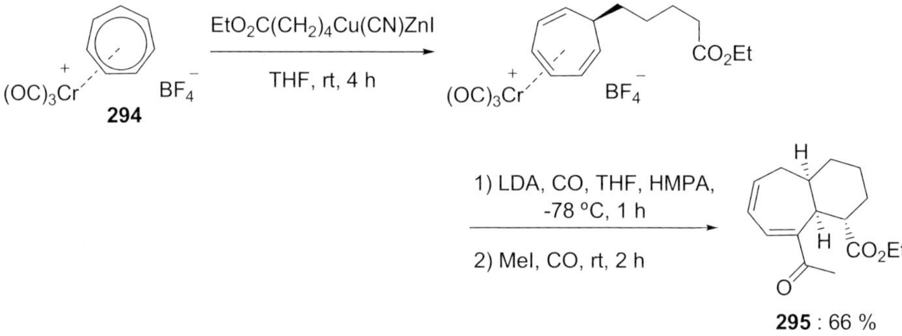

Scheme 7.84 Addition of functionalized zinc–copper reagents to cationic chromium complexes and subsequent cyclization.

Alkynyl iodides and bromides smoothly react with various zinc–copper organometallics at –60 °C leading to polyfunctional alkynes [48d]. Iodoalkynes, such as **296** [188] react at very low temperature, but lead in some cases to copper acetylides as byproducts (I/Cu-exchange reaction). 1-Bromoalkynes are the preferred substrates. Corey and Helel have prepared a key intermediate **297** of the side chain of

Cicaprost™ by reacting the chiral zinc reagent **298** with 1-bromopropyne leading to the functionalized alkyne **299** (Scheme 7.85) [189]. This cross-coupling has also been used to prepare a pheromone (**300**) [48d].

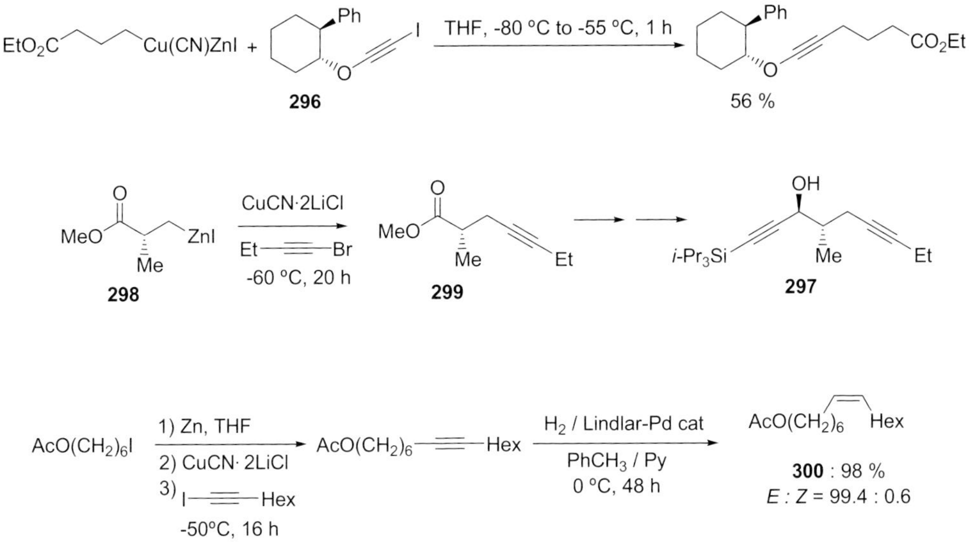

Scheme 7.85 Cross-coupling of zinc–copper organometallics with 1-haloalkynes.

Substitution at C_{sp2}-centers can also be accomplished as long as the haloalkene is further conjugated with an electron-withdrawing group at β-position. Thus, 3-iodo-2-cyclohexenone reacts with the zinc reagent **301** bearing a terminal alkyne affording the functionalized enone **302** [11]. Similarly the stepwise reaction of 3,4-dichlorocyclobutene-1,2-dione (**303**) with two different zinc–copper reagents furnishes polyfunctional squaric acid derivatives, such as **304** [190]. By using mixed diorganozinc reagents of the type FG-RZnMe [191], a catalytic addition-elimination can be performed with a wide range of β-keto-alkenyl triflates. Thus, the penicillin derivative **305** reacts with the mixed copper reagent **306** providing the desired product **307** in excellent yield (Scheme 7.86) [191].

Functionalized heterocycles such as **308** can be prepared in a one-pot synthesis in which the key step is the addition-elimination of a functionalized copper–zinc reagent **309** to the unprotected 3-iodoenone **310** producing the annelated heterocycle **308** in 41% (Scheme 7.87) [192].

Besides enones, several Michael acceptors having a leaving-group in β-position react with zinc–copper reagents. Thus, diethyl malonate [(phenylsulfonyl)methylene] (**311**) reacts with zinc–copper reagent **312** providing the addition substitution product **313** in 90% yield [193]. Similarly, the zinc–copper reagent **314** reacts with 2-phenylsulfonyl-1-nitroethylene (**315**) providing the intermediate triene **316** that cyclizes on silica gel at 25 °C within 4 h affording the Diels–Alder product **317** in 85% (Scheme 7.87) [194]. The cross-coupling reaction with unactivated alkenyl

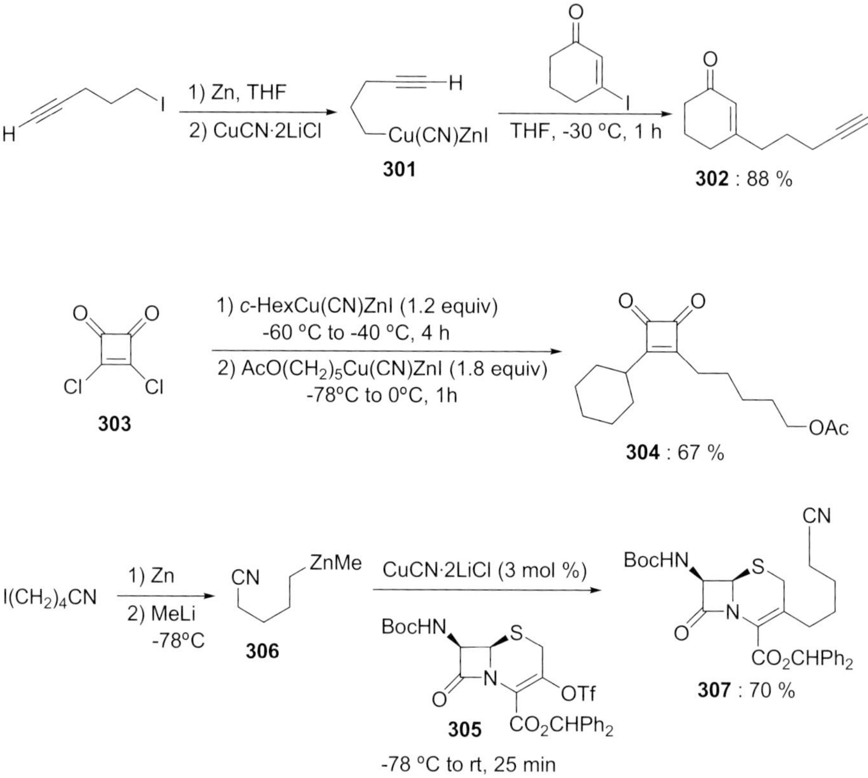

Scheme 7.86 Addition-elimination reaction with zinc–copper reagents.

iodides requires harsh conditions, but produces the desired products with retention of the double-bond configuration [166]. The alkylation of primary alkyl halides and benzylic halides can be readily performed with diorganozincs treated with one equivalent of Me$_2$CuMgCl in DMPU. This cross-coupling reaction tolerates a range of functionalities (ester, cyanide, halide and nitro group). The methyl group plays the role of a nontransferable group under our reaction conditions. Thus, the reaction of the dialkylzinc **318** with the nitro-substituted alkyl iodide **319** provides the cross-coupling product **320** in 83% yield [195]. A nickel-catalyzed cross-coupling reaction between an arylzinc reagent, such as **321** and a functionalized alkyl iodide can be successfully achieved using 3-trifluoromethylstyrene **322** as a promoter. The role of this electron-poor styrene will be to coordinate the nickel(II) intermediate bearing an aryl- and an alkyl residue and to promote the reductive elimination leading to the cross-coupling product **323** [196,197]. Interestingly, this cross-coupling reaction can be readily performed between two C$_{sp3}$ centers. Thus, the reaction of primary or secondary alkylzinc iodides with various primary alkyl iodides or bromides in the presence of catalytic amount of Ni(acac)$_2$ (10 mol%), Bu$_4$NI (3 equiv) and 4- fluorostyrene (20 mol%) provides the corresponding cross-coupling products in satisfactory yields [199]. More reactive secondary dialkylzincs and the mixed organozinc compounds $RZnCH_2SiMe_3$ undergo the cross-coupling in the absence of Bu_4NI [198].

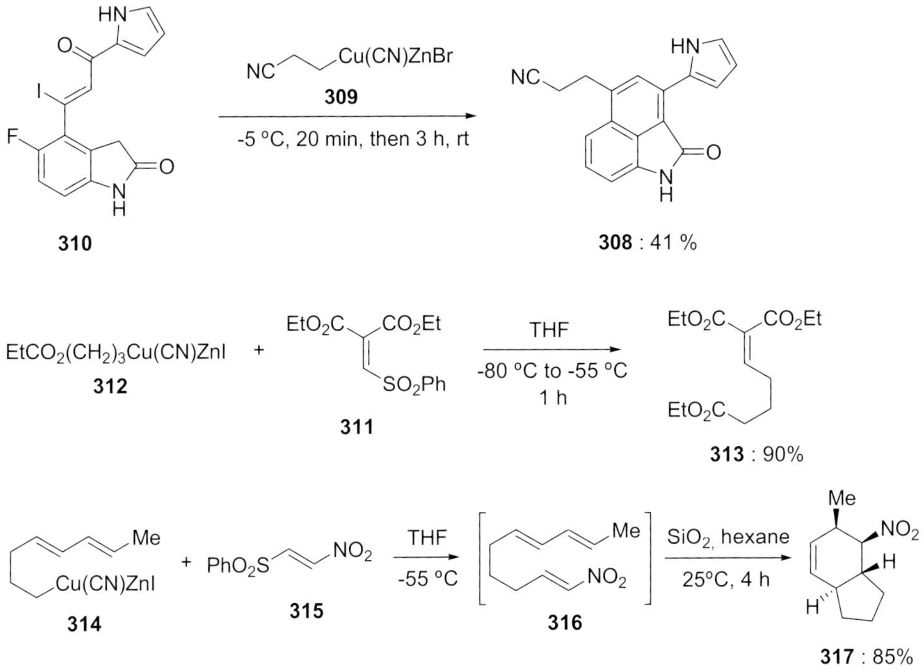

Scheme 7.87 Addition-elimination reactions of copper–zinc reagents with various Michael acceptors bearing a leaving group in β-position.

Thus, the secondary diorganozinc **324** obtained by the hydroboration of norbornene with Et_2BH and subsequent boron–zinc exchange undergoes a smooth cross-coupling with the iodoketone **325** at $-30\,°C$ (16 h reaction time) furnishing stereoselectively the exo-ketone **326** in 61% yield. Polyfunctional products, such as **327** are readily obtained by performing the cross-coupling of functionalized alkylzinc iodides with functionalized alkyl bromides (Scheme 7.88) [198].

7.3.2.2 Acylation Reactions

The uncatalyzed reaction of acid chlorides with organozincs is sluggish and inefficient. It is often complicated by side reactions and leads usually low yields of the desired acylation products. In contrast, the $CuCN \cdot 2LiCl$-mediated acylation of various zinc reagents affords ketones in excellent yields. The reaction of the zinc–copper reagent **328** obtained by the direct zinc insertion to the iodide **329** followed by a transmetallation with $CuCN \cdot 2LiCl$ reacts with benzoyl chloride at $25\,°C$ leading to the ketone **330** in 85% yield [52b]. The functionalized benzylic bromide **331**

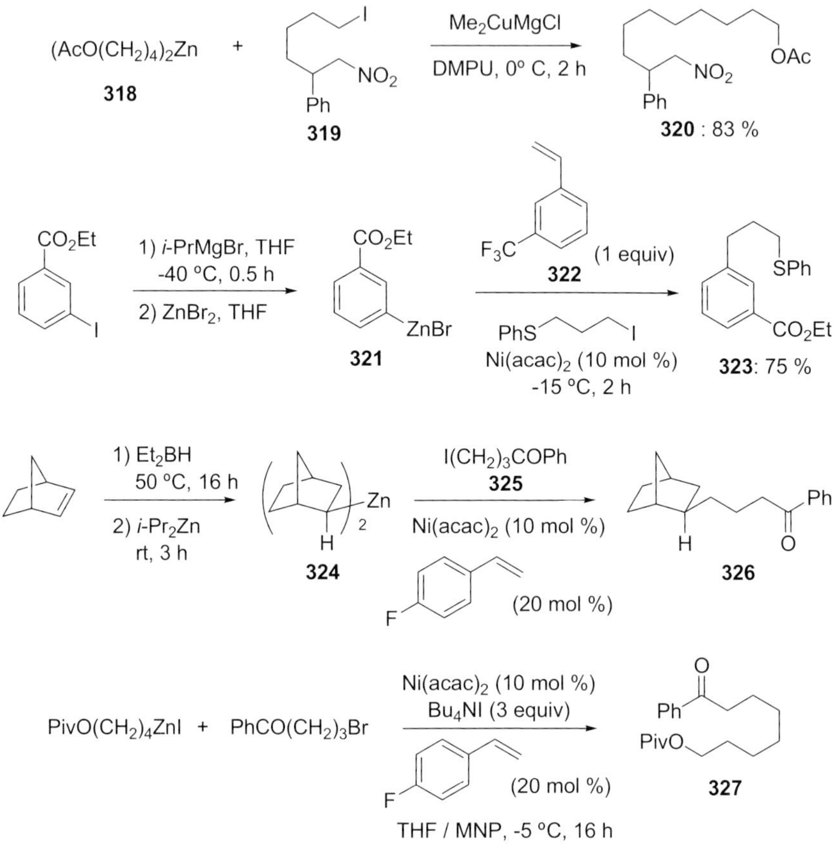

Scheme 7.88 Cross-coupling reactions with zinc organometallics.

is converted in the usual way into the zinc-copper derivative **332** that is readily acylated leading after aqueous workup to the functionalized indole **333** in 73% yield (Scheme 7.89) [50]. *Bis*-zinc organometallics, such as **334** and **335** are also acylated after transmetallation with CuCN·2LiCl leading to the corresponding diketones **336** [21] and **337** [200].

Allylic zinc reagents are highly reactive and add to acid chlorides and anhydrides. A double addition of the allylic moiety usually occurs leading to tertiary alcohols [15,20]. The double addition can be avoided by using a nitrile as substrate (Blaise reaction). By using Barbier conditions, it was possible to generate the zinc reagent corresponding to the bromide **338** and to add it to a nitrile. After acidic workup, the unsaturated ketone **339** is obtained in 82% yield (Scheme 7.90) [202,203].

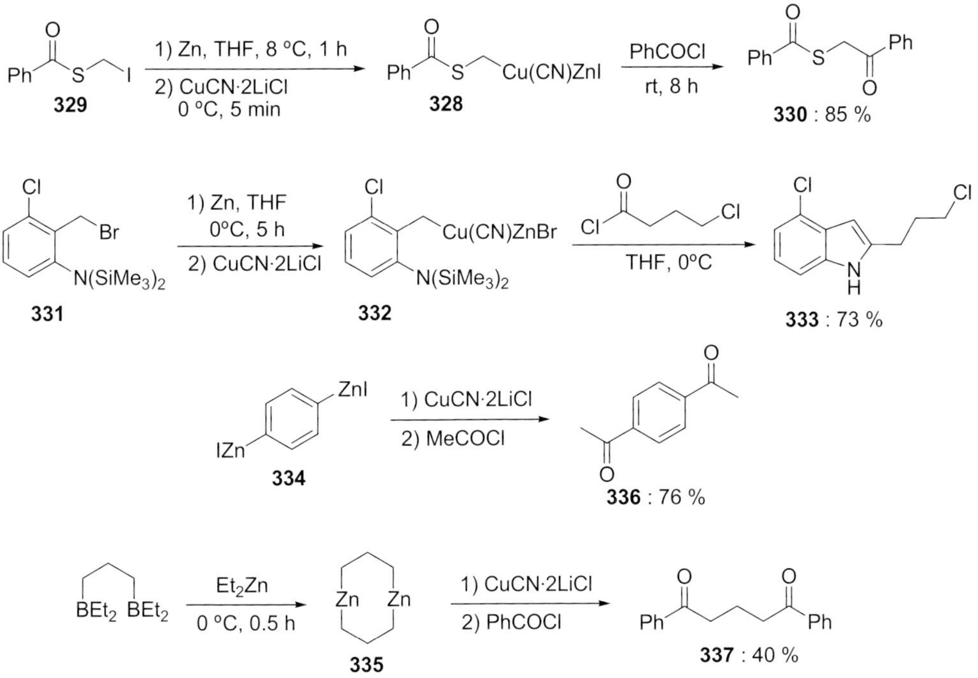

Scheme 7.89 Acylation of zinc–copper reagents.

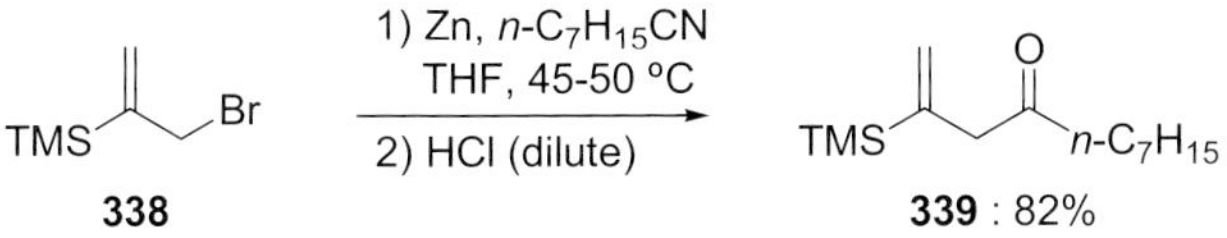

Scheme 7.90 Acylation by the addition of an allylic zinc reagent to a nitrile.

7.3.2.3 Addition Reactions

Allylic and to the same extent propargylic zinc reagents add to aldehydes, ketones and imines under mild conditions [203]. Thus, 2-carbethoxyallylzinc bromide (**340**) that is readily prepared by the reaction of ethyl (2-bromomethyl)acrylate [204] with zinc dust in THF (17–20 °C, 0.5 h) adds to a range of aldehydes and imines to provide α-methylene-γ-butyrolactones and lactams [205]. Chiral lactams can be prepared by adding the allylic zinc (**183**) to imines bearing amino alcohol substituents [206]. Functionalized allylic zinc reagents have been used to prepare a range of heterocycles, such as **341** [207]. The addition of propargylic zinc halides to aldehydes or ketones often provides mixtures of homopropargylic and allenic alcohols [208].

Interestingly, silylated propargylic zinc reagents, such as **342** may be better viewed as the allenic zinc reagent **343** that reacts with an aldehyde via a cyclic transition state affording only the *anti*-homopropargylic alcohol **344** with 90% yield

(Scheme 7.91) [137]. Alkylzinc halides react only sluggishly with aldehydes or ketones. This reactivity can be improved by activating the carbonyl derivative with a Lewis acid. Excellent results are obtained with titanium alkoxides [209], Me$_3$SiCl [210] or BF$_3 \cdot$OEt$_2$ [48c]; (Scheme 7.92). Depending on the nature of the catalyst either the 1,2-addition product (**345**) or the 1,4-addition product (**346**) is obtained by the addition of the zinc–copper reagent **347** to cinnamaldehyde [48c].

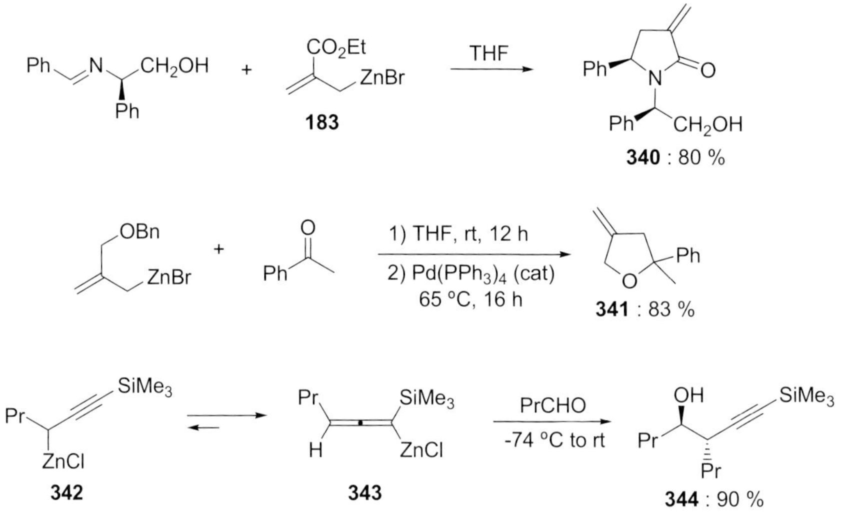

Scheme 7.91 Reaction of allenic, propargylic and allylic zinc reagents with carbonyl derivatives.

In a noncomplexing solvent like CH$_2$Cl$_2$, alkylzinc reagents, such as **348** and ent-**348** react with polyfunctional aldehydes leading to highly functionalized alcohols, such as **349a,b** with high diastereoselectivity in the "matched" case; Scheme 7.92 [158, 210]. Polyoxygenated metabolites of unsaturated fatty acids have been prepared by the addition of functionalized zinc–copper reagents to unsaturated aldehydes in the presence of BF$_3 \cdot$OEt$_2$ providing allylic alcohols of type **350** [211]. Allylic zinc reagents readily add to aldehydes with good stereoselectivity in some cases. Thus, the addition of dicrotylzinc [212a] to the amino-aldehyde **351** furnishes the alcohol **352** as one diastereomer in 82% yield. Interestingly, the corresponding Grignard reagent leads to a mixture of diastereoisomers (Scheme 7.93) [212b]. The addition of copper–zinc organometallics to imines is difficult. Benzylic zinc reagents display, however, an enhanced reactivity and react directly under well-defined conditions with *in situ* generated imines. A diastereoselective one-pot addition of functionalized zinc organometallics can be realized by performing the reaction in 5M LiClO$_4$ (in ether) in the presence of TMSCl. By using Reformatsky reagents and a chiral amine like (*R*)-phenylethylamine diastereoselectivities with up to 95% have been obtained. With the benzylic zinc reagent **353**, the secondary amine **354** is obtained in 60% yield and 70% *de* [213].

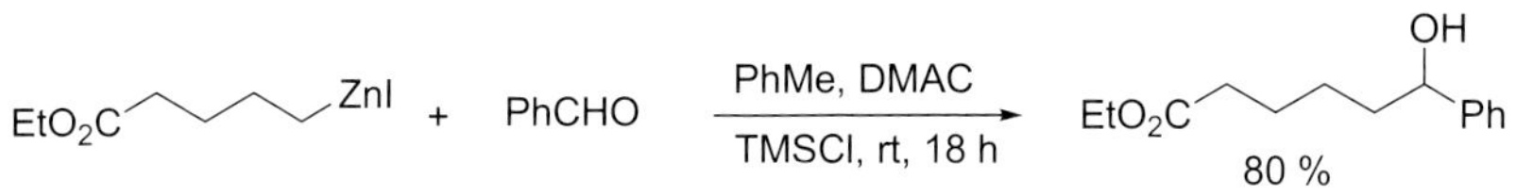

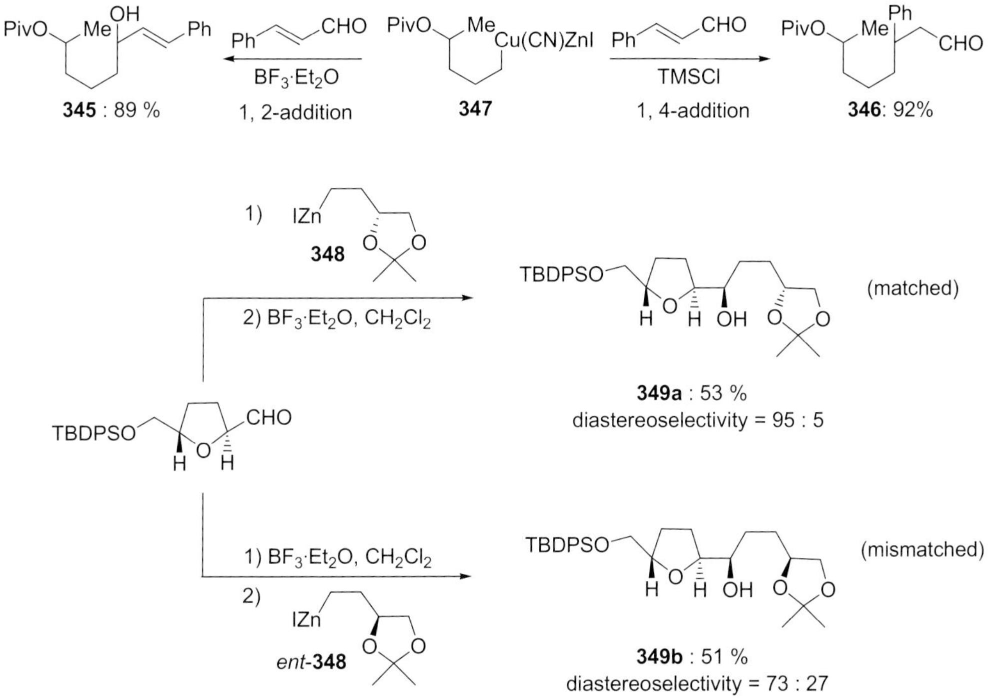

Scheme 7.92 Selectivity in the reaction of zinc and zinc–copper reagents with aldehydes.

Mangeney and coworkers have found that alkylzinc reagents add to reactive imine derivatives and have used this property to prepare chiral amino-acids (Scheme 7.94) [214]. The addition of *tert*-BuZnBr to chiral α-imino-esters bearing a phenylglycinol unit provides in ether the corresponding adduct **356** with a diastereoselectivity of 96:4. After removal of the chiral inductor by hydrogenolysis, the corresponding amino ester **357** is obtained in 85% *ee*.

Copper–zinc reagents add to various pyridinium salts leading either to the 1,2- or to the 1,4-adduct depending on the substituent pattern of the pyridine ring (Scheme 7.95) [150,215].

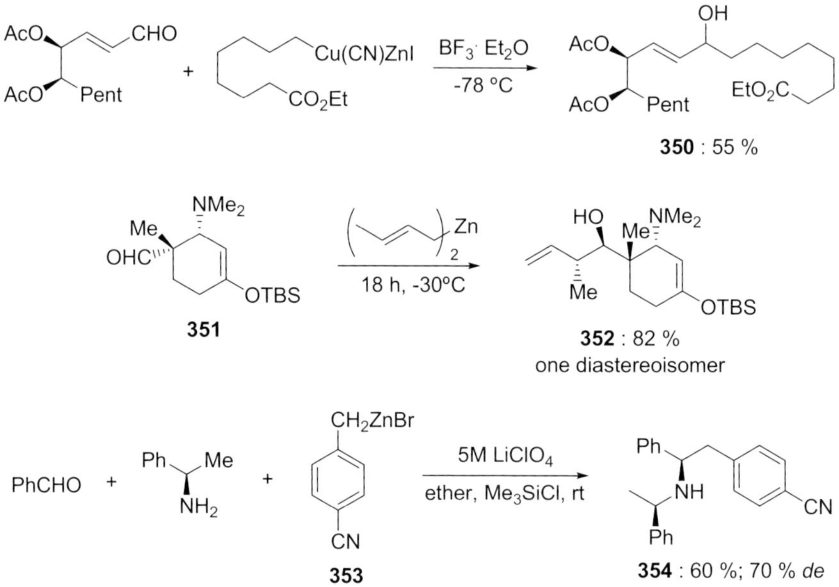

Scheme 7.93 Addition reactions of zinc and copper–zinc reagents to carbonyl compounds.

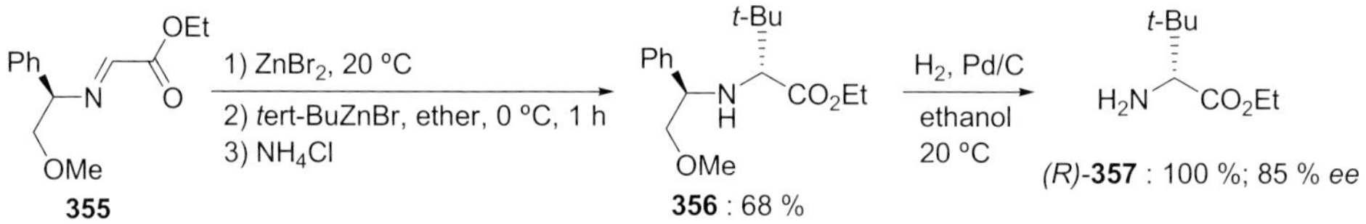

Scheme 7.94 Addition of a tertiary alkylzinc reagent to a chiral iminoester.

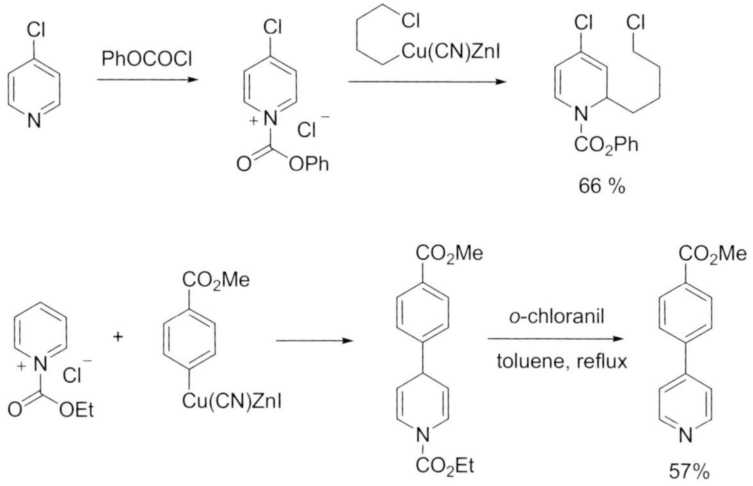

Scheme 7.95 Addition of copper–zinc reagents to pyridinium salts.

7.3.2.4 Michael Additions

Organocopper reagents derived from organolithium or magnesium derivatives add readily to various Michael acceptors [64]. The zinc–copper reagents obtained by reacting organozinc halides with CuCN · 2LiCl undergo numerous 1,4-addition reactions with various Michael acceptors. The presence of TMSCl is especially useful and effective [216]. It assures high yields of the conjugate adducts. Thus, the addition of various polyfunctional zinc–copper reagents with cyclohexenone provides the desired 1,4-addition product **358** in the presence of TMSCl in 95% yield [10, 217]. Sterically hindered enones, such as 3-cyclohexyl-2-cyclohexen-1-one **359** undergoes a conjugate addition in the presence of BF$_3$·OEt$_2$ affording the ketone **360** bearing a quaternary center in β-position (Scheme 7.96) [48a]. Unsaturated copper reagents are best added via alkenylzirconium species that are readily prepared by hydrozirconation. In the presence of catalytical amounts of Me$_2$Cu(CN)Li$_2$, they add to various enones affording the unsaturated ketones of type **361** in satisfactory yield (Scheme 7.96) [21].

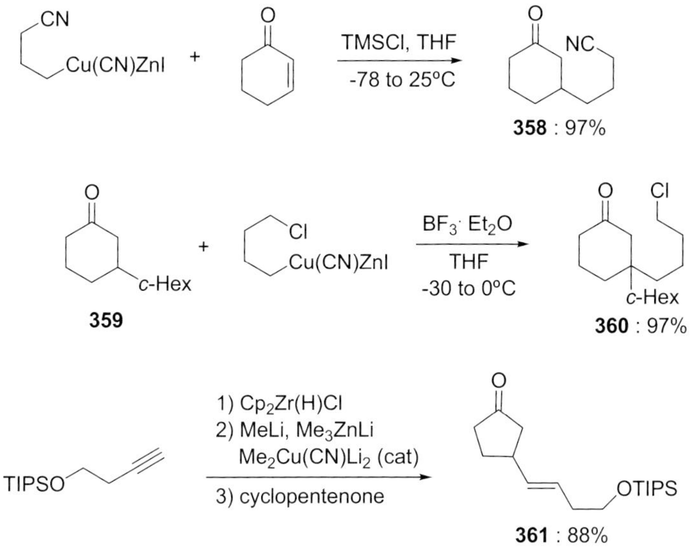

Scheme 7.96 Michael addition of zinc–copper reagents to enones.

Arylzinc reagents, obtained by the electrochemical reduction of the corresponding aromatic chloride or bromide using a sacrificial zinc electrode, allows the preparation of zinc reagents bearing a keto group [36]. *Exo*-methylene ketones such as **362** add various copper–zinc reagents. This methodology has been applied for the synthesis of various prostaglandins, such as **363** [218]. Enantioselective Michael additions have been pioneered by Feringa and coworkers [219] and Alexakis and coworkers [220]. Remarkably, only a catalytic amount of the chiral ligand **364** (4 mol%) and of Cu(OTf)$_2$ (2 mol%) is required. The 1,4-addition product **365** is obtained with an enantiomeric excess of 93% *ee* [219a]. It has been applied

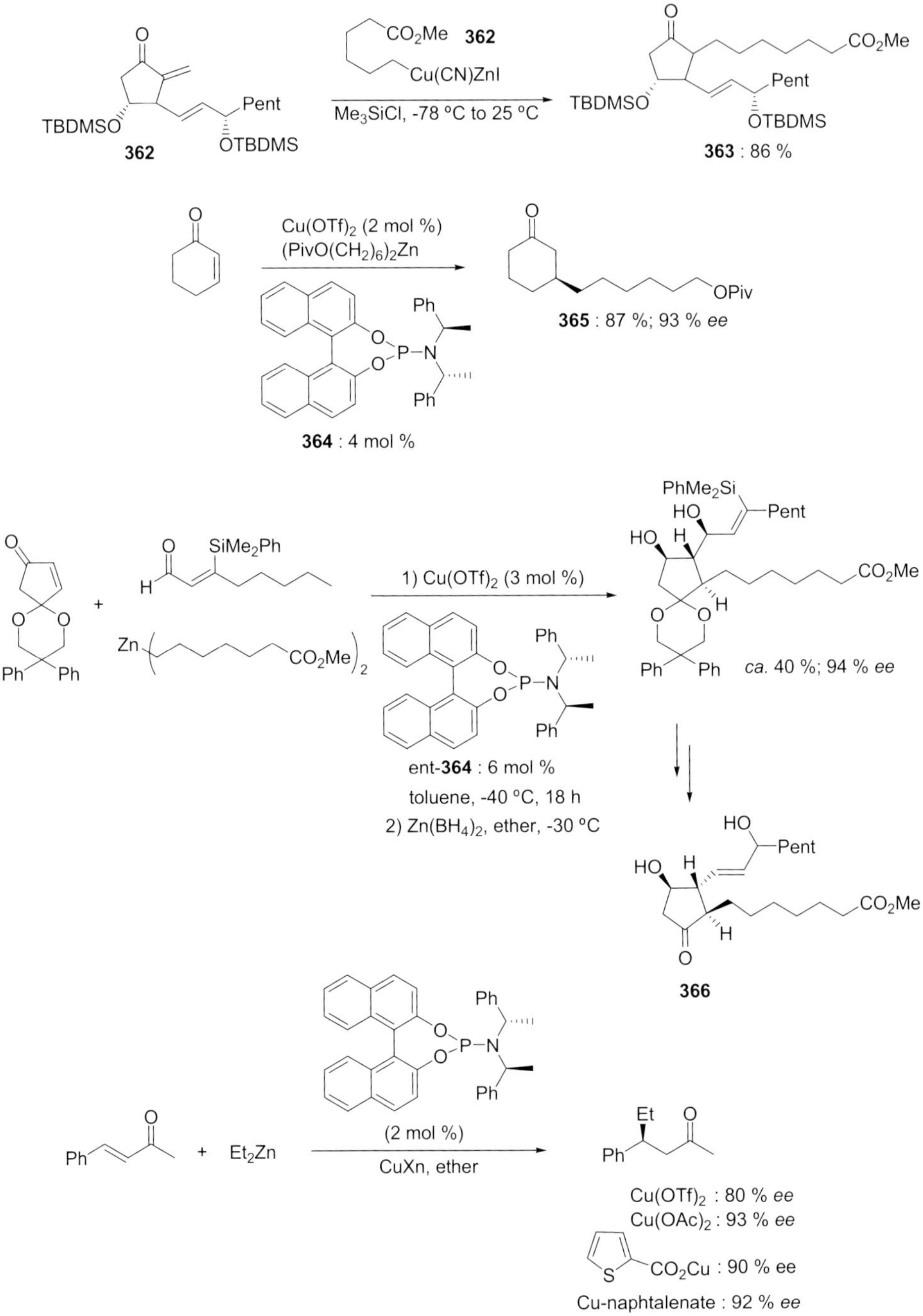

Scheme 7.97 Stereoselective Michael addition of zinc reagents to enones.

for an enantioselective synthesis of prostaglandin E_1 methyl ester (**366**) [221] and can be used for the performance of a highly regiodivergent and catalytic parallel kinetic resolution [222]. The nature of the copper salt strongly influences the enantioselectivity and copper carboxylates proved to be especially efficient (Scheme 7.97) [220c]. Hird and Hoveyda have developed a very efficient modular ligand based on various amino acids. The modular ligand **367** has been optimized for the enantioselective addition of Et_2Zn to the oxazolidinone **368** [223]. The resulting products of type **369** can be converted into other carbonyl compounds (ketones, Weinreb amides, carboxylic acids) by standard methods. A stereoselective synthesis of substituted pyrrolidines has been achieved by a sequential domino Michael addition and intramolecular carbozincation. Thus, the reaction of the acyclic ester **370** with a copper–zinc reagent such as **371** provides with high stereoselectivity the pyrrolidine **372** (d.r. = 95:5). The intermediate zinc–copper reagent obtained after cyclization can be trapped with an electrophile such as allyl bromide providing a product of type **373** in satisfactory yield [224].

The addition to α,β-unsaturated esters is usually difficult. However, under appropriate conditions, the 1,4-addition of diorganozincs to enoates is possible [225]. As mentioned above, Michael addition reactions can also be catalyzed by Ni(II)-salts [99]. The 1,4-addition of functionalized organozinc iodides to enones

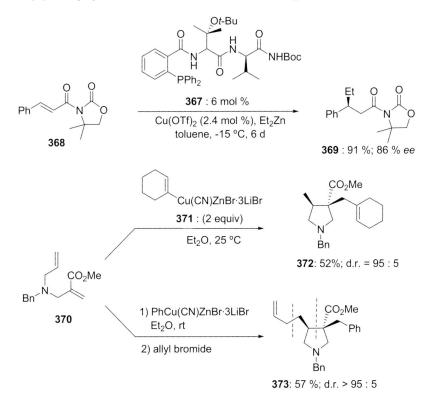

Scheme 7.98 Stereoselective additions of zinc–copper reagents to carbonyl compounds.

in the presence of Ni(acac)$_2$ in the presence of a diamine as ligand and TMSCl provides after hydrolysis the Michael adducts in satisfactory yields [226].

Nitroolefins are excellent Michael acceptors that react with a broad range of nucleophiles in a Michael fashion. The resulting functionalized nitroalkanes can be readily converted into amines by reduction reactions or to carbonyl compounds by a Nef reaction [227]. The addition of nucleophiles to nitroolefins is complicated by the subsequent addition of the resulting nitronate to remaining nitroolefin. Whereas such a side-reaction is quite fast for lithium and magnesium [228] nitronates, it is slow for zinc nitronates. Thus, various functionalized zinc–copper reagents add to nitroalkenes leading to Michael adducts, such as **374** in good yields [11,30]. The reaction with 1-acetoxy-2-nitro-2-propene (**375**) [172] and functionalized zinc–copper reagents provides an access to terminal nitroolefins, such as **376** [30]. The highly reactive reagent [229] undergoes the allylic substitution reaction already at –55 °C. Similarly, (2-phenylsulfonyl)nitroethylene (**377**) undergoes an addition-elimination reaction with the copper–zinc species **378** at –60 °C leading to the triene **379** in 92 % yield (Scheme 7.99) [30,229].

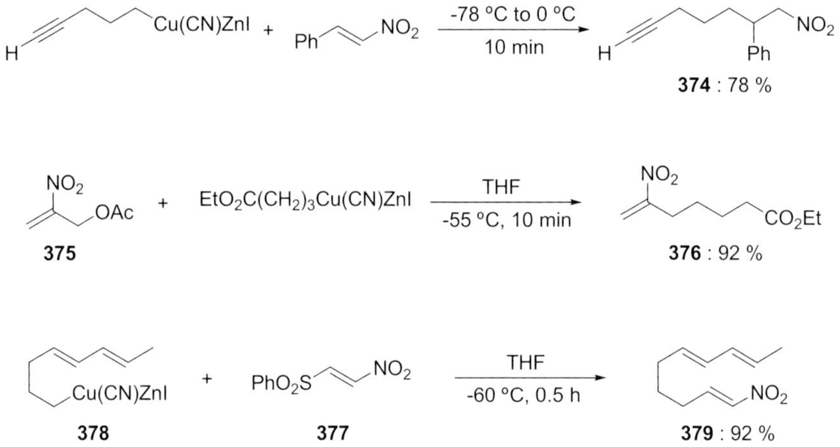

Scheme 7.99 Addition of zinc–copper reagents to nitroolefins.

Interestingly, *bis*(methylthio)-1-nitroethylene (**380**) reacts with dimetallic zinc–copper species leading to the corresponding *exo*-methylene cycloalkenes, such as **381** (Scheme 7.100) [30]. β-Disubstituted nitroolefins are especially difficult to prepare by nitroaldol condensation. The addition of zinc–copper reagents to nitroolefins followed by a reaction with phenylselenyl bromide produces after H$_2$O$_2$-oxidation E/Z-mixtures of β-disubstituted nitroalkenes, such as **382** (Scheme 7.100) [230].

Interestingly, the intermediate zinc nitronate resulting from the Michael addition, the oxidative Nef reaction can be directly performed leading to a polyfunctional ketone, such as **383** in high yield; Scheme 7.100 [30]. The addition of copper organometallics to triple bonds (carbocupration) [231] can be realized with copper reagents derived from zinc organometallics. The *syn*-addition of zinc–copper reagents to activated triple bonds, such as acetylenic esters is especially easy. The

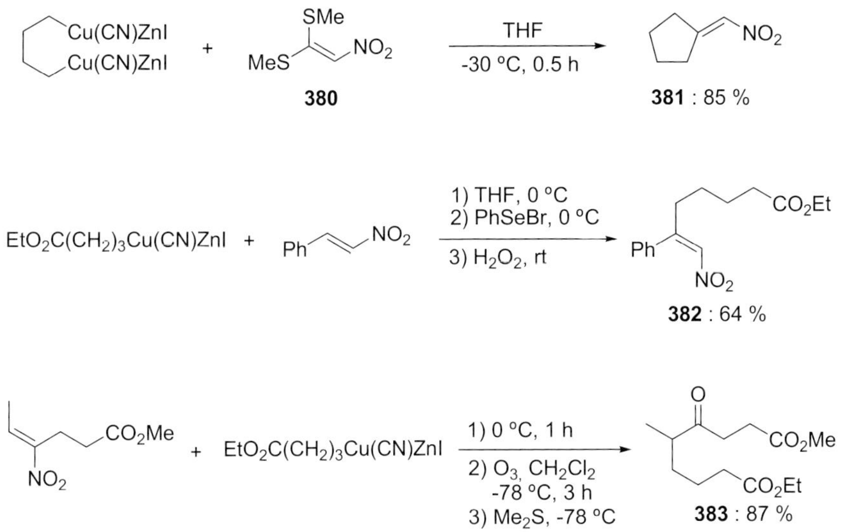

Scheme 7.100 Reaction of functionalized zinc–copper reagents with polyfunctional nitroolefins.

carbocupration of ethyl propiolate at –60 °C to –50 °C produces the *syn*-addition product **384** after protonation at low temperature. Interestingly, if the reaction mixture is warmed up in the presence of an excess of TMSCl, an equilibration occurs and the C-silylated unsaturated product **385** is obtained [48d]. The presence of acidic hydrogens of the amid group does not interfere with the carbocupration reaction. Thus, the unsaturated amide **386** affords, after addition, the *E*-unsaturated amide **387** in 53% as 10% *E*-isomer (Scheme 7.101) [11].

A formal [3+2]-cycloaddition can be accomplished by adding *bis*-(2-carbethoxy-ethyl) zinc to acetylenic esters [232c]. This reaction allows the construction of complex cyclopentenones, such as **388** which is a precursor of (±)-bilobalide (Scheme 7.101) [232c]. The allylzincation of trimethylsilylacetylenes can be performed intramolecularly providing a functionalized alkenylzinc that cyclizes in the presence of Pd(PPh₃)₄ at 25 °C within 3.5 h leading to the bicyclic product **389** in 84% yield (Scheme 7.102) [233]. The addition of allylic zinc halides to various alkynes occurs in the absence of copper salts. The related addition to 1-trimethylsilyl alkynes [234], unsaturated acetals [235] and cyclopropenes [236] occurs readily. Functionalized allylic zinc reagents can be prepared by the carbozincation of a dia-lkoxymethylenecyclopropane **390** with dialkylzincs. Thus, the reaction of **390** with Et₂Zn provides the allylic zinc reagent **391**. After the addition of an electrophile, the desired adduct **392** is obtained in good yield. [237].

Allylic zinc reagents are highly reactive reagents that are prone to undergo carbozincation of weakly activated alkenes. [143]. Thus, the addition of the mixed methallyl(butyl)zinc **393** with the vinylic boronate **394** provides the intermediate zinc reagent **395**. After the addition of an extra equivalent of ZnBr₂, CuCN · 2LiCl and allyl bromide, the reaction mixture was worked up oxidatively, providing the alcohol **396** in 83% yield. [238].

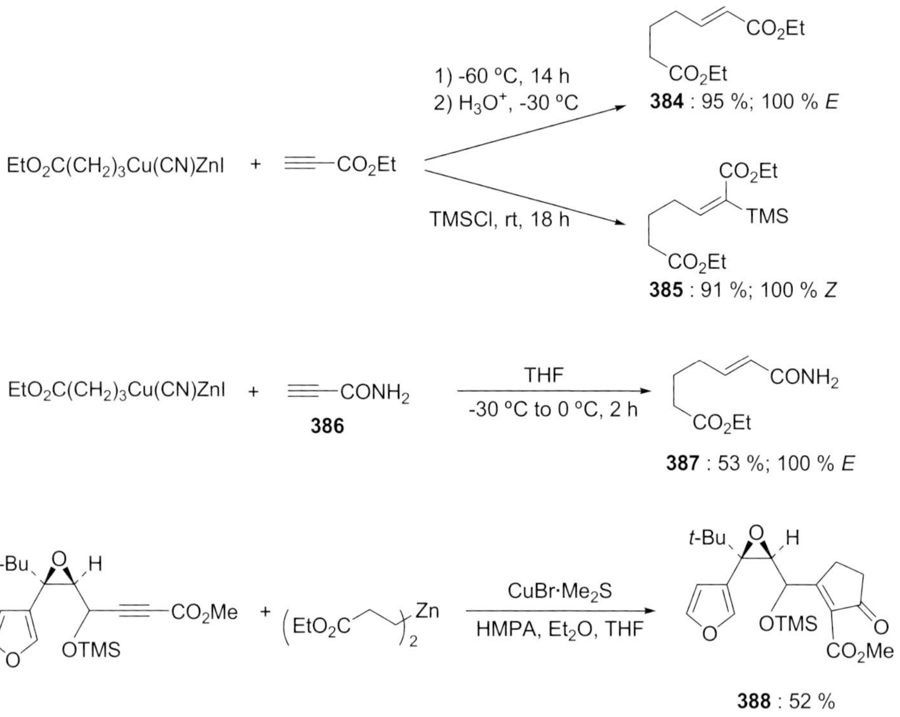
1) -60 °C, 14 h
2) H3O+, -30 °C
CO2Et
CO2Et
384 : 95 %; 100 % E
EtO2C(CH2)3Cu(CN)ZnI
+
CO2Et
TMSCl, rt, 18 h
CO2Et
TMS
CO2Et
385 : 91 %; 100 % Z
EtO2C(CH2)3Cu(CN)ZnI
+
CONH2
386
THF
-30 °C to 0 °C, 2 h
CONH2
CO2Et
387 : 53 %; 100 % E
t-Bu
O
H
OTMS
CO2Me
+
EtO2C
Zn
2
CuBr·Me2S
HMPA, Et2O, THF
t-Bu
O
H
O
OTMS
O
CO2Me
388 : 52 %

Scheme 7.101 Carbocupration of acetylenic carbonyl derivatives.

Functionalized allylic zinc reagents can also be generated *in situ*. The direct cyclization of allenyl aldehydes such as **397** with diorganozincs in the presence of Ni(cod)$_2$ (10–20 mol%) provides cyclic homoallylic alcohols such as **398** with good diastereoselectivity [239]. The allylzincation of alkenylmagnesium reagents is a very convenient synthesis of 1,1-bimetallic reagents of magnesium and zinc [240]. Thus, the addition of the ethoxy-substituted allylic zinc reagent (**399**) to the alkenylmagnesium reagent **400** providing after the addition of Me_3SnCl (1 equiv) the α-stannylated alkylzinc species **401** that is readily oxidized with O_2 leading to the aldehyde **402** with an excellent transfer of the stereochemistry (Scheme 7.103) [241].

The activation of primary zinc–copper reagents with $Me_2Cu(CN)Li_2$ allows to carbocuprate weakly activated alkynes. The carbocupration of the alkynyl thioether **403** is leading to the *E*-alkene **404** with high stereoselectivity [242]. Although, primary alkylzinc regents do not add to unactivated alkynes, the addition of the more reactive secondary copper–zinc reagents affords the desired product **405** with high *E*-stereoselectivity [242]. The intramolecular carbocupration proceeds also with primary copper organometallics leading after allylation of the *exo*-alkenylidene to cyclopentane derivative **406** in 60% yield (Scheme 7.104) [242]. The addition of zinc malonates, such as **407** produces the carbometallation product **408** in 50% yield (Scheme 7.104) [243].

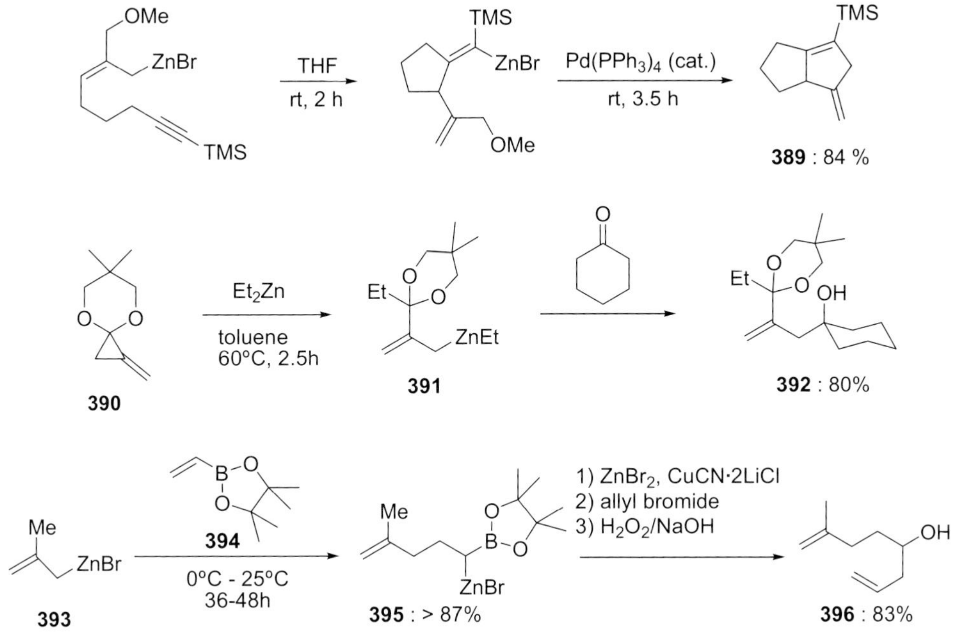

Scheme 7.102 Allylzincation of unsaturated compounds.

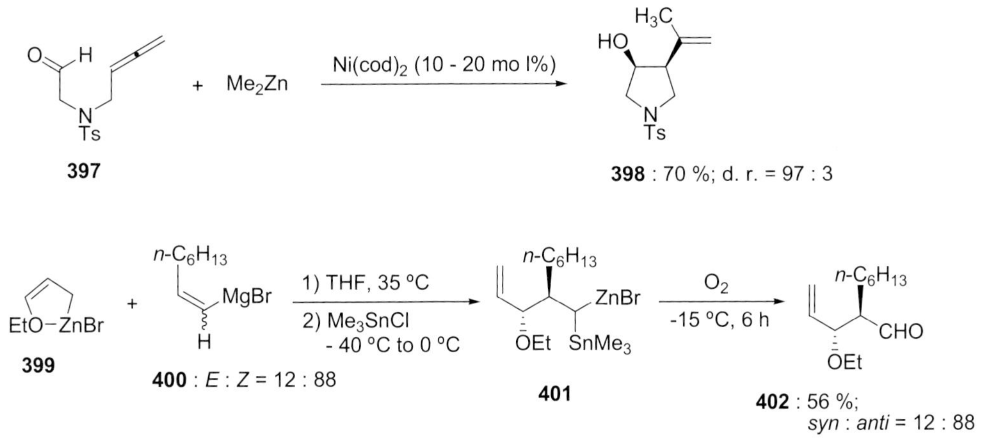

398 : 70 %; d. r. = 97 : 3

399

400 : *E* : *Z* = 12 : 88

401

402 : 56 %;
syn : *anti* = 12 : 88

Scheme 7.103 Allylzincation reactions.

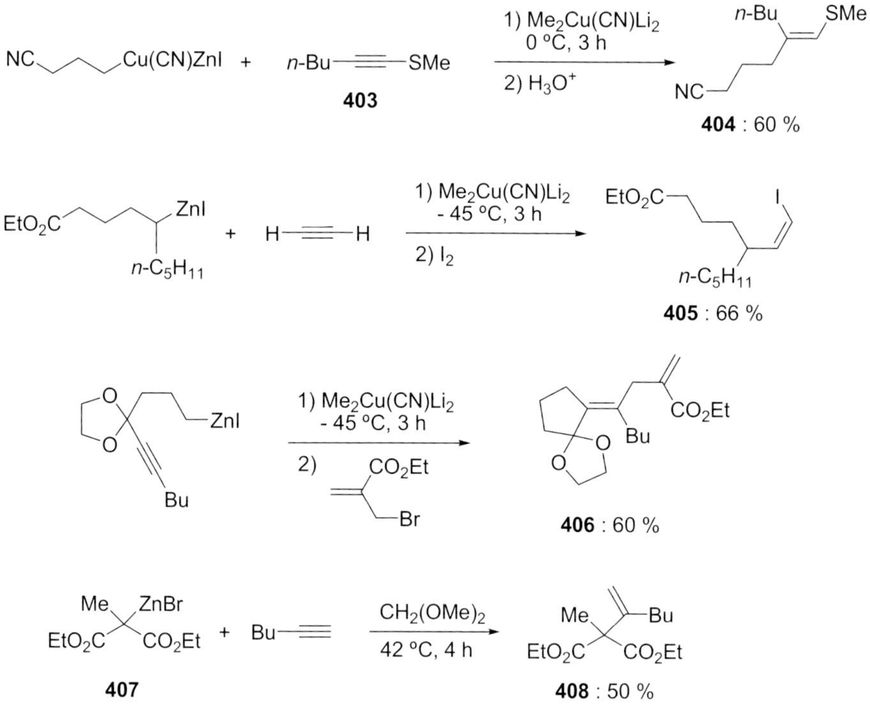

Scheme 7.104 Carbozincation and carbocupration of alkynes.

7.3.3
Palladium- and Nickel-catalyzed Reactions

Negishi and coworkers discovered 25 years ago that organozinc halides undergo smooth Pd(0)-catalyzed cross-coupling reactions with aryl, heteroaryl and alkenyl halides as well as acid chlorides [7, 244]. These cross-coupling reactions have a broad scope and have found many applications [7]. They have been performed with a range of polyfunctional organozinc halides [245]. These zinc organometallics may contain a silylated acetylene [246], an alkenylsilane [247], an allylic silane [248], an alkoxyacetylene [249], a polythiophene [250], polyfunctional aromatics [251], heterocyclic rings [252], an ester [21, 253], a nitrile [21, 58], a ketone [254], a protected ketone [252a], a protected aminoester [255], a stannane [256] or a boronic ester [53b]. The cross-coupling reaction with homoenolates proceeds especially well with *bis-(tris-o*-tolylphosphine)palladium dichloride [253a] leading to the desired cross-coupling products, such as **409**. Biphenyls, such as **410**, have been often prepared by using Pd(PPh₃)₄ as a catalyst [21]. Recently, Dai and Fu have demonstrated that Negishi cross-coupling reactions can be efficiently performed by using the sterically hindered phosphine (*t*-Bu)₃P as a ligand that results in very active catalytic species (*t*-Bu₃P)Pd [257]. In these cases, the cross-coupling can be performed with aryl chlorides and tolerates the presence of some functionalities. Thus, the cross-coupling between cheap 2-chlorobenzonitrile and *p*-tolylzinc chloride proceeds with exceptionally high turnovers (TON>3000) leading to the biphenyl **411** in 97% yield (Scheme 7.105) [257].

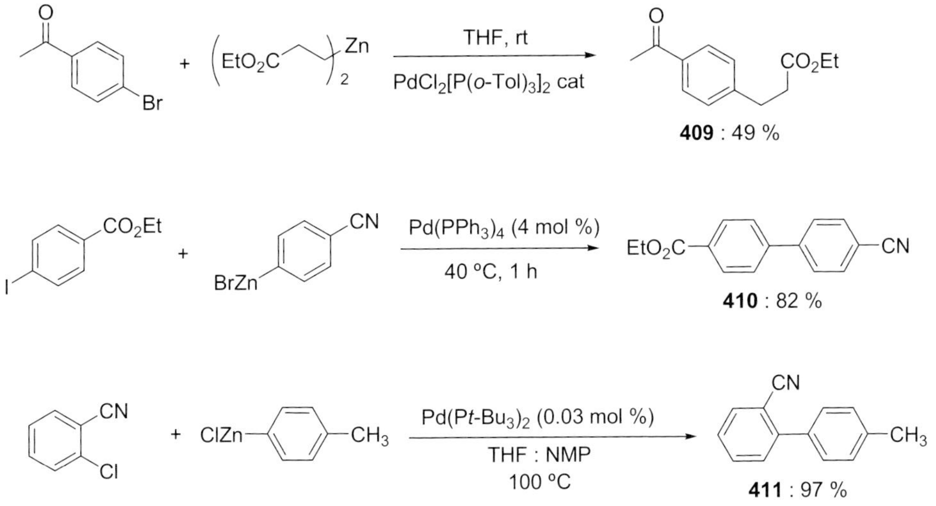

Scheme 7.105 Palladium-catalyzed cross-coupling reactions.

New α-amino acids such as **412** have been prepared in high optical purity by using the reaction of pyridyl bromide **413** with Jackson reagent **15** [25e] Fmoc-protected amino acids are routinely used in automated solid-phase peptide synthesis. The Fmoc-protected zinc reagent **414** is readily prepared from the corresponding alkyl iodide **415**. The Pd-catalyzed cross-coupling with various aryl iodides (Pd$_2$dba$_3$ (2.5 mol%), P(*o*-tol)$_3$ (10 mol%)) in DMF at 50 °C furnishes the corresponding arylated amino acid derivatives **416** in 25–59% yield [258]. Removal of the *tert*-butyl ester is readily achieved with Et$_3$SiH and TFA leading to Fmoc protected amino acids [259].

The reaction of the zinc homoenolate **417** with ketene acetal triflates such as **418** in the presence of a palladium(0)-catalyst (Pd(PPh$_3$)$_4$, 5 mol%) leads to the corresponding cross-coupling product **419**. This key sequence has been used for the iterative synthesis of polycyclic ethers [260]. A smooth cross-coupling is observed between (2*E*, 4*E*)-5-bromopenta-2,4-dienal **420** and various zinc reagents leading to dienic aldehydes of type **421** in good to excellent yields and high stereomeric purity. Using the isomeric (2*E*, 4*Z*)-5-bromopenta-2,4-dienal furnishes the corresponding diene with a partial isomerization of the double bonds (Scheme 7.107) [261]. Negishi cross-coupling reaction can be performed with complex alkenyl iodides such as **422** leading to the steroid derivative **423** in 88% yield. The palladium(0)-catalyst (Pd(PPh$_3$)$_4$) was generated *in situ* from Pd(OAc)$_2$ (10 mol%) and PPh$_3$ (40 mol%) (Scheme 7.107) [262].

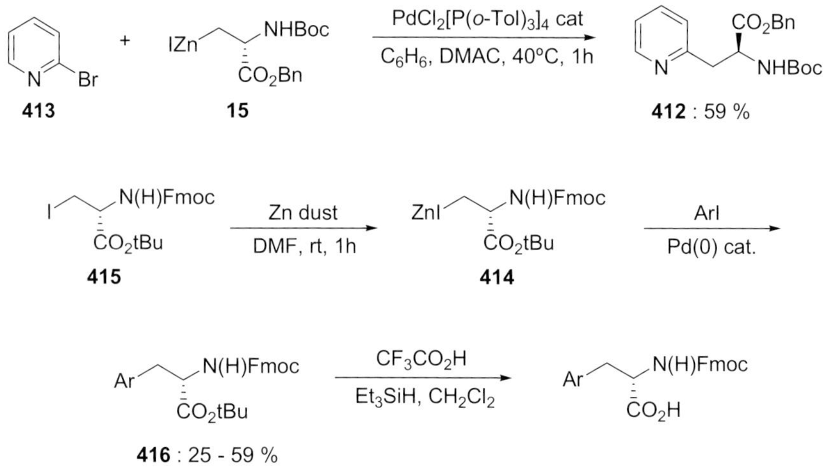

Scheme 7.106 Synthesis of α-amino acids by Negishi cross-couplings.

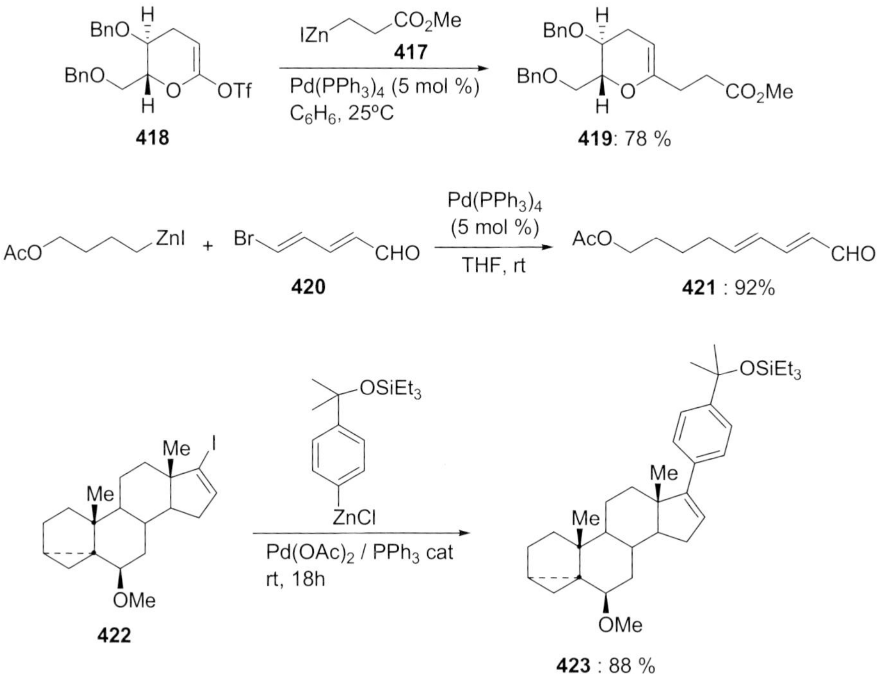

Scheme 7.107 Negishi cross-coupling of zinc organometallics with polyfunctional unsaturated derivatives.

Nickel salts are also excellent catalysts, however, the nature of the ligands attached to the nickel metal center is very important and often needs to be carefully chosen to achieve high reaction yields. Interestingly, nickel on charcoal

(Ni/C) proved to be an inexpensive heterogeneous catalyst for the cross-coupling of aryl chlorides with functionalized organozinc halides. Thus, the cross-coupling of 4-chlorobenzaldehyde with 3-cyanopropylzinc iodide produces in refluxing THF, the desired cross-coupling product **424** in 80% yield [263]. Tucker and de Vries developed an alternative homogeneous cross-coupling. Thus, the use of NiCl$_2$(PPh$_3$)$_2$ (7.5 mol%) and PPh$_3$ (15 mol%) was found to be an excellent cross-coupling catalyst for performing the cross-coupling between aryl chlorides such as **425** and functionalized arylzinc bromides like **426**. A complete conversion was obtained at 55 °C after a reaction time of 5 h affording the cross-coupling product **427** in 75% yield (Scheme 7.108) [264].

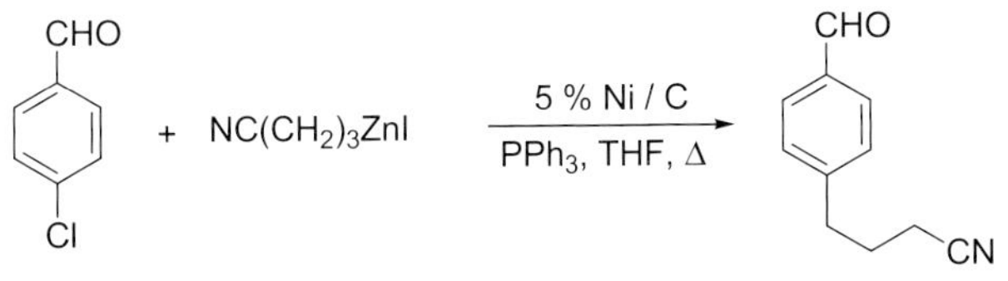

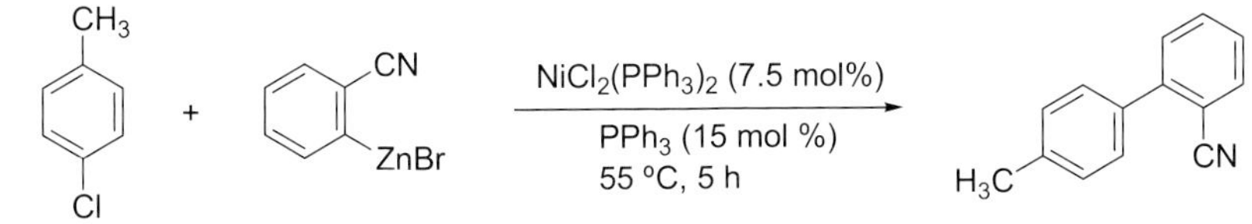

Scheme 7.108 Nickel-catalyzed cross-coupling reactions.

The cross-coupling of alkynylzinc halides or fluorinated alkenylzinc halides with fluorinated alkenyl iodides allows the preparation of a range of fluorinated dienes or enynes [265]. Functionalized allylic boronic esters can be prepared by the cross-coupling of (dialkylboryl)methylzinc iodide **428** with functionalized alkenyl iodides. The intramolecular reaction provides cyclized products, such as **429** (Scheme 7.109) [53c–e]. In some cases, reduction reactions [266] or halogen–zinc exchange reactions [40] are observed.

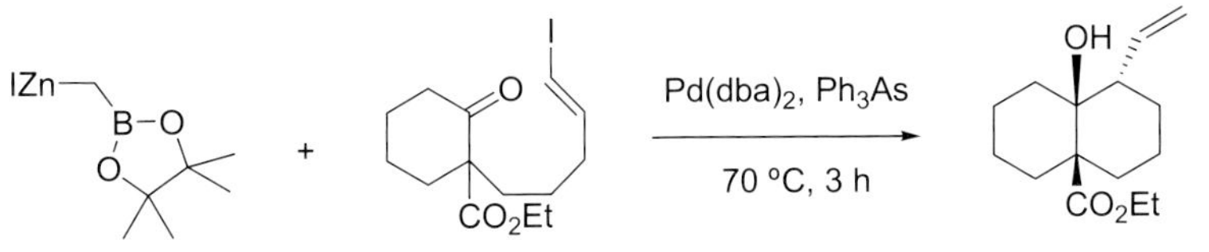

Scheme 7.109 *In situ* generation of allylic boronic esters via cross-coupling reactions.

Functionalized heterocyclic zinc reagents are very useful building blocks for the preparation of polyfunctional heterocycles, as shown with the pyridylzinc derivative **430** prepared by reductive lithiation followed by a transmetallation with zinc bromide [267]. The cross-coupling of the zinc reagent **430** with a quinolyl chloride provides the new heterocyclic compound **431**. The selective functionalization of positions 4 and 3 of pyridines is possible starting from 3-bromopyridine **432** that is selectively deprotonated in position 4 by the reaction with LDA in THF at −95 °C followed by a transmetallation with ZnCl$_2$. The resulting zinc species **433** undergoes a Pd-catalyzed cross-coupling with various aryl halides (Ar^1-X) affording products of type **434**. The cross-coupling of **434** with a zinc reagent Ar^2-ZnX in the presence of a Pd(0)-catalyst provides 3,4-diarylpyridines of type **435**. [268]. The functionalized iodoquinoline **436** reacts with *i*-PrMgCl at −30 °C providing an intermediate heteroarylmagnesium species that, after transmetallation to the corresponding zinc derivative **437**, undergoes a smooth cross-coupling reaction with

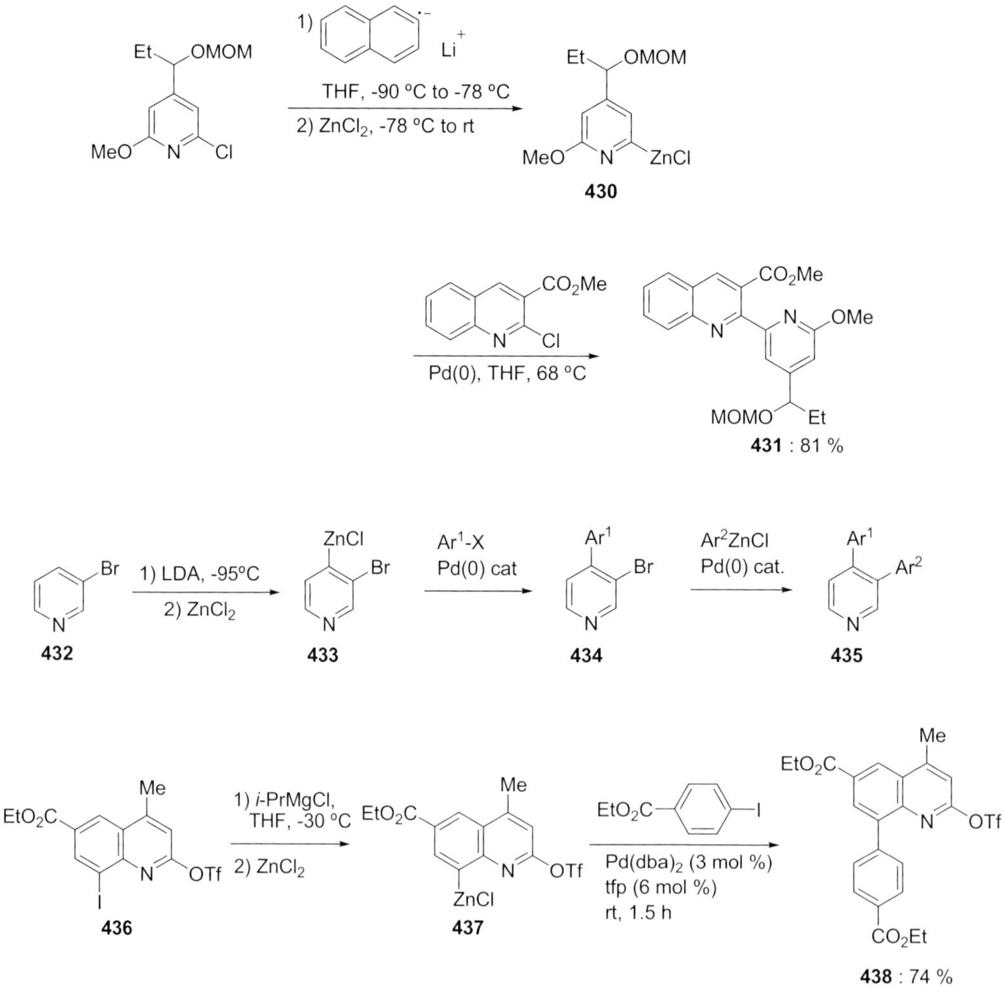

Scheme 7.110 Negishi cross-coupling with heterocyclic zinc reagents.

ethyl 4-iodobenzoate in the presence of Pd(dba)$_2$ (3 mol%) and tfp (6 mol%) in THF at rt providing the desired cross-coupling product **438** in 74% yield (Scheme 7.110) [269].

The selective functionalization of heterocycles is an important synthetic goal. Purines display multiple biological activities such as antiviral or cytostatic properties. The synthesis of analogs such as **441** can be achieved by a regioselective cross-coupling of functionalized benzylic zinc reagents such as **440** with the dichloropurine derivative **439** in the presence of Pd(PPh$_3$)$_4$ (5 mol%). [270]. L-Azatyrosine **442** is an anticancer lead compound that can be readily prepared using Jackson's reagent **15** and a Negishi cross-coupling reaction with 2-iodo-5-methoxy-pyridine **443** affording the amino-ester **444** in 59% yield. Its deprotection according to a procedure of Ye and Burke [271] produces L-azatyrosine **442** [272]. Functionalized alkenylzinc species such as **445** can be prepared via an iodine–magnesium exchange followed by a transmetallation with ZnBr$_2$. The Pd-catalyzed cross-coupling reaction with 4-iodobenzonitrile proceeds at 60 °C in THF leading to the arylated product **446** in 57% yield [273]. 1,1-Dibromo-1-alkenes of type **447** undergo a highly *trans*-selective Pd-catalyzed cross-coupling with alkylzinc reagents using *bis*-(2-diphenylphosphinophenyl) ether (DPE-phos) as a ligand. The use of Pd(*t*-Bu$_3$P)$_2$ is crucial for achieving stereospecific methylation with nearly 100% retention of configuration (Scheme 7.111) [274].

Ferrocenyl groups are useful tools for stereoselective syntheses. Hayashi and coworkers have discovered a novel class of ferrocenyl catalysts allowing the kinetic resolution of benzylic zinc derivatives, such as **448**. [275]. The racemic mixture of the benzylic zinc reagent **448** obtained by transmetallation from the corresponding Grignard reagent reacts with vinyl bromide in the presence of a Pd catalyst and the ferrocenyl aminophosphine **449** leading to the asymmetric cross-coupling product **450** with up to 85–86% *ee*. Complex ferrocenyl derivatives such as **453** being of interest as molecular materials with large second-order nonlinear optical properties (NLO) have been prepared by the cross-coupling of the ferrocenylzinc reagent **452** with 1,3,5-tribromobenzene. The starting zinc reagent has been prepared starting from the chiral ferrocenyl derivative **451**. Selective metallation of **451** with *tert*-BuLi followed by a transmetallation with ZnCl$_2$ furnishes the zinc reagent **452**. Negishi cross-coupling was best performed using PdCl$_2$(PPh$_3$)$_2$ (5 mol%) as catalyst (Scheme 7.112) [276].

Further applications of Negishi's cross-coupling reactions for the synthesis of new chiral ferrocenyl ligands have been reported [277]. Interestingly, cross-coupling reactions can also be performed using triorganozincates, such as **454** [278]. Chloroenyne **455** reacts with various magnesium zincates such as **454** generated *in situ* in the presence of PdCl$_2$(dppf) (5 mol%) in THF at 66 °C showing that alkenyl chlorides insert readily Pd(0)-complexes. Especially easy is the reaction of conjugated chloroenynes such as **455** leading to the enyne **456** (Scheme 7.113) [278]. A selective cross-coupling reaction of (*Z*)-2,3-dibromopropenoate **457** with organozinc compounds allows the preparation of highly functionalized enoates such as **458** (Scheme 7.113).

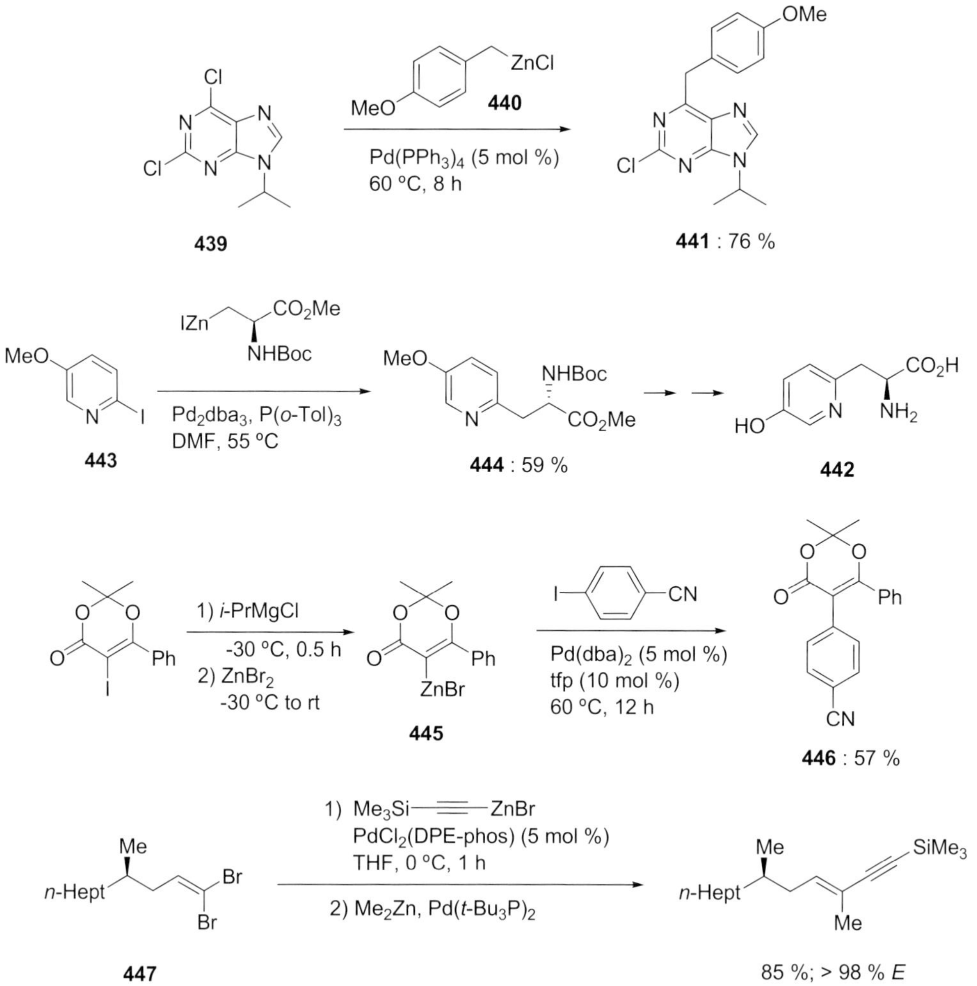

Scheme 7.111 Negishi cross-coupling reactions.

A flexible and convergent access to 2,3-disubstituted benzo [b]thiophenes, such as **459** has been developed. The most concise approach involves a sequential coupling of an *o*-bromoiodobenzene, such as **460** with benzylmercaptan and zinc acetylides leading to the adduct **461**. Treatment with iodine followed by an iodine–magnesium exchange and acylation provides the polyfunctional benzofuran derivatives like **459** [279]. Zeng and Negishi have recently developed a novel highly selective route to carotenoids and related natural products via Zr-catalyzed carboalumination and Pd-catalyzed cross-coupling of unsaturated zinc reagents. Starting from *β*-ionone (**462**), the first cross-coupling sequence affords the unsaturated product (**463**) that by an iterative reaction sequence and dimerization affords *β*-carotene (**464**) with high stereoisomeric purity (Scheme 7.114) [280].

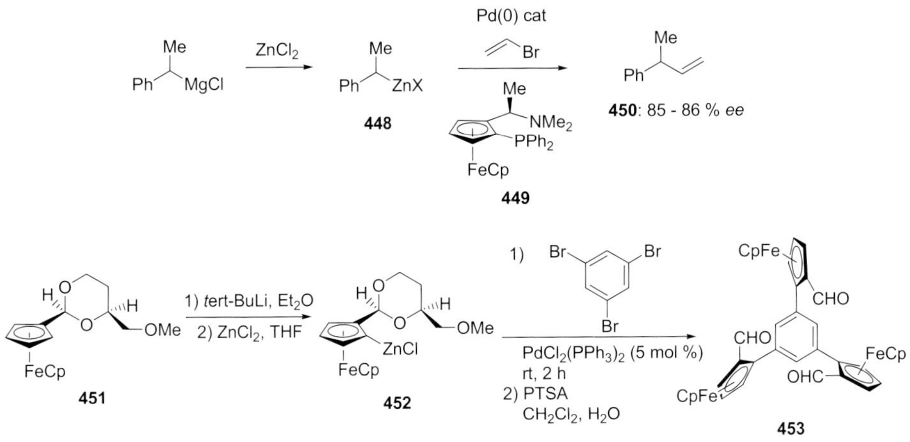

Scheme 7.112 Ferrocenyl groups in cross-coupling reactions.

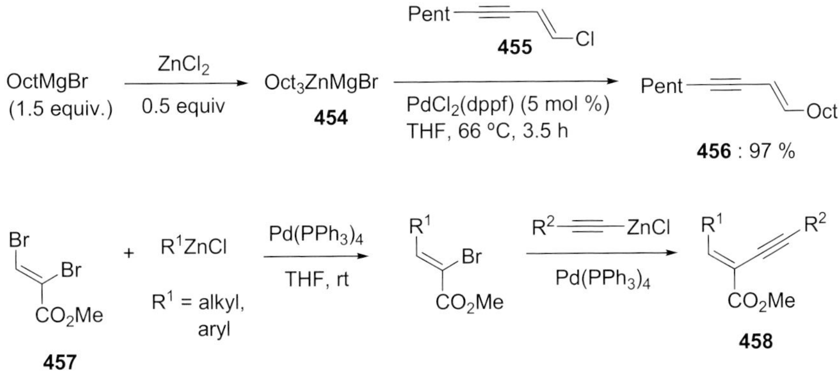

Scheme 7.113 Synthesis of enynes using zinc organometallics.

Acylation of organozincs with acid chlorides is efficiently catalyzed by palladium(0)-complexes [244c,d, 21]. Many functional groups are tolerated in this reaction. The cross-coupling of 3-carbethoxypropylzinc iodide with methacryl chloride provides an expeditive approach to polyfunctional enones, such as **465** [281]. The acylation of serine or glutamic acid-derived zinc species as developed by Jackson [25–27] provides chiral γ-keto-α-amino acids in good yields. The acylation of the Jackson reagent with phenyl chloroformate or the direct reaction of an organozinc reagent with carbon monoxide under sonication in the presence of catalytic amounts of $(PPh_3)_2PdCl_2$ leads to the C_2-symmetrical ketone **466** [25k]. In a related reaction, organozinc halides are treated with carbon monoxide and an allylic benzoate in the presence of a catalytic amount of palladium(0) complex and provides S-ketoesters (Scheme 7.115) [282]. The reaction of *syn*- or *anti*-3-iodo-2-methylbutanamide **467** with zinc powder furnishes the zinc reagent **468** that reacts with benzoyl chloride in the presence of Pd(0)-catalyst leading to the *syn*-product **469** (Scheme 7.115) [283].

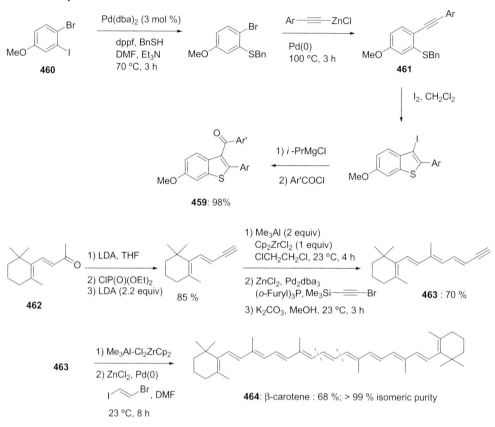

Scheme 7.114 Further applications of the Negishi cross-coupling reaction.

The reaction of thioesters such as **470** in the presence of nonpyrophoric Pd(OH)$_2$/C (Pearlman's catalyst) with various functionalized organozinc halides leads to functionalized unsymmetrical ketones such as **471** in high yield (Fukuyama reaction) [284]. This catalyst can also be used for Sonogashira and Suzuki reactions [285]. The Ni(acac)$_2$-catalyzed cross-coupling of the functionalized zinc reagent with various thioesters such as **472** provides under mild conditions the acylation product **473** in 74% yield. Interestingly, the addition of the organometallic zinc species to the thiolactone **474** furnishes after acidic treatment the vinylic sulfide **475** in 81% yield. The use of bromine for the activation of zinc dust for the preparation of the zinc reagent was found to be advantageous (Scheme 7.116) [286].

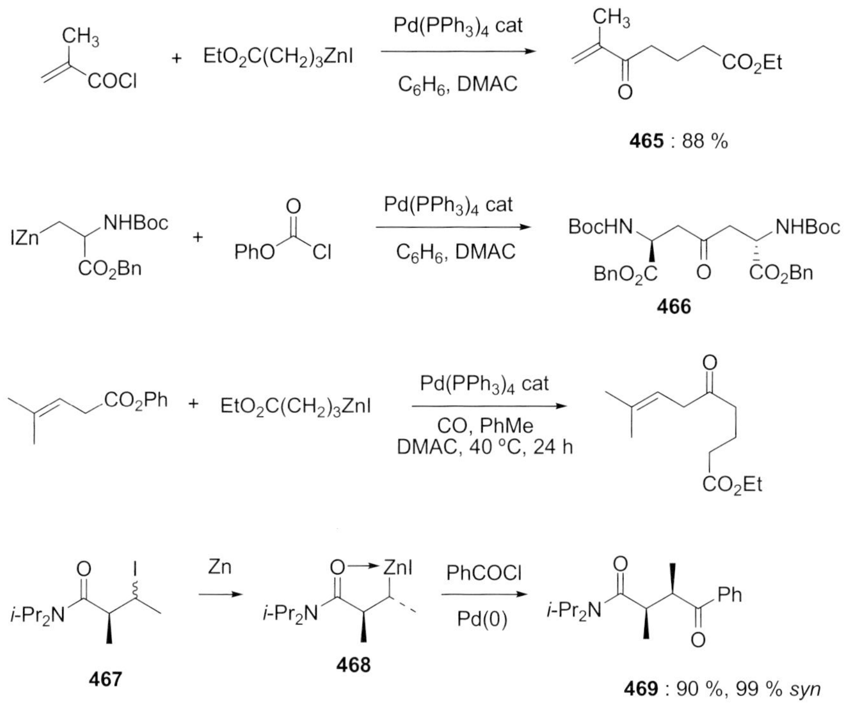

Scheme 7.115 Palladium-catalyzed acylation reactions of organozincs.

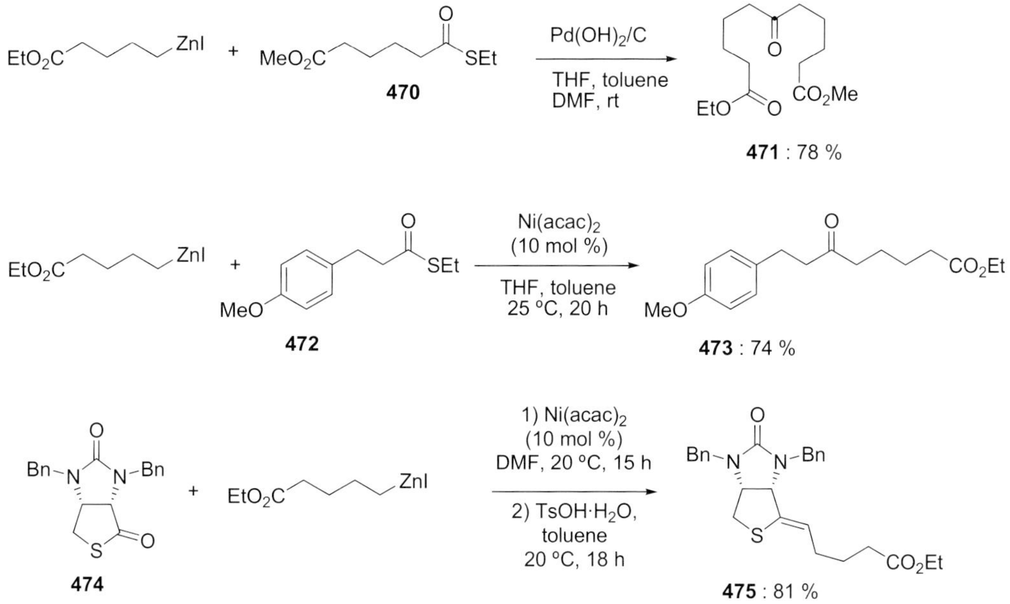

Scheme 7.116 Acylation of thioesters with zinc organometallics.

7.3.4
Reactions Catalyzed by Titanium and Zirconium(IV) Complexes

As mentioned previously, Lewis acids accelerate the addition of zinc organometallics to carbonyl derivatives. Titanium and zirconium(IV) salts are especially efficient catalysts. Oguni and coworkers [287] have shown in pioneering work that various chiral amino-alcohols catalyze the addition of diethylzinc to aldehydes [288]. Yoshioka and coworkers have shown that the 1,2-*bis*-sulfonamide **476** is an excellent ligand for the asymmetric addition of Et$_2$Zn to various aldehydes [289]. The catalysts of the TADDOL-family such as **477a** and **477b** are remarkable catalysts that tolerate many functional groups in the aldehyde as well as in the diorganozinc. These catalysts were discovered by Seebach and coworkers and they have found numerous applications in asymmetric catalysis [290]. The convenient preparation of diorganozincs starting from alkylmagnesium halides using ZnCl$_2$ in ether as the transmetallating reagent followed by the addition of 1,4-dioxane constitutes a practical method for the enantioselective addition of dialkylzincs [290f]. The enantioselective addition of polyfunctional diorganozincs is especially interesting. Thus, the reaction of the zinc reagent **478** with benzaldehyde in the presence of TADDOL **477a** provides the functionalized benzylic alcohol **479** in 84% *ee* [290f]. The more bulky ligand **477b** allows the addition of diorganozincs such as **480** to acetylenic aldehydes leading to propargylic alcohols like **481** in 96% *ee* (Scheme 7.117) [290g].

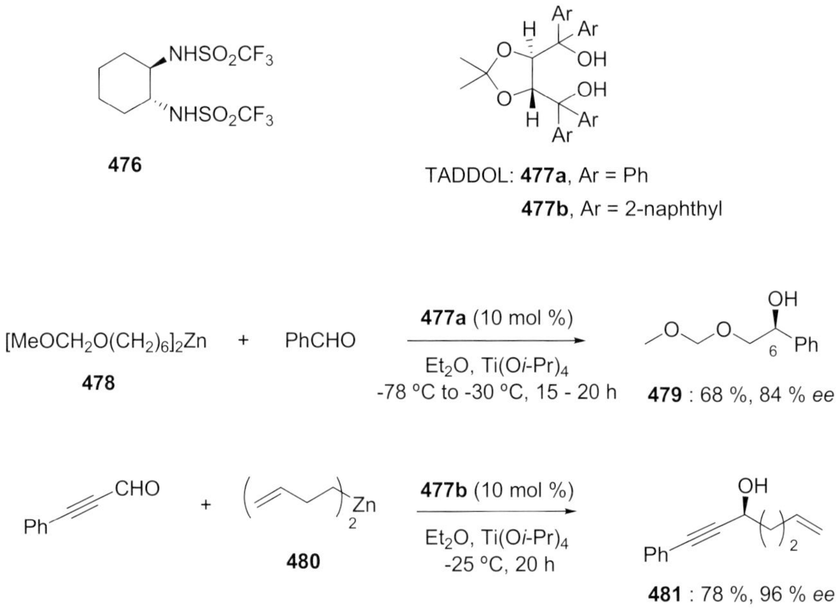

476

TADDOL: **477a**, Ar = Ph
        477b, Ar = 2-naphthyl

[MeOCH$_2$O(CH$_2$)$_6$]$_2$Zn + PhCHO

478

477a (10 mol %)
───────────────→
Et$_2$O, Ti(O*i*-Pr)$_4$
-78 °C to -30 °C, 15 - 20 h

479 : 68 %, 84 % *ee*

480

477b (10 mol %)
───────────────→
Et$_2$O, Ti(O*i*-Pr)$_4$
-25 °C, 20 h

481 : 78 %, 96 % *ee*

Scheme 7.117 Enantioselective addition of functionalized diorganozincs to aldehydes catalyzed by TADDOL.

This approach has been extended to the conjugate addition of primary dialkyl-zincs to 2-aryl- and 2-heteroaryl-nitroolefins and allows the preparation of enantio-enriched 2-arylamines [291]. Dendric styryl TADDOLS [292, 293] and polymer-bound Ti-TADDOLates [294] have proved to be very practical chiral catalysts for the enantioselective addition to aldehydes. Likewise, the immobilization of BINOL by cross-linking copolymerization of styryl derivatives has allowed several enantio-selective Ti and Al Lewis-acid-mediated additions to aldehydes [294]. Dialkylzincs obtained via an I/Zn-exchange or a B/Zn-exchange have also been successfully used for the enantioselective additions to a variety of aldehydes [295]. The use of *trans*-(1R,2R)-*bis*(trifluoromethanesulfamido)cyclohexane (**476**) is an excellent chi-ral ligand. The presence of an excess of titanium tetraisopropoxide (2 equiv) is, however, required [295]. The addition to unsaturated α-substituted aldehydes gives excellent enantioselectivities leading to polyfunctional allylic alcohols, such as **482** [93b]. Replacement of Ti(O*i*-Pr)$_4$ with the sterically hindered titanium alkoxide (Ti(O*t*-Bu)$_4$) leads to higher enantioselectivities [296].

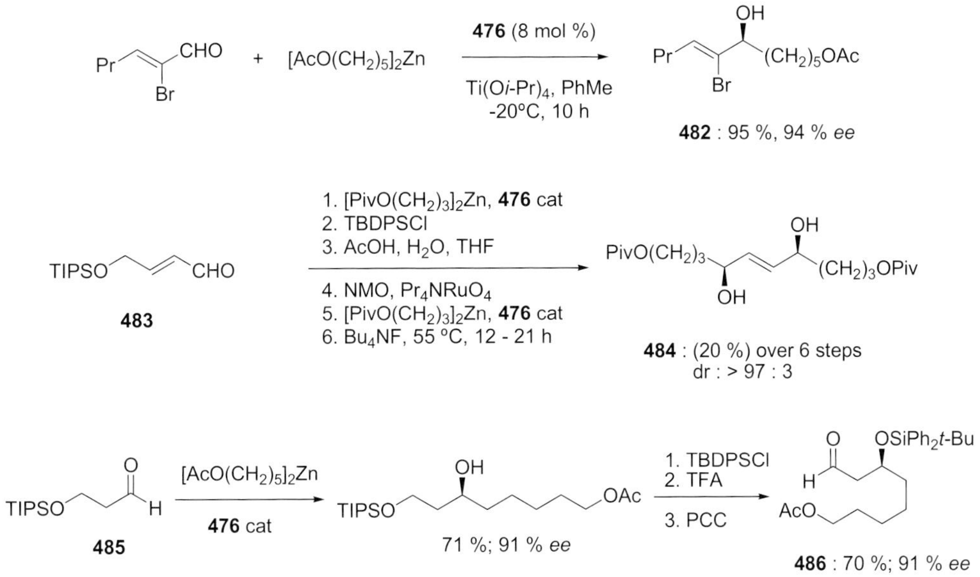

Scheme 7.118 Enantioselective addition of functionalized dialkylzincs to functionalized aldehydes.

This enantioselective preparation of allylic alcohols has been applied to the syn-thesis of the side chain of prostaglandins [297]. The addition to functionalized aldehydes, such as **483** allows the synthesis of C_2-symmetrical 1,4-diols, such as **484** with excellent diastereoselectivity and enantioselectivity [93b, 298]. An exten-sion of this method allows the synthesis of C_3-symmetrical diol [299]. Aldol-type products result from the catalytic enantioselective addition of functionalized di-alkylzincs to 3-TIPSO-substituted aldehydes, such as **485**, followed by a protection-deprotection and oxidation sequence affording **486** in 70% yield and 91% *ee* (Scheme 7.118) [300]. The addition to α-alkoxyaldehydes provides a general approach to monoprotected 1,2-diols that can be converted to epoxides, such as **487** in excellent enantioselectivity [301]. The configuration of the new chiral center does not depend on the configuration of the ligand **476**. Thus, the reaction of the chiral aldehyde **488** with a functionalized zinc reagent ([PivO(CH$_2$)$_3$]$_2$Zn) in the presence of the ligand **476** and **ent-476** gives the two diastereomeric allylic alcohols (**489** and **490**) with high stereoselectivity (Scheme 7.119) [298a].

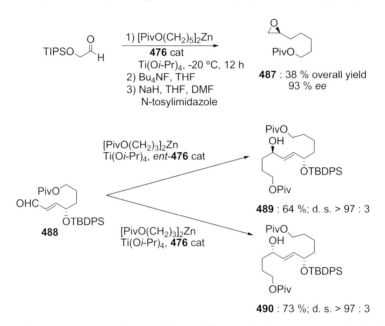

Scheme 7.119 Enantioselective additions of functionalized zinc reagents to chiral aldehydes.

Further applications to the preparation of chiral polyoxygenated molecules [302] and to the synthesis of the natural product (–)-mucocin [303] have been reported. The enantioselective addition of the functionalized diorganozinc reagent **491** to dodecanal in the presence of Ti(O*i*-Pr)$_4$ and the chiral ditriflamide **476** provides the desired chiral alcohol **492** with 99% of diastereomeric excess and 65% yield. The diorganozinc reagent was obtained by treating the alkyl iodide **493** with an excess of Et$_2$Zn in the presence of CuCN (3 mol%). The chiral building block **492** was used in the total synthesis of cycloviracin B1 of interest for its selective antiviral activity [304]. An elegant synthesis of (*R*)-(–)-muscone has been reported by Oppolzer et al. [305]. The alkenylzinc reagent was prepared by hydroboration of the acetylenic aldehyde **494** with (*c*-Hex)$_2$BH at 0 °C, followed by a B/Zn-exchange with Et$_2$Zn. The intramolecular addition in the presence of the chiral amino-alcohol **495** provides the allylic alcohol **496** with 72% yield and 92% *ee* [104c,305]. The enantioselective addition of dialkylzincs to aromatic ketones is especially difficult. Yus et al. have developed a new chiral ligand **497** allowing the addition of dialkylzincs to various aromatic ketones in good yields. The enantioselective addition is promoted by Ti(OiPr)$_4$ and tolerates the presence of several functional groups. Thus, the addition of Et$_2$Zn to phenacyl bromide provides the chiral alcohol **498** in 95% yield and 92% *ee* (Scheme 7.120) [306].

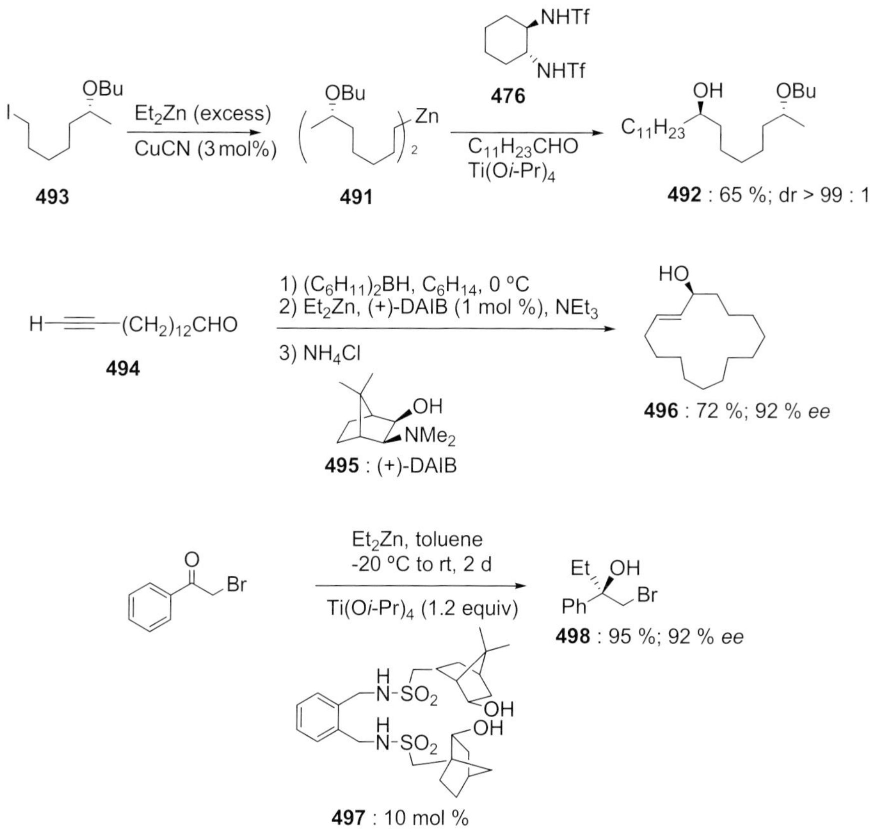

Scheme 7.120 Enantioselective additions mediated by titanium salts.

Alkynylzinc species generated *in situ* catalytically add to various aldehydes in very high enantioselectivities using reagent-grade toluene [307]. Chiral propargylic alcohols such as **499** are obtained in this way. As an application, the functional-ized alkyne **500** has been added to (R)-isopropylidene glyceraldehyde **501** in the presence of Zn(OTf)$_2$ and *N*-methylephedrine providing the propargylic alcohol **502** in 75% and a diastereomeric ratio of 94:6 [308]. The asymmetric addition of alkynylzinc reagents to aldehydes and ketones has recently been reviewed [309]. Especially important has been the stoichiometric addition of alkynylzinc deriva-tives leading to Efavirenz, a new drug for the treatment of AIDS [310]. The reac-tion of the chiral alkynylzinc reagent **503** obtained by the reaction of the aminoal-cohol **504** successively with Me$_2$Zn, *neo*-PentOH and the alkynyllithium **505** with the activated trifluoromethyl ketone **506** provides the tertiary alcohol **507** in 91% yield and 97% *ee*. A catalytic version of this reaction has been described by Anand and Carreira (Scheme 7.121) [311].

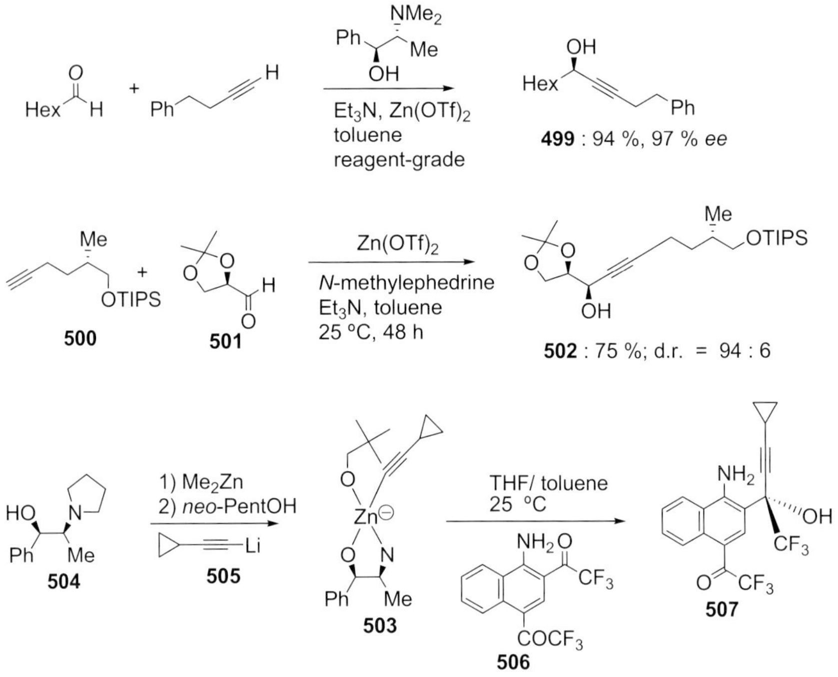

Scheme 7.121 Enantioselective preparation of propargylic alcohols.

Interestingly, the chiral diamine **508** catalyzes the enantioselective addition of boronic acids to aromatic ketones like acetophenone. The reaction produces interesting tertiary diarylcarbinols such as **509** in up to 93% *ee* [312]. Bolm and coworkers have shown that this approach can also be used for a simple preparation of chiral diarylcarbinols such as **510** in the presence of the chiral ferrocenyl ligand **511** (Scheme 7.122). [313]. The addition of diorganozincs to imines is a difficult reaction. However, Hoveyda and coworkers have found that the *in situ* generation of imines in the presence of Zr(O*i*-Pr)$_4$·HO*i*Pr, an excess of a dialkylzinc and a catalytic amount of the amino-acid derivative **512** allows an enantioselective addition leading to various amines such as **513** with high enantioselectivity. In the presence of the modular catalyst **514** and Zr(O*i*-Pr)$_4$·HO*i*-Pr (11 mol%), the imine **515** reacts with *bis*-alkynylzincs in the presence of (Me$_3$SiCH$_2$)$_2$Zn that is a zinc reagent with nontransferable Me$_3$SiCH$_2$ groups. [314]. Under these conditions, an efficient addition reaction proceeds affording the propargylic amine **516** in 83% yield and 90% *ee* (Scheme 7.122) [315].

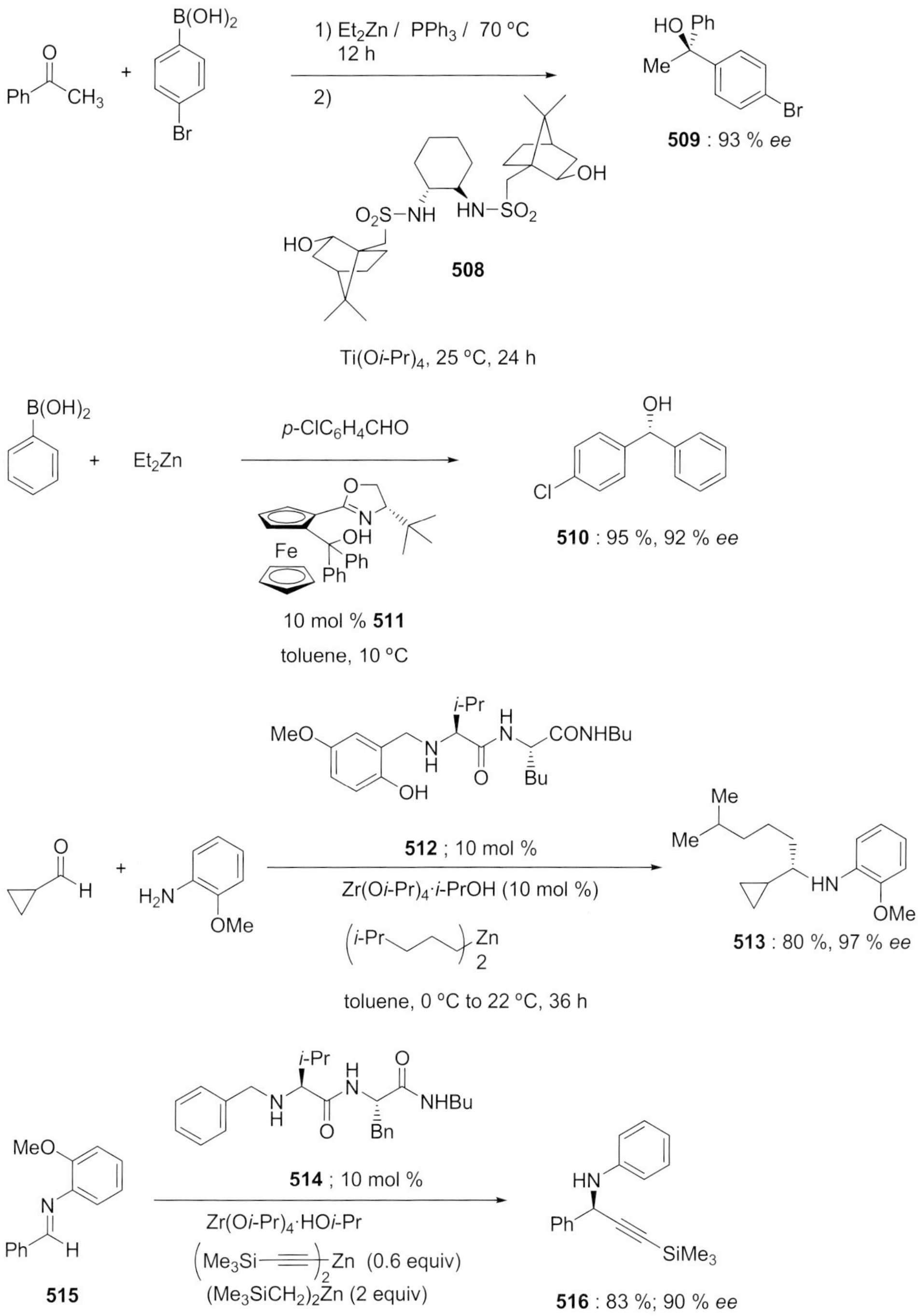

Scheme 7.122 Enantioselective additions of diorganozincs to unreactive carbonyl derivatives.

7.3.5
Reactions of Zinc Organometallics Catalyzed by Cobalt, Iron or Manganese Complexes

The catalysis of reactive main-group organometallics, such as organolithiums or organomagnesiums by transition-metal salts is often complicated by the formation of highly reactive transition-metal-centered ate-species ($R_n Met^-$) of transition metal complexes. These are especially prone to undergo β-hydrogen elimination reactions. Consequently, such transmetallations or catalysts have found little applications in organic synthesis [316]. Organozinc reagents, due to their lower reactivity, do not have a tendency to produce ate-species ($R_n Met^-$) of transition metal complexes, therefore the resulting transition-metal complexes of the type $RMetX_n$ have much higher thermal stability and a number of alkyl transition-metal complexes displaying synthetically useful properties can be accessed in this way. Thus, the reaction of cobalt(II) bromide with dialkylzincs in THF:NMP furnishes blue solutions of organocobalt intermediates that have a half-life of ca. 40 min. at –10 °C. Similarly the reaction of $FeCl_3$ with dipentylzinc produces a gray solution of an organoiron intermediate with a half-life of 2.5 h at –10 °C [317]. Interestingly, these new organocobalt(II) species undergo carbonylations at room temperature under mild conditions affording symmetrical ketones, such as **517** in satisfactory yield [318]. The stoichiometric preparation of organocobalt species is not necessary and catalytic amounts of cobalt(II) salts are sufficient to promote the acylation of diorganozincs [317]. Also, allylic chlorides react with zinc organometallics in a stereoselective manner. Thus, geranyl chloride (**518**) provides the S_N2-substitution product (*E*-**519**) with 90% yield and neryl chloride (**520**) affords the corresponding diene *Z*-**519** in 90% yield; (Scheme 7.123) [317].

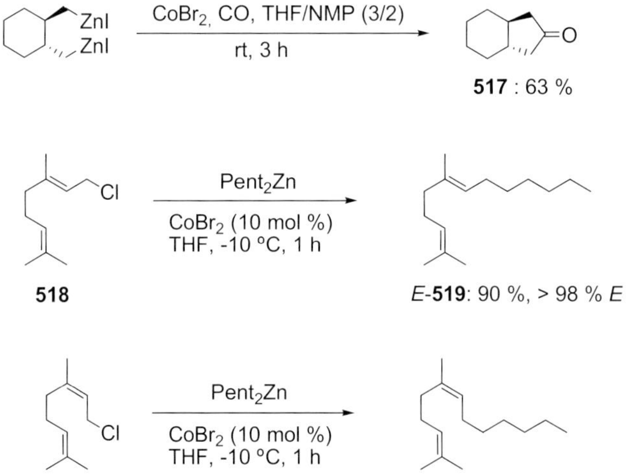

Scheme 7.123 Cobalt-mediated reactions of diorganozincs.

The reaction of manganese(II) salts with organozinc reagents does not provide the corresponding organomanganese reagents [319]. However, functionalized bromides can be metallated by Et_2Zn in the presence of catalytic amounts of $MnBr_2$ [320] leading to cyclized product, such as **520** in 82% yield (see Scheme 7.124) [321].

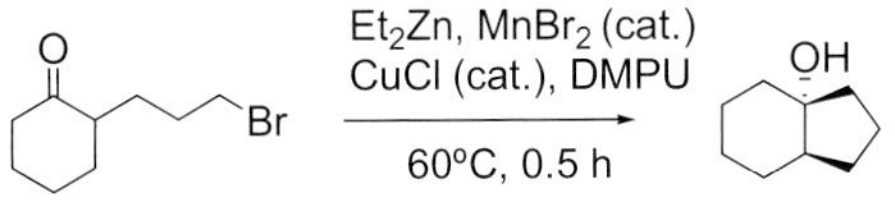

Scheme 7.124 Mn/Cu-mediated reaction of diorganozinc.

7.4
Conclusion

The synthesis of functionalized zinc organometallics can be accomplished with a variety of methods that have been developed in recent years. The intrinsic moderate reactivity of organozinc reagents can be dramatically increased by the use of the appropriate transition-metal catalyst or Lewis acid. Furthermore, the low ionic character of the carbon–zinc bond allows the preparation of a variety of chiral zinc organometallics with synthetically useful configurational stability. These properties make organozinc compounds ideal intermediates for the synthesis of complex and polyfunctionalized organic molecules.

References and Notes

1 E. Frankland, *Liebigs Ann. Chem.* **1849**, *71*, 171 and 213.

2 V. Grignard *Compt. Rend. Acad. Sci. Paris*, **1900**, *130*, 1322.

3 (a) S. Reformatsky, *Chem. Ber.* **1887**, *20*, 1210; **1895**, *28*, 2842; (b) A. Fürstner, *Synthesis*, **1989**, 571; (c) A. Fürstner, *Angew. Chem. Int. Ed.* **1993**, *32*, 164.

4 (a) H. E. Simmons, R. D. Smith *J. Am. Chem. Soc.* **1958**, *80*, 5323; (b) H. E. Simmons, T. L. Cairns, A. Vladuchick, C. M. Hoiness, *Org. React.* **1972**, *20*, 1; (c) J. Furukawa, N. Kawabata, J. Nishimma, *Tetrahedron Lett.* **1966**, *7*, 3353; (d) M. Nakamura, A. Hirai, E. Nakamura, *J. Am. Chem. Soc.* **2003**, *125*, 2341; (e) E. Nakamura, K. Sekiya, I. Kuwajima, *Tetrahedron Lett.* **1987**, *28*, 337.

5 H. Hunsdiecker, H. Erlbach, E. Vogt, German patent 722467, **1942**; *Chem. Abstr.* **1943**, *37*, 5080.

6 (a) P. Knochel, R. D. Singer, *Chem. Rev.* **1993**, *93*, 2117; (b) P. Knochel, *Synlett* **1995**, 393; (c) P. Knochel, S. Vettel, C. Eisenberg, *Appl. Organomet. Chem.* **1995**, *9*, 175; (d) P. Knochel, F. Langer, M. Rottländer, T. Stüdemann, *Chem. Ber.* **1997**, *130*, 387; (e) P. Knochel, J. Almena, P. Jones, *Tetrahedron* **1998** *54*, 8275; (f) P. Knochel, P. Jones, *Organozinc Reagents. A Practical Approach*, Oxford University Press, **1999**; (g) P. Knochel, N. Millot, A. L. Rodriguez, C. E. Tucker, *Org. React.* **2001**, *58*, 417; (h) A. Boudier, L. O. Bromm, M. Lotz, P. Knochel, *Angew. Chem. Int. Ed.* **2000**, *39*, 4415.

7 (a) E. Negishi, L. F. Valente,
M. Kobayashi, *J. Am. Chem. Soc.* **1980**,
102, 3298; (b) E. Erdik, *Tetrahedron*
1992, *48*, 9577

8 I. Antes, G. Frenking, *Organometallics*
1995, *14*, 4263.

9 E. Hupe, P. Knochel, K. J. Szabó, *Organometallics* **2002**, *21*, 2203.

10 P. Knochel, M.C.P. Yeh, S.C. Berk,
J. Talbert, *J. Org. Chem.* **1988**, *53*, 2390.

11 H. P. Knoess, M. T. Furlong, M. J.
Rozema, P. Knochel, *J. Org. Chem.*
1991, *56*, 5974.

12 S. C. Berk, M. C. P. Yeh, N. Jeong,
P. Knochel, *Organometallics* **1990**, *9*,
3053.

13 P. Knochel, C. J. Rao, *Tetrahedron*, **1993**,
49, 29.

14 (a) T. M. Stevenson, B. Prasad,
J. R. Citineni, P. Knochel, *Tetrahedron
Lett.* **1996**, *37*, 8375; (b) A. S. B. Prasad,
T. M. Stevenson, J. R. Citineni,
V. Nyzam, P. Knochel, *Tetrahedron* **1997**,
53, 7237.

15 M. Gaudemar, *Bull. Soc. Chim. Fr.* **1962**,
974.

16 E. Erdik, *Tetrahedron*, **1987**, *43*, 2203.

17 (a) J. K. Gawronsky, *Tetrahedron Lett.*
1984, *25*, 2605; (b) G. Picotin,
P. Miginiac, *Tetrahedron Lett.* **1987**, *28*,
4551.

18 (a) A. Fürstner, R. Singer, P. Knochel,
Tetrahedron Lett. **1994**, *35*, 1047;
(b) H. Stadtmüller, B. Greve, K. Lennick,
A. Chair, P. Knochel, *Synthesis* **1995**, 69.

19 (a) R. D. Rieke, *Science* **1989**, *246*, 1260;
(b) R. D. Rieke, S.-H. Kim, X. Wu,
J. Org. Chem. **1997**, *62*, 6921; (c) L. Zhu,
R. D. Rieke, *Tetrahedron Lett.* **1991**, *32*,
2865; (d) R. D. Rieke, M. S. Sell,
H. Xiong, *J. Am. Chem. Soc.* **1995**, *117*,
5429; (e) R. D. Rieke, P. M. Hudnall,
S. T. Uhm, *J. Chem. Soc., Chem. Com-
mun.* **1973**, 269; (f) R. D. Rieke,
P. T.-J. Li, T. P. Burns, S. T. Uhm, *J. Org.
Chem.* **1981**, *46*, 4323; (g) R. D. Rieke,
S. T. Uhm, *Synthesis* **1975**, 452;
(h) M. V. Hanson, J. D. Brown,
R. D. Rieke, *Tetrahedron Lett.* **1994**, *35*,
7205; (i) M. V. Hanson, R. D. Rieke,
Synth. Commun. **1995**, *25*, 101;
(j) R. D. Rieke, M. V. Hanson,
J. D. Brown, *J. Org. Chem.* **1996**, *61*,

2726 (k) A. Guijarro, R. D. Rieke, *Angew.
Chem. Int. Ed. Engl.* **1998**, *37*, 1679.

20 (a) M. V. Hanson, R. D. Rieke, *J. Am.
Chem. Soc.* **1995**, *117*, 10775; (b) L. Zhu,
R. M. Wehmeyer, R. D. Rieke, *J. Org.
Chem.* **1991**, *56*, 1445.

21 (a) B. H. Lipshutz, M. R. Wood, *J. Am.
Chem. Soc.* **1994**, *116*, 11689.
(b) B. H. Lipshutz, M. R. Wood, *J. Am.
Chem. Soc.* **1993**, *115*, 12625.

22 M. V. Hanson, J. D. Brown, R. D. Rieke,
O. J. Niu, *Tetrahedron Lett.* **1994**, *35*,
7205.

23 T. Ito, Y. Ishino, T. Mizuno, A. Ishikawa,
J. Kobayashi, *Synlett* **2002**, 2116.

24 (a) P. Knochel, J. F. Normant, *Tetra-
hedron Lett.* **1984**, *25*, 4383; (b) D. Peters,
R. Miethchen, *J. Prakt. Chem.* **1995**, *337*,
615; (c) T. Kitazume, N. Ishikawa, *J. Am.
Chem. Soc.* **1985**, *107*, 5186; (d) H. Tso,
T. Chou, Y. Lai, *J. Org. Chem.* **1989**, *54*,
4138; (e) J. P. Sestelo, J. L. Mascarenas,
L. Castedo, A. Mourino, *J. Org. Chem.*
1993, *58*, 118; (f) L. A. Sarandeses,
A. Mourino, J. L. Luche, *J. Chem. Soc.
Chem. Commun.* **1992**, 798.

25 (a) M. J. Dunn, R. F. W. Jackson,
J. Pietruszka, N. Wishart, D. Ellis,
M. J. Wythes, *Synlett* **1993**, 499;
(b) R. F. W. Jackson, N. Wishart,
M. J. Wythes, *Synlett* **1993**, 219;
(c) M. J. Dunn, R. F. W. Jackson,
J. Pietruszka, D. Turner, *J. Org. Chem.*
1995, *60*, 2210; (d) C. Malan, C. Morin,
Synlett **1996**, 167; (e) R. F. W. Jackson,
M. J. Wythes, A. Wood, *Tetrahedron Lett.*
1989, *30*, 5941; (f) R. F. W. Jackson,
A. Wood, M. J. Wythes, *Synlett* **1990**,
735; (g) M. J. Dunn, R. F. W. Jackson,
J. Chem. Soc., Chem. Commun. **1992**,
319; (h) R. F. W. Jackson, N. Wishart,
A. Wood; K. James, M. J. Wythes, *J. Org.
Chem.* **1992**, *57*, 3397; (i) C. S. Dexter,
R. F. W. Jackson, *J. Chem. Soc., Chem.
Commun.* **1998**, 75; (j) J. L. Fraser,
R. F. W.Jackson, B. Porter, *Synlett* **1994**,
379. (k) R. F. W. Jackson, W. Neil,
M. J. Wythes, *J. Chem. Soc., Chem. Com-
mun.* **1992**, 1587.

26 R. F. W. Jackson, I. Rilatt, P. J. Murray,
Chem. Commun. **2003**, 1242.

27 R. F. W. Jackson, I. Rilatt, P. J. Murray,
Org. Biomol. Chem. **2004**, *2*, 110.

28 C. Hunter, R. F. W. Jackson, H. K. Rami, *J. Chem. Soc., Perkin Trans.* 1, **2001**, 1349.

29 C. Jubert, P. Knochel, *J. Org. Chem.* **1992**, *57*, 5425.

30 C. Jubert, P. Knochel, *J. Org. Chem.* **1992**, *57*, 5431.

31 S. Huo, *Org. Lett.* **2003**, *5*, 423.

32 (a) B. H. Lipshutz, P. A. Blomgren, S. K. Kim, *Tetrahedron Lett.* **1999**, *40*, 197; (b) B. H. Lipshutz, P. A. Blomgren, *J. Am. Chem. Soc.* **1999**, *121*, 5819; (c) C. Dai, G. C. Fu, *J. Am. Chem. Soc.* **2001**, *123*, 2719.

33 (a) H. Blancou, A. Commeyras, *J. Flourine Chem.* **1982**, *20*, 255; (b) H. Blancou, A. Commeyras, *J. Flourine Chem.* **1982**, *20*, 267; (c) S. Benefice-Malouet, H. Blancou, A. Commeyras, *J. Flourine Chem.* **1985**, *30*, 171.

34 (a) R. N. Haszeldine, E. G. Walaschewski, *J. Chem. Soc.* **1953**, 3607; (b) W. T. Miller, E. Bergman, A. H. Fainberg, *J. Am. Chem. Soc.* **1957**, *79*, 4159; (c) R. D. Chambers, W. K. R. Musgrave, J. Savoy, *J. Chem. Soc.* **1962**, 1993; (d) D. J. Burton, *Tetrahedron* **1992**, *48*, 189; (f) D. J. Burton, Z.-Y. Yang, P. A. Morken, *Tetrahedron* **1994**, *50*, 2993.

35 (a) D. J. Burton, D. M. Wiemers, *J. Am. Chem. Soc.* **1985**, *107*, 5014; (b) D. M. Wiemers, D. J. Burton, *J. Am. Chem. Soc.* **1986**, *108*, 832; (c) B. Jiang, Y. Xu, *J. Org. Chem.* **1991**, *56*, 733; (d) P. A. Morken, H. Lu, A, Nakamura, D. J. Burton, *Tetrahedron Lett.* **1991**, *32*, 4271; (e) P. A. Morken, D. J. Burton, *J. Org. Chem.* **1993**, *58*, 1167.

36 (a) S. Sibille, V. Ratovelomanana, J. Périchon, S. Sibille, M. Troupel, *Synthesis* **1990**, 369; (b) M. Durandetti, M. Devaud, J. Périchon, *New J. Chem,* **1996**, *20*, 659; (c) S. Sibille, V. Ratovelomanana, J.-Y. Nédélec, J. Périchon, *Synlett* **1993**, 425; (d) C. Gosmini, J.-Y. Nédélec, J. Périchon, *Tetrahedron Lett.* **1997**, *38*, 1941, (e) N. Zylber, J. Zylber, Y. Rollin, E. Dunach, J. Périchon, *J. Organomet. Chem.* **1993**, *444*, 1; (f) Y. Rollin, S. Derien, E. Dunach, C. Gébéhenne, J. Périchon, *Tetrahedron* **1993**, *49*, 7723.

37 (a) P. Gomes, C. Gosmini, J. Périchon, *Synlett* **2002**, *10*, 1673; (b) P. Gomes, H. Fillon, C. Gosmini, E. Labbé, J. Périchon, *Tetrahedron* **2002**, *58*, 8417; (c) H. Fillon, E. Le Gall, C. Gosmini, J. Périchon, *Tetrahedron Lett.* **2002**, *43*, 5941; (d) C. Gosmini, Y. Rollin, J.-Y. Nédélec, J. Périchon, *J. Org. Chem.* **2000**, *65*, 6024.

38 S. Seka, O. Buriez, J.-Y. Nédélec, J. Périchon, *Chem. Eur. J.* **2002**, *8*, 2534.

39 (a) M. Mellah, E. Labbé, J.-Y. Nédélec, J. Périchon, *New J. Chem.* **2001**, *25*, 318; (b) C. Gosmini, J.-Y. Nédélec, J. Périchon, *Tetrahedron Lett.* **1999**, *41*, 201.

40 (a) H. Stadtmüller, R. Lentz, C. E. Tucker, T. Stüdemann, W. Dörner, P. Knochel, *J. Am. Chem. Soc.* **1993**, *115*, 7027; (b) H. Stadtmüller, A. Vaupel, C. E. Tucker, T. Stüdemann, P. Knochel, *Chem. Eur. J.* **1996**, *2*, 1204; (c) A. Vaupel, P. Knochel, *J. Org. Chem.* **1996**, *61*, 5743.

41 (a) H. Stadtmüller, C. E. Tucker, A. Vaupel, P. Knochel, *Tetrahedron Lett.* **1993**, *34*, 7911; (b) A. Vaupel, P. Knochel, *Tetrahedron Lett.* **1994**, *35*, 8349.

42 (a) A. Vaupel, P. Knochel, *Tetrahedron Lett.* **1995**, *36*, 231; (b) H. Stadtmüller, P. Knochel, *Synlett* **1995**, 463.

43 K. A. Agrios, M. Srebnik, *J. Org. Chem.* **1993**, *58*, 6908.

44 (a) I. Klement, P. Knochel, K. Chau, G. Cahiez, *Tetrahedron Lett.* **1994**, *35*, 1177; (b) E. Riguet, I. Klement, C. K. Reddy, G. Cahiez, P. Knochel, *Tetrahedron Lett.* **1996**, *37*, 5865.

45 (a) H. Fillon, C. Gosmini, J. Périchon, *J. Am. Chem. Soc.* **2003**, *125*, 3867; (b) P. Gomes, C. Gosmini, J. Périchon, *Synthesis,* **2003**, 1909; (c) I. Kazmierski, M. Bastienne, C. Gosmini, J.-M. Paris, J. Périchon, *J. Org. Chem.* **2004**, *69*, 936; (d) I. Kazmierski, C. Gosmini, J. M. Paris, J. Périchon, *Tetrahedron Lett.* **2003**, *44*, 6417; (e) S. Seka, O. Buriez, J. Périchon, *Chem. Eur. J.* **2003**, *9*, 3597.

46 (a) T. N. Majid, P. Knochel, *Tetrahedron Lett.* **1990**, *31*, 4413; (b) A. S. B. Prasad, P. Knochel, *Tetrahedron Lett.* **1997**, *53*, 16711; (c) S. C. Berk, P. Knochel, M. C. P. Yeh, *J. Org. Chem.* **1988**, *53*,

5789; (d) J. L. van der Baan, H. Stichter, G. J. J. Out, F. Bickelhaupt, G. W. Klumpp, *Tetrahedron Lett.* **1988**, *29*, 3579; (e) J. van der Louw, J. L. van der Baan, H. Stichter, G. J. J. Out, F. J. J. de Kanter, F. Bickelhaupt, G. W. Klumpp, *Tetrahedron* **1992**, *48*, 9877; (f) J. van der Louw, J. L. van der Baan, G. J. J. Out, F. J. J. de Kanter, F. Bickelhaupt, G. W. Klumpp, *Tetrahedron* **1992**, *48*, 9901; (g) T. A. van der Heide, J. L. van der Baan, F. Bickelhaupt, G. W. Klumpp, *Tetrahedron Lett.* **1992**, *33*, 475; (h) G. A. Molander, D. C. Shubert, *J. Am. Chem. Soc.* **1986**, *108*, 4683.

47 (a) C. E. Tucker, S. A. Rao, P. Knochel, *J. Org. Chem.* **1990**, *55*, 5446; (b) P. Knochel, T. S. Chou, H. G. Chen, M. C. P. Yeh, M. J. Rozema, *J. Org. Chem.* **1989**, *54*, 5202; (c) T. S. Chou, P. Knochel, *J. Org. Chem.* **1990**, *55*, 4791; (d) P. Knochel, T. S. Chou, C. Jubert, D. Rajagopal, *J. Org. Chem.* **1993**, *58*, 588.

48 (a) M. C. P. Yeh, P. Knochel, W. M. Butler, S. C. Berk, *Tetrahedron Lett.* **1988**, *29*, 6693; (b) H. G. Chen, J. L. Gage, S. D. Barrett, P. Knochel, *Tetrahedron Lett.* **1990**, *31*, 1829; (c) M. C. P. Yeh, P. Knochel, L. E. Santa, *Tetrahedron Lett.* **1988**, *29*, 3887; (d) M. C. P. Yeh, P. Knochel, *Tetrahedron Lett.* **1989**, *30*, 4799; (e) C. Retherford, M. C. P. Yeh, I. Schipor, H. G. Chen, P. Knochel, *J. Org. Chem.* **1989**, *54*, 5200.

49 M. C. P. Yeh, C.-N. Chuang, *J. Chem. Soc. Chem. Commun.* **1994**, 703.

50 H. G. Chen, C. Hoechstetter, P. Knochel, *Tetrahedron Lett.* **1989**, *30*, 4795.

51 M. C. P. Yeh, H. G. Chen, P. Knochel, *Org. Synth.* **1991**, *70*, 195.

52 S. AchyuthaRao, C. E. Tucker, P. Knochel, *Tetrahedron Lett.* **1990**, *31*, 7575; (b) S. AchyuthaRao, T.-S. Chou, I. Schipor, P. Knochel, *Tetrahedron*, **1992**, *48*, 2025.

53 (a) P. Knochel, *J. Am. Chem. Soc.* **1990**, *112*, 7431; (b) G. Kanai, N. Miyaura, A. Suzuki, *Chem. Lett.* **1993**, 845. (c) T. Watanabe, N. Miyaura, A. Suzuki, *J. Orgamomet. Chem.* **1993**, *444*, C1;

(d) T. Watanabe, M, Sakai, N. Miyaura, A. Suzuki, *J. Chem. Soc. Chem. Commun.* **1994**, 467.

54 (a) C. Janakiram Rao, P. Knochel, *J. Org. Chem.* **1991**, *56*, 4593; (b) P. Knochel, C. Janakiram Rao, *Tetrahedron*, **1993**, *49*, 29; (c) B. S. Bronk, S. J. Lippard, R. L. Danheiser, *Organometallics*, **1993**, *12*, 3340.

55 C. Retherford, T.-S. Chou, R. M. Schelkun, P. Knochel, *Tetrahedron Lett.* **1990**, *31*, 1833.

56 W. E. Parham, C. K. Bradsher, *Acc. Chem. Res.* **1982**, *15*, 300.

57 J. F. Cameron, J. M. J. Frechet, *J. Am. Chem. Soc.* **1991**, *113*, 4303; (b) P. Wiriyachitra, S. J. Falcone, M. P. Cava, *J. Org. Chem.* **1979**, *44*, 3957.

58 C. E. Tucker, T. N. Majid, P. Knochel, *J. Am. Chem. Soc.* **1992**, *114*, 3983.

59 E. M. E. Viseux, P. J. Parsons, J. B. J. Pavey, *Synlett* **2003**, 861.

60 B. A. Anderson, N. K. Harn, *Synthesis*, **1996**, 583.

61 S. Superchi, N. Sotomayor, G. Miao, B. Joseph, V. Snieckus, *Tetrahedron Lett.* **1996**, *37*, 6057.

62 (a) E. Lorthiois, I. Marek, J.-F. Normant, *Tetrahedron Lett.* **1997**, *38*, 89; (b) P. Karoyan, G. Chassaing, *Tetrahedron Lett.* **1997**, *38*, 85; (c) P. Karoyan, J. Quancard, J. Vaissermann, G. Chassaing, *J. Org. Chem.* **2003**, *68*, 2256; (d) P. Karoyan, G. Chassaing, *Tetrahedron: Asymmetry* **1997**, *8*, 2025; (e) P. Karoyan, A. Triolo, R. Nannicini, D. Giannotti, M. Altamura, G. Chassaing, G. E. Perrotta, *Tetrahedron Lett.* **1999**, *40*, 71; (f) P. Karoyan, G. Chassaing, *Tetrahedron Lett.* **2002**, *43*, 253; (g) J. Quancard, H. Magella, S. Lavielle, G. Chassaing, P. Karoyan, *Tetrahedron Lett.* **2004**, *45*, 2185.

63 (a) E. Nakamura, K. Kubota, *J. Org. Chem.* **1997**, *62*, 792; (b) K. Kubota, E. Nakamura, *Angew. Chem. Int. Ed. Engl.* **1997**, *36*, 2491; (c) M. Nakamura, T. Hatakeyama, K. Hara, E. Nakamura, *J. Am. Chem. Soc.* **2003**, *125*, 6362; (d) M. Nakamura, H. Isobe, E. Nakamura, *Chem. Rev.* **2003**, *103*, 1295; (e) M. Nakamura, N. Yoshikai, E. Nakamura, *Chem. Lett.* **2002**, 146;

(f) M. Nakamura, T. Inoue, A. Sato, E. Nakamura, *Org. Lett.* **2000**, *2*, 2193.

64 (a) J. P. Gillet, R. Sauvetre, J.-F. Normant, *Synthesis* **1986**, 538; (b) J.-F. Normant, *J. Organomet. Chem.* **1990**, *400*, 19.

65 (a) T. Shigeoka, Y. Kuwahara, K. Watanabe, K. Satro, M. Omote, A. Ando, I. Kumadaki, *J. Fluorine Chem.* **2000**, *103*, 99; (b) T. Shigeoka, Y. Kuwahara, K. Watanabe, K. Satro, M. Omote, A. Ando, I. Kumadaki, *Heterocycles*, **2000**, *52*, 383.

66 J. M. Bainbridge, S. J. Brown, P. N. Ewing, R. R. Gibson, J. M. Percy, *J. Chem. Soc. Perkin Trans 1*, **1998**, 2541.

67 R. Anilkumar, D. J. Burton, *Tetrahedron Lett.* **2002**, *43*, 6979.

68 J. Ichikawa, M. Fujiwara, H. Nawata, T. Okauchi, T. Minami, *Tetrahedron Lett.* **1996**, *37*, 8799.

69 (a) P. Wipf, W. Xu, *J. Org. Chem.* **1996**, *61*, 6556; (b) P. Wipf, W. Xu, *Tetrahedron Lett.* **1994**, *35*, 5197.

70 Y. Ni, K. K. D. Amarasinghe, J. Montgomery, *Org. Lett.* **2002**, *4*, 1743.

71 E. Negishi, N. Okukado, A. O. King, D. E. Van Horn, B. I. Spiegel, *J. Am. Chem. Soc.* **1978**, *100*, 2254.

72 R. K. Dieter, R. T. Watson, R. Goswami, *Org. Lett.* **2004**, *6*, 253.

73 E. Frankland, D. Duppa, *Liebigs Ann. Chem.* **1864**, *130*, 117.

74 M. J. Rozema, D. Rajagopal, C. E. Tucker, P. Knochel, *J. Organomet. Chem.* **1992**, *438*, 11.

75 P. Knochel, W. Dohle, N. Gommermann, F. F. Kneisel, F. Kopp, T. Korn, I. Sapountzis, V. A. Vu, *Angew. Chem. Int. Ed.* **2003**, *42*, 4302.

76 A. E. Jensen, P. Knochel, *J. Organomet. Chem.* **2002**, *653*, 122.

77 (a) D. Seyferth, H. Dertouzos, L. J. Todd, *J. Organomet. Chem.* **1965**, *4*, 18; (b) D. Seyferth, S. B. Andrews, *J. Organomet. Chem.* **1971**, *30*, 151.

78 (a) P. Knochel, N. Jeong, M. J. Rozema, M. C. P. Yeh, *J. Am. Chem. Soc.* **1989**, *111*, 6474; (b) P. Knochel, S. Achyutha Rao, *J. Am. Chem. Soc.* **1990**, *112*, 6146; (c) M. J. Rozema, P. Knochel, *Tetrahedron Lett.* **1991**, *32*, 1855; (d) M. R. Burns, J. K. Coward, *J. Org. Chem.* **1993**, *58*, 528; (e) A. Sidduri,

M. J. Rozema, P. Knochel, *J. Org. Chem.* **1993**, *58*, 2694; (f) A. Sidduri, P. Knochel, *J. Am. Chem. Soc.* **1992**, *114*, 7579.

79 P. Knochel, M. J. Rozema, C. E. Tucker, C. Relhwford, M. Furlong, S. Achyutha Rao, *Pure. Appl. Chem.* **1992**, *64*, 361.

80 E. Negishi, K. Akiyoshi, *J. Am. Chem. Soc.* **1988**, *110*, 646; (b) E. Negishi, K. Akiyoshi, B. O'Connor, K. Takagi, G. Wu, *J. Am. Chem. Soc.* **1989**, *111*, 3089.

81 T. Katsuhira, T. Harada, K. Maejima, A. Osada, A. Oku, *J. Org. Chem.* **1993**, *58*, 6166.

82 (a) T. Harada, K. Hattori, T. Katsuhira, A. Oku, *Tetrahedron Lett.*, **1989**, *30*, 6035 and 6039; (b) T. Harada, T. Katsuhira, K. Hattori, A. Oku, *J. Org. Chem.* **1993**, *58*, 2958; (c) T. Harada, T. Katsuhira, A. Oku, *J. Org. Chem.* **1992**, *57*, 5805.

83 J. P. Varghese, I. Zouev, L. Aufauvre, P. Knochel, I. Marek, *Eur. J. Org. Chem.* **2002**, 4151.

84 S. Achyutha Rao, P. Knochel, *J. Org. Chem.* **1991**, *56*, 4591;

85 R. Duddu, M. Eckhardt, M. Furlong, H. P. Knoess, S. Berger, P. Knochel, *Tetrahedron*, **1994**, *50*, 2415.

86 (a) M. Sakurai, T. Hata, Y. Yabe, *Tetrahedron Lett.* **1993**, *34*, 5939; (b) S. Sakami, T. Houkawa, A. Asaoka, H. Takei, *J. Chem. Soc. Perkin Trans 1*, **1995**, 285; (c) T. Houkawa, T. Ueda, S. Sakami, M. Asaoka, H. Takei, *Tetrahedron Lett.* **1996**, *37*, 1045.

87 C. Darcel, F. Flachsmann, P. Knochel, *Chem. Commun.* **1998**, 205.

88 F. Wang, J. Tang, L. Labaudinière, I. Marek, J.-F. Normant, *Synlett* **1995**, 723.

89 J.-F. Poisson, J. F. Normant, *J. Am. Chem. Soc.* **2001**, *123*, 4639.

90 J.-F. Poisson, J. F. Normant, *Org. Lett.* **2001**, *3*, 1889.

91 I. Wilson, R. F. W. Jackson, *J. Chem. Soc. Perkin Trans 1*, **2002**, 2845.

92 (a) J. Furukawa, N. Kawabata, J. Nishimura, *Tetrahedron Lett.* **1966**, *7*, 3353; (b) J. Furukawa, N. Kawabata, *Adv. Organomet. Chem.* **1974**, *12*, 83.

93 M. J. Rozema, A. Sidduri, P. Knochel, *J. Org. Chem.* **1992**, *57*, 1956;

(b) M. J. Rozema, C. Eisenberg, H. Lütjens, R. Ostwald, K. Belyk, P. Knochel, *Tetrahedron Lett.* **1993**, *34*, 3115.

94 A. B. Charette, A. Beauchemin, J.-F. Marcoux, *J. Am. Chem. Soc.* **1998**, *120*, 5114.

95 E. Hupe, M. I. Calaza, P. Knochel, *J. Organomet. Chem.* **2003**, *680*, 136.

96 L. Micouin, P. Knochel, *Synlett* **1997**, 327.

97 F. F. Kneisel, M. Dochnahl, P. Knochel, *Angew. Chem. Int. Ed.* **2004**, *43*, 1017.

98 (a) S. Berger, F. Langer, C. Lutz, P. Knochel, T. A. Mobley, C. K. Reddy, *Angew. Chem. Int. Ed.* **1997**, *36*, 1496; (b) P. Jones, C. K. Reddy, P. Knochel, *Tetrahedron* **1998**, *54*, 1471.

99 C. Petrier, J. De Souza Barbosa, C. Dupuy, J.-L. Luche, *J. Org. Chem.* **1985**, *50*, 5761.

100 E. J. Corey, J. O. Link, Y. Shao, *Tetrahedron Lett.* **1992**, *33*, 3435.

101 L. I. Zakharin, O. Y. Okhlobystin, *Z. Obshch. Chim.* **1960**, *30*, 2134; *Engl. Trans*, p. 2109, *Chem. Abstr.* **1961**, *55*, 9319a.

102 (a) K.-H. Thiele, P. Zdunneck, *J. Organomet. Chem.* **1965**, *4*, 10; (b) K.-H. Thiele, G. Engelhardt, J. Köhler, M. Arnstedt, *J. Organomet. Chem.* **1967**, *9*, 385.

103 (a) F. Langer, J. Waas, P. Knochel, *Tetrahedron Lett.* **1993**, *34*, 5261; (b) F. Langer, L. Schwink, A. Devasagayaraj, P.-Y. Chavant, P. Knochel, *J. Org. Chem.* **1996**, *61*, 8229.

104 (a) M. Srebnik, *Tetrahedron Lett.* **1991**, *32*, 2449; (b) W. Oppolzer, R. N. Radinov, *Helv. Chim. Acta* **1992**, *75*, 170; (c) W. Oppolzer, R. N. Radinov, *J. Am. Chem. Soc.* **1993**, *115*, 1593.

105 R. Köster, G. Griasnow, W. Larbig, P. Binger, *Liebigs Ann. Chem.* **1964**, *672*, 1.

106 A. Devasagayaraj, L. Schwink, P. Knochel, *J. Org. Chem.* **1995**, *60*, 3311.

107 H. C. Brown, A. K. Mandal, N. M. Yoon, B. Singaram, J. R. Schwier, P. K. Jadhav, *J. Org. Chem.* **1982**, *47*, 5069.

108 (a) L. Micouin, M. Oestreich, P. Knochel, *Angew. Chem. Int. Ed. Engl.* **1997**, *36*, 245; (b) A. Boudier, F. Flachsmann,

P. Knochel, *Synlett* **1998**, 1438; (c) E. Hupe, P. Knochel, *Org. Lett.* **2001**, *3*, 127.

109 E. Hupe, P. Knochel, *Angew. Chem. Int. Ed. Engl.* **2001**, *40*, 3022.

110 A. Boudier, E. Hupe, P. Knochel, *Angew. Chem. Int. Ed. Engl.* **2000**, *39*, 2294.

111 I. Fleming, N. J. Lawrence, *J. Chem. Soc. Perkin Trans* 1, **1992**, 3309.

112 (a) E. Hupe, M. I. Calaza, P. Knochel, *Tetrahedron Lett.* **2001**, *42*, 8829; (b) E. Hupe, M. I. Calaza, P. Knochel, *Chem. Eur. J.* **2003**, *9*, 2789.

113 K. Burgess, M. J. Ohlmeyer, *J. Org. Chem.* **1991**, *56*, 1027.

114 (a) K. Burgess, M. J. Ohlmeyer, *Chem. Rev.* **1991**, *91*, 1179; (b) A. H. Hoveyda, D. A. Evans, G. C. Fu, *Chem. Rev.* **1993**, *93*, 1307.

115 (a) D. A. Evans, G. C. Fu, A. H. Hoveyda, *J. Am. Chem. Soc.* **1988**, *110*, 6917; (b) K. Burgess, M. J. Ohlmeyer, *Tetrahedron Lett.* **1989**, *30*, 395.

116 E. Hupe, M. I. Calaza, P. Knochel, *Chem. Commun.* **2002**, 1390.

117 (a) S. Vettel, A. Vaupel, P. Knochel, *Tetrahedron Lett.* **1995**, *36*, 1023; (b) S. Vettel, A. Vaupel, P. Knochel, *J. Org. Chem.* **1996**, *61*, 7473.

118 (a) Y. Tamaru, A. Tanaka, K. Yasui, S. Goto, S. Tanaka, *Angew. Chem. Int. Ed. Engl.* **1995**, *34*, 787; (b) M. Shimizu, M. Kimura, S. Tanaka, Y. Tamaru, *Tetrahedron Lett.* **1998**, *39*, 609.

119 (a) Y. Tamaru, *J. Organomet. Chem.* **1999**, *576*, 215; (b) M. Kimura, I. Kiyama, T. Tomizawa, Y. Horino, S. Tanaka, Y. Tamaru, *Tetrahedron Lett.* **1999**, *40*, 6795; (c) M. Kimura, S. Matsuo, K. Shibata, Y. Tamaru, *Angew. Chem. Int. Ed. Engl.* **1999**, *38*, 3386; (d) M. Kimura, T. Tomizawa, Y. Horino, S. Tanaka, Y. Tamaru, *Tetrahedron Lett.* **2000**, *41*, 3627; (e) K. Shibata, M. Kimura, M. Shimizu, Y. Tamaru, *Org. Lett.* **2001**, *3*, 2181; (f) M. Kimura, Y. Horino, R. Mukai, S. Tanaka, Y. Tamaru, *J. Am. Chem. Soc.* **2001**, *123*, 10401.

120 (a) P. Jones, N. Millot, P. Knochel, *Chem. Commun.* **1998**, 2405; (b) P. Jones, N. Millot, P. Knochel, *J. Org. Chem.* **1999**, *64*, 186; (c) P. Jones, P. Knochel *Chem. Commun.* **1998**, 2407;

(d) N. Millot, P. Knochel, *Tetrahedron Lett.* **1999**, *40*, 7779; (e) C. Piazza, N. Millot, P. Knochel, *J. Organomet. Chem.* **2001**, *624*, 88.

121 R. Benn, E. G. Hoffmann, H. Lehmkuhl, H. Nehl, *J. Organomet. Chem.* **1978**, *146*, 103.

122 W. Oppolzer, H. Bienayme, A. Genevois-Borella, *J. Am. Chem. Soc.* **1991**, *113*, 9660.

123 C. Meyer, I. Marek, J.-F. Normant, *Tetrahedron Lett.* **1996**, 37, 857.

124 (a) W. Tückmankel, K. Oshima, H. Nozaki, *Chem. Ber.* **1986**, *119*, 1581; (b) R. M. Fabicon, M. Parvez, H. G. Richey, *J. Am. Chem. Soc.* **1991**, *113*, 1412 and 6680; (c) A. P. Purdy, C. F. George, *Organometallics* **1992**, *11*, 1955; (d) H. Toshiro, H. Wada, A. Oku, *J. Org. Chem.*, **1995**, *60*, 5370.

125 (a) M. Isobe, S. Konda, N. Nagasawa, T. Goto, *Chem. Lett.* **1977**, 679; (b) R. A. Kjonaas, R. K. Hoffer, *J. Org. Chem*, **1988**, *53*, 4133; (c) R. A. Kjonaas, E. J. Vawter, *J. Org. Chem*, **1986**, *51*, 3933.

126 (a) J. F. G. A. Jansen, B. L. Feringa, *Tetrahedron Lett.* **1988**, *29*, 3593; (b) J. F. G. A. Jansen, B. L. Feringa, *J. Chem. Soc. Chem. Commun.* **1989**, 741; (c) J. F. G. A. Jansen, B. L. Feringa, *J. Org. Chem.* **1990**, *55*, 4168.

127 D. Seebach, W. Langen, *Helv. Chim. Acta* **1979**, *62*, 1701 and 1710.

128 (a) M. Uchiyama, M. Koike, M. Kameda, Y. Kondo, T. Sakamoto, *J. Am. Chem. Soc.* **1996**, *118*, 8733; (b) Y. Kondo, N. Takazawa, C. Yamazaki, T. Sakamoto *J. Org. Chem.*, **1994**, *59*, 4717.

129 Y. Kondo, T. Komine, M. Fujinami, M. Uchiyama, T. Sakamoto, *J. Comb. Chem.* **1999**, *1*, 123; (b) G. Reginato, M. Taddei, *Il Farmaco* **2002**, *57*, 373.

130 T. Harada, T. Kaneko, T. Fujiwara, A. Oku, *J. Org. Chem*, **1997**, *62*, 8966.

131 I. N. Houpis, A. Molina, I. Dorziotis, R. A. Reamer, R. P. Volante, P. J. Reider, *Tetrahedron Lett.* **1997**, *38*, 7131.

132 (a) P. Knochel, in *Comprehensive Organic Synthesis*, Vol. 1 (eds: B. M. Trost, I. Fleming, S. L. Schreiber) Pergamon, Oxford **1991**, p. 221; (b) L. Miginiac, in *The Chemistry of the Metal-Carbon Bond* (eds: F. R. Havey, S. Patai), Wiley, New York, **1985**, vol. 3, p. 99; (c) P. Knochel, J.-F. Normant, *Tetrahedron Lett.* **1984**, *25*, 1475; (d) G. Rousseau, and J. Drouin, *Tetrahedron* **1983**, *39*, 2307; (e) P. Knochel, J.-F. Normant, *Tetrahedron Lett.* **1984**, *25*, 4383.

133 Y. A. Dembélé, C. Belaud, P. Hitchcock, J. Villiéras, *Tetrahedron: Asymmetry*, **1992**, *3*, 351.

134 P. Knochel, J.-F. Normant, *J. Organomet. Chem.* **1986**, *309*, 1.

135 H. Yamamoto, in *Comprehensive Organic Synthesis*, Vol. 2 (eds: B. M. Trost, I. Fleming, S. L. Schreiber), Pergamon, Oxford, **1991**, 81.

136 (a) J. A. Marshall, *Chem. Rev.* **2000**, *100*, 3163; (b) J. A. Marshall, M. M. Yanik, *J. Org. Chem.* **2001**, 66, 1373; (c) B. W. Gung, X. Xue, N. Katz, J. A. Marshall, *Organometallics* **2003**, *22*, 3158.

137 G. Zweifel, G. Hahn, *J. Org. Chem*, **1984**, *49*, 4565.

138 K. Nützel, In *Methoden der Organischen Chemie, Metallorganische Verbindungen Be, Mg, Ca, Sr, Ba, Zn, Cd*, vol. 13/2a, Thieme : Stuttgart, **1973**, p. 552.

139 T. Harada, E. Kutsuwa, *J. Org. Chem*, **2003**, *68*, 6716; (b) T. Harada, A. Oku in *Organozinc Reagents. A Practical Approach*, P. Knochel, P. Jones, Eds. Oxford University Press; Oxford, NY, **1999**, Chapter 6.

140 J. G. Giess, M. Le Blanc, *Angew. Chem.* **1978**, *90*, 654.

141 (a) I. Klement, P. Knochel, *Synlett* **1995**, 1113; (b) I. Klement, H. Lütjens, P. Knochel, *Tetrahedron* **1997**, *53*, 9135.

142 I. Klement, K. Lennick, C. E. Tucker, P. Knochel, *Tetrahedron Lett.* **1993**, *34*, 4623.

143 (a) M. Gaudernar, *C. R. Acad. Sci. Paris* **1971**, *273*, 1669; (b) P. Knochel, J.-F. Normant, *Tetrahedron Lett.* **1986**, *27*, 1039, 1043 and 4427; (c) I. Marek, J.-F. Normant, *Chem. Rev.* **1996**, *96*, 3241; (d) I. Marek, J.-F. Normant, *J. Org. Chem.* **1994**, *59*, 4154; (d) D. Beruben, I. Marek, J.-F. Normant, N. Platzer, *J. Org. Chem.* **1995**, *60*, 2488.

144 (a) F. Langer, P. Knochel, *Tetrahedron Lett.* **1995**, *36*, 4591; (b) F. Langer, K. Püntener, R. Stürmer, P. Knochel, *Tetrahedron: Asymmetry*, **1997**, *8*, 715.

145 A. Longeau, F. Langer, P. Knochel, *Tetrahedron Lett.* **1996**, *37*, 2209.

146 (a) S. J. Greenfield, A. Agarkov, S. R. Gilbertson, *Org. Lett.* **2003**, *5*, 3069; (b) H. Stadtmueller, P. Knochel, *Organometallics* **1995**, *14*, 3163.

147 N. Millot, C. Piazza, S. Avolio, P. Knochel, *Synthesis* **2000**, 941.

148 N. Gommermann, C. Koradin, P. Knochel, *Synthesis* **2002**, 2143.

149 F. Lamaty, R. Lazaro, J. Martinez, *Tetrahedron Lett*, **1997**, *38*, 3385.

150 (a) D. L. Comins, S. O'Connor, *Tetrahedron Lett.* **1987**, *28*, 1843; (b) D. L. Comins, M. A. Foley, *Tetrahedron Lett.* **1988**, *29*, 6711.

151 E. Nakamura, S. Aoki, K. Sekiya, H. Oshino, I. Kuwajima, *J. Am. Chem. Soc.* **1987**, *109*, 8056.

152 M. Ludwig, K. Polborn, K. T. Wanner, *Heterocycles* **2003**, *61*, 299.

153 A. R. Katritzky, Z. Luo, Y. Fang, P. J. Steel, *J. Org. Chem.* **2001**, *66*, 2858.

154 (a) S. N. Thorn, T. Gallagher, *Synlett* **1996**, 185; (b) S. D. Rychnovsky, N. A. Powell, *J. Org. Chem.* **1997**, *62*, 6460.

155 (a) S. Xue, K.-Z. Han, L. He, Q.-X. Guo, *Synlett* **2003**, 870; (b) J. D. Rainier, J. M. Cox, *Org. Lett.* **2000**, *2*, 2707.

156 W. J. Thompson, T. J. Tucker, J. E. Schwering, J. L. Barnes, *Tetrahedron Lett.* **1990**, *31*, 6819.

157 A. E. Decamp, A. T. Kawaguchi, R. P. Volante, I. Shinkai, *Tetrahedron Lett.* **1991**, *32*, 1867.

158 U. Koert, H. Wagner, U. Pidun, *Chem. Ber.* **1994**, *127*, 1447.

159 J. J. Almena Perea, T. Ireland, P. Knochel, *Tetrahedron Lett.* **1997**, *38*, 5961.

160 (a) C. K. Reddy, A. Devasagayaraj, P. Knochel, *Tetrahedron Lett.* **1996**, *37*, 4495; (b) M. Kitamura, T. Miki, K. Nakano, R. Noyori, *Tetrahedron Lett.* **1996**, *37*, 5141.

161 B. S. Bronk, S. J. Lippard, R. L. Danheiser, *Organometallics* **1993**, *12*, 3340.

162 S. Kim, J. M. Lee, *Tetrahedron Lett.* **1990**, *31*, 7627.

163 M. Arisawa, Y. Torisawa, M. Kawahara, M. Yamanaka, A. Nishida, M. Nakagawa, *J. Org. Chem.* **1997**, *62*, 4327.

164 (a) G. Posner, *Org. React.* **1972**, *19*, 1; (b) B. H. Lipshutz, *Synthesis* **1957**, 325; (c) B. H. Lipshutz, S. Seugepta, *Org. React.* **1992**, *41*, 135; (c) G. Posner, *Org. React.* **1975**, *22*, 253.

165 T. Stemmler, J. E. Penner-Hahn, P. Knochel, *J. Am. Chem. Soc.* **1993**, *115*, 348.

166 S. Marquais, G. Cahiez, P. Knochel, *Synlett* **1994**, 849.

167 H. T. Teunissen, F. Bickelhaupt, *Tetrahedron Lett.* **1992**, *33*, 3537.

168 (a) M. C. P. Yeh, P. Knochel, *Tetrahedron Lett.* **1988**, *29*, 2395; (b) H. Ochiai, Y. Tamaru, K. Tsubaki, Z. Yoshida, *J. Org. Chem.* **1987**, *52*, 4418; (c) M. Arai, T. Kawasuji, E. Nakamura, *J. Org. Chem.* **1993**, *58*, 5121; (d) E. Nakamura, K. Sekiya, M. Arai, S. Aoki, *J. Am. Chem. Soc.* **1989**, *111*, 3091; (e) K. Sekiya, E. Nakamura, *Tetrahedron Lett.* **1988**, *29*, 5155.

169 H. J. C. Deboves, U. Grabowska, A. Rizzo, R. F. W. Jackson, *J. Chem. Soc. Perkin Trans* 1, **2000**, 4284.

170 A. Hirai, A. Matsui, K. Komatsu, K. Tanino, M. Miyashita, *Chem. Commun.* **2002**, 1970.

171 B. H. Lipshutz, K. Woo, T. Gross, D. J. Buzard, R. Tirado, *Synlett* **1997**, 477.

172 (a) P. Knochel, D. Seebach, *Tetrahedron Lett.* **1982**, *23*, 3897; (b) D. Seebach, P. Knochel, *Helv. Chim. Acta* **1984**, *67*, 261.

173 S. Ma, A. Zhang, *J. Org. Chem.* **2002**, *67*, 2287.

174 T. Konno, M. Tanikawa, T. Ishihara, H. Yamanaka, *Collect. Czech. Chem. Commun.* **2002**, *67*, 1421.

175 A. Yanagisawa, Y. Noritake, N. Nomura, H. Yamamoto, *Synlett* **1991**, 251.

176 M. I. Calaza, E. Hupe, P. Knochel, *Org. Lett.* **2003**, *5*, 1059.

177 R. W. Hoffmann, *Chem. Rev.* **1989**, *89*, 1841.

178 N. Harrington-Frost, H. Leuser, M. I. Calaza, F. F. Kneisel, P. Knochel, *Org. Lett.* **2003**, *5*, 2111.

179 D. Soorukram, P. Knochel, *Org. Lett.* **2004**, *6*, 2409.

180 B. Breit, P. Demel, *Adv. Synth. Catal.* **2001**, *343*, 429; (b) B. Breit, P. Demel, *Tetrahedron* **2000**, *56*, 2833.

181 (a) A. S. E. Karlström, F. F. Huerta, G. J. Meuzelaar, J. E. Bäckvall, *Synlett* **2001**, 923; (b) G. J. Meuzelaar, A. S. E. Karlström, M. van Klaveren, E. S. M. Persson, A. del Villar, G. van Koten, J.-E. Bäckvall, *Tetrahedron* **2000**, *56*, 2895; (c) A. Alexakis, C. Malan, L. Lea, C. Benhaim, X. Fournioux, *Synlett* **2001**, 927; (d) A. Alexakis, K. Croset, *Org. Lett.* **2002**, *4*, 4147; (e) H. Malda, A. W. van Zijl, L. A. Arnold, B. L. Feringa, *Org. Lett.* **2001**, *3*, 1169; (f) O. Equey, E. Vrancken, A. Alexakis, *Eur. J. Org. Chem.* **2004**, 2151; (g) K. Tissot-Croset, D. Polet, A. Alexakis, *Angew. Chem. Int. Ed.* **2004**, *43*, 2426.

182 C. A. Luchaco-Cullis, H. Mizutani, K. E. Murphy, A. H. Hoveyda, *Angew. Chem. Int. Ed.* **2001**, *40*, 1456.

183 (a) F. Dübner, P. Knochel, *Tetrahedron Lett.* **2000**, *41*, 9233; (b) F. Dübner, P. Knochel, *Angew. Chem. Int. Ed.* **1999**, *38*, 379.

184 W. Blankenfeldt, J.-W. Liao, L.-C. Lo, M. C. P. Yeh, *Tetrahedron Lett.* **1996**, *37*, 7361.

185 (a) M. C. P. Yeh, M.-L. Sun, S.-K. Lin, *Tetrahedron Lett.* **1991**, *32*, 113; (b) M. C. P. Yeh, S. J. Tau, *J. Chem. Soc. Chem. Commun.* **1992**, 13; (c) M. C. P. Yeh, C.-J. Tsou, C.-N. Chuang, H.-C. Lin, *J. Chem. Soc. Chem. Commun.* **1992**, 890; (d) M. C. P. Yeh, B.-A. Shen, H.-W. Fu, S. I. Tau, L. W. Chuang, *J. Am. Chem. Soc.* **1993**, *115*, 5941.

186 M. C. P. Yeh, C.-C. Hwu, A.-T. Lee, M.-S. Tsai, *Organometallics* **2001**, *20*, 4965.

187 J. H. Rigby, M. Kirova-Snover, *Tetrahedron Lett.* **1997**, *38*, 8153.

188 H. Sörensen, A. E. Greene, *Tetrahedron Lett.* **1990**, *31*, 7597.

189 E. J. Corey, C. J. Helel, *Tetrahedron Lett.* **1997**, *38*, 7511.

190 A. Sidduri, N. Budries, R. M. Laine, P. Knochel, *Tetrahedron Lett.* **1992**, *33*, 7515.

191 B. H. Lipshutz, R. W. Vivian, *Tetrahedron Lett.* **1999**, *40*, 2871.

192 J.-J. Liu, F. Konzelmann, K.-C. Luk, *Tetrahedron Lett.* **2003**, *44*, 3901.

193 C. E. Tucker, P. Knochel, *Synthesis* **1993**, 530.

194 C. Retherford, P. Knochel, *Tetrahedron Lett.* **1991**, *32*, 441.

195 C. E. Tucker, P. Knochel, *J. Org. Chem.* **1993**, *58*, 4781.

196 (a) R. Giovannini, T. Stüdemann, G. Dussin, P. Knochel, *Angew. Chem. Int. Ed.* **1998**, *37*, 2387; (b) A. Devasagayaraj, T. Stüdemann, P. Knochel, *Angew. Chem. Int. Ed.* **1995**, *34*, 2723.

197 R. Giovannini, P. Knochel, *J. Am. Chem. Soc.* **1998**, *120*, 11186.

198 A. E. Jensen, P. Knochel, *J. Org. Chem.* **2002**, *67*, 79.

199 M. Piber, A. E. Jensen, M. Rottländer, P. Knochel, *Org. Lett.* **1999**, *1*, 1323.

200 (a) H. Eick, P. Knochel, *Angew. Chem. Int. Ed.* **1996**, *35*, 218; (b) A. S. Bhanu Prasad, H. Eick, P. Knochel, *J. Organomet. Chem.* **1998**, *562*, 133; (c) Y. Tamaru, H. Ochiai, F. Sanda, Z. Yoshida, *Tetrahedron Lett.* **1985**, *26*, 5529.

201 (a) K. Soai, T. Suzuki, T. Shono, *J. Chem. Soc. Chem. Commun.* **1994**, 317; (b) N. El Alami, C. Belaud, J. Villiéras, *J. Organomet. Chem.* **1987**, *319*, 303.

202 (a) P. Knochel, J.-F. Normant, *Tetrahedron Lett.* **1984**, *25*, 4383; (b) G. Rousseau, J. M. Conia, *Tetrahedron Lett.* **1981**, *22*, 649; (c) G. Rousseau, J. Drouin, *Tetrahedron* **1983**, *39*, 2307.

203 (a) L. Miginiac in *The Chemistry of Metal-Carbon Bond*, F. R. Hartley, S. Patai; Eds., Wiley, New York, **1985**, vol. 3, p. 99; (b) P. Knochel, in *Comprehensive Organic Synthesis*, B. M. Trost, I. Fleming, S. L. Schreiber, Eds.; Pergamon, Oxford **1991** vol. 1, p. 211.

204 (a) J. Villiéras, M. Rambaud, *Synthesis*, **1982**, 924; (b) J. Villiéras, M. Villiéras, *Janssen Chimica Acta* **1993**, *11*, 3.

205 (a) N. E. Alami, C. Belaud, J. Villiéras, *Tetrahedron Lett.* **1987**, *28*, 59; (b) N. El Alami, C. Belaud, J. Villiéras, *J. Organomet. Chem.* **1988**, *348*, 1.

206 Y. A. Dembélé, C. Belaud, J. Villiéras, *Tetrahedron : Asymmetry*, **1992**, *3*, 511.

207 (a) J. van der Louw, J. L. van der Baan, F. Bickelhaupt, G. W. Klumpp, *Tetrahedron Lett.* **1987**, *28*, 2889; (b) J. van der Louw, J. L. van der Baan, H. Stieltjes, F. Bickelhaupt, G. W. Klumpp, *Tetrahedron Lett.* **1987**,

28, 5929; (c) J. van der Louw,
J. L. van der Baan, *Tetrahedron* **1992**, *48*,
9877.

208 (a) J. L. Moreau, in *The Chemistry of
Ketenes, Allenes and Related Compounds*,
S. Patai, Ed.; Wiley, New York, **1980**,
p. 363; (b) M. Suzuki, Y. Morita,
R. Noyori, *J. Org. Chem.* **1990**, *55*, 441.

209 H. Ochiai, T. Nishihara, Y. Tamaru,
Z. Yoshida, *J. Org. Chem.* **1988**, *53*, 1343.

210 J. Berninger, U. Koert, C. Eisenberg-
Höhl, P. Knochel, *Chem. Ber.* **1995**, *128*,
1021.

211 P. Quinton, T. Le Gall, *Tetrahedron Lett.*
1991, *32*, 4909.

212 (a) H. Lehmkuhl, I. Döring, R. McLane,
H. Nehl, *J. Organomet. Chem.* **1981**, *221*,
1; (b) R. L. Soucy, D. Kozhinov, V. Behar,
J. Org. Chem. **2002**, *67*, 1947.

213 M. R. Saidi, N. Azizi, *Tetrahedron: Asymmetry* **2002**, *13*, 2523.

214 K. P. Chiev, S. Roland, P. Mangeney,
Tetrahedron: Asymmetry **2002**, 13, 2205.

215 (a) M.-J. Shiao, K.-H. Liu, L.-G. Lin, *Synlett* **1992**, 655; (b) W.-L. Chia, M.-J. Shiao,
Tetrahedron Lett. **1991**, *32*, 2033;
(c) T.-L.Shing, W. L. Chia, M.-J. Shiao,
T.-Y.Chau, *Synthesis* **1991**, 849;
(d) M.-J. Shiao, W. L. Chia; T.-L. Shing;
T. J. Chow, *J. Chem. Res. (S)* **1992**, 247;
(e) C. Agami, F. Couty, J.-C. Daran,
B. Prince, C. Puchot, *Tetrahedron Lett.*
1990, *31*, 2889; (f) C. Agami, F. Couty,
M. Poursoulis, J. Vaissermann, *Tetra-
hedron* **1992**, *48*, 431; (g) C. Andrés,
A. González, R. Pedrosa, A. Pérez-
Encabo, S. García-Granda, M. A. Salvadó,
F. Gómez-Beltrán, *Tetrahedron Lett.*
1992, *33*, 4743; (h) J.-L.Bettiol,
R.-J. Sundberg, *J. Org. Chem.* **1993**, *58*,
814; (i) J. Yamada, H. Sato,
Y. Yamamoto, *Tetrahedron Lett.* **1989**, *30*,
5611.

216 (a) C. Chuit, J. P. Foulon, J.-F. Normant,
Tetrahedron **1981**, *37*, 1385 and *Tetra-
hedron* **1980**, *36*, 2305; (b) M. Bourgain-
Commercon, J. P. Foulon, J. F. Normant,
J. Organomet. Chem. **1982**, *228*, 321;
(c) E. J. Corey, N. W. Boaz, *Tetrahedron
Lett.* **1985**, *26*, 6015 and 6019;
(d) A. Alexakis, J. Berlan, Y. Besace,
Tetrahedron Lett. **1986**, *27*, 1047;
(e) Y. Horiguchi, S. Matsuzawa,
E. Nakamura, I. Kuwajima, *Tetrahedron*

Lett. **1986**, *27*, 4025; (f) E. Nakamura,
S. Matsuzawa, Y. Horiguchi,
I. Kuwajima, *Tetrahedron Lett.* **1986**, *27*,
4029; (g) Y. Horiguchi, S. Matsuzawa,
E. Nakamura, I. Kuwajima, *Tetrahedron
Lett.* **1986**, *27*, 4025; (h) M. Bergdahl,
E.-L. Lindtstedt, M. Nilsson, T. Olsson,
Tetrahedron **1988**, *44*, 2055;
(i) M. Bergdahl, E.-L. Lindtstedt,
M. Nilsson, T. Olsson, *Tetrahedron* **1989**,
45, 535; (j) M. Bergdahl, E.-L. Lindtstedt,
T. Olsson, *J. Organomet. Chem.* **1989**,
365, C11; (k) M. Bergdahl, M. Nilsson,
T. Olsson, *J. Organomet. Chem.* **1990**,
C19; (l) M. Bergdahl, M. Eriksson,
M. Nilsson, T. Olsson, *J. Org. Chem.*
1993, *58*, 7238.

217 Y. Tamaru, H. Tanigawa, T. Yamamoto,
Z. Yoshida, *Angew. Chem. Int. Ed. Engl.*
1989, *28*, 351.

218 (a) H. Tsujiyama, N. Ono, T. Yoshino,
S. Okamoto, F. Sato, *Tetrahedron Lett.*
1990, *31*, 4481; (b) T. Yoshino,
S. Okamoto, F. Sato, *J. Org. Chem.* **1991**,
56, 3205; (c) K. Miyaji, Y. Ohara,
Y. Miyauchi, T. Tsuruda, K. Arai, *Tetrahedron Lett.* **1993**, *34*, 5597.

219 (a) B. L. Feringa, M. Pineschi,
L. A. Arnold, R. Imbos, A. H. M. de Vries,
Angew. Chem Int. Ed. **1997**, *36*, 2620;
(b) E. Keller, J. Maurer, R. Naasz,
T. Schader, A. Meetsma, B. L. Feringa,
Tetrahedron Asymmetry **1998**, *9*, 2409;
(c) R. Naasz, L. A. Arnold, M. Pineschi,
E. Keller, B. L. Feringa, *J. Am. Chem.
Soc.* **1999**, *121*, 1104; (d) R. Imbos,
M. H. G. Britman, M. Pineschi,
B. L. Feringa, *Org. Lett.* **1999**, *1*, 623;
(e) L. A. Arnold, R. Imbos, A. Mandoli,
A. H. M. De Vries, R. Naasz,
B. L. Feringa, *Tetrahedron* **2000**, *56*,
2865; (f) F. Bertozzi, P. Crotti,
B. L. Feringa, F. Macchia, M. Pineschi,
Synthesis **2001**, 483.

220 (a) A. Alexakis, J. Burton, J. Vastra,
P. Mangeney, *Tetrahedron: Asymmetry*
1997, *8*, 3987; (b) A. Alexakis,
G. P. Trevitt, G. Bernadelli, *J. Am.
Chem. Soc.* **2001**, *123*, 4358;
(c) A. Alexakis, C. Benhaim, S. Rosset,
M. Humann, *J. Am. Chem. Soc.* **2002**,
124, 5262; (d) O. Knopff, A. Alexakis,
Org. Lett. **2002**, *4*, 3835; (e) A. Alexakis,
C. Benhaim, *Eur. J. Org. Chem.* **2002**,

3221; (f) A. Alexakis, S. March, *J. Org. Chem.* **2002**, *67*, 8753.

221 L. A. Arnold, R. Naasz, A. J. Minnaard, B. L. Feringa, *J. Am. Chem. Soc.*, **2001**, *123*, 5841.

222 R. Naasz, L. A. Arnold, A. J. Minnaard, B. L. Feringa, *Angew. Chem. Int. Ed.* **2001**, *40*, 927;

223 A. W. Hird, A. H. Hoveyda, *Angew. Chem. Int. Ed.* **2003**, *42*, 1276.

224 F. Denes, F. Chemla, J. F. Normant, *Eur. J. Org. Chem.* **2002**, 3536.

225 G Cahiez, P. Verregas, C. E. Tucker, T. N. Majid, P. Knochel, *J. Chem. Soc. Chem. Commun.* **1992**, 1406.

226 D. H. Yang, J. X. Wang, Y. Hu, *J. Chem. Research (S)* **2002**, 79.

227 (a) D. Seebach, E. W. Colvin, F. Lehr, T. Weller, *Chimia* **1979**, *33*, 1; (b) A. Yoshikoshi, M. Miyashita, *Acc. Chem. Res.* **1985**, *18*, 284; (c) A. G. M. Barrett, G. G. Graboski, *Chem. Rev.* **1986**, *86*, 751; (d) R. Tamura, A. Kamimura, N. Ono, *Synthesis* **1991**, 423.

228 (a) S. B. Bowlus, *Tetrahedron Lett.* **1975**, 3591; (b) A.-T. Hansson, M. Nilsson, *Tetrahedron*, **1982**, *38*, 389; (c) S. Stiver, P. Yates, *J. Chem. Soc. Chem. Commun.* **1983**, 50.

229 C. Rutherford, P. Knochel, *Tetrahedron Lett.* **1991**, *32*, 441.

230 S. E. Denmark, L. R. Marcin, *J. Org. Chem.* **1993**, *58*, 3850.

231 J.-F. Normant, A. Alexakis, *Synthesis* **1981**, 841.

232 (a) M. T. Crimmins, P. G. Nantermet, *J. Org. Chem.* **1990**, *55*, 4235 ; (b) M. T. Crimmins, P. G. Nantermet, B. W. Trotter, I. M. Vallin, P. S. Watson, L. A. McKerlie, T.-L. Reinhold, A. W.-H. Cheung, K. A. Steton, D. Dedopoulou, J. L. Gray, *J. Org. Chem.* **1993**, *58*, 1038; (c) M. T. Crimmins, D. K. Jung, J. L. Gray, *J. Am. Chem. Soc.* **1993**, *115*, 3146.

233 J. van der Louw, C. M. D. Komen, A. Knol, F. J. J. de Kanter, J. L. van der Baan, F. Bickelhaupt, G. W. Klumpp, *Tetrahedron Lett.* **1989**, *30*, **4453**.

234 (a) E. Negishi, J. A. Miller, *J. Am. Chem. Soc.* **1983**, *105*, 6761; (b) G. A. Molander, *J. Org. Chem.* **1983**, *48*, 5409.

235 A. Yanagisawa, S. Habane, H. Yamamoto, *J. Am. Chem. Soc.* **1989**, *111*, 366.

236 K. Kubota, M. Nakamura, M. Isaka, E. Nakamura, *J. Am. Chem. Soc.* **1993**, *115*, 5867.

237 (a) M. Nakamura, N. Yoshikai, E. Nakamura, *Chem. Lett.* **2002**, 146; (b) K. Kubota, E. Nakamura, *Angew. Chem.* **1997**, *109*, 2581.

238 M. Nakamura, K. Hara, T. Hatakeyama, E. Nakamura, *Org. Lett.* **2001**, *3*, 3137.

239 (a) J. Montgomery, M. Song, *Org. Lett.* **2002**, *4*, 4009; (b) S.-K. Kang, S.-K. Yoon, *Chem. Commun.* **2002**, 2634.

240 (a) M. Gaudemar, *C.R. Hebd. Sceances Acad. Sci., Ser. C* **1971**, *273*, 1669; (b) Y. Frangin, M. Gaudemar *C.R. Hebd. Sceances Acad. Sci., Ser C* **1974**, *278*, 885; (c) M. Bellasoued, Y. Frangin, M. Gaudemar, *Synthesis* **1977**, 205; (d) P. Knochel, J.-F. Normant, *Tetrahedron Lett.* **1986**, *27*, 1039; (e) P. Knochel, J.-F. Normant, *Tetrahedron Lett.* **1986**, *27*, 1043; (f) P. Knochel, J.-F. Normant, *Tetrahedron Lett.* **1986**, *27*, 4427; (g) P. Knochel, J.-F. Normant, *Tetrahedron Lett.* **1986**, *27*, 5727; (h) D. Beruben, I. Marek, L. Labaudinière, J.-F. Normant, *Tetrahedron Lett.* **1993**, *34*, 2303; (i) J.-F. Normant, J.-C. Quirion, *Tetrahedron Lett.* **1989**, *30*, 3959; (j) I. Marek, J.-M. Lefrançois, J.-F. Normant, *Synlett* **1992**, 633; (k) I. Marek, J.-M. Lefrançois, J.-F. Normant, *Tetrahedron Lett.* **1992**, *33*, 1747; (l) I. Marek, J.-M. Lefrançois, J.-F. Normant, *Tetrahedron Lett.* **1991**, *32*, 5969; (m) I. Marek, J.-F. Normant, *Tetrahedron Lett.* **1991**, *32*, 5973.

241 P. Knochel, C. Xiao, M. C. P. Yeh, *Tetrahedron Lett.* **1988**, *29*, 6697.

242 S. A. Rao, P. Knochel, *J. Am. Chem. Soc.* **1991**, *113*, 5735.

243 M. T. Bertrand, G. Courtois, L. Miginiac *Tetrahedron Lett.* **1974**, 1945.

244 (a) M. Kobayashi, E. Negishi, *J. Org. Chem.* **1980**, *45*, 5223; (b) E. Negisihi, *Acc. Chem. Res.*, **1982**, *15*, 340; (c) E. Negisihi, V. Bagheri, S. Chatterjee, F.-T. Luo, J. A. Miller, A. T. Stoll, *Tetrahedron Lett.* **1983**, *24*, 5181; (d) R. A. Grey, *J. Org. Chem.* **1984**, *49*, 2288.

245 (a) E. Negishi, A. O. King, N. Okukado, *J. Org. Chem.* **1977**, *42*, 1821; (b) R. Rossi, F. Bellina, A. Carpita, R. Gori, *Synlett* **1995**, 344; (c) E. Negishi, H. Matsushita, M. Kobayashi, C. L. Rand, *Tetrahedron Lett.* **1983**, *24*, 3823; (d) J. G. Millar, *Tetrahedron Lett.* **1989**, *30*, 4913; (e) S. Hyuga, N. Yamashina, S. Hara, A. Suzuki, *Chem. Lett.* **1988**, 809; (f) A. Löffler, G. Himbert, *Synthesis* **1992**, 495; (g) R. C. Borner, R. F. W. Jackson, *J. Chem. Soc. Chem. Commun.* **1994**, 845; (h) T. Sakamoto, Y. Kondo, N. Takazawa, H. Yamanaka, *Tetrahedron Lett.* **1993**, *34*, 5955; (i) A. Minato, K. Suzuki, K. Tamao, M. Kumada, *Tetrahedron Lett.* **1984**, *25*, 2583; (j) C. A. Quesnelle, O. B. Familoni, V. Snieckus, *Synlett* **1994**, 349; (k) T. Sakamoto, S. Nishimura, Y. Kondo, H. Yamanaka, *Synthesis* **1988**, 485; (l) T. Sato, A. Kawase, T. Hirose, *Synlett* **1992**, 891.

246 E. Negishi, H. Matsushita, M. Kobayashi, C. L. Raud, *Tetrahedron Lett.* **1983**, *24*, 3823;

247 S. Hyuga, N. Yamayshina, S. Hara, A. Suzuki, *Chem. Lett.* **1988**, 809.

248 A. Minota, K. Suzuki, K. Tamao, M. Kumada, *Tetrahedron Lett.* **1984**, *25*, 2583.

249 A. Löffler, G. Himbert, *Synthesis* **1992**, 495.

250 G. Mignani, F. Leising, R. Meyrueix, H. Samson, *Tetrahedron Lett.* **1990**, *31*, 4743.

251 C. A. Quesnelle, O. B. Familoni, V. Snieckus, *Synlett* **1990**, 349.

252 (a) T. Sakamoto, Y. Kondo, N. Murata, H. Yamanaka, *Tetrahedron Lett.* **1992**, *33*, 5373; (b) T. Sakamoto, Y. Kondo, N. Takazawa, H. Yamanaka, *Heterocycles* **1993**, *36*, 941; (c) D. Trauner, J. B. Schwartz, S. J. Danishefsky, *Angew. Chem. Int. Ed. Engl.* **1999**, *38*, 3543; (d) A. S. Bell, D. A. Roberts, K. S. Ruddock, *Synthesis* **1987**, 843.

253 (a) E. Nakamura, I. Kuwajima, *Tetrahedron Lett.* **1986**, *27*, 83; (b) T. Sakamoto, S. Nishimura, Y. Kondo, H. Yamanaka, *Synthesis* **1988**, 485.

254 Y. Tamaru, H. Ochiai, T. Nakamura, Z. Yoshida, *Angew. Chem. Int. Ed.* **1987**, *99*, 1193.

255 R. F. W. Jackson, M. J. Wylhes, A. Wood, *J. Chem. Soc., Chem. Commun.* **1989**, 644.

256 T. Sato, A. Kawase, T. Hirose, *Synlett* **1992**, 891.

257 C. Dai, G. C. Fu, *J. Am. Chem. Soc.* **2001**, *123*, 2719.

258 H. J. C. Deboves, C. A. G. N. Montalbetti, R. F. W. Jackson, *J. Chem. Soc. Perkin Trans.* 1, **2001**, 1876.

259 A. Mehta, R. Jaouhari, T. J. Benson, K. T. Douglas, *Tetrahedron Lett.* **1992**, *33*, 5441.

260 (a) I. Kadota, H. Takamura, K. Sato, Y. Yamamoto, *J. Org. Chem.* **2002**, *67*, 3494; (b) Y. Tamaru, H. Ochial, T. Nakamura, Z. Yoshida, *Tetrahedron Lett.* **1986**, *27*, 955; (c) Y. Tamaru, H. Ochial, T. Nakamura, K. Tsubaki, Z. Yoshida, *Tetrahedron Lett.* **1985**, *26*, 5559.

261 N. Vicart, G.-S. Saboukoulou, Y. Ramondenc, G. Plé, *Synth. Commun.* **2003**, *33*, 1509.

262 A. Przezdziecka, A. Kurek-Tyrlik, J. Wicha, *Collect. Czech. Chem. Commun.* **2002**, *67*, 1658.

263 B. H. Lipshutz, P. A. Blomgren, *J. Am. Chem. Soc.* **1999**, *121*, 5819.

264 (a) J. A. Miller, R. P. Farrell, *Tetrahedron Lett.* **1998**, *39*, 6441. (b) C. E. Tucker, J. G. de Vries, *Top. Catal.* **2002**, *19*, 111.

265 (a) J. P. Gillet, R. Sauvêtre, J.-F. Normant, *Synthesis* **1986**, 538; (b) J.-P. Gillet, R. Sauvêtre, J.-F. Normant, *Tetrahedron Lett.* **1985**, *26*, 3999; (c) F. Tellier, R. Sauvêtre, J.-F. Normant, *J. Organomet. Chem.* **1986**, *303*, 309; (d) F. Tellier, R. Sauvêtre, J.-F. Normant, Y. Dromzee, Y. Jeannin, *J. Organomet. Chem.* **1987**, *331*, 281; (e) P. Martinet, R. Sauvêtre, J.-F. Normant, *J. Organomet. Chem.* **1989**, *367*, 1.

266 (a) K. Yuan, W. J. Scott, *Tetrahedron Lett.* **1989**, *30*, 4779; (b) K. Yuan, W. J. Scott, *J. Org. Chem.* **1990**, *55*, 6188; (c) K. Park, K. Yuan, W. J. Scott, *J. Org. Chem.* **1993**, *58*, 4866.

267 N. Murata, T. Sugihara, Y. Kondo, T. Sakamoto, *Synlett* **1997**, 298.

268 (a) G. Karig, J. A. Spencer, T. Gallagher, *Org. Lett.* **2001**, *3*, 835; (b) C. G. V. Sharples, G. Karig,

G. L. Simpson, J. A. Spencer, E. Wright, N. S. Millar, S. Wonnacott, T. Gallagher, *J. Med. Chem.* **2002**, *45*, 3235.

269 A. Staubitz, W. Dohle, P. Knochel, *Synthesis* **2003**, 233.

270 M. Hocek, I. Votruba, H. Dvoraka, *Tetrahedron* **2003**, *59*, 607.

271 B. Ye, T. R. Burke Jr., *J. Org. Chem.* **1995**, *60*, 2640.

272 A. W. Seton, M. F. G. Stevens, A. D. Westwell, *J. Chem. Res. (S)*, **2001**, 546.

273 J. Thibonnet, V. A. Vu, L. Berillon, P. Knochel, *Tetrahedron* **2002**, *58*, 4787.

274 J.-C. Shi, X. Zeng, E.-I. Negishi, *Org. Lett.* **2003**, *5*, 1825.

275 (a) T. Hayashi, T. Hagihara, Y. Katsuro, M. Kumada, *Bull. Chem. Soc. Jpn.* **1983**, *56*, 363; (b) T. Hayashi, *J. Organomet. Chem.* **2002**, *653*, 41.

276 V. Mamane, T. Ledoux-Rak, S. Deveau, J. Zyss, O. Riant, *Synthesis* **2003**, 455.

277 (a) M. Lotz, G. Kramer, P. Knochel, *Chem. Commun.* **2002**, 2546; (b) R. J. Kloetzing, M. Lotz, P. Knochel, *Tetrahedron: Asymmetry*, **2003**, *14*, 255; (c) T. Bunlaksananusorn, K. Polborn, P. Knochel, *Angew. Chem. Int. Ed.* **2003**, *42*, 3941; (d) T. Bunlaksananusorn, A. P. Luna, M. Boniu, L. Micouin, P. Knochel, *Synlett* **2003**, 2240.

278 (a) J.-F. Peyrat, E. Thomas, N. L'Hermite, M. Alami, J.-D. Brion, *Tetrahedron Lett.* **2003**, *44*, 6703; (b) M. Alami, S. Gueugnot, E. Doingues, G. Linstrumelle, *Tetrahedron* **1995**, *51*, 1209.

279 B. L. Flynn, P. Verdier-Pinard, E. Hamel, *Org. Lett.* **2001**, *3*, 651.

280 F. Zeng, E. Negishi, *Org. Lett.* **2001**, *3*, 719.

281 Y. Tamaru, H. Ochiai, T. Nakamura, Z. Yoshida, *Org. Synth.* **1989**, *67*, 98.

282 Y. Tamaru, K. Yasui, H. Takanabe, S. Tanaka, K. Fugami, *Angew. Chem. Int. Ed.* **1992**, *31*, 645.

283 M. Asaoka, M. Tamaka, T. Houkawa, T. Ueda, S. Sakami, H. Takei, *Tetrahedron* **1998**, *54*, 471.

284 H. Tokuyama, S. Yokoshima, T. Yamashita, T. Fukuyama, *Tetrahedron Lett.* **1998**, *39*, 3189.

285 Y. Mori, M. Seki, *J. Org. Chem.* **2003**, *68*, 1571.

286 (a) T. Shimizu, M. Seki, *Tetrahedron Lett.* **2002**, *43*, 1039; (b) T. Shimizu, M. Seki, *Tetrahedron Lett.* **2001**, *42*, 429; (c) see also M. Kimura, M. Seki, *Tetrahedron Lett.* **2004**, *45*, 1635.

287 (a) N. Oguni, T. Omi, Y. Yamamoto, A. Nakamura, *Chem. Lett.* **1983**, 841; (b) N. Oguni, T. Omi, *Tetrahedron Lett.* **1984**, *25*, 2823; (c) N. Oguni, Y. Matsuda, T. Kaneko, *J. Am. Chem. Soc.* **1988**, *110*, 7877.

288 (a) K. Soai, S. Niwa, *Chem. Rev.* **1992**, *92*, 833; (b) D. A. Evans, *Science* **1988**, *240*, 420; (c) R. Noyori, M. Kitamura, *Angew. Chem. Int. Ed. Engl.* **1991**, *30*, 49.

289 (a) M. Yoshioka, T. Kawakita, M. Ohno, *Tetrahedron Lett.* **1989**, *30*, 1657; (b) H. Takahashi, T. Kawakita, M. Yoshioka, S. Kobayashi, M. Ohno, *Tetrahedron Lett.* **1989**, *30*, 7095; (c) H. Takahashi, T. Kawakita, M. Ohno, M. Yoshioka, S. Kobayashi, *Tetrahedron* **1992**, *48*, 5691.

290 (a) A. K. Beck, B. Bastani, D. A. Plattner, W. Petter, D. Seebach, H. Braunschweiger, P. Gysi, L. VaVecchia, *Chimia* **1991**, *45*, 238; (b) B. Schmidt, D. Seebach, *Angew. Chem. Int. Ed. Engl.* **1991**, *30*, 99; (c) D. Seebach, L. Behrendt, D. Felix, *Angew. Chem. Int. Ed. Engl.* **1991**, *30*, 1008; (d) B. Schmidt, D. Seebach, *Angew. Chem. Int. Ed. Engl.* **1991**, *30*, 1321; (e) J. L. von dem Bussche-Hünnefeld, D. Seebach, *Tetrahedron* **1992**, *48*, 5719; (f) D. Seebach, D. A. Plattner, A. K. Beck, Y. M. Wang, D. Hunziker, W. Petter, *Helv. Chim. Acta* **1992**, *75*, 2171; (g) D. Seebach, A. K. Beck, B. Schmidt, Y. M. Wang, *Tetrahedron* **1994**, *50*, 4363.

291 H. Schafer, D. Seebach, *Tetrahedron* **1995**, *51*, 2305.

292 D. Seebach, A. K. Beck, A. Hechel, *Angew. Chem. Int. Ed.* **2001**, *40*, 92.

293 B. P. Rheiner, H. Sellner, D. Seebach, *Helv. Chim. Acta* **1997**, *80*, 2027.

294 H. Sellner, C. Faber, P. B. Rheiner, D. Seebach, *Chem. Eur. J.* **2000**, *6*, 3692.

295 (a) L. Schwink, P. Knochel, *Chem. Eur. J.* **1998**, *4*, 950; (b) W. Brieden, R. Ostwald, P. Knochel, *Angew. Chem., Int. Ed. Engl.* **1993**, *32*, 582.

296 S. Nowotny, S. Vettel, P. Knochel, *Tetrahedron Lett.* **1994**, *35*, 4539.

297 (a) R. Noyori, M. Suzuki, *Angew. Chem. Int. Ed. Engl.* **1984**, *23*, 847; (b) M. Suzuki, A. Yanagisawa, R. Noyori, *J. Am. Chem. Soc.* **1990**, *307*, 3348; (c) Y. Morita, M. Suzuki, R. Noyori, *J. Org. Chem.* **1989**, *54*, 1785; (d) R. Noyori, S. Suga, K. Kawai, S. Okada, M. Kitamura, *Pure Appl. Chem.* **1988**, *60*, 1597.

298 (a) S. Vettel, P. Knochel, *Tetrahedron Lett.* **1994**, *35*, 5849; (b) R. Ostwald, P.-Y. Chavant, H. Stadtmüller, P. Knochel, *J. Org. Chem.* **1994**, *59*, 4143; (c) W. R. Roush, K. Koyama, *Tetrahedron Lett.* **1992**, *33*, 6227.

299 H. Lütjens, P. Knochel, *Tetrahedron: Asymmetry* **1994**, *5*, 1161.

300 P. Knochel, W. Brieden, M. J. Rozema, C. Eisenberg, *Tetrahedron Lett.* **1993**, *34*, 5881.

301 C. Eisenberg, P. Knochel, *J. Org. Chem.* **1994**, *59*, 3760.

302 A. Fürstner, T. Müller, *J. Org. Chem.* **1998**, *63*, 424.

303 S. Bäurle, S. Hoppen, U. Koert, *Angew. Chem.* **1999**, *111*, 1341.

304 A. Fürstner, M. Albert, J. Mlynarski, M. Matheu, E. DeClercq, *J. Am. Chem. Soc.* **2003**, *125*, 13132.

305 W. Oppolzer, R. N. Radinov, J. de Bradander, *Tetrahedron Lett.* **1995**, *36*, 2607.

306 M. Yus, D. J. Ramón, O. Prieto, *Tetrahedron: Asymmetry* **2003**, *14*, 1103.

307 D. Boyalli, E. Frantz, M. Carreira, *Org. Lett.* **2002**, *4*, 2605.

308 A. Fettes, E. M. Carreira, *Angew. Chem. Int. Ed.*, **2002**, *41*, 4098.

309 L. Pu, *Tetrahedron* **2003**, *59*, 9873.

310 L. S. Tan, C. Y. Chen, R. D. Tillyer, E. J. J. Grabowski, P. J. Reider, *Angew. Chem. Int. Ed.* **1999**, *38*, 711.

311 N. K. Anand, E. M. Carreira, *J. Am. Chem. Soc.* **2001**, *123*, 9687.

312 O. Prieto, D. J. Ramon, M. Yus, *Tetrahedron: Asymmetry* **2003**, *14*, 1955.

313 (a) J. Rudolph, T. Rasmussen, C. Bolm, P.-O. Norrby, *Angew. Chem. Int. Ed.* **2003**, *42*, 3002; (b) J. Rudolph, N. Hermanns, C. Bolm, *J. Org. Chem.* **2004**, *69*, 3997; (c) C. Bolm, J. Mueller, *Tetrahedron*, **1994**, *50*, 4355; (d) C. Bolm, J. Mueller, G. Schlingloff, M. Zehnder, M. Neuburger, *J. Chem. Soc., Chem. Commun.* **1993**, *2*, 182; (e) C. Bolm, G. Schlingloff, K. Harms, *Chem. Ber.* **1992**, *124*, 1191; (f) C. Bolm, N. Hermanns, A. Classen, K. Muniz, *Bioorg. Med. Chem. Lett.* **2002**, *12*, 1795.

314 J. R. Porter, J. F. Traverse, A. H. Hoveyda, M. L. Snapper, *J. Am. Chem. Soc.* **2001**, *123*, 10409.

315 J. F. Traverse, A. H. Hoveyda, M. L. Snapper, *Org. Lett.* **2003**, *5*, 3273.

316 T. Yamamoto, A. Yamamoto, S. Ikeda, *J. Am. Chem. Soc.* **1971**, *93*, 3350.

317 C. K. Reddy, P. Knochel, *Angew. Chem. Int. Ed. Engl.* **1996**, *35*, 1700.

318 A. Devasagayaraj, P. Knochel, *Tetrahedron Lett.* **1995**, *36*, 8411.

319 (a) G. Cahiez, *Actual Chim.* **1984**, (Sept), 24; (b) B. Weidmann, D. Seebach, *Angew. Chem. Int. Ed. Engl.* **1983**, *22*, 31; (c) G. Cahiez, B. Laboue, *Tetrahedron Lett.* **1989**, *30*, 3545; (d) G. Cahiez, B. Laboue, *Tetrahedron Lett.* **1989**, *30*, 7369; (e) G. Cahiez, M. Alami, *Tetrahedron* **1989**, *45*, 4163; (f) G. Cahiez, P.-Y. Chavant, E. Metais, *Tetrahedron Lett.* **1992**, *33*, 5245; (g) G. Cahiez, B. Figadère, P. Cléry, *Tetrahedron Lett.* **1994**, *35*, 3065; (h) G. Cahiez, K. Chau, P. Cléry, *Tetrahedron Lett.* **1994**, *35*, 3069.

320 E. Riguet, I. Klement, C. K. Reddy, G. Cahiez, P. Knochel, *Tetrahedron Lett.* **1996**, *37*, 5865.

321 T. Stüdemann, M. Ibrahim-Ouali, G. Cahiez, P. Knochel, *Synlett* **1998**, 143.

Index

Organometallics. Paul Knochel
Copyright © 2005 WILEY-VCH Verlag GmbH & Co. KGaA, Weinheim
ISBN: 3-527-31131-9